Nanomaterials and Nanotechnology

Shengjie Peng · Peng Li

Nanomaterials and Nanotechnology

Basic, Preparation and Applications

Springer

Shengjie Peng
College of Materials Science
and Technology
Nanjing University of Aeronautics
and Astronautics
Nanjing, Jiangsu, China

Peng Li
College of Materials Science
and Technology
Nanjing University of Aeronautics
and Astronautics
Nanjing, Jiangsu, China

ISBN 978-981-97-8432-5 ISBN 978-981-97-8433-2 (eBook)
https://doi.org/10.1007/978-981-97-8433-2

This Springer imprint is published by the registered company Springer Nature Singapore Pte Ltd.
The registered company address is: 152 Beach Road, #21-01/04 Gateway East, Singapore 189721, Singapore

Preface

Nanomaterials are the most important part of nanoscience and technology, and they are also hot spots and difficulties in international competition. Nanomaterials generally refer to materials that have one dimension in space on the order of nanometers. Nanomaterials have surface effect, small size effect and macroscopic quantum tunneling effect. Therefore, in actual use, it will show many exotic properties, that is, its optical, thermal, magnetic, mechanical and chemical properties will be significantly different from those of coarse-grained materials. Therefore, the main purpose of this book is to enable students to understand the concept, structure, dimension, characterization method, performance and application of nanomaterials through reading.

Nanotechnology not only contains rich scientific connotations, but also provides people with a broad space for innovation, thus becoming a new territory for the development of physics, chemistry, material science, life science and information science. People are using nanotechnology to understand and transform nature at the nanoscale and create new materials by directly manipulating and arranging atoms and molecules. The emergence of nanotechnology marks that human science and technology has entered a new era, the era of nanotechnology. The speed of penetration of nanotechnology into various fields and the wide range of influence are far beyond people's expectations. In nanomaterials, quasi-zero-dimensional nanoparticle materials and nanopowder materials are the most dynamic components of ultrafine powder materials; nanotubes, nanowires, nanorods, etc., are called one-dimensional nanomaterials; nanoparticle films, nanofilms and nanomultilayers are quasi-two-dimensional nanomaterials; three-dimensional materials composed of nanoparticles and nanofibers are usually called nano-bulk materials. Nanostructure research will bring new opportunities to the synthesis and application of nanomaterials and has aroused people's great interest in understanding and mastering new achievements, new theories, new ideas and new technologies in the field of nanometers at home and abroad.

Based on this idea, we first introduced the development history, research content and development trend of nanomaterials and then described the preparation methods, characterization methods, physical, chemical and mechanical properties of nanomaterials, so that students have an overall understanding of nanomaterials. On this basis, the application of nanomaterials and the widely studied carbon nanomaterials will be taught. Finally, the latest research progress of nanomaterials will be introduced. Whole book enumerates a few real examples as far as possible, making it easy to understand and master.

Due to the limited level and ability of the editor, it is inevitable that there will be mistakes and omissions in the book. I sincerely hope that readers will criticize and correct me.

Nanjing, China Shengjie Peng

Contents

About the Authors

Prof. Shengjie Peng is currently Full Professor in Nanjing University of Aeronautics and Astronautics. He completed his Ph.D. (2010) in Nankai University and B.S. (2004) in Anhui Normal University. Then, he went to Nanyang Technological University and National University of Singapore as Postdoctoral Fellow with Prof. Qingyu Yan and Prof. Seeram Ramakrishna till 2017. His current research interests focus on development of rationally designed functional materials with finely tailored nanoscale architecture to tackle critical problems (such as hydrogen production, energy density, power density, cycle and calendar life, safety and cost) in diverse energy-related applications, including water splitting, ORR, CO_2 reduction, batteries, fuel cells, as well as clean and renewable energy. To date, he has co-authored over 150 peer-reviewed publications, more than 20 Chinese patents, one chapter in a book and one English book. Based on his research achievements in electrochemical catalysis and storage, he got several honors, including Fellow of the Royal Society of Chemistry, FRSC, China One Thousand Young Talent, China Jiangsu Specially Appointed Professor, High-Level Entrepreneurial and Innovative Talents Program of Jiangsu, etc.

Peng Li is Associate Professor at Nanjing University of Aeronautics and Astronautics. He received his bachelor's and master's degrees from Nanjing University in 2006 and 2009, respectively, and earned his PhD from Hokkaido University in 2013. His research mainly focuses on functional nanomaterials for energy conversion and storage.

Chapter 1
Basic Concepts of Nanomaterials

1.1 Definition of Nanomaterials

The naming of nanomaterials appeared in 1980s, which refers to any material with at least one dimension of 0.1–100 nm or composed of them as matrix units. Nanomaterials can be crystalline or quasicrystalline or amorphous. The basic unit or constituent unit of nanomaterials can be composed of atomic clusters, nanoparticles, nanowires or nano-films, which can be metal materials, inorganic non-metallic materials and polymer materials.

People have different opinions about the definition of the size range of nanomaterials. Some define the size of nanoparticles as 1–10 nm, some as 1–50 nm, and others as 1–100 nm [1]. In fact, the division of the average particle size of nano-materials is not very strict, but the critical size must be considered. When the particle size is reduced to a certain size in nanometer scale, the properties of the material suddenly change, even completely different from those of the conventional materials with the same composition. This size is defined as the critical size. The critical size of the mutation of different properties of the same nanomaterial is different, and the critical size of different nanomaterials with the same performance is also very different. That is to say, the properties of nanomaterials vary with the size, and this strong size effect is rare in conventional materials, which is the characteristic of nanomaterials. Therefore, the size range of nanomaterials cannot be defined by a fixed scale but is wider. In addition, nanomaterials are materials defined by size, which involve a wide range of materials. Conventional materials have corresponding nanomaterials, and nanoparticles of various materials generally include 10^4–10^5 atoms. Due to the quantum size effect, the energy levels of such atomic groups split, causing many changes in properties. For metals, the size of nanoparticles containing so many atoms can be very small, but for substances with large unit cells, the size of particles containing such atomic numbers may become very large. For example, the cell size of ferroelectrics is much larger than that of pure metals. Generally speaking, when

S. Peng and P. Li, *Nanomaterials and Nanotechnology*,
https://doi.org/10.1007/978-981-97-8433-2_1

the size of various substances is reduced to 1–100 nm, they all have different properties from conventional materials. Therefore, it is appropriate to divide the size of nanomaterials into 1–100 nm. In recent years, most reports on nanomaterials have adopted this range.

We call the materials whose constituent phases or grain structures are controlled below 100 nm as nanomaterials, that is, the materials or structures with at least one structural unit in the three-dimensional space in the nanometer scale range (1–100 nm) or constructed by them according to certain rules are defined as nanomaterials.

Nanomaterials are composed of two structural components: one is crystal component; The second is the interface component (grain boundary phase). Crystal components are composed of atoms in all nano-grains, and each atom is arranged in an orderly manner in its lattice position. Interface components are composed of interface atoms between grains formed by surface atoms of each nano-grain. When the grain size decreases, the relative volume of the grain boundary phase increases, and for nanocrystals, the atoms in the grain boundary phase can reach 15–50%. The structure of the grain boundary phase is determined by the relative orientation of adjacent grains and the inclination of the interface. If the grain orientation is random, all the grain boundary phases will show different structures, and the atomic spacing of all the grain boundary phases will be different. Statistically speaking, there is no preferred atomic spacing. Therefore, the average result will lead to various possible values of atomic spacing, that is to say, the structure of interface components has neither long-range ordered arrangement of crystals nor short-range order of amorphous. However, from another point of view, interface atoms are transformed from atoms on the surface of grains, so it has a special order, which is the special feature of nano-material structure.

1.2 Origin and Research Progress of Nanotechnology

1.2.1 The Origin of Nanotechnology

Nanotechnology refers to the understanding and transformation of nature within the nanometer scale. It studies the motion law and functional characteristics of the material composition system between 1 and 1–100 nm, and its application technology in actual production and life. Its main feature is "small". Imagine the possibility: "A material that is 10 times stronger than steel and weighs only a fraction of it; Indenting all the information of the Library of Congress into a device the size of a sugar cube; It can detect tumors that are only a few cells in size," is the propaganda slogan made by former US President Bill Clinton for the "New National Nanotechnology". Nano-disciplines include nanophysics, nano-electronics, nanomaterials, nano-mechanics, nano-biology, nanomedicine, nano-information technology and nano-manufacturing.

The scientist who first raised the question of science and technology on the nanometer scale was the famous physicist and Nobel Prize winner Richard Feynman (1918–1988). In 1959, Feynman predicted that human beings could make smaller machines with small machines, and finally they would arrange atoms one by one to make "products" according to human wishes. This was the earliest dream about nanotechnology. In the 1970s, scientists began to put forward ideas about nanotechnology from different angles. In 1974, scientist Donnie Gucci first used the term nanotechnology to describe precision machining. In 1982, German scientist (G. Binnig, 1947~) and Swiss scientist Basil (1933~) invented the scanning tunneling microscope (STM), which made it possible for human beings to see atoms and molecules in the atmosphere and at room temperature. It was a milestone for human beings to explore the nano-world and technology. In 1989, Stanford University moved the atomic group to "write" the name of Stanford University. In 1990, IBM used a scanning probe to move 35 atoms to form the three letters of IBM (see Fig. 1.1), creating the most "minimal" advertisement for human beings, and the nano myth shocked the world again. In 1993, the Beijing Vacuum Physics Laboratory of Chinese Academy of Sciences was also able to manipulate atoms freely and successfully wrote the word "China", which indicated that China began to occupy a place in the field of international nanotechnology. In July 1990, the first international conference on nanotechnology was held in Baltimore, USA, which marked the formal birth of nanotechnology.

In 1991, carbon nanotubes were discovered by human beings. Its mass is one sixth of that of steel with the same volume, but its strength is ten times that of steel. Professor Smalley, winner of the Nobel Prize in Chemistry, believes that carbon nanotubes will be the first choice for the best fibers in the future, and will also be widely used in ultramicro wires, ultramicro switches and nanoscale electronic circuits. Therefore, carbon nanotubes have become the research focus of nanotechnology.

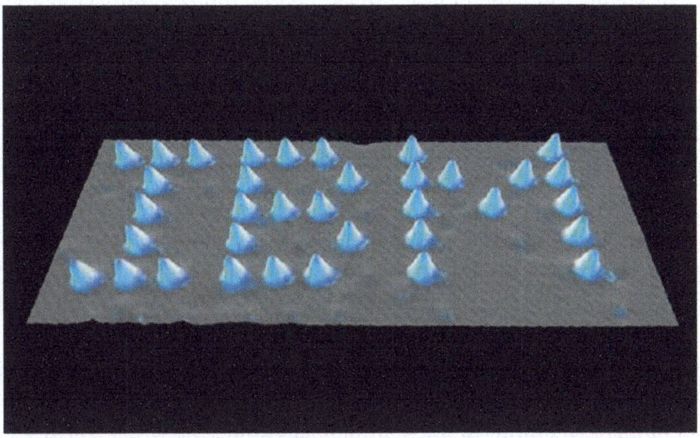

Fig. 1.1 IBM image written by atomic relocation [2]

In 1997, American scientists successfully moved a single electron for the first time. It is expected that a quantum computer with thousands of times higher speed and storage capacity will be developed in 20 years. In 1999, Brazilian and American scientists invented the world's smallest "scale" when conducting experiments on carbon nanotubes, which can weigh 10^{-9} g of objects, which is equivalent to the weight of a virus; shortly thereafter, German scientists developed a scale that can weigh a single atom, breaking the record jointly created by American and Brazilian scientists [3].

In 1999, nanotechnology gradually entered the market, and the annual turnover of nano products reached 50 billion US dollars. Throughout the world, governments in both developed and developing countries have realized that the development of nanotechnology will become a new driving force for economic growth in the twenty-first century, so they have continuously increased their investment in nanotechnology research and development. At the same time, some countries have formulated relevant strategies or plans and invested huge sums of money to seize the strategic highland of nanotechnology. Japan has set up a research center for nanomaterials and put nanotechnology into the research and development focus of the new 5-year science and technology basic plan; Germany has established a nanotechnology research network; the United States regards the nano-plan as the core of the next industrial revolution, and American government departments have increased their investment in basic scientific and technological research from $116 million in 1997 to $470 million in 2001. According to the report published by the Third World Academy of Sciences in July 2005, the three world economic entities, the United States, Japan and the European Union, plan to invest more than $3.7 billion, $3 billion and $1.7 billion in nanotechnology from 2005 to 2008, respectively, and the research projects cover many fields such as energy, medicine, microelectronics industry, materials and environment [4].

In the face of the world's nano-research fever, our government also attaches great importance to and supports the development and research of nanotechnology. In 2001, the Ministry of Science and Technology, the State Planning Commission, the Ministry of Education, the Chinese Academy of Sciences and the National Natural Science Foundation of China set up the National Nanotechnology Guidance and Coordination Committee to make overall plans for the research direction of nanotechnology in China. In the National Outline for the Development of Nanotechnology issued jointly in July 2001, it is pointed out that the main goals of the development of nanotechnology in China in the next 5–10 years are: to make significant progress in the frontier of nanotechnology, lay a foundation for development, and make major breakthroughs in the development and application of nanotechnology; gradually form a lean, cross-integrated and sustainable innovation backbone team of nanotechnology. In this development plan, it is also planned to establish a national nanotechnology research and development center and an industrialization base with enterprises as the main body to promote the coordinated development of basic research, applied research and industrialization. In the National Medium- and Long-Term Science and Technology Development Plan for 2006–2020 formulated by the State Council in early 2006, nanoscience was listed as one of the four main directions of basic scientific research during this period, and nanomaterials and nano-devices were regarded as the key targets for developing advanced materials. The key research and development projects related to nanotechnology are: the core technologies of nanoelectronics and nanobiology; R&D and industrialization of new functional materials; development of micro-nano electromechanical system on submicron scale [5]. During the Tenth Five-Year Plan period, China has made great progress in the research and development of nanotechnology and nanotechnology. The number of papers related to nanotechnology written by researchers in China is increasing at an annual rate of about 30%. In recent years, according to the statistics of the number of papers published in the core journals of the American Science Citation Index, the total number of nano-science papers in China ranks among the top in the world, and there are many research results published in world-famous scientific journals such as Science and Nature. According to incomplete statistics, there are more than 50 universities in China, as well as more than 20 research institutes and more than 300 enterprises of Chinese Academy of Sciences engaged in research and development related to nanotechnology, and there is a research team of more than 3000 people from research institutes, universities and enterprises.

1.2.2 Research Progress of Nanotechnology

1.2.2.1 Nanomaterials

The development of nanotechnology depends on nanomaterials, so nanomaterials are the foundation and core of nanotechnology. At the same time, nanotechnology is fundamentally changing the manufacturing and production methods of materials and devices in the future. The application of nanotechnology can make materials and products from atoms and molecules, that is, from atoms and molecules to nano-powder fibers and other small structural components, and then to materials and products [6]. This small-scale manufacturing method requires less materials, consumes less energy and causes lower pollution. On the nanometer scale, nano-structural units are synthesized by accurately controlling the size and composition, and then they are combined into larger structures with unique properties and functions to prepare lighter, stronger and designable materials [7].

Since 1990s, the development of quasi-one-dimensional nanomaterials has been the frontier field of nanotechnology. In January 1991, the carbon nanotubes, which are multilayer coaxial tubes, were first observed by Sumio Iijima of NEC Laboratory in Tsukuba, Japan. In October 2000, researchers from the University of Pennsylvania in the United States published an article in Science, saying that the mass of carbon nanotubes is one-sixth of that of steel with the same volume, but it is 100 times stronger than steel. It not only has good electrical conductivity, but also is the best thermal conductivity material at present. The excellent thermal conductivity of carbon nanotubes makes it a heat sink for computer chips in the future and can also be used as a protective material for various high-temperature components such as engines and rockets. The latest research shows that the cavity in carbon nanotubes can not only serve as a miniature test tube, mold or template, but also seal the second substance in this confined space and induce it to have a structure and behavior that can't be seen in macro materials. Computer simulation shows that the water sealed in carbon nanotubes can exist as a new ice phase. Under suitable conditions, the obvious boundary between liquid phase and solid phase in carbon nanotubes will disappear, and liquid substances will be continuously transformed into solids without obvious solidification process.

Modification of traditional materials by nano-addition is a basic method and idea to prepare nanomaterials. For example, adding 3–5% of 27 nm nanometer alumina to alumina ceramic substrate material can improve thermal stability by 2–3 times and thermal conductivity by 10–15%. Nanomaterials are added to plastics to enhance their anti-aging ability and prolong their life. Adding it to rubber can improve dielectric and wear resistance. Hiroichi Shinhara of Japan summarized the performance improvement of several nano-composite ceramics and found that nano-composite technology improved the strength and toughness of ceramic matrix materials by 2–5 times, and the working temperature increased by 25–133%.

The excellent properties of nanoparticles, such as large specific surface area, high surface reactivity, many surface-active centers, high catalytic efficiency and strong adsorption capacity, make them have important applications in chemical catalysis. Nano-powders such as platinum black, silver, aluminum oxide and iron oxide have been directly used as catalysts for oxidation, reduction and synthesis of polymer, which greatly improves the reaction efficiency. For example, 30 nm nano-Ni powder can increase the reaction degree of organic chemical hydrogenation and dehydrogenation by 15 times. The catalyst made of Ni and Cu–Zn alloy ultrafine powder with particle size less than 300 nm can make the hydrogenation reaction efficiency of organic matter 10 times higher than that of traditional Ni catalyst. Ultra-fine mixed light sintered body of Fe, Ni and γ-Fe_2O_3 can replace precious metal Pt powder and WC powder as purifying agent for automobile exhaust.

Since the discovery of semiconductor photocatalytic effect, people have been paying attention to it. The so-called semiconductor photocatalytic effect means that under the irradiation of light, valence band electrons jump to conduction band, and the holes in valence band take away hydroxyl electrons from the surrounding environment, and hydroxyl groups become free radicals and strong oxidants. Commonly used photocatalytic semiconductor nanoparticles include TiO_2 (anatase), Fe_2O_3, CdS, ZnS, PbS, PbSe, $ZnFe_2O_4$, etc. Photocatalytic effect of semiconductors has important applications in environmental protection, water quality treatment, organic matter degradation and ineffective pesticide degradation. For example, the United States and Japan make the above materials into hollow spheres, which float on the surface of wastewater containing organic matter or seawater polluted by oil spill and degrade organic matter or oil by sunlight. Coating a nano-TiO_2 film on the surface of automobile windshield and rearview mirror can play an anti-fouling and anti-fog role. Nano-TiO_2 powder can also be added to ceramic materials to make them have cleaning and sterilization functions, and can also be added to man-made fibers to make sterilization fibers [6]. The surface of anatase nano-TiO_2 particles is modified with Cu and Ag, which has better sterilization effect than single nano-TiO_2 or Cu and Ag, and has broad application prospects in refrigerators, air conditioners, medical devices, hospital operating rooms and so on. After killing bacteria, commonly used fungicides such as Ag and Cu may cause typhoid fever and cholera due to the release of pyrogenic and toxic components such as endotoxin. However, TiO_2 photocatalytic degradation of bacteria into CO_2, H_2O and organic acids does not exist this problem. Hydrogen can be extracted from methanol aqueous solution by photocatalytic effect of nano-TiO_2. Propyne can react with water vapor to produce combustible methane, ethane and propane by using Pt-doped nano-TiO_2 particles. Pt-doped nano-TiO_2 particles decompose acetic acid into methane and CO_2 by photocatalysis. In order to improve the photocatalytic efficiency, people also try to assemble nano-TiO_2 into porous solids to increase the specific surface, or compound zinc ferrite with nano-TiO_2 to improve the utilization rate of sunlight. Using porous ordered array Al_2O_3, anatase nano-TiO_2 wire array was synthesized in the micro of nano-cylindrical holes, and then the composite was glued to the bottom of epoxy resin village. After the template was removed, nano-TiO_2 wire array was formed on the bottom of ring resin village. Because of the large specific surface area of nanowires, the light receiving capacity of TiO_2 films with the same plane area is increased by 100 times, and the maximum photocatalytic efficiency can be increased by more than 300 times.

The future development direction of nanotechnology is to realize the "bottom-up" method to construct nano-devices. At present, there are two kinds of attempts in this field. One is to artificially realize single-atom manipulation and molecular surgery. Researchers of Osaka University in Japan have synthesized three-dimensional nano-cows and nano-springs in polymer materials by using two-photon absorption technology. The other is the molecular self-assembly technology of various systems. At present, the research object mainly focuses on nano-array system, Nano-inlay system, mesoporous and nano-particle composite system and nano-particle film. The purpose is to design a new material system and explore or improve the properties of materials according to the needs. The goal is to make preliminary preparations for the fabrication of nano-devices, such as high-brightness solid-state electronic display screens, nanocrystalline diodes, vacuum ultraviolet to near-infrared photovoltaic power generation and electronic light-emitting tubes controlled by blue, green and red light, etc. Nanocrystals can be used as the main materials, which are called "quantum" nanocrystals internationally. At present, nano-devices have been designed in the laboratory, including Si–SiO_2 light-emitting diodes, Si–Ni-doped nano-particle light-emitting diodes, and red, green and blue tunable diodes made of CdSe with different nanoscales. Scientists in bell laboratory used the self-assembly technology of organic molecule mercaptan to prepare single-layer field effect transistors with a diameter of 1–2 nm.

In addition, scientists expect to realize complex life-like functions in materials, create compounds with life-like behaviors, and make materials smart. For example, concrete can detect signs of strength decline internally, or it can respond to external corrosion and release chemicals to resist corrosion. For another example, it is expected that building materials can feel the weather conditions and respond to the weather changes by changing their internal structure so that air and moisture can penetrate. In the future, the comfort and energy efficiency of buildings will be greatly improved, and sensitive hazards can be automatically corrected or avoided. We may also create a chameleon-like camouflage materials that can change its shape and color at any place and at any time so as to be close to its environment. We can also make alloys with self-repairing function, which will automatically fill, bridge and strengthen tiny cracks. The materials made by combining plastic molecular chains with ceramic nanoparticles will be more wear resistant. In the near future, we can use nanotechnology to develop many materials that have never been seen in nature before.

In recent years, China has made great achievements in the research of functional nanomaterials, which has attracted international attention. The first is the synthesis of large-area oriented carbon nanotube arrays: The aperture of the nanotubes synthesized by this technology is basically the same, about 20 nm, the length is about 100 pm, and the nanotube array area reaches 3 mm*3 mm. This large-area oriented carbon nanotube array has important application prospects in field emission cathode of flat panel display. This article was published in the American journal Science in 1996. The second is the preparation of ultra-long carbon nanotubes: for the first time, ultra-long aligned carbon nanotube arrays with a length of 2–3 mm were prepared in large quantities. The length of this super-long carbon nanotube is 1–2 orders of magnitude longer than that of the existing carbon nanotube. This achievement was published in the British journal Nature in August 1998. The Financial Times of the United Kingdom introduced the work on long nanotubes with the title "Carbon nanotubes enter the stage of human growth". Thirdly, preparation of nitride nanorods: for the first time, one-dimensional blue-emitting gallium nitride nanorods with a diameter of 3–40 nm and a length of micro-scale were successfully prepared by using carbon nanotubes as templates, and the concept of limited reaction of carbon nanotubes was put forward. This achievement was rated as one of the top ten science and technology news in China in 1998. Fourth, the carbon nanotube array on the bottom of the silicon village was successfully developed, and the application of carbon nanotubes in field emission plane and nano-devices was promoted. Fifthly, one-dimensional nanowires and cables: a new method combining sol–gel and carbothermal reduction was used to synthesize tantalum carbide (TaC) nanowires coated with insulator SiO_2 and TaC nanowires coated with stone for the first time. At present, only a few research groups in the world can synthesize this material. Many foreign scientists spoke highly of this research paper after it was read out at the Fourth International Nano Conference held in Sweden in 1998. Sixth, nano gallium nitride microcrystals were prepared by benzene thermal method; non-aqueous solvothermal synthesis technology was discovered, and zinc nitride microcrystals with particle size of 30 nm were made at about 300 °C for the first time. Nano-crystallites of chemical (CrN), phosphate drill (Co_2P) and antimony sulfide (Sb_2S_3) were also prepared by synthesis, and the paper was published in the journal Science in 1997. Seventh, nano-diamond is made by catalytic pyrolysis; Diamond nano-powder was prepared by the reaction of carbon tetrachloride and sodium by catalytic pyrolysis at medium temperature (70 °C) in an autoclave. The paper was published in the journal Science in 1998. It is highly praised as "straw turns into gold".

1.2.2.2 Nanoelectronics

A major driving force for the development of nanotechnology comes from the electronics industry. The goal of nano-electronics is to further reduce the geometric structure of integrated circuits, beyond the limits encountered in current development, so as to make their functional density and data throughput reach unimaginable levels at present. To achieve this goal, it is necessary not only to innovate the concept of devices, but also to develop a brand-new manufacturing method of integrated circuit blocks in order to overcome the limitation of interconnection. On the nanometer scale, the physical laws that the traditional transistors follow are no longer applicable, and new physical effects will appear. Using nanoelectronics, novel quantum devices can be developed, such as resonant tunnel diodes, quantum lasers and quantum interference devices. By then, mankind may enter the "quantum kingdom". Another attractive research direction of nanoelectronics is the development of molecular electronic devices and biomolecular devices, which are still completely based on molecular combination and completely abandoned the silicon semiconductor-based electronics [8].

In the future, all nano-electronic devices will be smaller (higher integration), faster (faster response) and colder (lower power consumption and lower temperature rise of a single device). If the recording medium is in the form of nano-layers and nano-dots, the information on 1000 CDs may be stored in a watch-sized memory. In addition to a thousand times or even a million times the storage capacity, the speed of the computer will also be greatly improved. Devices that transmit electromagnetic signals (including radio signals and laser signals) will become smaller and more powerful. Anyone and anything will be connected to the future Internet at any time and any place. The future Internet is more like a ubiquitous information environment than just a computer network.

The Semiconductor Industry Association of the United States (SIA) has drawn up a technical development route for the continuous improvement of miniaturization, speed and power consumption of information processing devices. These information processing devices include nanosensors for signal acquisition, logic devices for signal processing, memories for data memory, displays for visualization and transmission devices for communication. According to SIA's prediction, by 2010, the precision of semiconductor chips can reach 100 nm, which is not far from nano-structured devices. In fact, in 1999, the University of California and Hewlett-Packard Company had successfully developed a 100 nm chip. In 1998, the University of Minnesota and Princeton University produced quantum disks.

At present, many kinds of nano-devices have been successfully developed by using nanotechnology. Single electron transistor, red, green and blue tunable nano light emitting diodes and micro detectors made of nano wires and nanorods have come out. Hitachi, Japan has successfully developed a single electronic transistor, which accomplishes a specific function by controlling the motion state of a single electron, that is, an electron is a multifunctional device. The University of Wisconsin has made quantum dots that can hold a single electron. Billions of such quantum dots can be accommodated on a needle tip. Quantum dots can be used to make single electronic devices with small volume and low energy consumption, which will be widely used in microelectronics and optoelectronics. In addition, if billions of quantum dots can be connected, each quantum dot is equivalent to a nerve cell in the brain, and combined with micro-processing technology, it will bring hope for the development of intelligent microcomputers. NEC research institute of Japan has the technology to fabricate fine quantum wire structures below 100 nm and has successfully fabricated quantum dot arrays with switching function on the bottom of GaAs semiconductor village. The United States has also successfully developed a nano-device with switching characteristics with a size of only 4 nm, which is driven by laser and has a fast opening and closing speed.

The reading head developed by taking advantage of the remarkable giant magnetoresistance (GMR) and tunneling magnetoresistance (TMR) phenomenon in nanomagnetism can improve the recording density of the magnetic disk by more than 30 times. Researchers in Zurich, Switzerland have prepared nanowires filled with Cu and Co alternately and used its giant magnetoresistance to prepare an ultramicro magnetic field sensor. Magnetic nanoparticles have small particle size, single domain structure and high coercivity, so they can be used as magnetic recording materials to improve signal-to-noise ratio and image quality. In 1997, the Nanostructure Laboratory of the Department of Electronic Engineering of the University of Minnesota successfully developed a nanostructured magnetic disk by nanolithography, and the Co rods with a length of 40 nm were arranged in a periodic array of quantum rods. Because nano-magnetic units are separated from each other, they are called quantum disks. It utilizes the storage characteristics of magnetic nanowire arrays, and the storage density can reach 400 Gb/in^2. Magnetic sensors prepared by using giant magneto-impedance effect of iron-based nanomaterials have come out, and magnetic liquids coated with superparamagnetic nanoparticles are also widely used in aerospace and some civil fields as long-life dynamic rotary seals.

1.2.2.3 Nanomedicine

Life systems are controlled by molecular behaviors on the nanometer scale, such as RNA–protein complexes in organisms, whose linearity is between 15 and 20 nm, and nucleic acids, lipids, hydrocarbons and viruses in organisms are also nanoparticles.

Cell organelles and other structural units in cells are "nano-machines" that perform certain functions. Cells are like "nano-workshops", and photosynthesis in plants are typical examples of "nano-factories". Every enzyme protein molecule in a cell is a living nano-robot. The change of the conformation of enzyme protein makes the actions between different domains of enzyme molecule like micro-human moving and rearranging the atomic arrangement of catalyzed molecules. All structural units in cells are micro-machines that perform some function, and ribosomes are processors that arrange amino acids in order to make protein molecules according to the instructions of genetic code. Golgi apparatus is a processing plant for processing and modifying newly manufactured protein molecules. The processed protein can be transported from the membrane vesicle to a certain position to function according to the instruction of the signal brain. Protein, which has completed its functional mission, will be labeled and sent to be hydrolyzed into amino acids and reused in the synthesis of new protein.

The size of nanoparticles is often smaller than cells and red blood cells in organisms, which provides a new opportunity for medical research. Injecting smart drugs coated with nanoparticles into blood and delivering them to the focus cells opens up a brand-new way for drug transmission and greatly enhances the efficacy of drug treatment. Nanotubes can absorb drug molecules and release them slowly in a certain period of time, making controllable drugs a reality. At present, examples that have been well applied include: the technology of cell separation by using nano-SiO_2 particles, the intracellular staining of nanoparticles, especially gold (Au) particles, and local targeted therapy by new drugs or antibodies coated with magnetic nanoparticles.

Biochips, including cell chips, protein chips (biomolecular chips) and gene chips (DNA chips), have the advantages of integration, parallelism and rapid detection, and have become the frontier technology of nano-bioengineering. It will be directly applied to clinical diagnosis, drug development and human genetic diagnosis. After being implanted into human body, people can enjoy medical treatment anytime and anywhere, and the premonitory information of diseases can be found in dynamic detection, making early diagnosis and prevention possible.

The application of nanotechnology in gene therapy is the most exciting field of nanotechnology, which mainly includes gene modification and bionic based on ordered assembly of DNA molecules and simulation of biological ordered structure. In gene modification therapy, protein and nucleic acid images can be obtained by tunnel scanning microscope, and DNA molecules can be deformed and rearranged in a small space.

In the aspect of DNA nano-bionic manufacturing, we use the specificity of base complementary rule, the simplicity of base, the diversity of genetic information and the topological targeting of double helix structure in the process of DNA replication and combine nanotechnology to manipulate a single atom and molecule to produce nano-organic–inorganic composite machines with various functions similar to every link in the life process. Using nanomaterials as gene delivery system has obvious advantages. It has been reported that using nanotechnology can make DNA locate in cells through active targeting and concentrate plasmid DNA to 50–200 nm with negative charge to promote its effective entry into the nucleus.

However, we can't ignore the negative effects of nanomaterials on organisms. In April 2003, Science first published an article to discuss the possible biological effects of the interaction between nanomaterials and biological environment. Subsequently, Nature and Science magazines published their editorial articles four times in 1 year, and the American Chemical Society and many academic magazines in Europe also published articles one after another, discussing with scientists in various fields whether nano-biological effects, especially whether nanoparticles have potential negative effects on human health, living environment and social security, that is, the environmental safety of nano-organisms. The research results of nano-biological effects put forward a new direction in the field of chemistry-modified chemistry to reduce or even eliminate nano-toxicity. Chemical modification of nano-molecules with negative nano-biological effect can eliminate their toxicity while maintaining their functional characteristics.

At present, there are few published research data on the negative biological effects of nanomaterials, and there is no systematic research data on any kind of nanomaterials. This work still needs a long time of accumulation and development, and a sound research system is established.

1.2.2.4 Nanomilitary

Nanotechnology will bring revolutionary influence on national defense and military field. For example, nano-electronic devices will be used for real-time communication between virtual training system and battlefield; nano-detection systems in chemistry, biology and nuclear weapons; new nano-materials can improve the strike and protection ability of conventional weapons; small robots made of nano-micro mechanical systems can complete special reconnaissance and strike tasks; nano-satellites can be launched by a small launch vehicle, forming a satellite network according to different orbits, monitoring every corner of the earth and making the battlefield more transparent. The application of nanomaterials in stealth technology is particularly noticeable.

The U.S. Department of Defense clearly recognized the importance of nanotechnology more than 10 years ago and has been playing an important role in supporting the development of nanotechnology. It has invested 35 million U.S. dollars every year to develop new nano weapons. Through the application of advanced nano-electronic devices in information control, the army will take advantage of information and make rapid response in early warning, missile interception and other fields. Through the application of nano-mechanics and micro-robots, unmanned combat aircraft will improve the flexibility of troops and increase the effectiveness of combat because there is no limitation of acceleration force on drivers. The performance of the national nuclear defense system will be greatly improved by controlling with nano-and micro-mechanical equipment. Through the application of nano-material technology, the corrosion resistance, wave absorption and concealment of weapons and equipment can be greatly improved, which can be used in ships, submarines and fighters. In the Gulf War in 1991, the fighter planes dispatched by the United States escaped Iraq's tight radar surveillance network on the first day and quickly reached the capital Baghdad, directly destroying the telegraph building and other military targets. During the 42-day battle, 1270 planes carried out the mission, which destroyed 95% of the important military targets of the Iraqi army, while none of the American fighters were damaged. This high-tech war once shocked the world. Why is Iraq's radar defense system helpless against American fighters? Why do American missiles hit Iraqi military targets so accurately? Why did Iraq's Iraq hit the air-to-surface missile? One reason is that the fuselage surface of American fighter F117A is covered with infrared and microwave stealth materials, which has excellent broadband microwave absorption ability and can escape radar monitoring. However, Iraqi military targets and tanks and other weapons have no stealth materials to defend against infrared detection, so they are easily found by sensitive infrared detectors on American fighter planes and hit accurately by advanced laser-guided weapons.

The stealth material on the skin of American F117A aircraft contains many kinds of ultrafine particles, which have strong absorption ability to electromagnetic waves in different bands. Why do ultrafine particles, especially nanoparticles, have stealth effect on infrared and electromagnetic waves? There are two main reasons: on the one hand, because the size of nanoparticles is much smaller than the wavelength of infrared and radar waves, the transmittance of nanoparticles is much stronger than that of conventional materials, which greatly reduces the reflectivity of waves and makes the reflected signals received by infrared detectors and radars very weak, thus achieving the function of stealth; on the other hand, the specific surface area of nanoparticles is 3–4 orders of magnitude larger than that of conventional coarse powder. The absorption rate of infrared light and electromagnetic wave is much higher than that of conventional materials, which greatly reduces the intensity of reflected signals obtained by infrared detectors and radars, so it is difficult to find the detected target and plays a stealth role. American scientists have used nanotechnology to develop smart combat suits for more than 10 years. In addition to hoping that the fabric of the combat suit has chemical protection function, they also envisage installing a microcomputer and a highly sensitive sensor in the combat suit. In this way, soldiers will be alerted in time and easily avoid incoming bullets. In their vision, the smart combat suit can also monitor the important changes in the surrounding environment, and it has the camouflage ability like a chameleon and blends in with the surrounding environment. The more advanced virtual reality system based on nanoelectronics makes military training more economical and efficient. Using insects as a platform, molecular robots are implanted into the nervous system of insects to control insects to fly to the enemy to collect information, so that the target loses its function.

In addition, the application of nanotechnology in the military has the following ideas, which will also become a reality in the near future:

(1) "Sparrow" satellite: The United States put forward the concept of nanosatellite in 1995. This kind of satellite is slightly bigger than sparrow, and its weight is less than 10 kg. All components are made of nanomaterials and integrated by the most advanced micro-electromechanical integration technology. It is reconfigurable and renewable, with low cost, good quality and strong reliability. A small rocket can launch hundreds of nanosatellites at a time. If 648 nanosatellites with different functions are arranged at equal intervals in the sun-synchronous orbit, it can ensure continuous monitoring of any point on the earth at any time. Even if a few satellites fail, the work of the whole satellite network will not be affected.

(2) "Mosquito" missile: Because the working speed of nano-devices is much faster than that of semiconductor devices, it can greatly improve the processing capacity of information transmission and storage of weapon control system and can create an intelligent micro-navigation system with new principles, which will make the concealment, maneuverability and survivability of guided weapons change qualitatively. Using nanotechnology to make shapes like mosquitoes and micro-missiles can play a magical combat effectiveness. Nano-missiles are directly controlled by radio waves, which can sneak into the target unnoticed, and their power is enough to blow up enemy artillery, tanks, planes, headquarters and ammunition depots.
(3) "Fly" aircraft: This is a pocket aircraft the size of a fly, which can carry all kinds of detection equipment and has the ability of information processing, navigation and communication. Its main function is to be secretly deployed in or near enemy information systems and weapon systems to monitor the enemy's situation. These nano-planes can hover and fly, and enemy radar can't find them at all. It is said that it can also adapt to all-weather operations and can transmit the information it has obtained back to its own missile launch base from hundreds of kilometers away, and directly guide missiles to attack targets.
(4) "Ant soldier": This is a miniature missile man controlled by sound waves. These robots are smaller than ants, but they have amazing destructive power. They can sneak into enemy weapons and equipment through various channels and stay hidden for a long time. Once activated, these "nano-soldiers" will show their magical powers: some specialize in destroying enemy electronic equipment, making it short-circuited and destroyed, and some act as blasters and detonate targets with special explosives; Some release various chemicals, which make the enemy metal brittle, oil coagulate or make the enemy personnel nervous and lose their combat effectiveness.

1.3 Classification of Nanomaterials

The material named after "nano" was in the 1980s. As a definition of a material, nanoparticles are limited to 1–100 nm, and nanomaterials refer to materials with nano-scale phase in microstructure. It includes three levels, namely nanoparticles, nanosolids and nano-assembly systems [9]. According to the nature, structure and performance of materials, there are different classification methods.

From different directions, nanomaterials can be divided into the following categories:

(1) According to the structure: Zero-dimensional nano-materials: the three dimensions of this material are nano-scale, that is, nanoparticles, atomic clusters, etc.; One-dimensional nano-materials: the dimensions of this material in two dimensions in space are nano-scale, that is, nanowires, nanorods, nanotubes, etc., or collectively referred to as nanofibers; Two-dimensional nano-materials:

the size of this material is nano-scale only in one dimension of space, that is, ultra-thin films, multi-layer films, superlattices, etc.; Three-dimensional nanomaterials, also known as nano-phase materials (for example, nano-mesoporous materials).

(2) According to chemical components, it can be divided into nano-metals, nanocrystals, nano-ceramics, nano-glasses, nano-polymers, and nano-composite materials.
(3) According to the physical properties of materials, it can be divided into nano-semiconductors, nano-magnetic materials, nano-nonlinear optical materials, nano-ferroelectrics, nano-superconducting materials and nano-thermoelectric materials.
(4) According to the application, it can be divided into nano electronic materials, nano optoelectronic materials, nano biomedical materials, nano sensitive materials and nano energy storage materials.
(5) According to the order of nanomaterials, it can be divided into crystalline nano-materials and amorphous nanomaterials. Nanomaterials can be single crystal or polycrystalline. It can be crystal structure, quasicrystal or amorphous phase (glassy state), metal, ceramics, oxides, nitrides, carbides or composite materials.

1.3.1 Nanoparticles

Nanoparticles refer to the aggregate of particles with linearity between 1 and 1–100 nm, which is the general name of all kinds of particle aggregates with this geometric size. The morphology of nanoparticles is not limited to sphere, but also sheet, rod, needle, star, mesh and so on [10]. Generally speaking, when the linearity of micro-particle polymer is less than 1 nm, it is called cluster, while the linearity of micro-powder is in micron level. The linearity of nanoparticles is just between the two, so they are also called ultrafine particles. Nanoparticles are invisible to ordinary microscopes, and the diameter of atoms is 0.08 nm; The diameter of nonmetallic atoms is generally 0.1–0.2 nm; The diameter of metal atoms is 0.3–0.4 nm; The size of red blood cells in blood is 200–300 nm; The virus is tens of nanometers in size. Therefore, nanoparticles are smaller than red blood cells and are about the same size as viruses.

Although the concept of nanomaterials is defined in modern times, the preparation and application of nanomaterials have existed in ancient China. The most typical particles are the ink in Four Treasures of the Study, which is carbon-containing nano-powder, and the antirust layer on the surface of ancient bronze mirrors in China has also been proved to be a thin film composed of nano-tin oxide particles. In September 2006, Nanocommunication published an article entitled "Early Use of Lead Sulfide Nanotechnology in Ancient Hair Dyeing Formulas". The research, which was jointly completed by researchers from French National Scientific Research Center, L 'Oré al Research Institute, Argonne National Laboratory and National Aviation Research Office, revealed an amazing discovery: as early as more than 2000 years ago, ancient

Greeks and Romans had used black lead sulfide nanocrystals formed on fiber cores to dye black and white hair and wool. It was just that people at that time did not clearly understand and realize the particle size of powder.

1.3.2 Nanosolid

Nano-solids are aggregates formed by the aggregation of nanoparticles. From the point of view of geometry, nano-solid can be divided into nano-block materials, nano-film materials and nano-fiber materials. These forms of nanosolids are also called nanostructured materials. Nano-particles, which constitute nano-solids, can be single-phase, different materials or different phases, which are called nano-phase materials and nano-composite materials respectively [11]. In 1963, Tian Liangji, a professor at Nagoya University in Japan, first obtained nanoparticles with clean surfaces by evaporation and condensation. In 1984, the team led by German Professor H Gleiter first developed the first artificial metal solid (CuPaAg and Fe). TiO_2 nano-solid was developed with Argonne Laboratory in the United States. At the end of 1980s, artificial nanosolids such as alloys, semiconductors and ceramic ionic crystals came out one after another.

The main feature of nano-solid materials is that they have huge inter-particle interfaces. For example, the solid composed of 5 nm particles will contain 10^{19} grain boundaries per cubic centimeter, and the diffusion coefficient of atoms is 10^{14}–10^{16} times higher than that of bulk materials, which makes nanomaterials have high toughness. Usually, ceramic materials have the advantages of high hardness, wear resistance and corrosion resistance, but they also have the disadvantages of brittleness and difficulty in machining. Nano-ceramics can increase toughness and improve brittleness to a certain extent. If the nano-ceramics are annealed to make the grain grow to micron size, the characteristics of ordinary ceramics will be restored. Therefore, the ceramics can be extruded and rolled by using the characteristics of nano-ceramics, and then heat-treated to transform them into ordinary ceramics, or surface heat-treated to keep the toughness inside the materials, but the surface shows high hardness, high wear resistance and corrosion resistance. The development trend of electronic ceramics is ultra-thin (only a few microns thick). In order to ensure homogeneity, the particle diameter should be about 1% of the thickness, so ultrafine particles should be used as raw materials. With the rapid development of integrated circuits, micro-components and high-power semiconductor devices, the demand for ceramic substrates with high thermal conductivity is increasing. The ceramic materials with high thermal conductivity include diamond, silicon carbide, aluminum nitride, etc. The thermal conductivity of compact sintered body made of ultrafine aluminum nitride is 100–220 W/(k m), which is 2.5–5.5 times higher than that of ordinary products. Fine ceramics made of ultrafine particles may be used in ceramic adiabatic turbine composite engines, ceramic turbines, high temperature and corrosion resistant bearings and rolling balls.

Composite nano-solid materials are also an important application field. For example, cermets containing 20% ultrafine cobalt particles are high temperature resistant materials for rocket nozzles. Metal aluminum contains a small amount of ceramic ultrafine particles, which can be made into a new structural material with light weight, high strength, good toughness and strong heat resistance. Ultrafine particles may also be used as raw materials for gradient functional materials. For example, the high-temperature resistant surface of the material is ceramic, the side contacting with the cooling system is metal with good thermal conductivity, and the composition between the ceramic and the metal changes slowly and continuously. This material can be used as the thermal insulation material of the space shuttle and the structural material of the nuclear fusion reactor with a temperature difference of 1000 °C. Graded functional materials are new materials developed in recent years. It is expected that they can be made into bioactive artificial teeth, human bones and artificial organs in medical biology, and can be made into composite electromagnetic functional materials and optical materials.

1.3.3 Nano-assembly System

The nano-structured system synthesized by artificial assembly is called nano-assembly system, which is also called nano-scale pattern material. Nanoparticles and their composed nanowires and tubes are used as basic units, which are assembled and arranged in one, two and three dimensions to form a system with nano-structure. Nanoparticles, wires and tubes can be arranged in order or disorder, which is characterized by being able to be designed according to people's wishes, and the whole system has the expected characteristics, so this field is considered as an important frontier topic in material chemistry and physics.

In recent 10 years, in the process of constructing ordered nano-structure system, people have developed nano-particle assembly technology. This technology takes nanoparticles (including zero-dimensional nanoparticles, one-dimensional nanowires, etc.) as basic structural units, combines molecular assembly technology and chemical modification technology, and controls the interaction between basic structural units, functional molecular assembly systems and basic structural units through structural design at all levels, thus constructing one-dimensional, two-dimensional and three-dimensional ordered assembly systems with special functions and properties. Self-assembly technology is a bottom-up, small-to-large manufacturing method, that is, the device is completely constructed from atomic or molecular level. Up to now, there is no mature view on the division of nanostructure self-assembly system. According to whether the driving force in the construction of nanostructure system is external or internal, it can be roughly divided into two categories: one is artificial nanostructure assembly system; The second is nanostructure self-assembly system and molecular self-assembly system.

Artificial nanostructure assembly system is a one-dimensional, two-dimensional and three-dimensional nanostructure system, including nano-ordered array system and mesoporous complex, which is artificially assembled and arranged by physical and chemical methods according to human will. People can also assemble their own nanoparticles, nanotubes and nanorods to create a new material system that does not exist in nature. Self-assembly system of nanostructures refers to connecting atoms, ions or molecules together to form a nanostructure through weak and less directional non-covalent bonds, such as hydrogen bonds, van der Waals bonds and weak ionic bonds. The key to the self-assembly process is not the simple superposition of weak forces among a large number of atoms, ions and molecules, but a whole and complex synergy. There are two important conditions for the formation of nanostructure self-assembly system. One is the existence of a sufficient number of noncovalent bonds or bonds, because noncovalent bonds such as Van der Waals bonds are very weak (211–412 kJ/mol), and only a sufficient number of weak bonds exist can a stable nanostructure system be constructed through synergy. The other is that the energy of self-assembly system is low, otherwise it is difficult to form a stable self-assembly system.

1.4 Basic Structural Units of Nanomaterials

The basic structural units that constitute nano-films, multilayer films, nano-phase solids and nanostructures are as follows: (1) clusters; (2) nanoparticles; (3) Quasi-one-dimensional nanomaterials (nanotubes, nanorods, nanowires, nano-spins, coaxial nano-cables, etc.); (4) artificial atoms.

1.4.1 Cluster

1.4.1.1 Definition of Clusters

Atomic, molecular or ionic clusters, referred to as clusters, are relatively stable aggregates formed by several to hundreds or even thousands of atoms, molecules or ions through physical and chemical binding forces.

Cluster is the transition state of various substances from atoms and molecules to bulk substances, or it represents the initial state of condensed matter, and it is a new level of material structure between atoms, molecules and macro-solids, sometimes called the "fifth state" of matter [12]. For example, C_{60}, C_{70}, etc. When evolving from atoms to macroscopic bulk materials, there is only one arrangement before four atoms, and for four atoms, it is quadrangular arrangement, that is, tetrahedron arrangement. When adding another atom to five atoms, there are two arrangements, that is, there may be two ways to grow up, as shown in Fig. 1.2.

One arrangement is that the fifth atom forms a triangle with an edge of the tetrahedron, that is, two included angles are formed. In this way, in the process of gradually

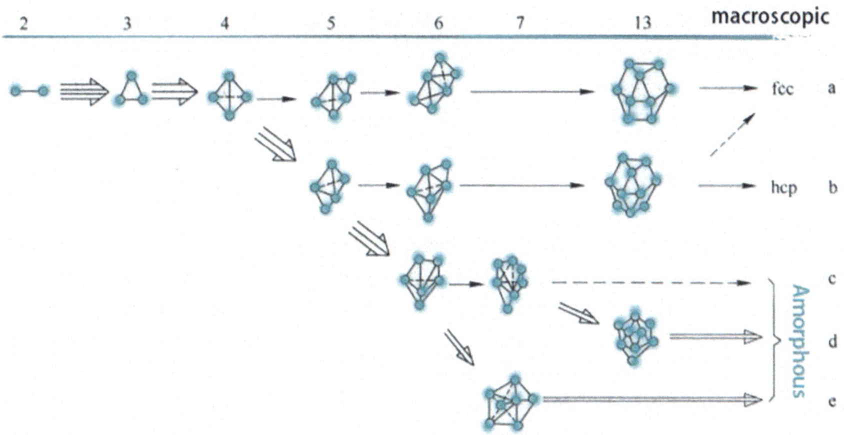

Fig. 1.2 Evolution from atoms to macroscopic bulk materials [13]

increasing atoms downwards, there is only one arrangement until the macroscopic crystal develops into the structural form of face-centered cubic crystal.

Another form of arrangement is the double triangular pyramid. Arranging in this way and adding another atom, that is, the number of atoms is 6, will produce two different forms of arrangement. In the process of gradually increasing atoms, this one is similar to the fcc above, that is, in the process of gradually growing up, there is only one way to arrange atoms until it develops into a macroscopic crystal and forms a hexagonal (hcp) crystal structure. In the process of increasing atoms, they will evolve into a variety of arrangement forms, until the macroscopic bulk materials, they all develop into non-crystal family. Clusters of several to hundreds or even thousands of atoms formed in the middle of these evolutions are called clusters. Clusters are smaller than nanoparticles and larger than atoms and molecules [14].

Atomic cluster is a newly discovered chemical species in 1980s, and its size is usually less than 1 nm. It is a thermodynamically unstable system, which is generally generated under non-equilibrium conditions, and is unstable compared with macroscopic substances and nanoparticles. In general, in-situ detection is used to study the properties of clusters, that is, instruments are used to detect them when they are just prepared, otherwise they will continue to grow, and clusters contain a limited number of atoms and molecules, which is less than that of nanoparticles.

1.4.1.2 Classification of Clusters

Atomic clusters can be divided into univariate atomic clusters, binary atomic clusters, polybasic atomic clusters and atomic cluster compounds. Univalent atomic clusters include metal clusters (such as Nan, Nin, etc.) and nonmetallic clusters. No nmetallic clusters can be divided into carbon (C_{60}, C_{70}, fullerene, etc.) and non-carbon clusters (such as B, P, S, Si clusters, etc.). Atomic cluster compounds are compounds formed by the combination of atomic clusters with other molecules by coordination chemical bonds.

1.4.1.3 Characteristics of Clusters

One of the basic research problems of clusters is to reveal the mechanism of clusters, that is, how clusters are gradually developed from atoms and molecules, and the changing laws of their structures and properties with this development. It includes the critical size and process change law of clusters developing into macroscopic solids. Clusters are often produced in non-equilibrium conditions and are difficult to produce in equilibrium gas phase. For smaller clusters, every time an atom is added, the cluster structure changes, which is called reconfiguration. However, when the cluster size reaches a certain level, it will become a massive solid structure. At this time, except for the relaxation of surface atoms, the reconstruction will no longer occur with the addition of atoms, and its properties will not change significantly. The corresponding cluster size is the critical size or called the joint point. This joint point may be different for different substances, and this difference can also show different growth characteristics for the same substance.

Atomic clusters are different from molecules with specific size and shape, aggregates formed by weak interaction between molecules and crystals with strong periodicity. Its shape can be varied, such as spherical, skeleton, onion, tubular, layered, linear and so on. They have not yet formed regular crystals, except inert gases, and they are all closely bonded aggregates by chemical bonds. Atomic clusters have many strange characteristics, such as extremely high chemical activity and catalytic activity, quantum size effect and nonlinear effect of light, geometric size effect of conductivity, conductivity and superconductivity of Ca-doped and doped atoms, and conductivity of carbon tubes and carbon onions.

1.4.1.4 Magic Numbers of Clusters

In the mass spectrometry analysis of various clusters, there is a common law: in the process that the abundance of clusters slowly decreases with the increase of the number of atoms N, when some specific values $n = N$, there is a sudden enhancement peak, which indicates that clusters with these specific numbers of atoms (molecules) have particularly high thermodynamic stability. This number n is called the magic number of clusters. This feature is very similar to the electronic state in atoms and the nucleon state in nuclei, indicating that clusters also have shell structure. This is closely related to symmetry and interaction potential. Magic cluster refers to a cluster with a specific number of atoms with a closed shell structure of electrons or atoms, so it is extremely stable. The specific number of atoms here is called magic number. Magic numbers are a series of separated numbers. For the two-dimensional close-packed system, these numbers are 1, 3, 6, 10, etc., that is, only when the number of atoms in the n(n + 1)/2 (positive integer) cluster is equal to the magic number, it will have high stability. The magic number sequence of the cluster is related to the bonding mode of the atoms that make up the cluster. Metal bonds are derived from free valence electrons, semiconductor bonds are oriented covalent bonds, alkali metal halides are ionic bonds, and the interaction between inert element atoms is van der Waals bonds.

The clusters composed of inert elements have a shell structure dominated by positional order. Carbon clusters are mainly covalent bonds, and their bonding methods are directional and saturated. Carbon clusters prepared by laser evaporation have high stability, especially those with atomic numbers of 20, 24, 28, 32, 36, 50, 60 and 70.

For metal clusters and metal clusters (large clusters), it is generally assumed that clusters are piled up according to the same (cubic ccp or hexagonal hcp) rule as bulk metals, that is, a central atom is surrounded by 12 outer atoms, forming a 13-atom cluster M_{13}, and then increasing layer by layer, which can form 2-layer, 3-layer and more layers of clusters respectively. The total number of atoms in each layer is related to the number of layers N, which is $10n^2$. Another way of stacking is to take M_{13} as the basic cluster unit and add the rest M to the basic cluster to form (M_{13}) and (n = 1–13) clusters respectively. With the decrease of metal particle diameter, the proportion of surface metal atoms will obviously increase. It can be predicted that if metal clusters and bulk metals are composed of the same number of atoms, the metal clusters have much larger surface atoms than bulk metals.

1.4.1.5 Properties and Research Significance of Clusters

The physical and chemical properties of clusters vary with the number of atoms, molecules or ions contained, and many of their properties are different from those of single atoms, molecules and bulk solids (or liquids), such as magic number and shell structure, quantum size effect and surface effect. The physical and chemical properties of clusters are a hot topic in current research, because they are aggregated by atoms and molecules, and they are a transition state from micro to macro. Therefore, they provide a new method to study the transition from atoms and molecules to condensed state system. By studying clusters, we can understand some complex condensed phenomena, such as nucleation, dissolution, adsorption, phase transition and other special systems. Just as embryology explains biological laws in its special way, cluster research helps us to understand some properties and laws of condensed matter (Fig. 1.3).

Cluster widely exists in nature and human practical activities, involving many materials movement processes and phenomena, such as catalytic combustion, crystal growth, nucleation and solidification, critical phenomena, phase transition, sol, photography, film formation and sputtering, etc. It constitutes the intersection of physics and chemistry, and is a chemical branch that spans synthetic chemistry, chemical kinetics, crystal chemistry, structural chemistry, atomic cluster chemistry, etc., as well as cross-atom, molecular physics, surface physics and so on. Clusters

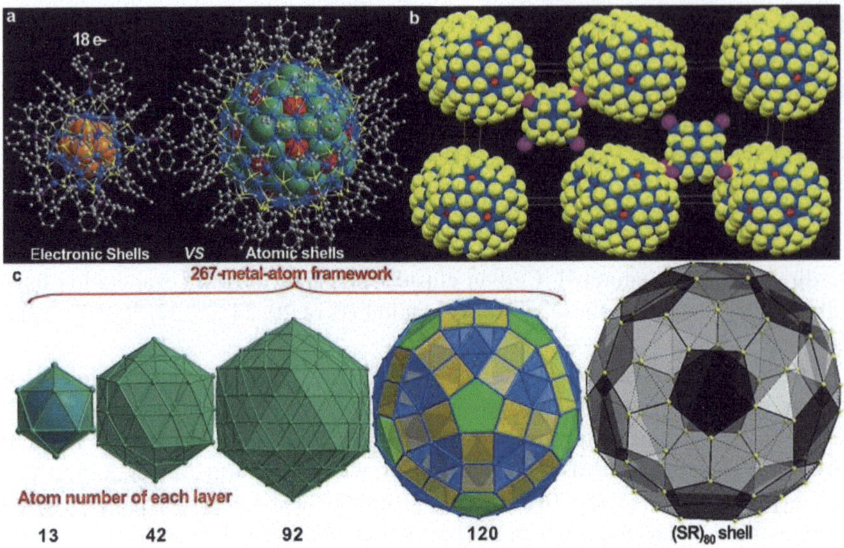

Fig. 1.3 Shell structure of metal clusters with different magic numbers [15]

have broad application prospects in quantum dot lasers and single electron transistors, especially in developing new materials as structural units.

1.4.2 Nanoparticles

Nanoparticles are also called nanoparticles, nanoparticles, ultrafine particles and nano-powder. Generally, it refers to granular substances with a particle size between 1 and 1–100 nm. Its scale is larger than atomic clusters and smaller than ordinary micropowder.

Nanoparticles are tiny particles invisible to the naked eye and general optical microscope. Therefore, the size of nanoparticles is a fraction of that of red blood cells and bacteria and is equivalent to or slightly smaller than that of viruses, so that such small objects can only be observed by high-power electron microscopes. Yoshiji Ueda of Nagoya University defines nanoparticles as: particles that can only be seen with an electron microscope are called nanoparticles [16].

As early as about 1861, with the establishment of colloid chemistry, scientists began to study the nanoparticle system (colloid), and the truly effective study of discrete nanoparticles began in the 1920s and 1960s. In the past decades, people have done a lot of work on the preparation, properties and application of various nanoparticles. In recent years, the research on the preparation, properties and applications of nanoparticles has become more popular, and a series of meaningful results

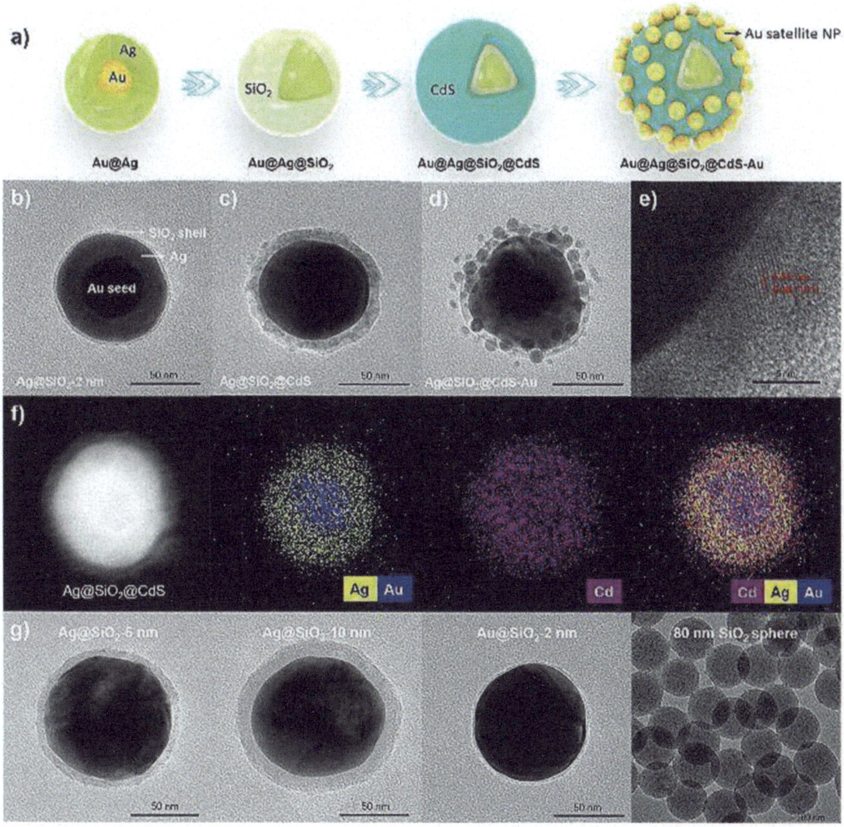

Fig. 1.4 Transmission electron microscope photos of nanoparticles from several materials [17]

have been obtained. Figure 1.4 is a transmission electron microscope photograph of nanoparticles of several materials.

1.4.3 Quasi-One-Dimensional Nanomaterials

Quasi-one-dimensional nano-materials refer to nano-materials in two-dimensional direction, whose length is much larger than that in the above-mentioned two-dimensional direction, and even macro-materials (such as millimeters and centimeters) [18]. According to the specific shape, it can be divided into tubes, rods, wires, silk rings, spirals, etc. A nanorod with a small aspect ratio (the ratio of length to diameter) and a circular cross section is a nanowire with a large aspect ratio and a rectangular cross section is a nanoribbon. Figure 1.5 is an electron microscope photograph of several quasi-dimensional nanomaterials.

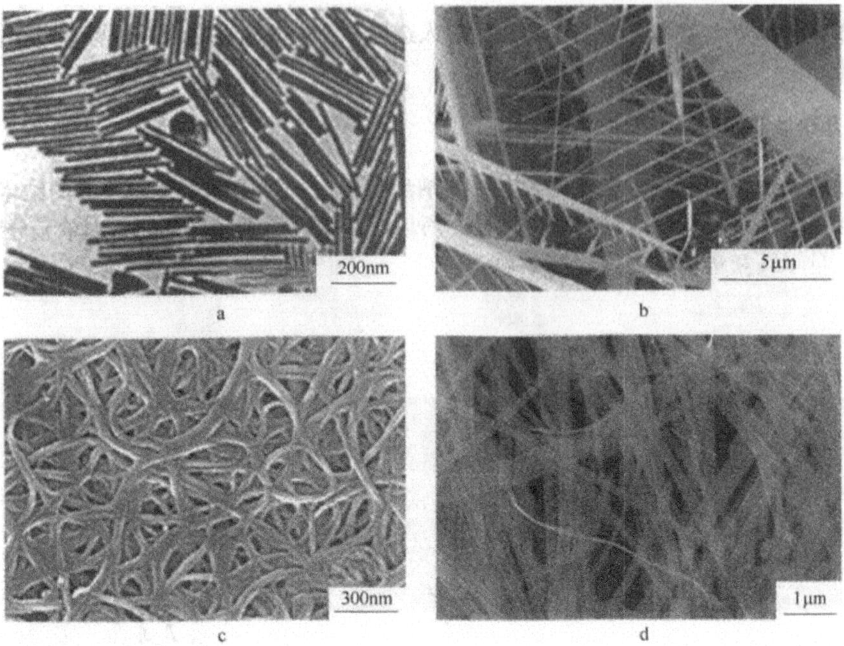

Fig. 1.5 Electron microscopy photos of several quasi one-dimensional nanomaterials [19]

In 1991, S. Iijima of Nippon Electric Company discovered carbon nanotubes (multi-walled tubes) while studying buckyballs. In 1993, single-walled carbon nanotubes were discovered again. The diameter of carbon nanotubes is generally from 0.33 nm to tens of nanometers, and the hollow space inside the tube can absorb many substances of appropriate size.

1.4.4 Artificial Atoms

Artificial atom is an aggregate composed of a certain number of actual atoms, the size of which is less than 100 nm, which was put forward by Professor Ashoori of Massachusetts Institute of Technology. Artificial atoms are sometimes called quantum dots. Artificial atoms include quasi-zero-dimensional quantum dots, quasi-one-dimensional quantum rods and quasi-two-dimensional quantum disks, and even quantum devices around 100 nm are regarded as artificial atoms. When the scale of the system is equivalent to the physical characteristic quantity, the quantum effect is very significant. When VLSI circuits are miniaturized to about 100 nm, the working principle of VLSI circuits based on traditional concepts and principles is challenged, and the principle of quantum mechanics will play an important role. The motion law of electrons in artificial atoms will appear new phenomena that are difficult to explain in classical physics. Dutch and British scientists have observed the quantization step phenomenon of electron transport in GaAs/GaAlAs artificial atoms.

In artificial atoms, when the coherence length of electron wave function is equivalent to the scale of artificial atoms, electrons can no longer be regarded as classical particles moving in the external field, and the fluctuation of electrons is fully exerted in transportation, which will lead to universal conductance fluctuation and non-local conductance. The unique quantum effect shown by the characteristics of electron transport in artificial atoms will lay a theoretical foundation for the design and manufacture of quantum effect principal devices and nanostructure devices.

1.4.4.1 Similarities Between Artificial Atoms and Real Atoms.

The similarities between man-made atoms and real atoms are as follows: (1) Man-made atoms have discrete energy levels and discontinuous charges, and electrons also move in an orbital manner in man-made atoms, which is very similar to real atoms. (2) When the electronic energy level of hydrogen atom is treated by quantum mechanics, the calculation of Schrodinger equation shows that the electronic energy level is quantized. (3) The law of electron filling is also similar to that of real atoms, and it obeys Hunde's rule.

1.4.4.2 The Difference Between Artificial Atoms and Real Atoms

The spectral measurement of a ring-shaped artificial atom containing a few electrons in a magnetic field reveals the triple structure of the first excited state, and it is found that the electrons in the artificial atom lack spin degeneracy. The differences between artificial atoms and real atoms are as follows: (1) Artificial atoms contain a certain number of real atoms. (2) There are various shapes and symmetries of artificial atoms. Real atoms can be described by simple spheres and cubes, but artificial atoms are not limited to these simple shapes. Except for quantum dots with high symmetry, tiny systems with low symmetry and complex shapes with sizes less than 100 nm can all be called artificial atoms. (3) The strong interaction between electrons in artificial atoms is much more complicated than that in real atoms. With the increase of the number of atoms in artificial atoms, the electron orbital spacing decreases, the strong Coulomb repulsion, the localized effect of the system and the Pauli incompatibility principle make the electron spins orderly arranged in the same direction. Artificial atom is the best object to study multi-electron system. (4) The electrons in the actual atom are attracted by the nucleus to do orbital motion, while the electrons in the artificial atom are in a parabolic potential well, which tends to fall to the bottom of the potential well. Due to Coulomb repulsion, some electrons are in the upper part of the potential well, which makes them have the characteristics of free electrons. (5) Another important feature of artificial atoms is that releasing an electron or taking out an electron can easily cause charge fluctuation, and releasing an electron is equivalent to charging the artificial atom. These phenomena are the physical basis for designing single-electron transistors.

1.4.5 Carbon Nanotubes, Nanorods, Nanowires and Nanowires

It is necessary to use high-resolution electron microscope technology to observe the related structure of carbon nanotubes. The research shows that multilayer carbon nanotubes are generally composed of several to dozens of single-walled carbon nanotubes, and the tube spacing is about 0.34 nm, which is equivalent to the [14] plane spacing of graphite. The diameter of carbon nanotubes is a few tenths to tens of nanometers, and the side of each single-walled tube is composed of hexagonal carbon atoms, and the length is generally tens of nanometers to microns, and the two ends are capped by pentagons of carbon atoms. Single-walled carbon nanotubes may have three types of structures, namely single-arm nanotubes, zigzag nanotubes and chiral nanotubes, as shown in Fig. 1.6 [20]. The formation of these types of carbon nanotubes depends on how the hexagonal lattice two-dimensional graphite sheet of carbon atoms is "rolled up" to form a cylinder. Different types can be explained according to the single cell of a carbon nanotube. In Fig. 1.7, the OAB'B box is a single cell of carbon nanotubes, the chiral vector $C_h = na_1 + ma_2$, a_1 and a_2 are unit vectors, n and m are integers, and the chiral angle θ is the included angle between the chiral vector and a1. In order to form nanotubes, the single cell OAB'B is rolled up, so that O and A, B and B' coincide, and the end is capped with 1/2 fullerene, and finally carbon nanotubes are formed. When the graphite sheet is rolled up to form the cylindrical part of the nanotube, the ends of the chiral vectors are weighted with each other, and the chiral vectors form the circumference of the circular cross section of the nanotube. Different M and N values lead to different nanotube structures. When $n = m$ and $\theta = 30°$, single-arm nanotubes are formed; when n or m is 0 and $\theta = 0°$, serrated nanotubes are formed. Theta is between 0 and 30 degrees to form chiral nanotubes.

The properties of carbon nanotubes are determined by their diameter and chiral angle θ, and these two parameters depend on the values of m and n. Diameter dt $= (\sqrt{3}/\pi)$ ac-c $(m^2 + mn + n^2)1/2$ac-c is the distance between adjacent carbon atoms on the graphite sheet, and $\theta = \tan^{-1}[\sqrt{3}\ n/(2\ m + n)]$ carbon nanotubes often exist in the form of nanorods or nanowires, which conforms to the definition of nanomaterials, that is, quasi-one-dimensional solid nanomaterials refer to two-dimensional directions. Generally speaking, the material with a small aspect ratio (the ratio of length to diameter) is called nanorod, and the material with a large aspect ratio is called nanowire. So far, there is no strict standard to distinguish between nanorod and nanowire. In many documents, nanowires with a length of less than 1 m are called nanorods, and those with a length of more than 1 pm are called nanowires or wires. Semiconductor and metal nanowires are often called quantum wires.

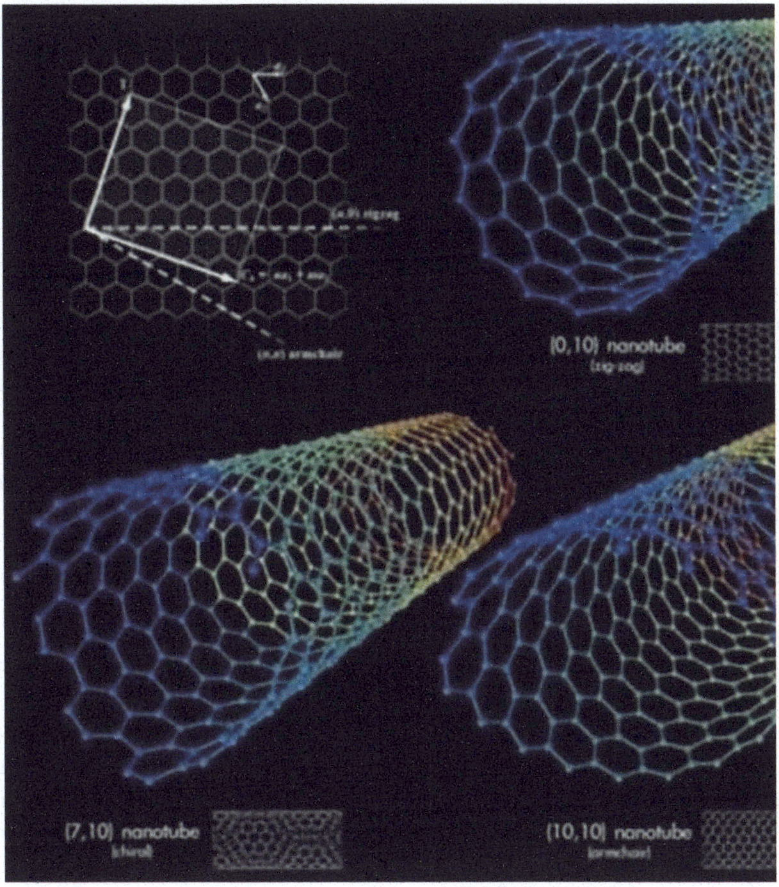

Fig. 1.6 Three types of carbon nanotubes [21]

Fig. 1.7 Schematic diagram of single cell OAB'B forming carbon nanotubes on a two-dimensional graphite sheet at (n, m) = (4, 2) [22]

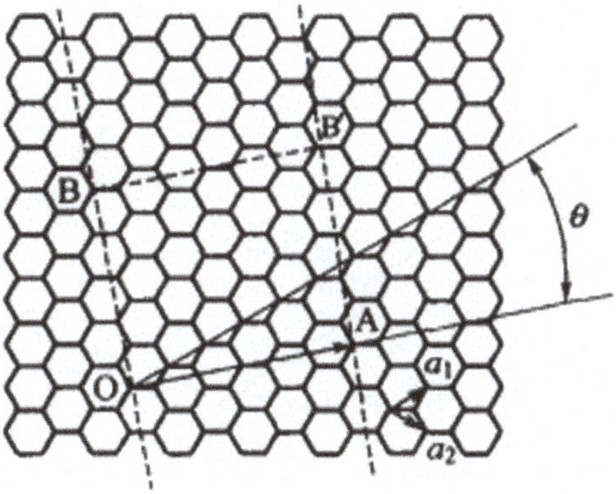

1.4.6 Nano-film

As a regularly arranged nano-film, the crystal structure of its surface can be described by TLK model, as shown in Fig. 1.8. The model holds that the surface is composed of a terrace with a low miller indices base and a ledge with a single atom height, and there are several kink positions on the ledge. The above model is called terrace-ledge-kink model, which is called TLK model for short [23].

In the process of forming nano-films, the atoms incident on the solid surface can stay on the bend, the ridge or the flat surface of the substrate and become adsorbed atoms. Then the atoms adsorbed on the surface can migrate from one location to another in the form of diffusion through the surface. It can be approximately considered that the binding energy of surface atoms is proportional to the number of atoms in the nearest neighbor and the next nearest neighbor, so the atoms attached to the edge step are more firmly bonded than those attached to the substrate platform. At equilibrium, these surface atoms in different states all have a certain concentration. Among them, the concentration of atoms attached to the position with the largest binding energy is the largest, while the concentration of atoms attached to the substrate platform is very small. The TLK model can be used to discuss the crystal growth, gasification, dissolution and surface phenomena conveniently [25].

The observation on the surface of the thin film shows that the actual surface is composed of the so-called "surface phase" of several atomic layers outside the solid, and its thickness is about several tenths to several nanometers. Atoms in the surface layer can have different degrees of displacement and leave the equilibrium position, which makes the surface structure different from the structure on the parallel plane in the body. This change of surface phase is called "reconstruction". If the distance between atoms on the surface compresses or expands perpendicular to the surface, it is called "relaxation". Surface atoms may also change in the height of atomic size on the surface to form a "step" structure. The atomic layer on the surface of the

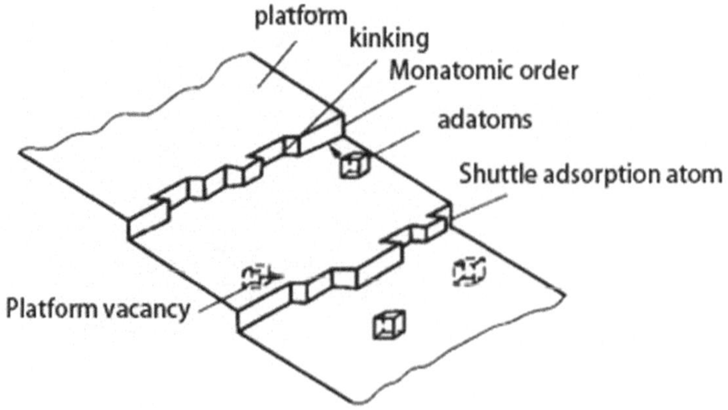

Fig. 1.8 TLK model of thin film crystal [24]

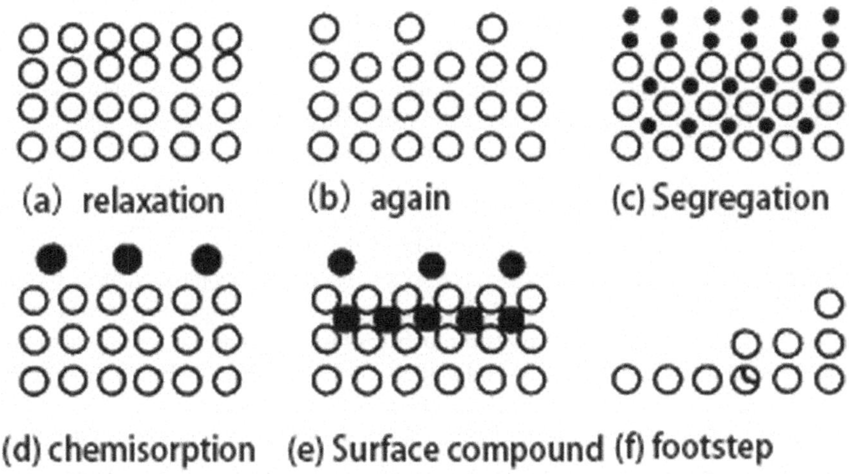

Fig. 1.9 Types of changes in surface state of nano thin films [26]

film may be different from the crystal form in the film, forming a "reconstruction", and the surface composition may also change, resulting in "segregation". In addition, "chemical adsorption" will occur on the surface, and even "surface compounds" may be generated. Figure 1.9 shows the types of changes in some states that may occur in the surface layer.

Generally speaking, the surface energy on the densely packed surface is the lowest, so it is called low energy surface. The higher the concentration of high-energy atoms on the crystal plane, the greater the surface energy value, so the higher the activity of the crystal plane when it is combined with other atoms.

1.4.7 Nanophase Particles

1.4.7.1 Definition of Nanophase Solid Materials

Nano-phase material refers to a three-dimensional solid material composed of a large number of particles or particles with a grain size of 1–100 nm. Some people call it nano-solid material, nano-crystalline material, nano-structured material and even nano-material [27].

Some people think that particles with a particle size of several nanometers can't be considered as traditional crystals with long-range order, so the solid that can't be pressed into nanoparticles is called nano-polycrystal but should be called nano-state solid or nano-phase solid.

1.4.7.2 Structure of Nanophase Solid Materials

In fact, the description of the structure of nano-phase solid should take into account the size, morphology and size distribution of nanoparticles that constitute nano-phase solid. The shape of interface, the type, quantity and composition of defects in particles and interface. Chemical composition and distribution of impurity elements in particles and interfaces, etc.

Among these factors, the microstructure of the interface has a very important influence on the properties of nano-phase solids. Therefore, people have been trying to study the interface structure. Although people have studied the interface of different kinds of nano-phase solids through various experimental testing methods and obtained many experimental results, their views on the structure of nano-phase solids are still not consistent. This is mainly because there are relatively few experimental data about nano-phase solid materials, and different nano-phase solids have different preparation methods, and their interfaces are quite different. In addition, when some testing methods are used for testing, the structure of the sample may be changed in the process of sample preparation. At present, the models of interface structure of nano-phase solid materials mainly include gas-like model, short-range order model, interface defect model, interface structure variable model and so on. People have different views on these models and have not yet formed a unified and systematic theory. Therefore, the theory about this part of nano-phase solid structure is not mature. Here, I will introduce some recognized conclusions to you. Studying the microstructure of nano-phase solid materials is very important for further understanding the characteristics of nano-phase solid materials. The following systematically introduces the structural characteristics of nano-phase solid materials. The basic composition of nano-phase solid materials is nanoparticles plus the interface between them. Due to the small size of nanoparticles, the volume fraction of interface is almost the same as that of nanoparticles, so the interface of nano-phase solid materials can not be simply regarded as a defect, it has become one of the basic components of nano-phase solid materials and plays an important role in its performance [28].

The microstructure of nano-phase solid materials was studied by means of TEM, XRD and other testing methods, and the structural units of nano-phase solid materials could be divided into two components. One is the particle component, in which all atoms or molecules are located in the particle. One is the interface component, in which all atoms or molecules are located on the interface between particles. Components here refer to the structural units that make up a substance. For example, the structural unit that makes up a metal is a metal atom, such as Pd, and its structural unit is a Pd atom; Oxides, such as TiO_2, are composed of TiO_2 molecules.

According to the arrangement state of particles in nano-phase solid, it can be divided into three kinds of nano-phase solid materials: (1) Nanocrystalline materials are composed of grain components (all atoms are located on the lattice points of grains) and grain boundary components. (2) Nano-amorphous materials are composed of amorphous components and interface components. (3) Nanocrystalline materials. Grain component, amorphous component and quasicrystal component, which are composed of quasicrystal component and interface component, are collectively referred to as particle component, while grain boundary component and interface component are collectively referred to as interface component. Interface component has the following two characteristics: (1) atomic density is reduced; (2) that coordination numb change of the nearest neighbor atom.

The structural characteristics of nano-phase materials are there are many defects such as grain boundaries, phase boundaries or dislocations, the proportion of atoms on the interface is very high (the volume fraction of atoms in the defect area is equivalent to that in the crystal), and lattice mismatch is concentrated at the boundary, forming a structure far from equilibrium. Therefore, the structure of nano-phase materials is characterized by a large proportion of atoms in the defect environment. For example, when the particle size in nanomaterials is 5 nm, there are 10 interfaces per cubic centimeter. If the small grains have an equiaxed shape, for a compressed body with a grain size of about 3–6 nm and an interface thickness of about 1 nm, according to simple geometric estimation, the volume fraction of the interface is about 50%.

1.4.7.3 Characteristics of Nanophase Solid Materials

The characteristics of nano-phase solid materials are determined by the interaction of particle size, phase composition and interface.

In 1984, H Geiter of the University of Salland, Germany, prepared Pd, Cu and Fe nanocrystals with clean surfaces for the first time by inert gas coagulation, and then pressed them into nano-phase solid materials in situ in a vacuum chamber and proposed an interface structure model of nano-phase materials. Subsequent research found that CaF_2 and TiO_2 nano-ceramics showed good toughness at room temperature, which made people see a new way to toughen ceramics.

Nano-ceramics refers to ceramic materials with nano-scale phase in microstructure. That is to say, grain size, grain boundary width, second phase distribution, defect size, etc. are all on the nanometer level.

Ceramic materials are brittle under normal circumstances, but nano-ceramic materials made of nanoparticles have good toughness, which is because the solid materials made of nanoparticles have large interfaces and the atomic arrangement at the interfaces is quite chaotic. Atoms migrate easily under the condition of external force deformation, so they show good toughness and certain ductility, which makes ceramic materials have novel mechanical properties. It is also based on the same reason that nano-phase ceramics can be deeply deformed without breaking. Under pressure, nano-sized particles tend to slide against each other more than millimeter-sized particles. This process is called particle boundary sliding, which is the basic way of nano-phase ceramics deformation. Therefore, once a small initial crack appears, the atoms in the surrounding material will move to fill people. The smaller the particle size, the shorter the distance that atoms need to move, so the faster the repair. But in ordinary ceramics, the particles are combined with each other. This is the so-called "unbreakable ceramic bowl" introduced at some current trade fairs.

Traditional ceramic materials are brittle, with poor toughness and strength, so their application is limited. Nano-ceramics may overcome the brittleness of ceramic materials and have the same flexibility and machinability as metals (ideal). Ryan Cahn, a famous British scientist, wrote in Nature magazine that "nano-ceramics are a strategic way to solve the brittleness of ceramics". Many experts believe that if we can solve the technical problem of grain growth in the sintering process of single-phase nano-ceramics, so as to control the grain size of ceramics below 50 nm, it will have incomparable advantages such as high hardness, high toughness, low temperature superplasticity and easy processing.

Nano-ceramics may be used in unbreakable ceramics and bulletproof glass. Although there are still many key technologies to be solved, nano-ceramics have excellent mechanical properties at room temperature and high temperature, bending strength and fracture toughness, which make them widely used in cutting tools, bearings, automobile engine parts and many other aspects, and play an irreplaceable role in many harsh environments such as ultra-high temperature and strong corrosion, and have broad application prospects. Nano-ceramics can also be made into nano-composite ceramic bearings, cutting tools, chemical high-temperature wear-resistant parts and so on.

1.5 Common Nanomaterials

1.5.1 Nanofibers

Porous materials with high surface area have been applied in many fields. Nanofibers, with their super-high specific surface area, have shown its powerful potential vitality in many fields [29]. Different from traditional rigid porous materials, nanofibers have good flexibility and ductility, which greatly enhances the controllability of their pore size and morphology (see Fig. 1.10).

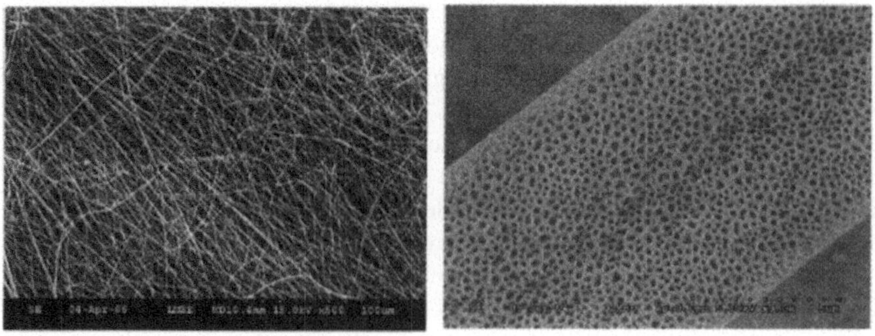

Fig. 1.10 Electrospun nanofibers [30]

Hollow nanofibers were obtained by electrospinning two incompatible liquids with coaxial capillary jet device, and then selectively removing the inner core. The diameter and wall thickness of the fiber can also be controlled by the flow rate of the liquid in the inner and outer layers. This kind of hollow nanofiber will be expected to be used in microfluidic devices and light wave transmission fields.

The application of polymer nanofibers in biomedicine is a hot research topic at present. There are mainly the following aspects.

The functional membrane is a vegetation nanofiber membrane based on electrospinning technology, which can be used as a drug sustained-release material. At the same time, changing the composition and fiber diameter of the material can effectively control the degradation rate of biomaterials, and the experiment proves that the nanofiber membrane prepared by electrospinning has an anti-adhesion effect. Smith Daniel and others designed a medical device based on electrostatic spinning principle, which can directly spray degradable polymer materials on the wound to form a fiber coating film. The membrane can not only promote skin growth, but also has no scar after wound healing [31].

The nanofiber structure formed by electrospinning of cell scaffold can basically meet the requirements of tissue engineering scaffold, that is, supporting and guiding cell proliferation. Li prepared glycolide/lactide copolymer cell scaffold by electrospinning. He thought that its structure was similar to that of natural tissue extracellular matrix, with porosity, wide pore size distribution and good mechanical properties.

The porosity and adaptability of bionic artificial blood vessels affect the tissue reaction. Porous artificial blood vessels are beneficial to the growth of host tissue and make its inner wall better intimated. Using electrospinning technology, using polyurethane, polytetrafluoroethylene and other raw materials, the multilayer artificial blood vessel has good flexibility, high porosity and excellent mechanical properties. From a biological point of view, almost all organs of human body, such as bone, periodontal tissue, collagen, skin and cartilage, exist in the form of nanofibers. Electrospinning can prepare nano-scale bionic fibers, which has broad development prospects.

Cell carrier Chu and others embedded bone cells in the double-layer polylactic acid fiber membrane prepared by electrospinning, forming a fiber membrane cell/fiber membrane structure. It was found that the structure remained intact and did not become brittle after being immersed in liquid nitrogen. At the same time, the porous structure of fiber membrane is beneficial to cell release and cell activity [32].

In addition, based on the unique properties of nanofibers, researchers put forward more than ten applications, including solar sails, optical sails, mirrors used in space, botanical insecticide, nano-conductors and nano-electrical applications (Fig. 1.11), such as field effect transistors, subminiature antennas, chemical catalyst devices and ammonia storage tanks of fuel cells.

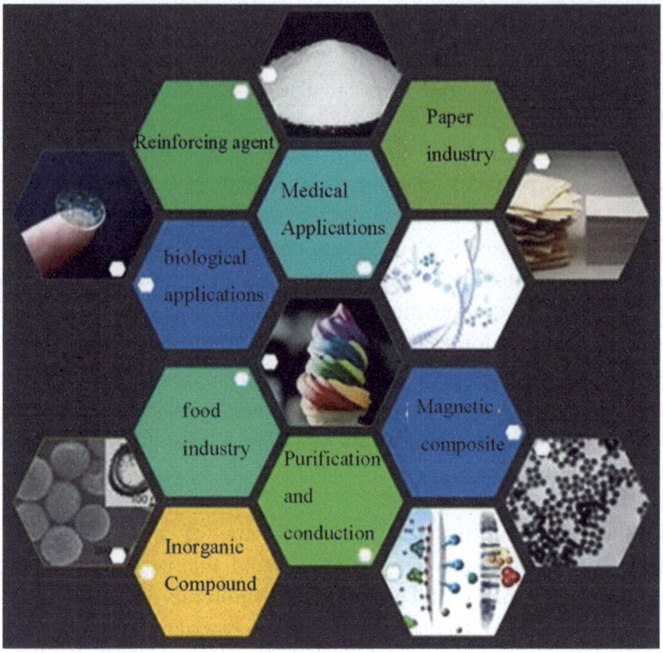

Fig. 1.11 Application fields of nanofibers [33]

1.5.2 Carbon Nanotubes

Carbon nanotubes (CNTs) are a new type of carbon materials. In recent decades, carbon nanotubes have been the focus of scientific research in the world. Carbon nanotubes (CNTs) are seamless hollow nano-scale coaxial cylinders rolled by graphite with lamellar structure. There are two "hats" formed by half fullerene spheres at both ends of the cylinder, which can be generally divided into single-walled carbon nanotubes composed of a layer of stone and multi-walled carbon nanotubes composed of multiple layers of graphite coaxially. Two kinds of CNTs have the characteristics of small tubes and large length-diameter ratio, which makes them have excellent properties. It is a new material that attracts worldwide attention. For example, the average Young's modulus of multi-walled carbon nanotubes is about 1800 GPa, which is 100 times that of steel, and the bending strength can reach 14.2 GPa, and the stored strain can reach 100 keV, showing superior mechanical properties, while the density is only 1/6, making it one of the materials with the best mechanical properties ever. Comwell et al. found that when carbon nanotubes are stressed, they can release stress through pentagonal and heptagonal pairs, showing good self-lubricating performance, which shows a bright future for the application of self-lubricating performance of carbon nanotubes. Limited by its geometric shape, the thermal expansion of carbon nanotubes in the direction perpendicular to the tube axis is almost zero. Large area directionally grown carbon nanotubes can be used to prepare field emission tubes. Carbon nanotubes have two unique and excellent electrical properties, one is the field emission property, and the other is the dual electrical properties of carbon nanotubes. These two unique electrical properties make this new material have a good application prospect in microelectronics. Field emission is one of the important technologies in electromagnetic research, because such devices have a wide range of applications, such as making kinescopes, scanning electron microscopes, high-energy electronic weapons, designing and manufacturing sensitive switches, UHF oscillators, and field emission flat panel displays. Because the top of carbon nanotubes can be made extremely sharp, it can emit electrons under the action of lower excitation electric field than other materials, and because of the strong carbon–carbon bond, carbon nanotubes can work for a long time without being damaged, which has excellent field electron emission performance. This performance can be used to make flat display devices thinner and more power-saving [34].

With such excellent mechanical properties, carbon nanotubes are an excellent fiber material, and its performance is superior to any current fiber. It has the inherent properties of carbon fiber, the electrical and thermal conductivity of metal materials, the heat and corrosion resistance of ceramic materials, the softness and plaidability of textile fibers, and the slight processability of polymer materials. It is a functional material and structural material with multiple functions and multiple purposes and is expected to be applied to many aspects in the field of materials. In nano-machinery, nanoscales have been made. Nanoscale is similar to a suspended pendulum, and its bending constant is known. By measuring the vibration frequency, the mass of particles bonded to one end of a cantilever beam can be measured. This principle is also applicable to measuring the tiny mass bonded to the top of the free end of carbon nanotubes. This is the newly discovered nanometer scale, and it is also the most sensitive and smallest weighing instrument in the world. Some experts believe that this nanoscale can be used to measure the mass of macromolecules and biological particles (such as viruses) and may also lead to the emergence of a nano-mass spectrometer. As a probe of probe electron microscope, carbon nanotubes are one of the closest commercial applications of carbon nanotubes. The nano-scale diameter of carbon nanotubes makes the microscope probe prepared by it better than the traditional Si or Si; N pyramid-shaped needle tip has higher resolution. Carbon nanotubes (CNTs) have a large aspect ratio, which is deeper than the traditional pyramid-shaped needle tip and can detect slits and deep-seated characteristics. In addition, carbon nanotubes have good elasticity and flexibility, which can avoid damaging samples and probe tips, and the end of carbon nanotubes can be selectively chemically modified to prepare probe tips for analyzing functional groups of organic and biological samples. Because of the excellent characteristics of carbon nanotube probe tips, the research in this field has become one of the hot topics in recent years.

At the same time, carbon nanotubes (CNTs) are a kind of potential hydrogen storage materials. Because of their pipeline structure and graphite-like gaps between multi-walled carbon tubes, CNTs have become the most potential hydrogen storage materials and become the focus of current research. Scientists all over the world have done a lot of work on the ammonia storage performance of carbon nanotubes. It is found that the hydrogen storage capacity of single-walled carbon nanotubes weighing about 500 mg can reach 41.2% (mass fraction) at room temperature, and 78.13% of the stored hydrogen can be released at room temperature, and the remaining hydrogen can also be released after heating. This single-walled carbon nanotube can be reused. This achievement has opened up a broad prospect for the research of hydrogen storage materials. Carbon nanotubes are a promising hydrogen storage material, which is expected to promote the utilization of hydrogen energy, especially the early realization of hydrogen fuel cell vehicles. However, the research on hydrogen storage of carbon nanotubes started late, and there are still many aspects, such as cycle characteristics, thermodynamic and kinetic behavior of hydrogen storage, how to further improve its mass hydrogen storage capacity and volume hydrogen storage capacity, hydrogen storage and release mechanism, etc., which need in-depth and detailed research. At present, scientists at home and abroad are also working hard for it. In a word, carbon nanotubes, as a new nanomaterial, have important application prospects in many fields and are one of the most promising nanomaterials in the twenty-first century. Professor Smalley, winner of the Nobel Prize in Chemistry, believes that carbon nanotubes will be the first choice for the best fibers in the future, and will also be widely used in ultramicro wires, ultramicro switches and nanoscale electronic circuits.

1.5.3 Nanometer Titanium Dioxide

Nano-titanium dioxide (TiO_2), as a new photocatalyst, anti-ultraviolet agent, photoelectric effect agent, etc., with its magical functions, will show broad application prospects in the fields of antibacterial and mildewproof, exhaust purification, deodorization, water treatment, antifouling, weather resistance and aging resistance, and automotive topcoat [35]. With the maturity of industrial production and functional application of its products, it will play an important role in the technological revolution in the fields of environment, information, materials, energy, medical care and sanitation. Nano-titanium dioxide is divided into anatase type and rutile type, and the appearance is white powder. Among them, anatase is mainly used as photocatalyst, and there are many studies on photocatalytic activity of anatase titanium dioxide in the literature. It is a nanometer powder made of nanometer titanium dioxide doped with some metals or metal oxides. When anatase TiO_2 is irradiated with ultraviolet light of less than 388 nm, electrons are excited from the valence band to the conduction band, leaving holes in the valence band, forming a photogenerated electron–hole pair: $e^- - h^+$. Electrons and holes can be captured by oxidant and reductant, respectively (see Fig. 1.12). For example, O_2 and H_2O adsorbed on the surface of TiO_2 react to generate superoxide anion free radicals, which have strong oxidation and can react with harmful gases at room temperature to decompose organic pollution and harmful bacteria. Rutile titanium dioxide has unique particle shape, good dispersibility and good shielding effect on ultraviolet rays, and can be widely used in cosmetics, protective paints, etc., which can improve the aging resistance, scouring resistance and self-cleaning function of coating films. Using pure nano-TiO_2

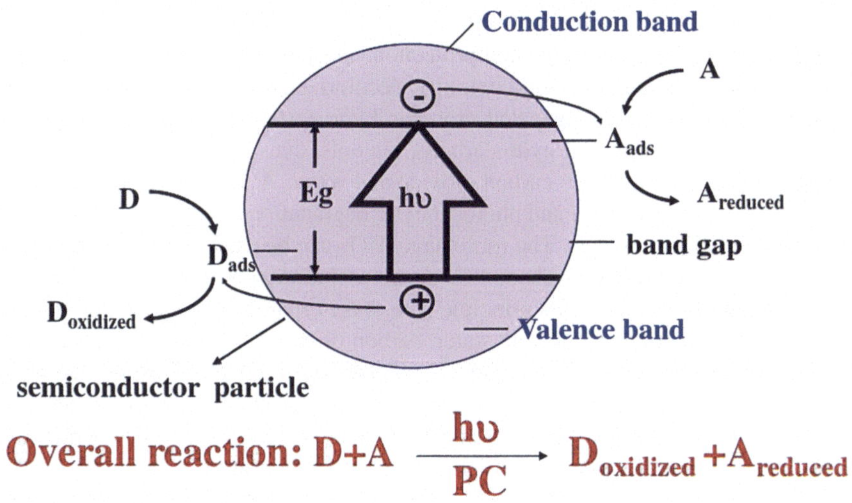

Fig. 1.12 Principle of TiO_2 photocatalysis [36]

as photocatalyst requires external ultraviolet light irradiation, which is not in line with the principle of energy saving. Sunlight is a continuous spectrum and an inexhaustible clean energy source. In order to make full use of solar energy and change the situation of traditional ultraviolet irradiation catalysis, scholars all over the world are committed to expanding the absorption wavelength range of TiO_2 and strive to make the catalytic reaction effectively under visible light. Therefore, various modification methods of nano-TiO_2 have emerged, such as manufacturing composite semiconductors, doping metal modification, and precious metal deposition modification on the surface of TiO_2 photocatalyst.

Antibacterial and mildew-proof mechanism of nano-TiO_2 Because of the characteristics of electronic structure of TiO_2, it generates superoxide anion free radicals and hydroxyl free radicals with strong chemical activity when exposed to light, attacking organic substances and degrading organic pollutants. When encountering bacteria, it directly attacks the cells of bacteria, causing the organic matter in the cells of bacteria to degrade, thereby killing bacteria and decomposing them. Commonly used fungicides such as silver and copper can inactivate bacterial cells, but after the bacteria are killed, the corpse releases harmful components such as endotoxin. Nano-TiO_2 can not only affect the reproduction of bacteria, but also destroy the cell membrane structure of bacteria, completely degrade bacteria and prevent endotoxin from causing secondary pollution. Nano-TiO_2 is a non-dissolving material, which can degrade organic pollutants and kill bacteria, but it does not decompose and dissolve itself, and has a lasting photocatalytic effect and a lasting effect of sterilization and degradation of pollutants.

TiO_2 photocatalytic technology is simple in process and low in cost. It can catalyze and decompose bacteria and pollutants by natural light at room temperature and pressure. It has the characteristics of high activity, no secondary pollution, no irritation, safety, no toxicity, good chemical stability and thermal stability, and is one of the most promising green environmental protection catalysts. Using nano-TiO_2 to treat organic wastewater can effectively detoxify, decolorize and mineralize halogenated aliphatic hydrocarbons, halogenated aromatic hydrocarbons, nitroaromatic hydrocarbons, polycyclic aromatic hydrocarbons, phenols, dyes and pesticides in water, and finally degrade them into carbon dioxide and water. At present, the research in this field has made progress, and photocatalytic degradation of sewage will become an effective treatment means. The rutile nano-TiO_2 can be used to degrade nitrogen oxides (NO_x, etc.) and effectively control the air pollution caused by industrial waste gas and automobile exhaust. Its principle is to make organic or inorganic pollutants undergo redox reaction to generate water, carbon dioxide, salt, etc., thus purifying the air. The results show that nano-TiO_2 photocatalytic air purification coatings, ceramics and other materials have good prospects in eliminating nitrogen oxides. It can be predicted that in the twenty-first century, nanotechnology and nanomaterials will be more and more closely related to human life in the future.

1.5.4 Nano-ceramics

Nano-ceramics is a new type of ceramic material, which is characterized by nanoscale microstructure and excellent properties. The development of nano-ceramics is the combination of nanotechnology and ceramic materials. Through the application of nano-technology, the properties of traditional ceramic materials have been comprehensively upgraded and improved [37].

The preparation methods of nano-ceramics mainly include sol–gel method, high-energy ball milling method, vapor deposition method and so on. Among them, sol–gel method is one of the most commonly used methods. Metal ions or organic substances are dissolved in a solvent to form a colloidal solution, and then the sol–gel is made into a solid through steps such as drying and heat treatment, and finally the nano-ceramic material is obtained.

Nano-ceramics have many excellent properties. First of all, nano-ceramics have high hardness and strength and can withstand high temperature and high-pressure environment. Secondly, nano-ceramics have good wear resistance and corrosion resistance and can maintain stable performance under harsh working conditions. In addition, nano-ceramics also have excellent insulating properties and thermal stability and can maintain the characteristics of electrical insulation and thermal stability at high temperature.

Nano-ceramics have a wide range of applications. First of all, nano-ceramics have important applications in the electronic field. Because of its excellent insulating properties and thermal stability, nano-ceramics are widely used as insulating layers of electronic components and protective layers of high-temperature components. Secondly, nano-ceramics are also widely used in the mechanical field. Because of its high hardness and strength, nano-ceramics are used to manufacture high-performance tools, bearings and mechanical seals. In addition, nano-ceramics can also be used to manufacture high-temperature stoves, ceramic catalysts and sensors.

Although nano-ceramics have many advantages, there are also some challenges and problems. Firstly, the preparation process of nano-ceramics is complicated, and it is necessary to control the concentration of sol, the choice of solvent and the conditions of heat treatment. Secondly, the cost of nano-ceramics is high, and the preparation and processing process require high energy and equipment investment. In addition, the stability and reliability of nano-ceramics need to be further improved and studied.

The development of nano-ceramics is of great significance for improving material properties, prolonging material service life, reducing energy consumption and reducing environmental pollution. However, the preparation and application of nano-ceramics still face some challenges, such as the complexity of preparation process, high cost and the aggregation and reunion of nanoparticles. Therefore, it is the current research direction to further study and develop the preparation technology and application methods of nano-ceramic materials.

Nano-ceramic is a new ceramic material with excellent properties and wide application prospects. With the continuous development and progress of nanotechnology, the preparation methods and properties of nano-ceramics will be further improved and upgraded. It is believed that in the future, nano-ceramics will play an important role in various fields and bring more scientific and technological innovation and development to mankind.

1.5.5 Nanometer Powder

Nano-powder is a kind of powder material with nanometer particle size. The preparation methods of nano-powder mainly include physical method, chemical method and biological method. Physical methods include gas phase method, solidification method and mechanical method. Chemical methods include sol–gel method, deposition method and hydrothermal method. The biological principle is to use organisms or biological processes to prepare nano-powders [38]. Nano-powder has a large specific surface area and special physical, chemical and biological properties, so it is widely used in many fields.

Nano powder has many unique properties and advantages. First of all, nano-powder has a large specific surface area. Due to the small particle size, the specific surface area of nano-powder is relatively large, which makes nano-powder interact with other materials better and improves the reaction rate and catalytic activity of materials. Secondly, nano-powder has excellent optical properties. Because the particle size is close to the wavelength of light, nano-powder can show special optical effects, such as quantum size effect and surface plasmon resonance effect. In addition, nano-powder has excellent magnetic, electrical, and thermal properties, and can play an important role in the fields of magnetic materials, electronic devices and heat conduction materials.

Nano-powder has been widely used in many fields. First of all, nano-powder has important applications in the field of materials science. Because of its large specific surface area and special properties, nano-powder is widely used to prepare high-performance ceramics, metal alloys and composite materials. Secondly, nano-powder also has important applications in the energy field. Nano-powder is used to prepare efficient solar cells, lithium-ion batteries, and fuel cells. In addition, nano-powder is also widely used in biomedical field. Nano-powder is used to prepare drug carriers, biosensors and biological imaging agents, which has important clinical application value.

However, the preparation and application of nano-powder also face some challenges and problems. Firstly, the preparation method of nano-powder is complicated, and it is necessary to control the reaction conditions and particle size distribution of powder. Secondly, the stability and safety of nano-powder also need further study and evaluation. Because of its large specific surface area and high surface activity, nano-powder is easy to react with other substances in the environment, which may cause potential risks to the environment and human health.

Generally speaking, nano-powder is a kind of powder material with unique properties and wide application prospects. With the continuous development and progress of nanotechnology, the preparation methods and properties of nano-powder will be further improved and promoted. It is believed that in the future, nano-powder will play an important role in various fields and bring more scientific and technological innovation and development to mankind.

1.5.6 Nano-film

Nano-film is a kind of thin film material with nanometer thickness, which is usually between 1 and 100 nm. Compared with traditional thin films, nano-films have higher specific surface area and more interface effects, so they have many unique properties and applications [39].

First of all, nano-films have excellent optical properties. Because its thickness is close to the wavelength of light, nano-films can show special optical effects, such as surface plasmon resonance and quantum size effect. These effects make nano-films widely used in optical devices, optical sensors and photocatalysis.

Secondly, nano-films have excellent electrical properties. Due to its nano-scale thickness, nano-films can confine and transport electrons, thus having higher conductivity and electron mobility. This makes nano-films have important application potential in electronic devices, conductive materials and energy storage.

In addition, nano-films have excellent magnetic, thermal and mechanical properties. Nanofilms can adjust their magnetism by controlling their thickness and composition, thus playing an important role in the fields of magnetic materials and magnetic storage devices. Nano-film also has excellent thermal conductivity due to its nanoscale thickness and can be used to prepare efficient thermal interface materials and thermal management devices. In addition, due to its high specific surface area and interfacial effect, nano-films have excellent mechanical properties, which can be used to prepare materials with high strength, toughness and flexibility.

Nanofilms are widely used in many fields. First of all, nano-films have important applications in nano-electronic devices and optoelectronic devices. Nanofilms can be used to prepare high-performance transistors, photodiodes and solar cells. Secondly, nano-films also have important applications in the energy field. Nanofilms can be used to prepare efficient catalysts, lithium-ion batteries and fuel cells. In addition, nano-membranes are widely used in biomedical field. Nano-membranes can be used to prepare drug carriers, biosensors and tissue engineering materials, and have important clinical application value.

However, the preparation and application of nano-films also face some challenges and problems. Firstly, the preparation method of nano-film is complicated, and it is necessary to control the thickness, composition and structure of the film. Secondly, the stability and reliability of nano-films need further study and evaluation. Due to the nano-scale thickness of nano-film, the defects and inhomogeneities of its surface and interface may affect the properties and applications of materials.

Generally speaking, nano-film is a kind of thin film material with unique properties and wide application prospects. With the continuous development and progress of nanotechnology, the preparation methods and properties of nano-films will be further improved and promoted. It is believed that in the future, nano-films will play an important role in various fields and bring more scientific and technological innovation and development to mankind.

1.5.7 Nanoblock

Nanoblock is a kind of block material with nanometer size. Compared with traditional bulk materials, nano-bulk materials have larger specific surface area and more interface effects, so they have many unique properties and applications [40].

Firstly, nano-blocks have excellent mechanical properties. Because of its nanometer size, nano-block can realize the size effect of crystal and the strengthening effect of interface, so it has higher strength, hardness and toughness. This makes nano-block widely used in material strength, wear resistance and corrosion resistance.

Secondly, nano-blocks have excellent optical properties. Because its size is close to the wavelength of light, nano-blocks can show special optical effects, such as surface plasmon resonance and quantum size effect. These effects make nano-blocks widely used in optical devices, optical sensors and photocatalysis.

In addition, nano-blocks have excellent electrical properties. Because of its nanoscale size and large specific surface area, nano-blocks can confine and transport electrons, thus having higher conductivity and electron mobility. This makes nano-block have important application potential in the fields of electronic devices, conductive materials and energy storage.

Nanoblocks are widely used in many fields. First of all, nano-blocks have important applications in materials science and engineering. Nano-blocks can be used to prepare materials with high strength, toughness and wear resistance, such as high strength steel, ceramic composites and nano-coatings. Secondly, nano-blocks also have important applications in the energy field. Nanoblocks can be used to prepare efficient catalysts, lithium-ion batteries and fuel cells. In addition, nano-blocks are also widely used in the biomedical field. Nanoblocks can be used to prepare drug carriers, biosensors and tissue engineering materials, and have important clinical application value.

However, the preparation and application of nano-blocks also face some challenges and problems. Firstly, the preparation method of nano-blocks is complicated, and it is necessary to control the size, shape and composition of the blocks. Secondly, the stability and reliability of nano-blocks also need further study and evaluation. Due to the nano-scale size of nano-blocks, the defects and inhomogeneities of their surfaces and interfaces may affect the properties and applications of materials.

Generally speaking, nano-block is a kind of block material with unique properties and wide application prospects. With the continuous development and progress of nanotechnology, the preparation methods and properties of nano-blocks will be further improved and upgraded. It is believed that in the future, nano-blocks will play an important role in various fields and bring more scientific and technological innovation and development to mankind.

References

1. J. Yang, S. Tighe, *Procedia Soc. Behav. Sci.* **2013**, *96*, 1269–1276.
2. A. Davoodabadi, H. Ghasemi, *Adv. ColloidK. Interface. Sci.* **2021**, *290*, 102385.
3. L. Bu, X. Zhu, Y. Zhu, C. Cheng, Y. Li, Q. Shao, L. Zhang, X. Huang, *CCS. Chem.* **2020**, *2*, 1972–1982.
4. C. Scherer, A. F. M. Figueiredoneto, *Brazilian J. Phys.* **2005**, *35*, 718–727.
5. B. Kalidasan, A. K. Pandey, R. Saidur, M. Samykano, V. V. Tyagi, *Int. Mater. Rev.* **2022**, *2*, 140–183.
6. L. Tian, L. Lu, J. Feng, M. P. Melancon, *Acta. Pharm. Sin.* B **2018**, *3*, 360–370.
7. J. Corbett, P. A. McKeown, G. N. Peggs, R. Whatmore, *CIRP. Ann.* **2000**, *49*, 523–545.
8. H. Tong, S. Ouyang, Y. Bi, N. Umezawa, M. Oshikiri, J. Ye, *Adv. Mater.* **2011**, *2*, 229–251.
9. B. W. Mwakikunga, S. Motshekga, L. Sikhwivhilu, M. Moodley, M. Scriba, G. Malgas, A. Simo, B. Sone, M. Maaza, S. Sinha Ray, *Sens. Actuat. B. Chem.* **2013**, *184*, 170–178.
10. G. Lespes, S. Faucher, V. I. Slaveykova, *Front. Environ. Sci.* **2020**, *8*, 71–74.
11. Q. Li, H. Zhu, L. Hu, J. Chen, X. Xing, *Acc.Chem.Res.* **2019**, *52*, 2694–2702.
12. P. Jena, ACS. Energy. Lett. **2020**, *5*, 428–429.
13. J. Valente, E. Plum, I. J. Youngs, N. I. Zheludev, *Adv. Mater.* **2016**, *26*, 5176–5180.
14. X. Yang, X. An, S. Ling, H. Huang, Y. Zhang, G. Chen, C. Li, Q. Wang, *CCS. Chem.* **2020**, *3*, 895–903.
15. C. Xu, M. Nasrollahzadeh, M. Selva, Z. Issaabadi, R. Luque, *Chem. Soc. Rev.* **2019**, *48*, 4791–4822.
16. Z. Lu, E. Linga Reddy, H. Moussab, S. Youssef, A.-T. Antonio, O.-C. Samy, H. Jean-louis, L. Jun, W. Nini, G. Daniel, D.-G. Pascal, B. Jean-Marie, *Appl. Catal. B. Environ.* **2017**, *5*, 44–59.

17. G. Ashrafi, M. Nasrollahzadeh, B. Jaleh, M. Sajjadi, H. Ghafuri, *Adv. Colloid. Interface. Sci.* **2022**, *306*, 102599.
18. C. Tian, W. Miao, L. Zhao, J. Wang, *Reviews in Physics* **2023**, *10*, 622–701.
19. J. Babak, N. Atefeh, E. Mahtab, N. Mahmoud, H. A. Jacky, F. Paolo, B. G. Manoj, J. *Mater. Chem.* A **2023**, *11*, 9333–9382.
20. K. Arivalagan, S. Ravichandran, K. Rangasamy, E. Karthikeyan, *Int. J. ChemTech. Res.* **2011**, *3*, 534–538.
21. A. Brouillard, N. Deshpande, A. A. Kulkarni, *Adv. Healthcare. Mater.* **2021**, *6*, 2001680.
22. L. Donaldson, *Mater. Today* **2021**, *50*, 3–6.
23. D. N. Bukharov, A. V. Osipov, S. M. Arakelyan, A. O. Kucherik, *J. Phys. Conf. Ser.* **2019**, *1331*, 012008.
24. M. Grossman, F. Bouville, F. Erni, K. Masania, R. Libanori, A. R. Studart, *Adv. Mater.* **2016**, *8*, 1605039.
25. C. Zhang, D. A. McAdams, J. C. Grunlan, *Adv. Mater.* 2016, *39*, 8566–8566.
26. F. Haase, P. Hirschle, R. Freund, S. Furukawa, Z. Ji, S. Wuttke, *Angew. Chem. Int. Ed.* **2020**, *50,* 22350–22370.
27. J. W. Burdon, P. D. Calvert, *Inter. SAMPE tech. conference*, **1995**, *6*, 170625.
28. X. Yang, C. Wang, R. Lu, Y. Shen, H. Zhao, J. Li, R. Li, L. Zhang, H. Chen, T. Zhang, X. Zheng, *Nano Energy* **2022**, *3*, 895–903.
29. Hartgerink, D. J., *Sci.* **2001**, *294*, 1684–1688.
30. C. Y. Hui, M. Liu, Y. Li, J. D. Brennan, *Angew. Chem. Int. Ed.* **2018**, *17*, 4549–4553.
31. Y. Nan, D. Tan, J. Shao, M. Willatzen, Z. L. Wang, *ACS. Energy. Lett.* **2021**, *6*, 2313–2319.
32. Q. Wang, K. Feng, H. Li, *Bioresource. Technol.* **2020,** *315*, 123804.
33. S. M. Jafari, C. Arpagaus, M. A. Cerqueira, K. Samborska, *Trends Food Sci. Tech.* **2021**,*109*, 632–646.
34. Y. Wei, W. Kong, Y. Wang, *Cement. Concrete. Comp.* **2021**, *119*, 104025.
35. D. Yue-Hua, Y. Yong-Sheng, Z. Wen, W. Ling-Ling, M. Yan-Li, X. U. Wan-Zhen, *Metall. Analysis* **2009**, *11*, 10640.
36. J. Li, J. Liu, W. Huo, J. Yu, X. Liu, M. J. Haslinger, M. Muehlberger, P. Kulha, X. Huang, *Mater. Today Nano*, **2022**, *18*, 100201.
37. X. Chang-Qing, G. U. Yi, Z. Fan-Hao, *J. Cent. South. Univ.* **2003**, *10*, 87–90.
38. P. G. Mccormick, T. Tsuzuki, J. S. Robinson, J. Ding, *Adv. Mater.* **2010**, *13*, 1008–1010.
39. S. Zheng, Y. Zhu, S. Krishnaswamy, *Pro. of SPIE Inter. Society for Optical Engineering* **2012,** *8346*, 7–11.
40. Y. Tanaka, S. Kaneda, K. Sasaki, *Nano. Lett.* **2013,** *13*, 2146–2150.

Chapter 2
Fundamental Effects of Nanomaterials

The structure of the material determines the performance of the material. The morphology, particle size and distribution of the microstructure of the material are directly related to the performance of the material. When people crush a macroscopic object into ultrafine particles, it will show many strange properties, that is, its optical, thermal, electrical, magnetic, mechanical and chemical properties will be significantly different from those of bulk solids. The appearance of these characteristics is mainly caused by its surface and interface effects, small size effects, quantum size effects, quantum tunneling effects, and dielectric confinement effects [1].

2.1 Surface and Interface Effects of Nanomaterials

In the long-term scientific research process of human beings, researchers have gradually realized that surface interface science actually studies the phase interface problem of materials or devices from micro to macro scales. In current scientific research and practical industrial applications, surface interface phenomena and the key scientific principles behind them have been widely used. The materials or industrial fields involved mainly include photodiode devices, solar cells, electrochemical energy storage devices, industrial adsorption and desorption, electrocatalysis, infiltration, etc. For example, lotus leaves usually show super-hydrophobic properties that are difficult to infiltrate. The reason is that the surface of lotus leaves has a complex nanostructure, which changes the state of lotus leaves when they are in contact with water. Based on this principle, self-cleaning clothing and self-cleaning solar cells can be designed. The team of Academician Jiang Lei of Beihang University has made many pioneering contributions to the development of this field [2].

As a booming frontier hot material, nanomaterials will introduce many surfaces and interfaces in both material preparation and device construction due to their small size characteristics, so it is very important to further study them. At the same time,

S. Peng and P. Li, *Nanomaterials and Nanotechnology*,
https://doi.org/10.1007/978-981-97-8433-2_2

the research on the surface and interface effects of nanomaterials will promote the further development of related fields.

2.1.1 Surface Atoms of Nanomaterials

Due to the extremely small size of nanomaterials, for the constituent atoms in nanomaterials, the characteristic is that the number of atoms in the surface state accounts for the majority of the total number of constituent atoms. The proportion of surface atoms of nanoparticles is not only related to particle size, but also affected by particle crystal form (or atomic packing method). When nanoparticles have the same size, the proportion of their surface atoms is determined by the particle crystal form.

Here we choose triangular pyramid crystals and cubic crystals for a brief description. Assuming that they are all atomic close-packed structures, the total number of atoms in the particles is [3]:

$$N = \sum 10n^2 + 2 \tag{2.1}$$

In the formula, N is the total number of atoms constituting the nanoparticle; n is the number of shells of atoms stacked.

For triangular pyramid crystals, when the number of atomic accumulation layers constituting the particles is only 2 and 3, the atoms are all on the surface, and their proportion is 100%.

Based on the above analysis, it can be known that the proportion of surface atoms in nanoparticles is related to its size and crystal form, and reducing the size can significantly increase the proportion of surface atoms. At the same time, under the condition of the same size, the crystal structure will also affect the proportion of surface atoms. Based on the above conclusions, in the research and development of materials, when some special properties of surface atoms need to be utilized, the proportion of surface atoms can be increased through size design and crystal form design. For example, in the field of catalytic science, the occurrence of catalytic reactions is often closely related to the surface structure of catalysts. Surface atoms contain more unsaturated atoms and dangling bonds, and having a large number of active sites can accelerate the progress of chemical reactions. By nano-size and crystal design of catalysts, the proportion of surface atoms can be increased, and the catalytic efficiency can be significantly improved [4].

2.1.2 Surface Energy of Nanomaterials

Surface energy refers to the non-volume work required to reversibly increase the surface area of a substance under the conditions of isothermal, isobaric, and composition constant. Surface energy is a measure of the energy to break the chemical bonds between molecules when building the surface of a material. The physical and chemical properties of materials are limited and regulated by surface energy. The surface energy is positively correlated with the surface area of the system. For the nanoparticle system, because more atoms are on the surface, the surface energy of the particle is higher, and it has extremely high physical and chemical activity.

As the size decreases, the specific surface area of the particle increases rapidly. When the size reaches the nanoscale, the atoms on the surface of the particle account for a considerable proportion, and the particle has a very high surface energy. The surface effect refers to the change in the properties of the material caused by the sharp increase in the ratio of the number of atoms in the surface layer of nanoparticles to the total number of atoms as the particle size becomes smaller. For example, the mass fraction of Au atoms in the surface layer of Au particles with a particle size of 20 nm (the atomic radius of Au atoms is 130 pm) is 7.6%, while that of Au particles with 5 nm is 26.4%. When the particle size is 10 nm, the specific surface area is 90 m^2/g, and when the particle size drops to 2 nm, the specific surface area increases to 450 m^2/g (the specific surface area is inversely proportional to the particle size). Nanoparticles are small in size and high in surface energy, and the atoms located on the surface account for a considerable proportion. In addition, Table 2.1 lists the relationship between the particle size of nano-Cu particles and the specific surface area, the ratio of surface atoms, surface energy, and the number of atoms in a particle. It can be seen from Table 2.1 that as the particle size decreases, the number of surface atoms increases rapidly. This is due to the sharp increase in surface area due to the small particle size. For example, when the particle size is 10 nm, the specific surface area is 66 m^2/g; when the particle size is 5 nm, the specific surface area is 180 m^2/g, when the particle size drops to 2 nm, the specific surface area increases sharply to 450 m^2/g. Such a high specific surface makes the number of atoms on the surface more and more, and at the same time, the surface energy increases rapidly. It can be seen from Table 2.1 that the particle size of Cu nanoparticles increases from 100 to 10 to 1 nm, and the specific surface area and surface energy of Cu particles increase by 2 orders.

The relationship between the ratio of the number of surface atoms to the total number of atoms and the particle size is shown in Fig. 2.1 The atoms on the surface of nanoparticles account for a considerable proportion, that is, the particles have very high specific surface and surface energy. Due to the increase in the number of surface atoms, insufficient atomic coordination and high surface energy, these surface atoms are highly active, extremely unstable, and easily combined with other atoms. For example, metal nanoparticles will burn in the air, and inorganic nanoparticles will absorb gas and react with the gas when exposed to air. The following examples illustrate the reasons for the high surface activity of nanoparticles. Figure 2.2 shows

Table 2.1 The relationship between the particle size of nano-Cu particles and the specific surface area, the ratio of surface atoms, surface energy and the number of atoms in a particle [5]

Particle size/nm	Specific surface area of Cu/m^2 g^{-1}	Proportion of surface atoms to all atoms/%	Number of atoms in a particle	Specific surface energy/J mol^{-1}
100	6.6		8.46×10^7	5.9×10^2
20		10		
10	66	20	8.46×10^4	5.9×10^3
5	180	40	1.06×10^4	
2	450	80		
1	660	99		5.9×10^4

a two-dimensional plan view of a single cubic crystal grain. Assuming that the particle is circular, the solid circle represents the atoms on the surface, and the hollow circle represents the internal atoms. The particle size is 3 nm, and the interatomic distance is about 0.3 nm. Obviously, the solid circle atoms have incomplete neighbor coordination. There are "E" atoms that lack one neighbor, "D" atoms that lack two neighbors, and "A" atoms that lack three neighbor coordination. Surface atoms like "A" are extremely unstable and quickly go to "B". As soon as these surface atoms meet other atoms, they combine quickly to stabilize them, which is why the surface activity of nanoparticles is high. In fact, this activity of surface atoms not only causes changes in the transport and configuration of the surface atoms of nanoparticles, but also changes the surface electron spin conformation and electron energy spectrum.

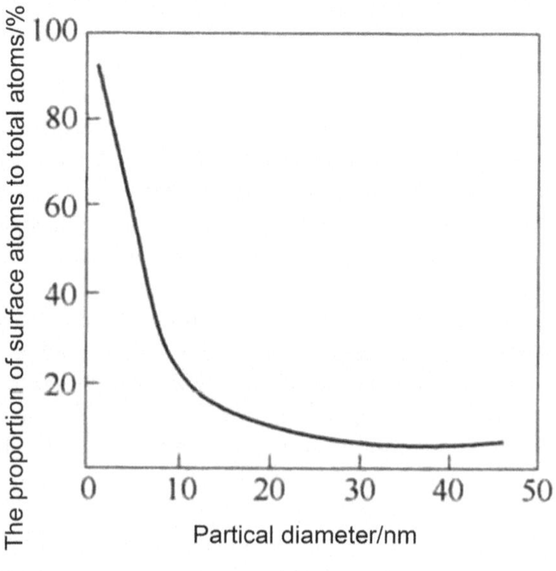

Fig. 2.1 The relationship between the proportion of surface atoms to the total number of atoms and particles [6]

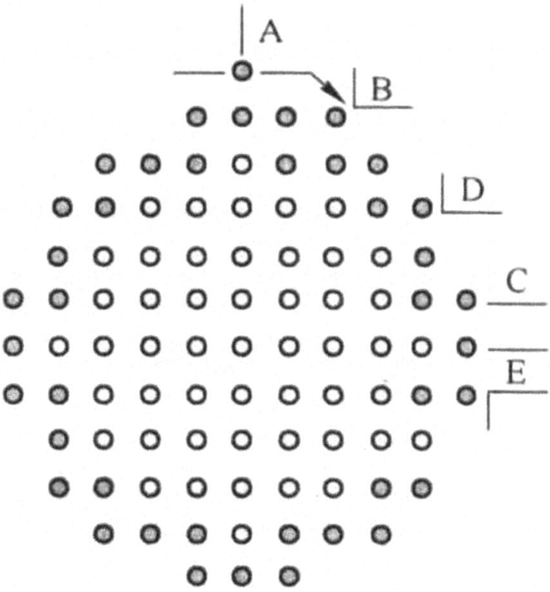

Fig. 2.2 Two-dimensional plan view of a single cubic grain [7]

With the decrease of particle size, the specific surface area, specific surface energy and specific interfacial binding energy of particles increase rapidly. This is because the surface layer atoms are in an asymmetric force field, causing the surface layer atoms to be in a high-energy state. The force on the surface layer atoms and the internal atoms is unbalanced, and the atoms on the surface layer are mainly subjected to the force from the internal side, so the atomic coordination number is unsaturated, and it is easy to combine with external atoms and tend to be stable. Therefore, the chemical activity of the surface layer atoms is high, and they have the so-called surface energy that is excess than the internal ones. The most common case where nanoparticles are extremely prone to agglomeration with each other is a clear example. Taking advantage of their high surface energy, nanoparticles can be widely used as catalysts and chemical additives.

In view of the large specific surface area and many surface activation centers of nanoparticles, the catalytic activity and selectivity are much better than conventional catalysts, and even some of the previously incapable reactions can be completely carried out. Nano-catalysts are internationally known as fourth-generation catalysts. According to reports, nano-metal catalysts catalyze NH_4ClO_4 and are widely used in the field of explosives; nano-catalysts have obvious photocatalytic activity for photolysis of water to produce H_2 and some organic synthesis reactions; they have high activity and selectivity for catalytic oxidation, reduction and cracking reactions. How to improve the reaction rate and efficiency, and how to optimize the reaction pathway has become the focus of research.

As an additive, it can improve and improve the performance of certain products. Nanomaterials have provided a variety of additives with superior performance due to their unique advantages.

In cosmetic processing, nanomaterials can replace organic compounds as new sunscreens and antibacterial agents. They have excellent properties such as non-toxic, odorless, non-irritating, heat-stable, light-stable and chemically stable. For example, nano-composite silver-based antibacterial agents are known for their rapid and long-term sterilization, and can be used in cosmetics, hygiene products, paints, water treatment, enamel, chemical fibers, plastics, and building materials.

In the rubber industry, materials such as nano-$CaCO_3$ are used to replace traditional reinforcing, wear-resistant and anti-aging additives such as carbon black and high-energy, high-cost silica, and will change the consistent black appearance of rubber products, making them rich in color, excellent in performance and attractive in cost performance.

In the production of plexiglass, adding surface-modified nano-SiO_2 can enhance the anti-ultraviolet radiation performance of plexiglass, improving service life and anti-aging ability; adding nano-Al_2O_3 to plexiglass can improve the high-temperature impact toughness of the product without affecting transparency.

2.1.3 Surface Reconstruction and Surface State

The structure of surface atoms is closely related to a series of physical and chemical properties of nanomaterials, such as the surface electronic state, work function, catalysis and oxidation of nanomaterials. Studies on the surface of many semiconductor materials have shown that the arrangement of atoms on the surface is significantly different from that in the atomic layer. The arrangement of atoms away from the surface has a certain periodicity, which can be described by a two-dimensional Bravais lattice. If the symmetry of the atomic layer in the body is described by basis vectors τ_1 and τ_2, the surface atomic layer is generally described by $n\tau_1$ and $n\tau_2$, which can be written as $(n \times m)$. This phenomenon is called surface reconstruction or surface remodeling. The reconstructed surface should minimize the Gibbs free energy G of the system:

$$G = U - TS + pV \tag{2.2}$$

In the formula, U is the thermodynamic energy or internal energy of the system; T is the temperature; S is the entropy; p is the pressure; V is the volume. Under vacuum $(p = 0)$ conditions, changes in thermodynamic energy due to structural changes are usually dominant. Therefore, the change in surface structure should minimize the energy of the system.

The surface reconstruction of representative (2×2) and Si (111) surfaces on the A-plane of GaAs (111) is introduced below. The entry-vacancy model proposed by Tong et al. in 1984 demonstrates the (2×2) surface reconstruction of GaAs

(111) A surface. This model is consistent with many experiments and theoretical calculations. According to this model, three quarters of the Ga atoms move down. Clamped between the three As atoms with which it is bonded. It is roughly located in a plane, which means that Ga forms a bond with As in the sp^2 hybrid orbital. Another quarter of the Ga is missing, forming vacancies to accommodate the entry of three Ga atoms. The As atoms form three near 90° p-type bonds and one s-type orbital. The bonding situation of Ga atoms and As atoms on the surface and the nearby local structure are very similar to those on the GaAs (110) surface. It can be seen that the above-mentioned electronic structures and the nearby geometric configurations of the surface Ga atoms and As atoms are the most favorable in terms of energy.

The Si (111) surface has various surface reconstructions. Figure 2.3a shows the surface reconstruction image of Si (111) (7 × 7) measured by scanning tunneling microscope, which is also one of the most complex surface structures recognized in the early days. The top view of this (7 × 7) surface reconstruction structure is shown in Fig. 2.3b, and the bright spot marked with a cross shows a good six-fold rotational symmetry. Such surface atoms have a large number of dangling bonds, so adatoms are needed to reduce the unsaturation effect of dangling bonds. Figure 2.3c shows the modified atomic adsorption model.

Nanomaterials have a very high proportion of surface atoms. Just as the atomic energy level forms a three-dimensional energy band after forming a crystal, a certain type of dangling bond or a certain restructured bond on the surface of a nanomaterial will also form a surface energy band. On the relaxed surface, there are several

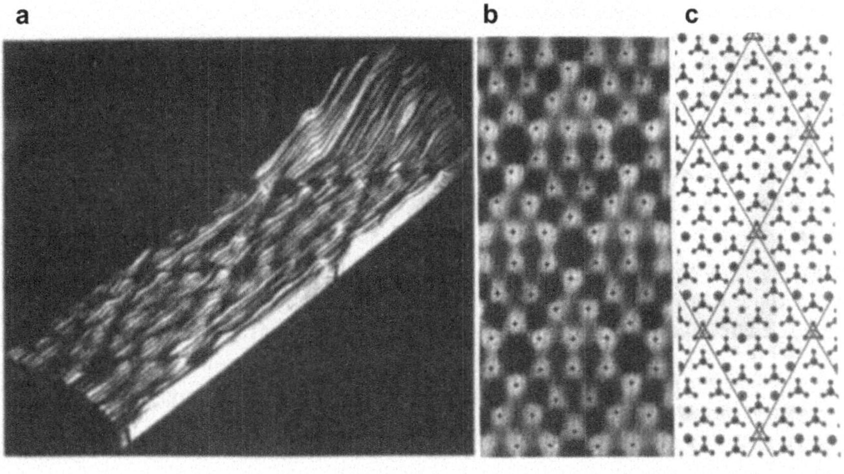

Fig. 2.3 Surface reconstruction of a typical semiconductor nanomaterial Si, **a** the earliest surface reconstruction image of Si (111) (7 × 7) measured by scanning tunneling microscope; **b** the top view of Figure (**a**); **c** the corrected adatom model (the positions of the top atoms of the bottom layer are represented by dots, the remaining atoms with unsatisfied dangling bonds are represented by encircled circles, the thickness of the circles represents the depth of the measurement, adatoms are represented by solid black dots with corresponding bond arms, empty potential adatom positions are represented by triangles with empty indicated by a circle) [8]

types of dangling bonds and restructured bonds in the original cell, and correspondingly there are several surface energy bands. Some of the energy bands can also be partially occupied, corresponding to different surface reconstructions, the surface energy bands should have different structures. The electrons in the surface energy bands constitute the so-called two-dimensional electron gas. The wave function of electrons in the belt can also be described by the two-dimensional Bloch equation, and the group velocity of electrons can still be written as [9]:

$$v = \frac{1}{\hbar}\frac{dE(k)}{dk} \tag{2.3}$$

In some cases, the bottom or top of the surface energy band has a parabolic Ek relationship:

$$E = E_0 + \frac{\hbar^2 k^2}{2m} \tag{2.4}$$

In the formula, and is the wave vector along the surface direction; E_0 is the energy of the band bottom or band top; m is the effective mass of the surface band, which is not directly related to the effective mass in the body.

The surface energy bands of nanomaterials determine their electrical conductivity. Surfaces with unsaturated dangling bonds may result in partially filled surface energy bands and metallic conductivity. Metallic surface energy bands do not necessarily consist of metal atoms adsorbed on the surface, and conversely, metal atoms adsorbed on the surface do not necessarily form surface energy bands.

In the Si (111) surface (7 × 7) reconstruction described above, there are 17 unpaired dangling bonds in each cell, so this surface should be metallic. The analysis of the surface energy band structure shows that the S_1 band and the empty band U_1 come from the dangling bonds of the added atoms, the S_2 band comes from the dangling bonds of other atoms, and the S_3 band comes from the back bonds of the added atoms. Since these dangling bonds were originally partially occupied, the S $_1$ band should be partially filled and thus the surface appears metallic. Of course, the conductance of the Si (111) plane (7 × 7) reconstructed surface is very small. Because the dispersion of the S_1 band is very insignificant, that is, it corresponds to a large effective mass. Considering that the S_1 band corresponds to the dangling bonds of the added atoms, and there is a large spatial separation between the added atoms in each (7 × 7) unit cell, the electrons in this band are strongly localized.

2.2 Small Size Effect of Nanomaterials

2.2.1 *Definition of Small Size Effect*

For extremely small-sized nanoparticles, when their size is reduced to the same size or smaller than the physical characteristic size of light wavelength (below 100 nm), de Broglie wavelength (10–100 nm), excitonic Bohr radius (1–10 nm), and superconducting coherence length (below a few nanometers), the periodic boundary conditions of the crystal inside the particle will be destroyed. Materials show excellent properties in mechanics, heat, optics and magnetism. The Kubo theory is a typical example of the small size effect. For example, light absorption is enhanced. When the particles of all metal materials are reduced to the state of ultrafine particles, they will lose their original luster and color, and appear black, and the finer the particle size, the blacker. Using this characteristic, it can be used to manufacture high-efficiency photothermal and photoelectric conversion materials. It can be used as high-efficiency solar energy collection equipment, wave-absorbing equipment for electron microscope-nuclear magnetic resonance spectrometers, infrared sensitive components and infrared stealth materials. In terms of magnetism, magnetically ordered state transitions to magnetically disordered state, and superconducting phase transitions to normal phase transition. People have produced magnetic recording powder with high storage density, which can be used in magnetic tapes, magnetic disks, magnetic cards and magnetic keys.

2.2.2 *Effect of Small Size Effect on Properties of Nanomaterials*

The small size effect will affect the properties of nanomaterials in many ways. In terms of physical properties, here is a brief description of the changes in light, heat, force, magnetic and electrical properties.

2.2.2.1 Special Optical Properties

Compared with conventional bulk materials, nanomaterials exhibit many abnormal optical properties. These properties have broad application prospects in civil and military fields, such as smart windows, smart roofs, etc., which can automatically adjust light transmittance and realize automatic adjustment of indoor temperature.

Here we take the optical properties of metals as an example. Due to their different absorption and reflection abilities to various wavelengths of visible light, bulk metals will show different metallic luster. However, when the size of the metal particles changes from bulk to nano-scale, the size of the metal particles is smaller than the wavelength of light, and the particles will completely absorb light. Nano-particles usually exhibit black characteristics, and the smaller the size, the deeper the black. For example, the reflectivity of silver-white platinum (Pt) is reduced to 1% after nanometerization, and becomes platinum black; the reflectivity of gold (Au) is also less than 10% after nanometerization, and it also appears black. High-efficiency photoelectric conversion materials can be developed by using the strong absorption of nano-metal particles to improve the efficiency of converting light energy into heat or electricity. In the field of photocatalysis, the catalytic efficiency can also be improved by nano-catalysts. At the same time, the strong absorption properties can be applied to infrared sensor materials on the one hand, and can also be used to prepare protective coatings with stealth functions to realize the stealth of military equipment such as fighter jets and tanks, and have a wide range of applications in the military field.

In addition to the strong absorption and low reflection characteristics of nanoparticles, there is an obvious blue shift phenomenon in the absorption and emission bands of the spectrum. Taking spherical particles as the research object, an external field of a certain frequency will cause resonance, resulting in surface plasmon oscillations. Assuming that there are N electrons in the unit volume of the system, and the displacement of the electrons relative to the positive charge is X, the electric polarization P can be expressed as [10]:

$$P = NqX \tag{2.5}$$

In the formula, q is the charge amount of the electron. Since the reverse electric field caused by polarization is $-P/\varepsilon$, the electron motion equation is:

$$m\frac{d^2X}{dt^2} = -Nq^2X/\varepsilon \tag{2.6}$$

In the formula, m is the equivalent mass of spherical particles; ε is the dielectric constant of the system. From this, the plasmon resonance frequency (w_p) can be obtained as:

$$w_p = \left[Nq^2/(cm)\right]^{\frac{1}{2}} \tag{2.7}$$

where c is the speed of light in vacuum. Usually, the plasmon resonance frequency is located in the visible ultraviolet or near-ultraviolet band, and the electronic energy level interval in ultrafine particles increases with the decrease of size, which will cause the light absorption peak to shift to the short wave direction, which is called blue shift.

For example, the infrared absorption peak of bulk silicon carbide (SiC) is 794 cm^{-1}, but when it is nano-sized, the absorption peak reaches 814 cm^{-1}, which shows an obvious blue shift. In terms of emission spectrum, taking Y_2O_3:Eu^{2+} powder as an example, by comparing the emission spectra of micron-scale and nano-scale powders, the emission spectrum also shows a significant blue shift phenomenon. Studying the optical properties of nanoparticles is of great significance to the development of high-efficiency photoelectric conversion materials, thermoelectric conversion materials, absorbing materials and photosensitive materials.

There are many reasons for the blue shift of the spectrum. On the one hand, as the particle size decreases, the width (energy gap) between the energy levels of molecular orbitals increases, and the characteristic band of the absorbed light undergoes a blue shift; on the other hand, the surface tension of nanoparticles is large, the crystal lattice is prone to shrinkage distortion, the bond length decreases, and the resonance frequency of the internal bonds in the crystal increases, resulting in a shift in the absorption band.

2.2.2.2 Special Thermal Properties

The thermal properties of a material are controlled by the motion behavior of its atoms and molecules. When the material size is equivalent to the characteristic size of hot carriers, its thermal properties will show obvious size dependence. As the temperature decreases further, the hot carriers will have a longer mean free path, and the thermal properties are more dependent on the dimensional characteristics of the material.

Table 2.2 lists the change of the corresponding melting point of some materials after the size is nanosized. Obviously, the melting point of the material in the bulk state is significantly higher than that of the nano-sized material, that is, the melting point decreases as the particle size decreases. The melting point of nanoparticles is lower than that of homogeneous macro-scale solid materials, because the number of atoms contained in the particles is not very large, and the thermal motion of the surface atoms is more intense than that of the internal atoms. Taking Au as an example, when its size is reduced to 2 nm, its melting point is about 1/3 of that in the bulk state. Further research on it shows that Au nanoparticles can continuously transform between multiple crystal forms, which may be the reason why the thermal behavior of nanoparticles is significantly different from that of bulk materials. According to the lattice vibration characteristics and the Debye assumption, the internal energy U of the crystal can be calculated, namely [11]:

$$U = 3\sum_{k} \frac{\hbar\theta k}{\exp\left(\frac{\hbar\theta k}{k_B T}\right) - 1} \tag{2.8}$$

Table 2.2 Comparison of melting points of metal particles in bulk state and after nanonization [12]

Material type	Normal melting point of bulk material/°C	Melting point of nanomaterials/°C
Cu	1053	750 (40 nm)
Ag	690	100
Au	1064	372 (2 nm)

where k is the wave vector; T is the system temperature; $\hbar$ is Planck's constant; θ is the Debye temperature; k_B is Boltzmann's constant.

Experiments show that the melting point of Au particles with a particle size of 2 nm drops from 1064 °C of bulk Au to 327 °C; the melting point of Cu with an average particle size of 40 nm drops from 1053 °C of ordinary solid Cu to 750 °C. The melting point drop of nanoparticle powder has greatly improved and benefited the production technology of the powder metallurgy industry and the ceramic industry, and is very attractive to related production departments. For example, adding 0.1–0.5% mass ratio of nano-Ni powder to aluminum powder can reduce the sintering temperature of the powder metallurgy process from the original 3000 °C to 1200–1300 °C, which improves energy consumption and production efficiency.

In conventional bulk materials, formula (2.8) can be simplified as [13]:

$$u_{bulk} = 9nk_BT\left(\frac{T}{\theta}\right)^3 \int_0^{x_D} \frac{x^3}{\exp(x) - 1}dx \tag{2.9}$$

In the formula, u_{bulk} is the internal energy per unit volume of the bulk crystal material; n is the atomic number density per unit volume; x_D is the integration limit corresponding to the Debye temperature. In traditional bulk materials, due to the large size of the material, the contribution of surface phonons can be ignored in the solution of the internal energy of the crystal. In formula (2.9), only the contribution of the phonon mode inside the material is considered. However, in nanomaterials, because the size of the material in some dimensions is on the order of nanometers, the number of atoms contained in it is reduced, and the contribution of surface phonons cannot be ignored, and the internal energy in the tiny lattice needs to be reconsidered. However, due to the size effect of the lattice, the basic thermal properties of nanomaterials show obvious size dependence.

The change in heat capacity and thermal expansion coefficient of nanoparticles is another special thermal property compared to bulk materials. For nanomaterials, due to the small size effect, a large part of the atoms that make up the crystal are located on the grain boundary, and these atoms have different atomic configurations from the atoms inside the crystal, thus making the thermodynamic properties of nanomaterials different from bulk materials. Theoretical calculations show that in nanomaterials, due to the contribution of interfacial atoms, the heat capacity of crystals increases with decreasing size.

For example, the heat capacity of (Pd) nanocrystals (6 nm) is increased by 29% to 53% compared with bulk Pd crystals, and the heat capacity of copper (Cu) nanocrystals is also nearly double that of bulk Cu. In addition, for the thermal expansion coefficient, the thermal expansion of the material actually comes from the anharmonic vibration generated by the internal lattice, and the lattice at the crystal interface often has stronger anharmonic vibration, and thus the crystal interface will significantly increase the thermal expansion coefficient of the material. Nanomaterials have more grain boundaries, resulting in higher thermal expansion coefficients of nanoscale crystalline materials than traditional bulk materials.

2.2.2.3 Special Mechanical Properties

At the same time, the small size effect significantly increases the strength and hardness of the material, because the structure of various materials in the nanophase is basically composed of fine grains and a large number of interfaces.

An example is a concept widely used in the field of metallic materials: fine-grain strengthening, the core of fine-grain strengthening lies in hindering the slippage of dislocations inside the crystal through the grain boundaries formed between the grains. For metal materials, its crystal structure is polycrystal composed of many microcrystals. In metals with fine-grain strengthening, when external force acts on the fine grains, the plastic deformation produced by the grains is dispersed by more fine grains, which can effectively alleviate the overall deformation of the metal, and the phenomenon of stress concentration is more difficult to occur. In addition, as the grain size decreases, the proportion of grain boundaries in the system increases, and complex grain boundaries make it difficult for internal cracks to expand under external forces. Therefore, the strength of materials can be significantly improved by reducing the grain size. At the same time, research on whether there is a limit size for fine-grain strengthening is currently very hot. When the grain size is refined to less than a certain critical value, even if the size continues to decrease, the strength will no longer increase, and even show a downward trend. This phenomenon is usually attributed to the transformation of the internal deformation mechanism of the material. However, it is still difficult to prepare high-quality and large-sized nanocrystals, which makes it difficult to elucidate their internal mechanism. The method of using fine-grain strengthening to prepare ultra-high-strength materials still needs to be further explored. The team of Academician Ke Lu of the Institute of Metal Research, Chinese Academy of Sciences has achieved a series of major achievements in the field of strengthening and toughening of metal materials around fine-grain strengthening, leading the development of this field.

Traditional ceramic materials are brittle, but ceramics made of nanoparticles have good toughness. The reason is that its organizational structure has a large number of dense interfaces, and the arrangement of interface atoms is quite chaotic. When encountering an external force that can cause deformation, these atoms are easy to migrate, therefore, showing toughness and certain ductility, so that the ceramic has novel mechanical properties. This is the so-called "unbreakable ceramic bowl" introduced at some current trade fairs. For example, nano-TiO_2 ceramics can be bent at room temperature, and its plastic deformation can reach 100%, and its hardness and strength are 3–4 times that of ordinary TiO_2 ceramics [14].

2.2.2.4 Special Magnetic Properties

The dependence of the magnetic properties of materials on the material particle size can also be attributed to the influence of the small size effect. In magnetic materials, when the size of the magnetic particles that make up the material continues to decrease, the internal structure of the particles will change from the original multi-domain state to a single-domain state. This transformation makes the mechanism of magnetization reversal change, from the domain wall displacement mode that originally existed in the multi-domain state to the magnetic domain rotation mode. The change of the transformation mechanism makes the coercive force of magnetization reversal significantly increased. This rule is widely used in the preparation of permanent magnetic powders. The critical size of multi-domain conversion to single domain varies greatly with different materials, for example, in iron ferrite. The critical size of single domain is about 1 μm, while its critical size is only about 17 nm in iron microparticles. Taking iron as an example, the coercive force of pure iron is about 80 A/m under bulk conditions, but when its size is reduced to less than the critical size (16 nm), its coercive force can be as high as 80,000 A/m. Due to the single magnetic domain structure and high coercive force of magnetic nanoparticles, magnetic nanoparticles can be used as devices for recording data. For example, the magnet of video tape is made of magnetic nanomaterials. It not only records a large amount of information, but also has high signal-to-noise ratio and high image quality. In addition, nanocrystalline magnetic materials are also widely used in power transformers, pulse transformers, transformers, magnetic switches and other devices.

When the size of the magnetic particles continues to decrease and the magnetic anisotropy Kv is equal to or smaller than the thermal energy kT of the system, the existence of thermal disturbance will cause the coercive force of the nanoparticles to decrease significantly. At this time, the material enters a paramagnetic state, and the coercive force decreases. The material can be used as a good soft magnetic material. For example, when the particle size of iron is lower than 4.5 nm, the coercive force decreases to 0, and iron at this time has superparamagnetism. Superparamagnetic materials have a wide range of applications in biomedicine. The magnetic energy of superparamagnetic particles can be controlled by an external magnetic field to make them move, so as to realize the delivery of certain drugs and targeted therapy, such as tumor targeted therapy.

2.2.2.5 Special Electrical Properties

The electrical properties of materials will change due to the existence of small size effects. For example, in terms of dielectric properties, since the size of nanoparticles will affect the mean free path of electron movement, for a spherical particle with a hypothetical diameter of d, the mean free path of its internal electrons can be considered as d/2. Therefore, the relaxation time of electrons caused by surface scattering can be obtained as $(d/2)v_F$, and v_F is the Fermi velocity. Assuming that the scattering time is τ_0, then for nanoparticles, the relaxation time τ satisfies the following formula [15]:

$$\frac{1}{\tau} = \frac{1}{\tau_0} + \frac{2v_F}{d} \tag{2.10}$$

As the particle size decreases, the relaxation time will decrease accordingly. When the particle size is smaller than a certain value, formula (2.10) can be simplified as:

$$\frac{1}{\tau} = \frac{2v_F}{d} \tag{2.11}$$

The dielectric constant of nanoparticles can be expressed as:

$$\varepsilon(\omega) = \varepsilon_b(\omega) - \frac{\omega_p^2}{\omega\left(\omega + \frac{i}{\tau}\right)} = \varepsilon_1 + i\varepsilon_2 \tag{2.12}$$

where ω_p is the sub-bulk resonance frequency of the bulk material, and ε_1 is the real part of the dielectric constant of the material. When $\omega\tau \geq 1$, $\varepsilon_2 \approx \omega_p{}^2/\omega^2\tau = 2\omega_p{}^2 v_F/\omega^2 d$, that is, the plasmon resonance frequency will move to the low frequency direction as the particle size decreases, and the particle dielectric loss (ε_2) increases as the size decreases. The regulation of the dielectric behavior can be achieved by adjusting the particle size.

On the other hand, due to the small size effect of nanoparticles, the nanometerization of the material will significantly increase the number of interface atoms inside it, resulting in a significant increase in resistance. When the size is small enough, the material will even exhibit completely different properties from the original bulk material, such as changing from a conductor to an insulator. Because of this, the small size effect has brought certain obstacles to the integration of modern semiconductor devices, slowing down the further continuation of Moore's Law, and the development of nanomaterials is expected to break through the current technical bottleneck of transistor development.

In addition to the effects of the above aspects, the small size effect can also cause changes in the acoustic properties, chemical properties, and superconductivity of nanomaterials.

2.3 Quantum Size Effect of Nanomaterials

When the size of nanoparticles is small to a certain value, the phenomenon that the electronic energy level near the metal Fermi level changes from quasi-continuous to discrete energy levels, and the phenomenon that nano-semiconductor particles have discontinuous highest occupied molecular orbital and lowest unoccupied molecular orbital energy levels, and the phenomenon of widening the energy gap are called the Quantum Size Effect. Due to the existence of quantum size effects, nanomaterials have different physical and chemical properties from traditional materials, and thus have a wide range of applications in the fields of catalysis, quantum dot luminescence, and preparation of superconducting materials.

2.3.1 Quantum Concepts and Energy Levels

In the theory of quantum mechanics, the physical quantity is no longer a continuous state in classical mechanics, but converted into a quantized state, and energy level is a classical concept to describe quantization. Taking the conductivity of materials as an example, conductors, semiconductors, and insulators can be qualitatively and quantitatively distinguished by analyzing the size of the energy level interval of the material. Quantitative distinction requires the use of the concept of energy gap. The difference in material conductance can be determined by the separation between the conduction band and the valence band, which is usually called the forbidden band width, or in the field of nanomaterials research, the energy gap.

The quantum effect of small-scale systems means that the energy of electrons is quantized to form discrete electronic state energy levels, and the movement of electrons in the system is constrained. The most direct result of the quantization of free electron energy is that when the voltage is not suitable across the metal particle, it will not conduct electricity. This is a wonderful phenomenon. When the metal nanoparticle gets an extra electron from the external circuit, the metal particle becomes negatively charged, and its Coulomb force is enough to repel the next electron from the external circuit to enter the metal particle, thus cutting off the continuity of the current.

2.3.2 The Effect of Quantum Size Effect on the Properties of Nanomaterials

Quantum size effects can lead to significant changes in the macroscopic physical properties of materials. Due to the limited number of conducting electrons in materials at the nanometer scale, the energy levels of nanoparticles show discretization characteristics at low temperatures. From a quantitative point of view, according to the free electron gas model, the Fermi level of metals can be expressed as [16]:

$$E_F = \frac{\hbar^2}{2m}\left(3\pi^2 n\right)^{\frac{2}{3}} \tag{2.13}$$

In the formula, the electron number density n = N/V does not change with the size, N is the total number of electrons in the system, V is the volume of the system, and the Fermi level (E_F) will not change with the size of the nanoparticles. The density of states near the Fermi surface is:

$$g(E_F) = \frac{1}{2\pi}\left(\frac{2m}{\hbar^2}\right)^{\frac{3}{2}} E_F^{\frac{1}{2}} = \frac{3}{2} \times \frac{n}{E_F} \tag{2.14}$$

Since there are two different spin states on each allowed energy level, the number of energy levels per unit volume at the Fermi surface is 1/2 $g(E_F)$. Therefore, according to Kubo theory, the energy level interval is:

$$\delta = 4E_f/3N \tag{2.15}$$

where δ is the energy level interval; E_F represents the Fermi potential energy; N represents the total number of electrons in the particle.

Equation (2.15) indicates that the average spacing of energy levels is inversely proportional to the total number of free electrons in the constituent particles. The energy band theory shows that the electronic energy levels near the Fermi level of metals are generally continuous, which is only true at high temperature or macroscopic size. For ultrafine particles with only a limited number of conducting electrons, the energy levels are discrete at low temperatures. For macroscopic substances containing infinite atoms (i.e., the number of conducting electrons $N \to \infty$), the energy level spacing $\delta \to \infty$ can be obtained from formula (2.15), that is, the energy level spacing for large particles or macroscopic objects is almost zero; while for nanoparticles, the number of atoms contained is limited, and the value of N is very small, which leads to a certain value for δ, that is, the splitting of the energy level spacing. Quantum size effects must be considered when the energy level spacing is greater than thermal, magnetic, magnetostatic, electrostatic, photonic, or condensation energies of superconducting states. Quantum size effects lead to magnetic, optical, acoustic, thermal, electrical, and superconductive properties of nanoparticles that differ significantly from their macroscopic properties. The fluctuation of electrons in discrete quantized energy levels at the same time brings a series of special properties to nanoparticles, such as high optical nonlinearity, specific catalytic and photocatalytic properties, strong oxidizing and reducing properties, and so on. For example, conductive metals may become semiconductors or insulators when made into nanoparticles; the size of the magnetic moment is related to whether the number of electrons in the particles is odd or even. The specific heat capacity of the material will also appear abnormal; the spectral line will shift to the short wavelength direction; the catalytic activity is delicately related to the number of atoms, and one more atom or one less atom will greatly affect the activity level. This is the objective manifestation of quantum effects. Therefore, quantum effects must be considered for nanoparticles at low temperatures, and the original macroscopic laws are no longer valid.

A classic case where quantum size effects affect material properties is the metal–insulator state transition of silver (Ag) nanoparticles at small sizes. The specific process of solving the critical dimension for transformation is as follows. Assuming that the temperature is 1 K, the electron density of Ag is $n = N/V = 6.0 \times 10^{22}\ cm^3$, according to the Kubo formula (2.15) and the Fermi level energy calculation formula (2.13), it can be obtained [17]:

$$\delta/k_B = \left(1.45 \times 10^{-18}\right)/V \tag{2.16}$$

When $\delta/k_B = 1$, $V = 1.45 \times 10^{-18}$ cm^3 can be obtained, and then the critical diameter (d_0) of Ag nanoparticles can be obtained as 14 nm according to the volume calculation formula of the sphere.

According to Kubo theory, only when $\delta > k_BT$ can the energy level split, that is, when the size of Ag nanoparticles is smaller than the critical size, the quantum size effect will appear. In addition, considering the limitation of electron lifetime, the condition of electron lifetime $\tau > \hbar/\delta$ needs to be satisfied. In summary, at a temperature of 1 K, when the size of Ag nanoparticles is less than 14 nm, it transforms from a conductor to an insulator. When the temperature is higher than 1 K, the Ag nanoparticles need to be smaller in size to be able to transform into an insulator.

Combined with the above theoretical inference, the researchers designed experiments to prove it. The results show that nano-Ag does exhibit high-resistance insulator properties. Making full use of the quantum size effect is expected to bring revolutionary innovations to the performance design of materials and the development of devices.

2.4 Energy Band Theory of Nanomaterials

The properties of materials are closely related to the movement behavior of electrons inside. In order to describe the movement characteristics of electrons in solid materials, researchers started from quantum mechanics to explain the conductance behavior of metals and then proposed the energy band theory, which is mainly used to explain the universal characteristics of electron movement in crystals. From the perspective of energy band theory, it can be stated that solids exhibit the characteristics of conductors, insulators, and semiconductors. Therefore, energy band theory is widely used in the research and development of semiconductor materials and devices. In nanomaterials, due to the small size of nanoparticles, the energy band structure exhibits discretization characteristics different from traditional materials, which makes the thermodynamic properties of nanomaterials significantly different from traditional bulk materials. In the post-Moore era, with the further reduction in the size of integrated circuits, when the chip size reaches the nanoscale, the movement of electrons is limited to the nanoscale. The study of electron movement behavior at the micro-nano scale is very necessary for the design of nano-devices. Therefore, clarifying the energy band theory of nano-materials is of great significance for promoting the development of nano-materials and devices.

2.4.1 Brief Introduction to Solid Energy Band Theory

The study of the energy band theory of solids should start from the study of the state of electron motion in solids, construct and solve the Schrödinger equations of all particle interaction systems. However, the number of ions and electrons in actual materials is almost infinite, and the movement of many electrons also affects each other. Therefore, it is impossible to solve the problem by constructing the Schrödinger equation. Only approximate methods can be used to solve it under reasonable assumptions. For crystalline materials, due to their periodicity, both the internal grid point arrangement and the real ion arrangement are periodic, so it can be assumed that the potential field where electrons are in during the movement is also periodic.

Here, the single-electron approximation method is adopted, assuming that the position of the atomic nucleus in the solid is fixed, and the electrons move in the potential field of the fixed nucleus and the average potential field of other electrons, and the motion problem of the multi-electron system in the solid is converted into a single-electron system for solution. For the potential field of electrons, it can be expressed as [18]:

$$U(x + na) = U(x) \tag{2.17}$$

In the formula, U represents the potential field where the electrons are located; x represents the spatial position coordinates of the electrons; a is the lattice constant of the crystal; n is an integer, representing the periodic number of repeated arrangement of electrons.

In order to describe the movement law of electrons in the crystal, it is necessary to solve the wave function of electron movement, that is, to find the functional formula corresponding to U(x), and combine this functional formula with the Schrödinger equation to solve it. In addition, in order to further simplify the problem, we make the following assumptions: (a) The lattice inside the crystal is complete, and the size of the crystal is infinite, and the influence of the crystal surface on the movement of electrons is not considered; (b) It is assumed that the thermal movement of ions in the crystal system does not affect the movement of electrons; (c) The moving electrons are independent of each other and there is no interaction; (d) Assume that the periodic potential field does not change with the change of space position. Under these assumptions, U(x) can be expressed as:

$$U(x) = U_0 + \sum_n U_n e^{i\pi nx/a} \tag{2.18}$$

In the formula, U_0 and U_n represent the potential energy of electrons in the initial state and n state, respectively; a is the lattice constant. Substitute it into the stationary Schrödinger equation of electron motion in one-dimensional space:

$$\frac{d^2\varphi}{dx^2} + \frac{8\pi^2 m}{h^2}(E - U)\varphi = 0 \tag{2.19}$$

In the formula, φ is the amplitude function; E is the total energy of electrons; U is the potential energy of electrons; m is the mass of electrons; $h = 2\pi\hbar$, $\hbar$ is Planck's constant.

Bloch proved that the solution of this equation has the following form:

$$\varphi(x) = e^{ikx} f(x) \tag{2.20}$$

In the formula, f(x) is a periodic function of the electron space coordinates; k is the period number, and its period is the same as the lattice arrangement and potential energy distribution:

$$f(x) = f(x + na) \tag{2.21}$$

This conclusion is called Bloch's theorem. Under the approximation of nearly free electrons, the motion of electrons satisfies the Schrödinger equation:

$$\left(-\frac{h}{2m}\nabla^2 + V\right)\Psi(r) = E\Psi(r) \tag{2.22}$$

In the formula, ∇ is the Laplace operator; E is the applied electric field strength; V is the volume of the object; $\Psi(r)$ is the wave function of free electrons, combined with Bloch's theorem, the solution of Schrödinger's equation can be realized.

Under the near-free electron approximation, solve the relationship between the energy of the electron and the wave vector (E − k), combined with the energy expression of the free electron, there is:

$$E = \frac{\hbar^2}{2m} k^2 \tag{2.23}$$

where E and k represent the energy and wave vector of the electron, respectively.

It can be seen from formula (2.23) that the E ~ k relationship under free electron approximation is parabolic. For some values of k, this parabolic relationship still holds true; but for other values of k, the energy E differs greatly from this parabolic relationship, that is, at $k = \pm n\pi/a$, the energy is no longer in a quasi-continuous state, and electrons can only occupy the high-energy $E_n + |U_n|$ energy level after occupying the lower-energy $E_n - |U_n|$ energy level, while the two energy levels cannot be occupied by electrons, showing the characteristics of prohibition.

Usually, we refer to the energy range that electrons can occupy under the approximate conditions of near-free electrons as the allowable band, and in the allowable band, the lower energy band that is bent and filled by electrons is defined as the valence band. Above the valence band, the energy band with higher energy that is not filled by electrons is called the conduction band. Between the top of the valence band filled by electrons and the bottom of the unfilled conduction band, electrons cannot be filled. This energy region is defined as the forbidden band. Solid materials can be divided into conductors, insulators and semiconductors according to the characteristics and bandwidth of their energy band structures.

For a conductor, a large number of electrons inside it are located in the conduction band that can move freely, and under the action of an external electric field, the electrons move directional to form a current. Common metallic materials are usually conductors. For insulators, the valence band in the energy band is completely filled, while the conduction band is completely empty without electrons. At the same time, the width between the conduction band and the valence band is very large. At room temperature, electrons cannot transition from the valence band to the conduction band only by means of thermal excitation. In a completely ideal insulator material, all electrons are directly bound by atoms, showing the property that it is difficult to move. Common inorganic non-metallic materials and polymer materials are usually insulators. For semiconductors, the electrons in its energy band structure completely fill the valence band, and the conduction band is completely empty, but the width between the conduction band and the valence band is small, and it only shows insulating properties at low temperatures. At high temperatures, some electrons can migrate from the valence band to the conduction band through thermal excitation, and can also form a current under a certain external field. The electronic equipment we use often contains a large number of semiconductor materials and devices.

In addition, in order to describe the distribution state of electrons in the energy band of solid materials, the concept of Fermi level is introduced to describe this characteristic. The physical meaning of the Fermi level means that electrons can fill all energy levels below the Fermi level in semiconductor materials under the condition of absolute zero, while electrons cannot fill the energy levels above the Fermi level. The Fermi surface is the equivalent energy surface of the highest energy level occupied by electrons at absolute zero. When the temperature is higher than absolute zero, the Fermi–Dirac distribution of electrons can be solved based on the Fermi level and temperature, so the Fermi level can completely describe the distribution state of electrons.

2.4.2 Energy Band Theory of Nanomaterials

At high temperatures, the energy band structure of macroscopic crystals can be regarded as a completely continuous state. As the size of the material decreases, the energy band structure will evolve from quasi-continuous energy bands to discrete energy levels at the atomic scale. Especially for nanoparticle systems, the number of conduction electrons that can move inside is limited, and the thermal excitation energy of electrons is k_BT, where k_B is the Boltzmann constant, and T is the temperature. Therefore, when the temperature is low, the free electron gas is affected by thermal excitation. The range is limited, and the energy bands of nanoparticles will show significant discretization characteristics, making their physical properties significantly different from those of bulk materials.

Assuming a metal nanoparticle with a diameter of d as the research object, under high temperature conditions, due to the thermal excitation of the free electron gas, taking the specific heat capacity as an example, the difference between the nanoparticle and the traditional bulk material is small. But as the temperature decreases, the physical properties of the material are determined almost entirely by the few energy levels around the Fermi level. Therefore, the grade interval model can be used to approximately describe the physical properties of a single metal nanoparticle at low temperature, and accordingly, the specific heat capacity of a single nanoparticle can be expressed as [19]:

$$C(T) = k_B \exp\left(-\frac{\delta}{k_B T}\right) \tag{2.24}$$

In the formula, C is the specific heat capacity; T is the temperature of the nanoparticles; δ is the energy level interval; k_B is the Boltzmann constant.

For nano-metal particles, due to the discretization of their energy levels, their thermodynamic properties at low temperature are significantly different from those of traditional materials. Although the theoretical solution of the specific heat capacity of a single metal nanoparticle at low temperature can be achieved through the equivalent energy level approximation model, it can only be characterized experimentally for the aggregate of nanoparticles, which makes it impossible to verify the correctness of the specific heat capacity solution experimentally. In an aggregate composed of nanoparticles, there are differences in the grain size of the particles, resulting in a statistical distribution of energy level intervals.

However, in aggregates, the statistical distribution of energy level intervals caused by particle size differences needs to be considered. Kubo theory takes the aggregate composed of metal nanoparticles as the object, and proposes to solve the electronic energy state of the particle aggregate based on the assumption of degenerate Fermi liquid and the assumption of electrical neutrality of nanoparticles.

Among them, the degenerate Fermi liquid hypothesis regards the electronic state near the Fermi surface as a degenerate electron gas limited by size, and assumes that the energy levels between them are discontinuous energy levels in the quasi-particle state, and assumes that there is no mutual exchange between quasi-particles. When the thermal excitation energy ($k_B T$) of electrons is much smaller than the average energy level interval (δ) between two adjacent energy levels, the energy level distribution of electrons close to the Fermi surface in this system obeys the Poisson distribution [20]:

$$P_n(\Delta) = \frac{1}{n!\delta}\frac{\Delta}{\delta}^{n} \exp\left(-\frac{\Delta}{\delta}\right) \tag{2.25}$$

In the formula, Δ is the energy state interval between two energy levels close to the Fermi surface; $P_n(\Delta)$ is the corresponding probability density; n is the number of energy levels between the two energy states. When Δ is the interval between adjacent energy levels, n = 0. The Kubo model improves the limitations of the traditional equal-level interval model, and can more accurately describe the physical properties of ultrafine particles at low temperatures.

The assumption of electrical neutrality of nanoparticles means that it is very difficult to extract or inject an electron from a single nanoparticle system through thermal fluctuations. For this reason, Kubo theory proposes that the work (W) done by overcoming the Coulomb force is required to realize this process:

$$k_BT \ll W\frac{1.5\times10^5 k_B}{d} = \frac{e^2}{4\pi\varepsilon_0 d} \approx \frac{e^2}{d} = \frac{1.5\times10^5 k_B}{d} \tag{2.26}$$

In the formula, d is the diameter of the nanoparticle; e is the electronic charge; ε_0 is the vacuum permittivity. It can be seen from formula (2.26) that as the diameter of the particle decreases, the work done to overcome the Coulomb force increases, and it is difficult to change the electrical neutrality of nanoparticles at low temperatures only by relying on thermal fluctuations.

Under the condition of sufficiently low temperature, when the particle size is only 1 nm, the work done by extracting or injecting an electron to overcome the Coulomb force (W) is two orders of magnitude smaller than the average energy level interval (δ). According to the above formula, it can be known that $k_BT \ll \delta$. Therefore, at low temperature, particles with a size of 1 nm will show discrete electronic energy levels. This discretization feature makes the properties of nanomaterials significantly different from bulk materials.

From the 1970s to the 1980s, the preparation and experimental characterization techniques of nanoparticles continued to develop, which facilitated the further study of the physical properties of nanoparticles, and thus made a series of progress related to theoretical prediction. For example, the test results based on the electronic motion characteristics, magnetic resonance characteristics and thermodynamic properties of nanomaterials verify the discretization of energy levels in nanoparticles, which provides experimental support for Kubo's theory. However, Kubo's theory still has limitations and needs to be further revised.

2.5 Dielectric Confinement Effect of Nanomaterials

With the continuous reduction of the particle size and the continuous increase of the specific surface area of the nanocrystalline particles, the change of the surface state will cause a significant change in the properties of the particles. For example, when a layer of medium with a small dielectric constant is modified on the surface of a semiconductor nanomaterial, compared to other media exposed around the semiconductor nanomaterial, the electric force lines of the charge carriers in the coated nanomaterial are more likely to pass through the coating film, resulting in a greater change in its optical properties than the bare nanomaterial, which is the dielectric confinement effect. When the dielectric constant of the nanomaterial and the medium differs greatly, a significant dielectric confinement effect occurs. At this time, the Coulomb force between charged particles is enhanced, resulting in enhanced binding energy and oscillator strength between electron–hole pairs, which weakens the main factor that produces the quantum size effect: the space confinement energy between electron–hole pairs, that is, the energy change caused by the surface effect is greater than the energy change caused by the space effect at this time, so that the energy band gap is reduced. Reflected in the optical properties, the absorption spectrum shows a significant red shift phenomenon.

In general, both transition metal oxides and semiconductor particles may produce dielectric confinement effects. The dielectric confinement of nanoparticles has important effects on light absorption, photochemistry, and optical nonlinearity. When analyzing the optical phenomena of nanomaterials, not only the quantum size effect, but also the dielectric confinement effect should be considered. Transition group metal oxides, such as Fe_2O_3, Co_2O_3, Cr_2O_3 and Mn_2O_3 nanoparticles dispersed in sodium dodecylbenzene sulfonate have optical third-order nonlinear enhancement effect. Measurements of Fe_2O_3 nanoparticles show that the third-order nonlinear coefficient is two orders of magnitude higher than that in water. This third-order nonlinear enhancement phenomenon is attributed to the dielectric confinement effect.

The Brus formula describes the effect of dielectric confinement on the shift (blue shift, red shift) of the light absorption band edge [21]:

$$E(r) = E_g(r = \infty) + h^2\pi^2/2\mu r^2 - 1.786e^2/\varepsilon r - 0.248E_{ry} \quad (2.27)$$

where E(r) is the absorption band gap of nanoparticles; $E_g(r = \infty)$ is the band gap of the bulk phase; r is the particle radius; $\mu = (1/m_{e-} + 1/m_{h+})^{-1}$ is the reduced mass of electrons and holes. The second term is the quantum confinement energy (blue shift), the third term indicates that the dielectric confinement effect leads to an increase in the permittivity, causing a red shift, and the fourth term is the effective Rydberg energy.

2.6 Macroscopic Quantum Tunneling Effect of Nanomaterials

2.6.1 Definition of Macroscopic Quantum Tunneling

Under certain conditions, electrons can cross the potential barrier like a train passing through a tunnel, and the effect of electrons penetrating through the potential barrier to participate in the conduction process is called tunneling effect. Electrons have both particle and wave properties, that is, wave-particle duality. In recent years, it has been found that some macroscopic quantities, such as the magnetization of microparticles and the magnetic flux in quantum coherent devices, also have tunneling effects, which are called macroscopic quantum tunneling effects. Macroscopic quantum tunneling is one of the basic quantum phenomena, that is, when the total energy of microscopic particles is less than the height of the barrier, the particle can still pass through the barrier. In the early days, it was used to explain the continued superparamagnetism of Ni nanoparticles at low temperature. In recent years, it has been found that the domain wall motion velocity in Fe–Ni films is basically independent of temperature below a certain critical temperature.

Quantum tunneling is a universally applicable law, which is the basis of many physical phenomena and can be realized in the field of electronics. Like tunnel diodes and the application of tunneling effects to measure small changes in magnetic fields, scanning tunneling microscopes are also based on the principle of quantum tunneling effects to directly observe the atomic structure of material surfaces, and use it to explain the essence of why nuclear radiation exists. In fact, the condensed force of neutrons and protons in the nucleus is so strong that it is impossible to break away from the nucleus. You must know that the protons in the nucleus have positive charges, while the neutrons have no charges. How do they gather together to form the nucleus? Scientific research has shown that the strong attraction between protons and neutrons is generated by the mutual exchange of mesons. According to the meson hypothesis, mesons can transmit the nuclear force, which is much stronger than the electromagnetic force, and thus can combine protons and protons. But this gravitational pull works only over extremely short distances. This particle was finally discovered later, and the force transmitted by the meson is called the "strong force". What causes protons to protons, and protons to neutrons? Quantum mechanics explains that there is a special force called the "strong force" that holds them together.

In a word, the study of the macroscopic quantum tunneling effect is of great significance to both basic research and practical applications. It limits the time limit for information storage on magnetic tapes and magnetic disks. Quantum size effect and tunnel effect will be the basis of future electronic devices, or it will establish the limit of further miniaturization of existing microelectronic devices. When electronic devices are further miniaturized, the above-mentioned quantum effects must be considered. For example, when designing and manufacturing semiconductor integrated circuits, when the size of the circuit is close to the wavelength of electrons, electrons will overflow the device through the tunnel effect, so that the device cannot work normally.

2.6.2 *Ballistic Transmission*

Classical transport theory tells us that the essence of material conductance is the movement of electrons in the lattice, and its transport law conforms to Ohm's law. For example, in a three-dimensional rectangular conductor, its conductance is proportional to the cross-sectional area S and inversely proportional to the length L [22]:

$$G = \sigma \frac{S}{L} \tag{2.28}$$

In the formula, G is the conductance of the material; σ is the conductivity, and at a certain temperature, the conductivity in the entire conductor is constant. According to the free electron gas model, the movement of electrons in the lattice has three degrees of freedom. According to Boltzmann's theorem, the thermal energy of each electron can be expressed as:

$$GE_T = \frac{3}{2} k_B T \tag{2.29}$$

For an electron with mass m and moving at an average velocity v without an external electric field, its total energy (kinetic energy) is $mv^2/2$. Therefore, $mv^2/2 = 3/2k_BT$ is established, and the average thermal velocity of electrons can be obtained as:

$$v_T = \sqrt{\frac{3k_BT}{m}} \tag{2.30}$$

In nanomaterials, when the length of material particles is reduced from the macroscopic level to the nanometer or atomic level, the transport characteristics of the internal electrons will change significantly. When the particle length is on a macroscopic scale, the transport characteristics of electrons conform to the diffusion equation. When electrons are transported in particles, they will encounter various obstacles and repeatedly scatter. That is, when electrons are transported in a macroscopic system, they have the characteristics of "random walk" from entering to leaving the system. When the length of the particle is reduced to nanometers or shorter, the size of the electronic device is smaller than the scattering length of the electron (i.e., the mean free path). At this time, there will be no scattering problem when the electron is transported from one end of the particle to the other end. The movement of the electron under this condition is called ballistic transmission. The transport law is that electrons transported ballistically will have quantized conductance when they pass through a short and narrow tunnel. This quantization effect of conductance exists in the mesoscopic physical system, that is, the scale is at the disk level of 0.01–1 μm. With the continuous reduction in the size of integrated circuits, the size of metal oxide–semiconductor devices is approaching the limit, and it is more important to study the electron transport behavior at this mesoscopic scale. The movement of electrons at the mesoscopic scale is governed by the quantum mechanism, and electrons are quantized in units of $2q^2/h$ Studying and mastering the ballistic transport mechanism will hopefully lead to better regulation of the physical properties of electronic and optical devices in the future.

2.6.3 *Inelastic Tunneling*

Another form of particle tunneling compared to ballistic transport is inelastic tunneling. In electronic devices, when the energy levels of the source and the drain are quite different, that is, when the energy levels of the two levels do not match very high, if electrons want to tunnel through the potential barrier, the transport of electrons will show another mode different from elastic transport: inelastic tunneling. In this process, electrons release excess energy by exciting phonons. When the energy of the phonon is exactly equal to the mismatched bandgap energy, electrons will tunnel across the potential barrier, which is inelastic tunneling. In order to achieve this process, it is usually necessary to use a third material as an intermediate transition layer. The researchers found in experiments that when oxide is used as the intermediate transition layer, it is more conducive to the occurrence of inelastic tunneling.

The significance and impact of the research on the macroscopic quantum tunneling effect on the development of nanomaterials and devices are various. In the field of nanomaterials research, scanning tunneling microscopy utilizes this effect. At the same time, the macroscopic quantum tunneling effect determines the integration limit of the device to a certain extent.

2.7 Coulomb Blockade and Single-Electron Tunneling Effect of Nanomaterials

The Coulomb blocking effect is one of the most important physical phenomena discovered in the mesoscopic field in the 1980s. When the scale of the system enters the nanoscale (several nanometers for metal particles and tens of nanometers for semiconductor particles), the system is charge "quantized", that is, the charging and discharging process is discontinuous, and the energy E_e required to charge an electron is $e^2/2C$, where e is the charge of an electron, and C is the capacitance of a small system. The smaller the system, the smaller C is, and the greater the energy E_e is. We call this energy Coulomb blockage energy.

In fact, the Coulomb blockade energy is the Coulomb repulsion energy of the previous electron to the next electron. This leads to the process of charging and discharging a small system, electrons cannot be transported collectively, but are transported one by one. This single-electron transport behavior in small systems is usually called the Coulomb blockade effect. If the nanoparticles are connected to the circuit through a very thin insulating layer to form a "tunneling junction-Coulomb island-tunneling junction" structure, when certain conditions are met to charge and discharge the system, the electrons cannot be transported collectively, but transported one by one, showing small steps one by one on the IV curve, which is usually called Coulomb blocking (Fig. 2.4a) or Coulomb step (Fig. 2.4b) effect. Due to the existence of the Coulomb blocking effect, the rise of the current with the voltage is no longer a straight line, but a sawtooth-shaped step on the IV curve. These are collectively known as the single-electron tunneling phenomenon, which is the physical basis of single-electron devices.

In each oscillation cycle, the amount of charge change is e, that is, the charge is quantized in units of e. Electrostatic energy, also known as Coulomb blockage energy, is the Coulomb electrostatic repulsion energy of the previous electron to the next electron, so that electrons cannot flow collectively in a small system such as

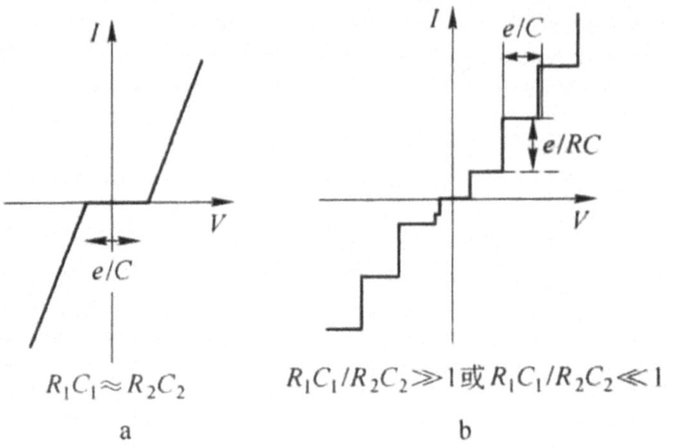

Fig. 2.4 **a** Coulomb blockade, **b** Coulomb step effect [23]

quantum dots, but are transmitted one by one. This is how single-electron transistors work.

If two quantum dots are connected by a "junction", the behavior of a single electron on one quantum dot across the energy barrier to the other quantum dot is called single-electron tunneling. The condition for single electron tunneling is [24]:

$$e^2/2C > k_B T \tag{2.31}$$

$$R_j > h/e^2 \tag{2.32}$$

In order to make a single electron tunnel from one quantum dot to another quantum dot, the voltage (V/2) applied on a quantum dot must overcome E_e, that is, $V > e/C$. Usually, both Coulomb blockade and quantum tunneling are observed at extremely low temperature, and the observed condition is $e^2/2C > k_B T$. It has been estimated that if the size of quantum dots is about 1 nm, the above effects can be observed at room temperature. When the size of quantum dots is in the range of tens of nanometers, the above effects must be observed at the temperature of liquid nitrogen. The reason is easy to understand. The smaller the size of the system, the smaller the capacitance C and the larger $e^2/2C$, which allows observation at higher temperatures. Using Coulomb blockade and quantum tunneling effects can design next-generation nanostructured devices, such as single-electron transistors and quantum switches [25]. The above-mentioned small size effect, surface interface effect, and quantum size effect are all basic characteristics of nanoparticles and nanosolids. It makes nanoparticles and nano-solids present many strange physical and chemical properties, and some "abnormal phenomena" appear [26].

References

1. N.-T. Suen, S.-F. Hung, Q. Quan, N. Zhang, Y.-J. Xu, H. M. Chen, *Chem. Soc. Rev.* **2017**, *46*, 337–365.
2. S. D. Ebbesen, S. H. Jensen, A. Hauch, M. B. Mogensen, *Chem. Rev.* **2014**, *114*, 10697–10734.
3. Y. Ding, F. Fan, Z. Tian, Z. L. Wang, *J. Am. Chem. Soc.* **2010**, *132*, 12480–12486.
4. P. Parthasarathy, K. N. Sheeba, *Int. J. Energ. Res.* **2015**, *39*, 147–164.
5. J. K. Nørskov, J. Rossmeisl, A. Logadottir, L. Lindqvist, J. R. Kitchin, T. Bligaard, H. Jónsson, *J. Phys. Chem. B* **2004**, *108*, 17886–17892.
6. J. C. Mitani, F. B. M. de Waal, K. Hosaka, W. C. McGrew, M. Nakamura, A. Nishimura, R. W. Wrangham, J. Yamigiwa, *Int. J. Primatol.* **2012**, *33*, 10–18.
7. Yan, Jiao, Yao, Zheng, Mietek, Jaroniec, S. Zhang, Q. J. C. S. Renew. Sust. Energ. Rev. 2015, 15, 12–18.
8. Y. Shi, B. Zhang, *Chem. Soc. Rev.* **2016**, *45*, 1529–1541.
9. L. Q. Zhang, M. Laato, *Ann. Chir. Gynaecol.* **2001**, *90*, 29–32.
10. J. Alazemi, J. Andrews, *Renew. Sust. Energ. Rev.* **2015**, *48*, 483–499.
11. J. E. Huber, J. W. Kolar, *Renew. Sust. Energ. Rev.* **2019**, *10*, 317–326.
12. S. E. Hosseini, M. A. Wahid, *Int. J. Energ. Res.* **2020**, *44*, 4110–4131.
13. B. C. Glavovic, T. F. Smith, I. White, *Clim. Dev.* **2022**, *14*, 829–833.
14. M. E. Mondejar, R. Avtar, H. L. B. Diaz, R. K. Dubey, J. Esteban, A. Gómez-Morales, B. Hallam, N. T. Mbungu, C. C. Okolo, K. A. Prasad, Q. She, S. Garcia-Segura, *Sci. Total Environ.* **2021**, *794*, 148539.
15. F. Razi, I. Dincer, *J. Clean. Prod.* **2020**, *264*, 121582.
16. E. A. Fletcher, R. L. Moen, *Science* **1977**, *197*, 1050–1056.
17. W.-H. Cheng, A. de la Calle, H. A. Atwater, E. B. Stechel, C. Xiang, *ACS Energy Lett.* **2021**, *6*, 3096–3113.
18. A. Kovač, M. Paranos, D. Marciuš, *Int. J. Hydrogen Energy* **2021**, *46*, 10016–10035.
19. S. U. M. Khan, M. Al-Shahry, W. B. Ingler, *Science* **2002**, *297*, 2243–2245.
20. Y. Zhang, J. Chen, C. Xu, K. Zhou, Z. Wang, J. Zhou, K. Cen, *Int. J. Hydrogen Energy* **2016**, *41*, 2215–2221.
21. A. Tofighi, F. Sibieude, *Int. J. Hydrogen Energy* **1984**, *9*, 293–296.
22. R. Nandi, S. Sengupta, *Crit. Rev. Microbiol.* **1998**, *24*, 61–84.
23. J.-F. Xu, N.-Q. Ren, A.-J. Wang, J. Qiu, Q.-L. Zhao, Y.-J. Feng, B.-F. Liu, *Int. J. Hydrogen Energy* **2010**, *35*, 13467–13474.
24. Q. Jin, N. Qureshi, H. Wang, H. Huang, *Fuel* **2019**, *244*, 536–544.
25. G. Davila-Vazquez, S. Arriaga, F. Alatriste-Mondragón, A. de León-Rodríguez, L. M. Rosales-Colunga, E. Razo-Flores, *Rev. Environ. Sci. Bio.* **2008**, *7*, 27–45.
26. C.-Y. Lin, C.-H. Cheng, *Int. J. Hydrogen Energy* **2006**, *31*, 832–840.

Chapter 3
Physical Properties of Nanomaterials

Nano-materials have a strong size effect, and their properties are determined by size. The size effect of nano-materials is based on quantum effect and surface-interface effect. Nano-materials also have macroscopic quantum tunneling effect.

3.1 Basic Dimensions of Nanomaterials

3.1.1 Size Effect

The so-called size effect means that the properties of nanomaterials will change when the size of the constituent phases of nanomaterials, such as the size of grains and the size of the second phase particles, decreases, and the properties of nanomaterials will change or suddenly change when the size of the constituent phases is as small as a certain crical size. The quantum effect of small-size (nm-level) system means that the energy of electrons is quantized to form discrete quantum state energy levels, and the movement of electrons in this system is constrained. The quantum size effect of control particles was studied theoretically as early as 1962. Japanese scientist Kubo defined the quantum size effect: with the decrease of the size of metal particles, the electron energy level near the Fermi level of metal changed from quasi-continuous to discrete. Kubo has long proposed the calculation method of energy level spacing and given a famous formula [1]:

$$\delta = \frac{1}{3} E_F / N \tag{3.1}$$

where "δ" is between energy levels; "E_F" is the Fermi level and "N" is the total number of electrons. Macroscopic objects contain infinite atoms (that is, the number of electrons contained in "N_∞"), so "$\delta 0$", that is, the energy level spacing of large

S. Peng and P. Li, *Nanomaterials and Nanotechnology*,
https://doi.org/10.1007/978-981-97-8433-2_3

particles or macroscopic objects is almost zero, and nanoparticles contain a limited number of atoms, and the N value is very small, resulting in a certain value, that is, when the energy level spacing is greater than electrostatic energy and light energy. This leads to the electrical and optical characteristics of nanoparticles and macroscopic characteristics are significantly different. This effect is called quantum size effect (small size effect). The size effect of nanomaterials also relates to the stability of nanostructures. The results of computer simulation calculation of the free energy of nanocrystalline materials and their corresponding amorphous state show that when the size of nanostructures is less than a certain critical size, the phase transition from nanocrystalline to amorphous state will occur [2].

3.1.2 Surface and Interface Effects

Surface-interface effect is another basic effect of nanomaterials. Because nanocrystalline materials contain a large number of grain boundaries, the atoms on the grain boundaries occupy a fairly high proportion. Nanoparticles are small in size and large in surface, and the atoms located on the surface account for a considerable proportion (see Table 3.1). The decrease of the particle size of nanoparticles will eventually lead to the increase of atomic activity on the surface, which will not only cause changes in atomic transport and configuration on the surface of nanoparticles, but also cause changes in surface electron spin conformation and electron energy spectrum. The above situation is called surface and interface effect [3].

As can be seen from Table 3.1 and Fig. 3.1, with the decrease of particle size, the number of surface atoms increases rapidly. This is due to the small particle size and the sharp increase in surface area. For example, when the particle size is 10nm, the specific area is 90 $m^2\ g^{-1}$, and when it is 5 nm, the specific surface area is 180 $m^2\ g^{-1}$, the particle size decreases to 2 nm, and the specific surface area increases sharply to 450 $m^2\ g^{-1}$. Such a high specific surface area makes the number of atoms on the surface more and more, and at the same time the surface energy increases rapidly, as shown in Table 3.2. It can be seen from the table that the specific surface area and surface energy of Cu nanoparticles increase by two orders of magnitude from 100 to 10 to 1 nm [5].

Table 3.1 Nanoparticle size versus surface atomic number

Size (nm)	Total number of atoms	The proportion of surface atoms (%)	Specific surface area ($m^2\ g^{-1}$)
10	3 × 104	20	90
4	4 × 103	40	225
2	2.5 × 102	80	450
1	30	99	900

Fig. 3.1 The relationship between the ratio of surface atomic number to total atomic number and particle size [4]

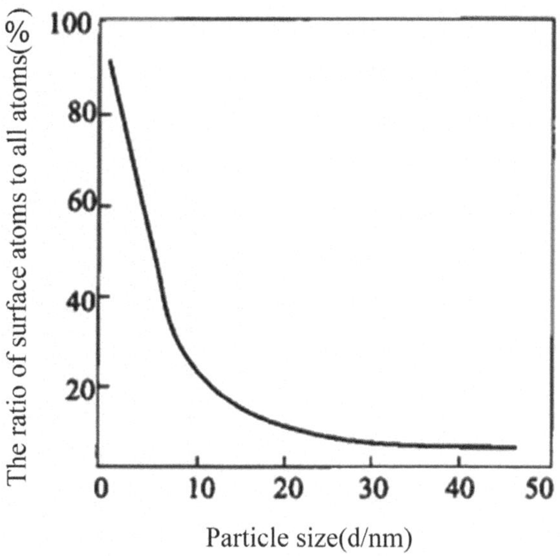

Table 3.2 The relationship between the particle size and specific surface area, the proportion of the number of surface atoms, the surface energy and the number of atoms in a particle

Particle size (d/nm)	The specific surface area of Cu ($m^2\ g^{-1}$)	The proportion of surface atoms (%)	The number of atoms in a particle	Specific surface energy ($J\ mol^{-1}$)
100	6.6		8.46×107	5.9×102
20		10		
10	66	20	8.46×104	5.9×103
5		40	1.06×104	
2		80		
1	660	99		5.9×104

Due to the increase in the number of surface atoms, insufficient coordination of atoms and high surface energy, these surface atoms are highly active, extremely unstable and easily combined with other atoms. For example, metal nanoparticles will burn in the air, and inorganic nanoparticles will absorb gas and react with gas when exposed to the air. Here is an example to illustrate the reasons for the high surface activity of nanoparticles. Figure 3.2 is a two-dimensional plan view of a single cubic crystal grain. Assuming that the grain is round (on the plane), the solid circle represents the atoms located on the surface, and the hollow circle represents the internal atoms. The grain size is 3nm, and the atomic spacing is about 0.3 nm. Obviously, there are "E" atoms missing one neighbor, "D" atoms missing two neighbors and "A" atoms missing three neighbors. Surface atoms such as "A" are extremely unstable and quickly move to the "B" position. When these surface atoms meet other

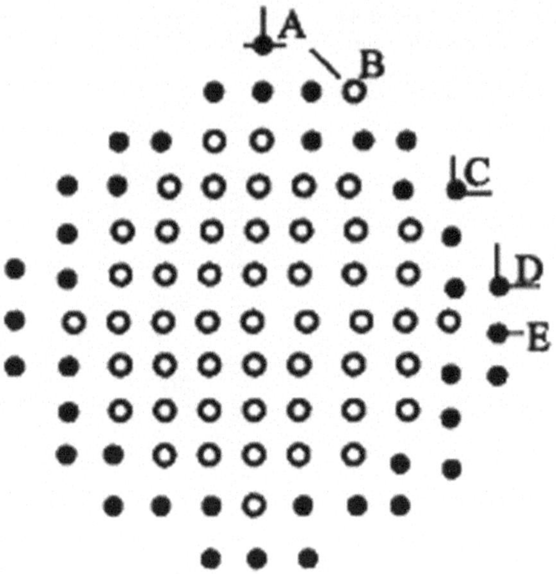

Fig. 3.2 Ultrafine pattern diagrams in which atoms with a single cubic lattice structure are configured as closely as possible (or) shaped [6]

atoms, they quickly combine and stabilize them, which is the reason for their activity. This activity of surface atoms not only causes changes in the surface atomic transport and configuration of nanoparticles, but also causes changes in the surface electron spin conformation and electron energy spectrum.

The surface and interface effects make nano-materials have a high diffusion rate. Through various methods, it has been observed that A1, Cu and Bi have high diffusion coefficients in nanocrystalline Pd. For polycrystalline materials, diffusion can be carried out along the free surface, grain boundary and lattice, in which the diffusion coefficient along the surface is the largest and along the lattice is the smallest. Generally, the interfacial volume fraction of metals is very small, so the grain boundary diffusion is not easy to show. In nanocrystals, because of the large proportion of grain boundaries, grain boundary diffusion may be absolutely dominant. The activation energy of nano-grain boundary diffusion along grain boundary is equivalent to that along free surface.

3.1.3 Macro Quantum Tunneling Effect

The ability of microscopic particles to penetrate the barrier is called tunneling effect. In recent years, it has been found that some macroscopic quantities, such as the magnetization of microparticles and the magnetic flux in quantum coherent devices, also have tunneling effect, which is called macroscopic quantum tunneling effect. It has been used to explain that ultrafine nickel particles continue to maintain superparamagnetism at low temperature. In recent years, it has been found that the velocity in Fe–Ni film is basically independent of temperature when it is below a certain critical temperature. Therefore, it has been suggested that the zero-point vibration of quantum mechanics can play a similar thermal fluctuation effect at low temperature, so that the reorientation of the magnetization vector of micro-particles near zero temperature can maintain a limited relaxation time, that is, there is still a non-zero magnetization reversal rate at absolute zero. The similar viewpoint explains that the high magnetocrystalline anisotropic single crystal produces a stepped inversion magnetization mode at low temperature and some effects in quantum interference devices. The above-mentioned size effect, surface interface effect, quantum size effect and quantum tunneling effect are the basic characteristics of nanoparticles and nanosolids. It makes nanoparticles and nanosolids present many strange physical and chemical properties, and some "abnormal phenomena" appear. For example, metal is a conductor, but nano-metal particles will show electrical insulation at low temperature due to quantum size effect. Generally, $PbTiO_3$, $BaTiO_3$ and $SrTi_3$ are typical ferroelectrics, but they will become paraelectric when their size reaches nanometer order. Ferromagnetic substances enter the nanometer scale (about 5 nm), and show strong paramagnetic effect because they change from multi-domain to single domain; When silicon nitride particles with a particle size of more than ten meters constitute nano-ceramics, they have no typical covalent bond characteristics, and the interface bond structure appears partial polarity, and the resistance is very small under alternating current; After being made into nano-micro-platinum, the chemically emotional metal platinum has become an excellent catalyst.

3.2 Electrical Properties

Due to the increase of atomic volume fraction in the nano-material boundary, the resistance of nano-materials is higher than that of similar coarse-grained materials, and even size induction occurs, and the metal transforms into insulator, and the resistance of materials decreases obviously in the magnetic field. The electrical properties change strangely because the transmission process of electrons in nano-materials is constrained by spatial dimensions, thus showing quantum confinement effect. In nanoparticles, or in a very thin short metal wire, the kinetic energy or energy of electrons is quantized because the movement of electrons in particles is limited. The electrical properties of nanomaterials depend on their structures. For example, with the different structural parameters of carbon nanotubes, carbon nanotubes can be metallic or semiconducting.

3.2.1 Dielectric Constant and Dielectric Loss

Dielectric materials or dielectrics are materials whose basic electrical properties are polarization. The so-called polarization refers to the relative movement (electric displacement) of the positive and negative centers of atoms or ions in materials under the action of electric field, which leads to the phenomenon of electric moment (electric dipole). The main mechanisms of polarization are:

1) Electron displacement polarization: under the action of external electric field, the electron cloud and nucleus of atoms are relatively displaced.
2) Particle displacement polarization: under the action of external electric field, there is a relative position between positive ions and ions.
3) When there is no external electric field, the positive and negative charge centers of some substances do not coincide, and there is an inherent electric dipole moment. However, due to thermal motion, the orientation of the electric dipole moment of molecules is randomly distributed, and the total electric moment is zero. Under the action of an external electric field, the electric dipole moment partially turns to the direction of the electric field and is aligned.
4) Spontaneous polarization: in the volume of 32 point groups, 20 point groups do not have central symmetry, but have piezoelectric characteristics due to elastic deformation polarization. Ten of these 20 point groups have a unique polar axis (spontaneous polarization axis) and spontaneous polarization can occur. Usually, spontaneous polarization can change with the change of temperature, which is called pyroelectricity. In pyroelectric crystals, the spontaneous polarization direction of some crystals can be changed under external electric field, and these crystals are called ferroelectrics. Obviously, ferroelectrics are pyroelectric, piezoelectric and dielectric at the same time; On the other hand, it is not necessarily true.

In an electrostatic field, the electric displacement "$D = \varepsilon_0 \varepsilon E$", where "$\varepsilon_0$" and "$\varepsilon$" are the relative permittivity of the vacuum and the medium, respectively, and E is the electric field. If the medium has no conductance in an electrostatic field, there is no dielectric loss. In alternating electric fields, the polarization changes with the change of the electric field. When the electric field changes relatively quickly, the polarization will lag behind the change of the electric field, resulting in a phase difference of "δ" between the electric field and the polarization. In fact, the various polarizations in the medium are a relaxation process, and a certain relaxation time is passed from the initial state to the final state. This relaxation of the medium will cause dielectric loss, also known as dielectric loss, in the alternating electric field. The dielectric loss is expressed as the tangent of the phase difference, i.e. the dielectric loss is equal to "$\tan \delta$".

In a dynamic electric field, the electric displacement "$D = \varepsilon^* E$", where "E" is the dynamic electric field and "ε^*" is the dynamic permittivity, which is a complex number related to the angular frequency of the electric field "ω" [7].

$$\varepsilon^*(\omega) = \varepsilon'(\omega) + i\varepsilon''(\omega) \tag{3.2}$$

If the medium has no loss (no conductance) in the electrostatic field, then when "ω_0", the second term in Eq. (3.2) tends to zero, and the dynamic permittivity tends to the static permittivity. In an alternating electric field, the dielectric loss factor is:

$$\tan\delta = \varepsilon''/\varepsilon' \tag{3.3}$$

The dielectric loss is related to the relaxation process of polarization. For example, the potential barrier must be overcome in the electric moment steering polarization, and relaxation will lead to loss. The transition of ions from one equilibrium state to another depends on thermal fluctuation, which is related to nonlinear vibration and will also lead to loss. Dielectric constant and dielectric loss are two important parameters to characterize dielectric properties. Their frequency and temperature changes are usually represented by frequency spectrum and temperature spectrum.

Nano-dielectric materials have size effect and interface effect, which will strongly affect their dielectric properties. These effects are mainly manifested in:

1) Interface polarization caused by space charge, because the interface with large volume fraction of nano-materials can produce strong interface polarization or space charge polarization on both sides of the interface under the action of external electric field;
2) The dielectric constant or dielectric loss has a strong scale effect. For example, if there is an electric domain in a ferroelectric, that is, a region with consistent spontaneous polarization orientation, the structure of the electric domain will directly affect the piezoelectric and dielectric properties of the ferroelectric. With the decrease of the size, the ferroelectric single domain will undergo a ferroelectric-paraelectric phase transition driven by the size, which will weaken the spontaneous polarization and lower the Curie point, which will affect the orientation polarization and dielectric properties.
3) The alternating current conductance of nanodielectric materials is often much larger than that of conventional dielectrics. For example, the conductance of nano-α-Fe_2O_3 solid is 3–4 orders of magnitude higher than that of conventional materials, and nano-silicon nitride also has obvious alternating current conductance with the decrease of size. The increase of the conductance of nano-dielectric materials will lead to the increase of dielectric loss, so nano-dielectric materials will show many dielectric characteristics different from conventional dielectrics.

3.2.2 Electrical Properties of Nano-oxides

Many oxides have low dielectric constant and low dielectric loss. For example, Al_2O_3, MgO, CaO, BeO, TiO_2, Y_2O_3, ZrO_2 etc., their relative dielectric constants are generally below 10, and their dielectric losses are in the order of 10^{-4}–10^{-3}. The electrical constant of TiO_2 is slightly higher, such as the relative dielectric constant of rutile structure TiO_2 is between 100 and 200, and that of anatase structure is about 30.

The study of dielectric constants of nano-metal oxides such as Al_2O_3, TiO_2 and Fe_2O_3 shows that frequency and particle size of oxides have great influence on dielectric constants. Figure 3.3 shows the effects of frequency and particle size of anatase TiO_2 on dielectric constant. As can be seen from the figure, the dielectric constant increases rapidly with the decrease of measurement frequency. At the same

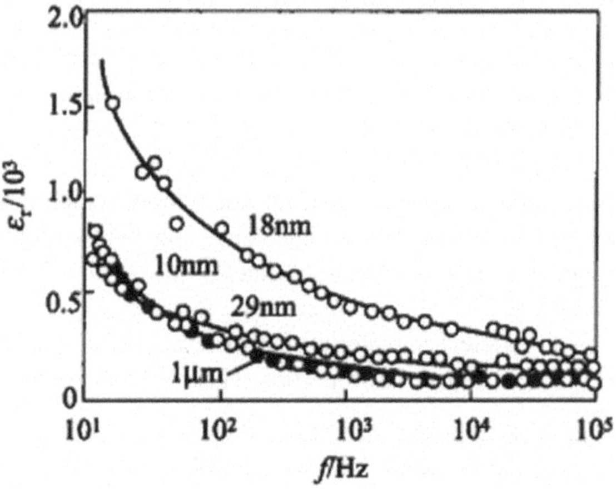

Fig. 3.3 Relationship between particle size, permittivity and frequency of anatase phase TiO_2 at room temperature [8]

time, the dielectric constant has a size effect, that is, TiO_2 with particle sizes of 10, 29 nm and 1 μm have very similar dielectric constants, while the dielectric constant of TiO_2 with particle size of 18 nm is significantly higher than that of the first three. Temperature also has a significant effect on the dielectric properties of anatase TiO_2. For example, when the frequency is 96 kHz, TiO_2 with a particle size of 18 nm has a peak at 75 °C, while the dielectric loss has a peak of 1.0 at 50 °C, and the dielectric loss is less than 0.1 at 0 °C. In the low frequency range (f < 0.1 kHz), the dielectric constant of rutile TiO_2 with an average particle size of 12 nm prepared by oxidation process is 1–2 orders of magnitude higher than that of conventional coarse crystals, but the dielectric loss is also as high as 5–20.

Rutile-phase TiO_2 powder with an average grain size of 8.5 nm was prepared by reactive ion plating, and the powder was pressed into a 10 mm sample under the pressure of 890 MPa, which can be used to measure the dielectric properties of rutile-phase TiO_2. Figure 3.4 shows the relationship between the dielectric constant at room temperature and the measurement frequency of the sample in vacuum and in air. It can be seen from the figure that the measurement frequency and air pressure have significant effects on the dielectric constant of the sample. When the measurement frequency is near zero in vacuum, the change of dielectric constant is as high as 2–3 orders of magnitude, and at the same time, the dielectric constant in air is three orders of magnitude higher than that in vacuum. Figure 3.5 shows the frequency spectrum of imaginary part of dielectric constant of nano-γ-Fe_2O_3 sample in air and vacuum (inset). The sample is made of Fe_2O_3 powder with an average particle size of 12 nm under a pressure of 20 MPa. The spectrum of the real part of the sample is similar to that of the imaginary part, showing that the dielectric constant decreases sharply with the increase of the number of measurements near the frequency of zero. The reason why the dielectric constant drops sharply with the increase of experimental times near zero frequency is unclear, which may be related to the existence of a large number of dangling bonds in the sample because it is only pressed without sintering. With the increase of measurement times or the extension of time, the dangling bonds in the sample decrease greatly and finally reach a stable value. During vacuum measurement, it was found that the capacitance of the sample would be stabilized after 48 h. Therefore, it can be considered that it is of no practical value not to reach a stable dielectric constant value. Besides dangling bonds, there are a lot of defects in the interface of pressed samples. In the atmosphere, the dielectric constant of nano-rutile TiO_2 increases abnormally, which is mainly caused by the defects in the material interface and the physical or chemical adsorption of polarized molecules in the atmosphere by dangling bonds. In vacuum, the dielectric constant of nano-rutile TiO_2 is less than 10, which is much lower than that of ordinary coarse-grained TiO_2 because it can not absorb polarized molecules in the atmosphere [9].

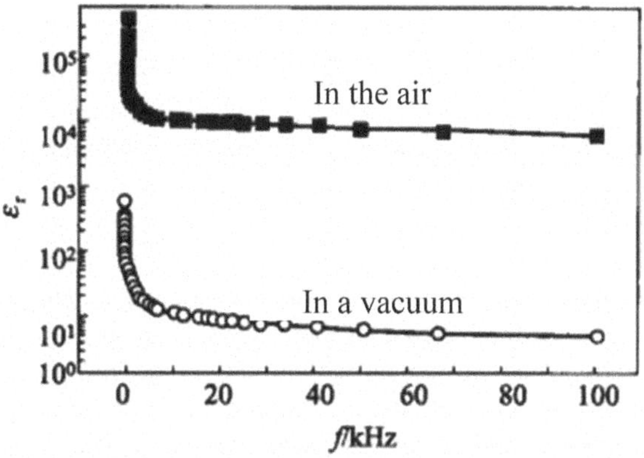

Fig. 3.4 Relationship between dielectric constant and frequency of TiO_2 in the nanometer rutile phase [8]

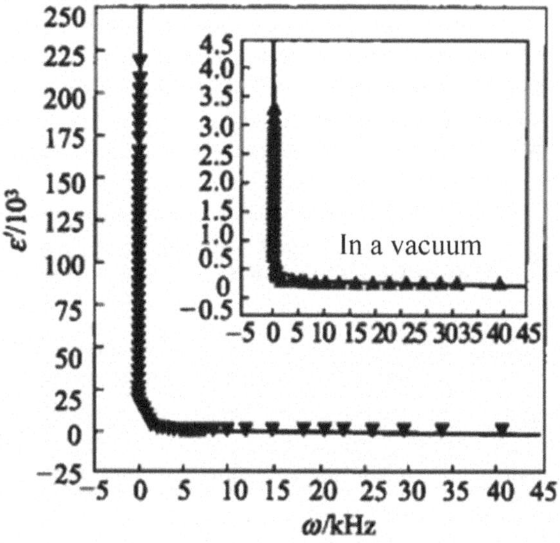

Fig. 3.5 Frequency spectrum of the imaginary part of the dielectric constant of nano γ-Fe_2O_3 in air and vacuum (inset) [10]

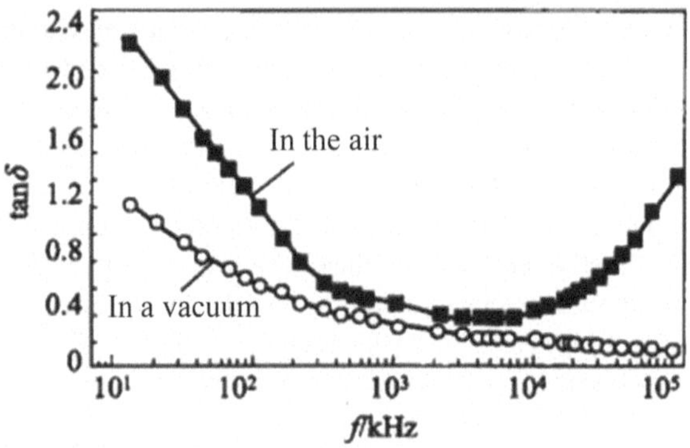

Fig. 3.6 Dielectric loss spectrogram of nanorutile phase TiO_2 in vacuum and air [10]

Figure 3.6 shows the dielectric loss spectrum of nano rutile TiO_2 in vacuum and air. It can be seen from the figure that the dielectric loss in air is much greater than that in vacuum in low frequency and high frequency regions, and the lowest value appears around 1000 Hz, while the dielectric loss in vacuum decreases monotonously with the increase of frequency, and there is no lowest value of loss. The results of measuring the equivalent impedance spectrum of the sample show that the impedance of the sample in vacuum is obviously greater than that in air. Therefore, the main reason for the increase of dielectric loss in air is the loss caused by the increase of conductance. At the same time, the dielectric loss of nano-rutile TiO_2 in vacuum is mainly conductivity loss, which decreases monotonously with the increase of frequency, and there is no peak. The dielectric loss of nano-γ-Fe_2O_3 samples in vacuum is also much less than that in air, which is also because the conductance of samples in vacuum is smaller than that in air, and the dielectric loss in vacuum is mainly caused by conductance loss. The loss of nano-TiO_2 in the atmosphere is not only dielectric loss, but also loss caused by polarization relaxation, such as electric moment steering and potential barrier that ions must overcome when they transition from one equilibrium position to another, so part of energy is consumed in the form of heat energy.

3.3 Magnetic Properties

When the grain size is reduced to nanometer level, the ferromagnetic interaction between grains begins to have an important influence on the macroscopic magnetism of materials. Nanoparticles are generally single domain particles because of their ultra-fine size, and their technical magnetization process is determined by the magnetic anisotropy of grains and the magnetic interaction between grains. The magnetic anisotropy of nano-grains is related to the shape, crystal structure, internal stress and atoms on the surface of grains, which is significantly different from coarse-grained materials and shows obvious small-size effect.

When the particle size or grain size of a magnetic substance enters the nanometer range, its magnetic properties have obvious size effect. Therefore, nano-materials have many magnetic properties that coarse-grained or micro-grained materials do not have. For example, nanowires have strong shape anisotropy because of the large ratio of length to diameter. When their diameter is less than a certain critical value, they have the characteristic of magnetization along the axis of the filament under zero magnetic field. In addition, magnetic parameters such as coercivity, saturation magnetization and Curie temperature are all related to grain size.

3.3.1 Coercivity

In magnetic properties, the coercivity is most strongly influenced by the change of grain size. For roughly spherical grains, the coercivity increases with the decrease of grain size, and after reaching a maximum, the coercivity decreases with the further decrease of grain size. The grain size corresponding to the maximum coercivity is equivalent to the size of a single domain, and its size ranges from ten to several hundred nanometers for different alloy systems. When the grain size is larger than the single domain size, the relationship between coercivity H_c and average grain size D is as follows:

$$H_c = C/D \tag{3.4}$$

where “C” is a constant with respect to materials. It can be seen that when the grain size of the nanomaterial is larger than the single domain size, the coercivity also increases with the decrease of the grain size “D”, which corresponds to Eq. (3.4). When the grain size of the nanomaterial is smaller than a certain size, the coercivity decreases sharply with the decrease of the grain. At this time, the relationship between coercivity and grain size is:

$$H_c = C''/D^6 \tag{3.5}$$

Where “C′′” is a constant with respect to the material. Equation (3.5) is in good agreement with the measured data. Figure 3.7 shows the H_c of some Fe-based alloys versus grain size D. Figure 3.8 complements the relationship between the coercivity Hc and the average particle size D of Fe and Fe–Co alloy particles in the range of 1–1000 nm, and the relationship between the residual magnetic ratio “M_r/M_S”and “D” is also shown.

The size effect of coercivity can be qualitatively explained using Fig. 3.9. There are three critical dimensions of diameter D on the abscissa of the figure. When D > D_{crit}, the particles are multi-domain, and their antimagnetization is the domain wall displacement process, and the H_c is relatively small; When D < D_{crit}, the particles are single domains, but when $d_{crit} < D < D_{crit}$, non-uniform rotation occurs, and the H_c increases with the decrease of D. When $d_{th} < D < d_{crit}$, it is the average rotation

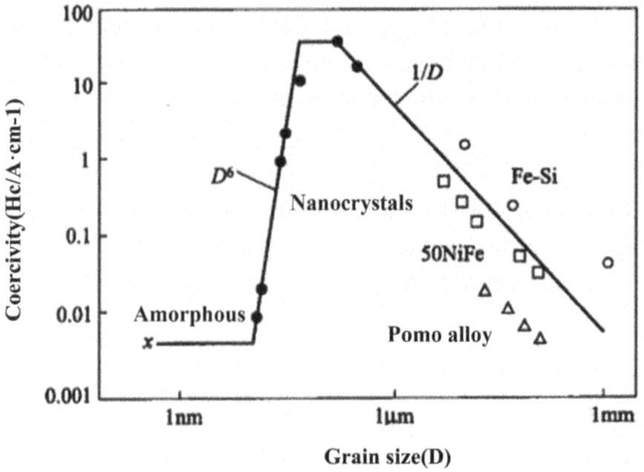

Fig. 3.7 Relationship between coercivity (H_c) and grain size (D) [11]

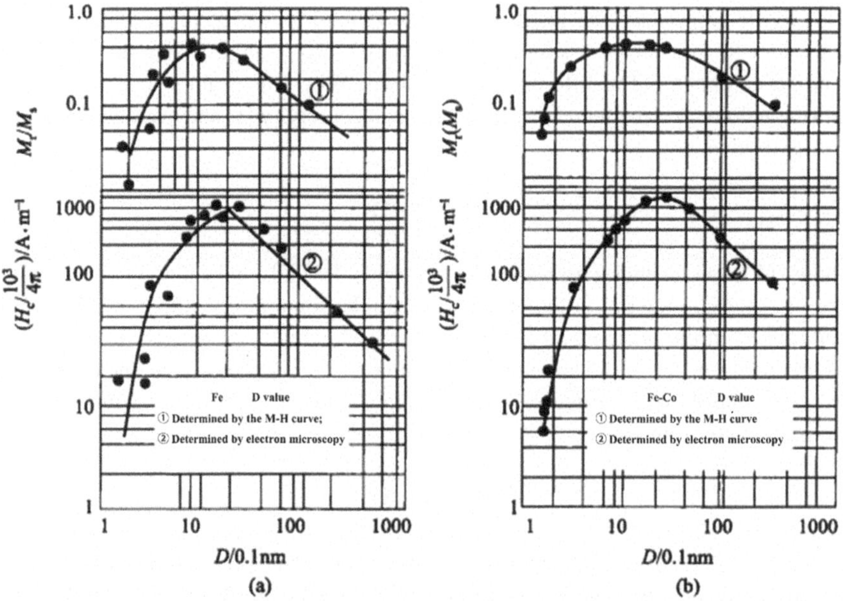

Fig. 3.8 Relationship between Fe and Fe–Co alloy particles H_c and D **a** Fe; **b** Fe–Co [12]

region and H_c reaches a maximum value. When $D < d_{th}$, Hc decreases sharply with the decrease of D, which is caused by superparamagnetism caused by the magnetization direction of the particles as "magnetic brownian motion" when the thermal kinetic energy k_BT is greater than the barrier that needs to be overcome for magnetization reversal [13].

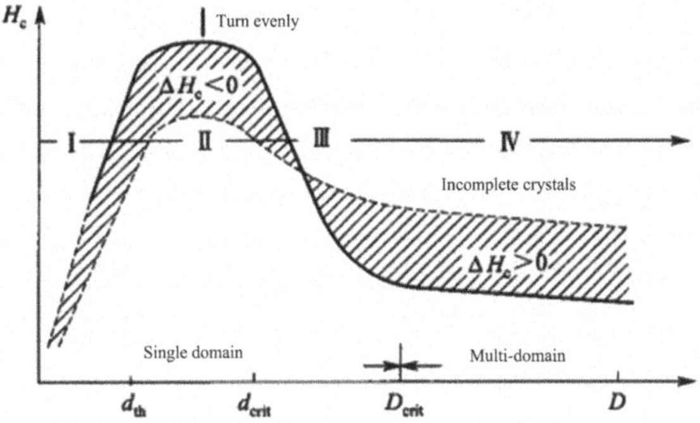

Fig. 3.9 Relationship between H_c and diameter D of particles [14]

3.3.2 *Superparamagnetism*

Superparamagnetism is the magnetism caused by the influence of thermal motion energy on the spontaneous magnetization direction of particles when the particle volume is small enough. Superparamagnetism can be defined as the phenomenon that the magnetization of magnetic materials reaches equilibrium after time "t" when an arbitrary field changes. Materials in superparamagnetic state have two characteristics:

1) No hysteresis loop;
2) Coercive force is equal to zero. Figure 3.10 shows the magnetization curve of strong magnetic phase $Co_{90}Cu_{10}$(2.7 nm) in Co–2%Cu alloy after dissolution and decomposition, which shows that the particle is in superparamagnetic state. The size of the material is the decisive factor whether the material is in superparamagnetic state, and superparamagnetism has a strong size effect. Meanwhile, superparamagnetism is also related to time and temperature.

For a single-axis single-domain particle aggregate, the easy magnetization direction of each particle is parallel, and the magnetic field magnetizes it in the easy magnetization direction. When the magnetic field is cancelled, the residual magnetism

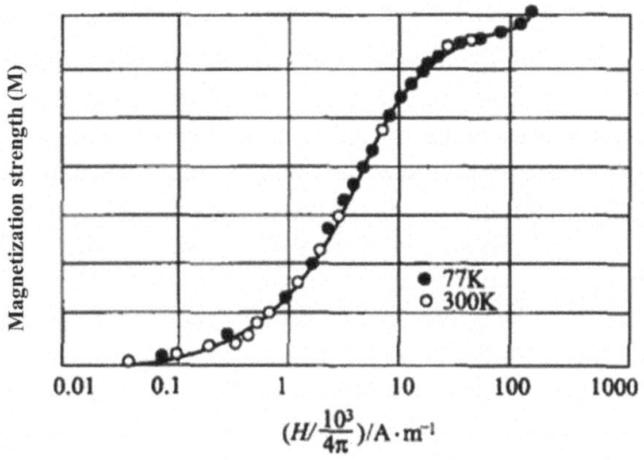

Fig. 3.10 Superparamagnetic of Co-rich particles in Co–Cu alloy [15]

"$M_r(0) = M_s$", and "M_s" is the saturation magnetization. The magnetization reversal is hindered by the barrier "$\Delta E = KV$" in the difficult-to-magnetize direction, and demagnetization can only be achieved if the applied magnetic field is sufficient to overcome the barrier. If the particle size is small enough, there can be a probability that thermal motion can cause "M_s" to cross the barrier "ΔE", the macroscopic quantum tunneling effect occurs, and the tunneling probability "$p \approx \exp[-KV/(k_BT)]$", where "K" is the anisotropic constant and "V" is the volume of the particle. If the remanence "M_r" tends to zero after a sufficiently long time, the attenuation process is:

$$M_r(t) = M_r(0)\exp(-t/\tau) \tag{3.6}$$

where "τ" is the relaxation time. It can be expressed as follows:

$$\tau = \tau_0 \exp\left(\frac{KV}{k_BT}\right) = f_0^{-1} \exp\left(\frac{KV}{k_BT}\right) \tag{3.7}$$

where "f_0" is the frequency oscillator, and its value is about $10^{-9}s^{-1}$.

Superparamagnetic limits are critical for magnetic storage materials. If 1 bit of information is to be stored in a spherical particle for 10 years, the volume of the particle $V > 4\ k_BT/K$ is required. For a typical thin film recording medium, the effective anisotropy constant $K_{eff} = 0.2$ J cm^{-3}. At room temperature, the volume of particles should be greater than 828 nm^3; For cubic grains, their side lengths should be greater than 9 nm. In addition, superparamagnetism is a condition for the preparation of magnetic liquids.

3.3.3 *Saturated Magnetization, Curie Temperature and Magnetic Susceptibility*

The saturation magnetization of microcrystals is insensitive to the size of grains or particles. However, when the size is reduced to 20 nm or less, the saturation magnetization will be strongly reduced because the atoms located on the surface or interface occupy a considerable proportion, and the atomic structure and symmetry of the surface atoms are different from those inside. For example, the content of 6 nm Fe is nearly 40% lower than that of coarse-grained bulk Fe. Figure 3.11 shows the magnetization curves of ferrite soft nickel soft magnetic materials with different grain sizes. In the figure, the vertical axis represents the specific saturation magnetization, and a, b, c and d represent the samples of 8, 13, 23 and 54 nm respectively. It can be seen from the figure that the specific saturation magnetization of the sample decreases sharply with the decrease of grain size. In the figure, the specific surface areas of samples a, b, c and d are 153.5, 103.2, 55.8 and 23.7 $m^2\ g^{-1}$, respectively. Therefore,

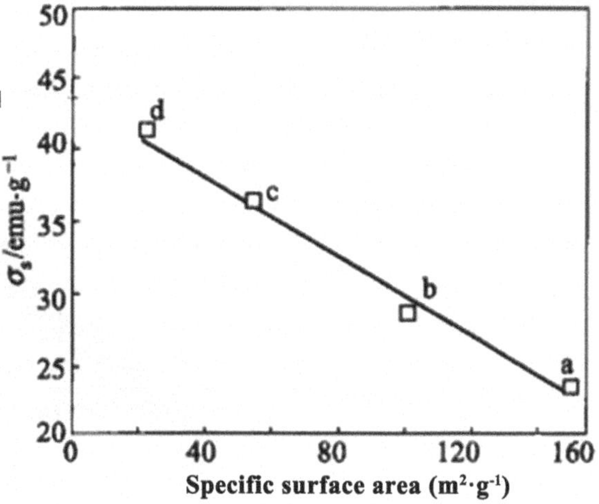

Fig. 3.11 Magnetization curves of nickel ferrite with different grains ($emu \cdot g^{-1} = 4\pi \times 10^{-7} Wb \cdot m \cdot 1$ [16]

the smaller the grain, the larger the specific surface area and the more it decreases. Therefore, the huge surface is very unfavorable to magnetization.

Nanomaterials usually have a lower Curie temperature, for example, the Curie temperature of 70 nm Ni is 40 °C lower than that of crystalline Ni. It has also been reported that $MnFeO_4$ and T_c of particles increase when the diameter is 2–25 nm. The huge surface or interface in nano-materials is the main reason for the decrease of "T_c". The decrease of "T_c" is unfavorable to the application of nano-magnetic materials.

The magnetic susceptibility "X" of nanoparticles is related to the temperature and the singularity of the number of electrons in the particles. Generally speaking, the total number of conduction electrons "N" of bivalent simple metal particles is even; Simple monovalent metal particles may be half odd and half even. Statistical theory shows that when "N" is odd, "X" obeys Curie–Weiss law, and "X" is inversely proportional to "T"; When "N" is an even number, the magnetic susceptibility of particles increases with the increase of temperature. Figure 3.12 shows the relationship between "X" and particle size of $MgFe_2O_4$ particles at different temperatures, which directly shows the influence of particle size on "X". The curves in the figure represent the measured values of particle sizes of 6, 7, 8, 11, 13 and 18 nm from bottom to top respectively. As can be seen from the figure, particles with each particle size have a temperature corresponding to the maximum value "X", which is called "freezing or ending" temperature T_B [17]. Above T_B, the value of "X" begins to decline. T_B corresponds to the threshold value of thermal activation energy. When the temperature is higher than T_B, the crystal anisotropy of nanoparticles can be overcome by thermal activation energy, showing superparamagnetic characteristics. The smaller the particle size, the lower the freezing temperature.

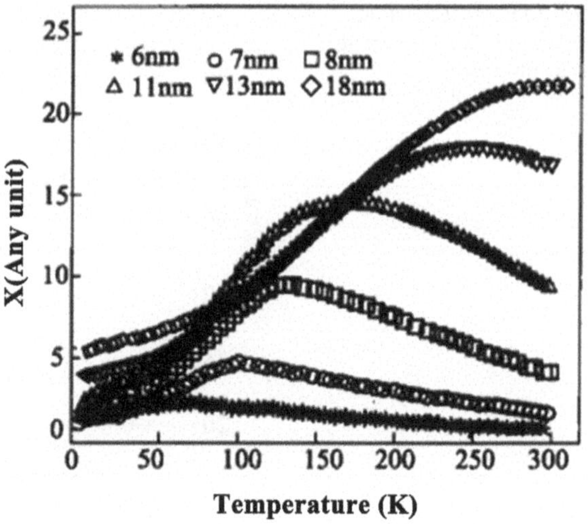

Fig. 3.12 $MgFe_2O_4$ particle X versus temperature and particle size [16]

3.4 Optical Properties

The particle size of nanoparticles (10–100 nm) is smaller than the wavelength of light wave, so it will have complex interaction with human light. Under proper evaporation and deposition conditions, black metal ultrafine particles which can easily absorb light can be obtained, which is called metal black, which is in sharp contrast with the high reflectivity glossy surface formed by metal during vacuum coating. Due to the quantum size effect, the absorption luster of nano-semiconductor particles is generally blue-shifted, and nano-materials can be used as infrared sensor materials because of their high light absorption rate.

3.4.1 Basic Concepts

The quantum effect, large specific surface effect, irregular atomic arrangement and bond configuration of nano-materials have great influence on the optical properties of nano-particles, which makes nano-materials very different from homogeneous bulk materials. The theoretical basis of studying the optical properties of nano-materials is quantum mechanics. This chapter will not elaborate on this specific theory, but in the process of understanding the optical characteristics of nano-materials, we often encounter the following concepts, which are introduced here first.

3.4.1.1 Excitation

Excitation can be simply understood as a bound electron–hole pair. Electrons excited from valence band to conduction band are usually free, and holes moving freely in valence band and electrons moving freely in conduction band are bound together through Coulomb interaction to form bound electron–hole pairs to form excitons. When electrons and holes recombine, they emit light, that is, they release energy in the form of photons, as shown in Fig. 3.13. According to the strength of the interaction between electrons and holes, excitons can be divided into Wannier excitons (loosely bound) and Frenke excitons (tightly bound). Vannell excitons are usually encountered in nano-materials such as semiconductors and metals. The relationship between this exciton energy "E_n" and the wave vector "$\vec{K}$" can be written as:

$$E_n(K) = E_g + \frac{\hbar^2 K^2}{2m} - \frac{R^*}{n^2}(n = 1,\ 2,\ 3,\ \ldots) \tag{3.8}$$

where "E_g" is the energy gap of the corresponding material; "$m = m_e^* + m_h^*$" is the sum of the effective masses of the electron space; "R*" is the equivalent Rydberg energy of the exciton. "$R^* = 13.6\frac{\mu}{\varepsilon^2}eV$", "$\varepsilon$" is the dielectric constant (sometimes directly called the permittivity), "μ" is the equivalent mass of the electron and the hole, "$\frac{1}{\mu} = \frac{1}{m_e^*} + \frac{1}{m_h^*}$" If $K = 0$ in Eq. (3.8), then the exciton energy is:

$$E_n(K) = E_g - \frac{R^*}{n^2}(n = 1,\ 2,\ 3,\ \ldots) \tag{3.9}$$

$E_n(K)$ the specific energy gap is small, so when the direct transition between bands is allowed, the energy of the required photon absorbed by exciton light is smaller than that of intrinsic absorption, that is, there is an absorption process corresponding to exciton light absorption in the long wave direction of intrinsic absorption edge.

Because the intrinsic equation of exciton is similar to that of hydrogen-like atom, the radius of exciton is quantized, and the smallest exciton radius is called exciton Bohr radius, which is expressed as [19]:

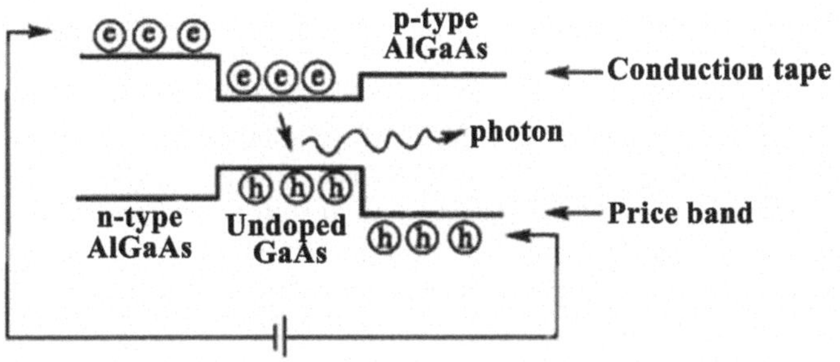

Fig. 3.13 Schematic diagram of semiconductor excitons and luminescence [18]

$$a_B = 0.053 \frac{m_0 \varepsilon}{\mu} \tag{3.10}$$

where "m_0" is the static mass of the electron. In semiconductor light-emitting materials, when the size of the material system is similar to the exciton Bohr radius "a_B", a quantum confinement effect occurs, that is, a series of discrete values appear in the energy levels in the system, and the movement of electrons in the system where the energy levels are quantized is constrained and limited.

In the periodic table, the exciton Bohr radius "a_B" of group I–VII element ions is small, such as "$a_B \approx 0.7$ nm" of CuCl; For group II–VI semiconductors, such as "$a_B \approx 3.0$ nm. " of CdS, "$a_B \approx 3.5$ nm" of CdSe. They have a strong quantum confinement effect at small sizes (less than 2 nm). However, because I–VII and II–VI semiconductors are difficult to make small-sized microcrystals, they are not suitable as strong quantum-confined materials, and III–V group materials are ideal strong quantum-confined materials, they have small electron–hole effective mass and large dielectric constant. From Eq. (3.10), it can be seen that groups III–V have large exciton Bohr radii, such as "$a_B \approx 31.6$ nm" of InAs; At room temperature, InSb has the narrowest band gap, the smallest electron–hole effective mass, the largest permittivity, and the exciton Bohr radius of InSb is also the largest, "$a_B \approx 67.8$ nm", so InSb is widely used to study strong quantum confinement.

3.4.1.2 Spectral Line and Its Shift

(1) Blue shift and broadening

When the size of semiconductor particles is close to its exciton Bohr radius, with the decrease of particle size, the effective band gap of semiconductor particles increases, and its corresponding absorption spectrum and fluorescence spectrum shift blue, thus forming a series of discrete energy levels in the energy band. The absorption band E(r) of nano-semiconductor particles can be described by the following formula:

$$E(r) = E_g + \frac{\hbar^2 \pi^2}{2\mu r^2} - \frac{1.8e^2}{4\pi \varepsilon_0 \varepsilon_r} + 0.248R^* \tag{3.11}$$

where "r" is that radius of the nanoparticles; "E(r)" is a function of r and is the energy gap of the corresponding semiconductor material. The second term on the right side of the equation is the quantum confinement energy, which is the blue shift; The third term is the Coulomb interaction energy of electron–hole pairs, which is the red shift that causes the wavelength of the spectral line to shift to the long wave direction; The fourth term is the spatial correlation energy due to the proximity of electrons and holes, which "R^*" is the equivalent Rydberg energy of excitons.

Compared with bulk materials, the absorption band of nanoparticles generally moves to the short wave direction, that is, the blue shift phenomenon. As the size of nanoparticles decreases, the energy gap widens, which leads to the light absorption band moving to the short wave direction. The occupied energy level and unoccupied width (energy gap) of electrons increase with the decrease of particles, so the quantum confinement effect is the fundamental reason for the blue shift and infrared absorption bandwidth of nano-materials Due to the small size of nanoparticles, large surface tension distorts the lattice and reduces the lattice constant. For example, the study of nano-oxide and nitride particles shows that the distance between the first neighbor and the second neighbor becomes shorter. The shortening of bond length leads to the increase of bond intrinsic vibration frequency of nanoparticles, which leads to the shift of infrared absorption band to high wave number and the blue shift of spectral lines of nanomaterials due to interface effect.

(2) Red shift

In some cases, when the particle size is reduced to nanometer level, it can be observed that the light absorption band is "red-shifted" relative to coarse-grained materials, that is, the absorption band moves to the long-wave direction. In terms of the energy level transition of the spectral line, the red shift of the spectral line is that the energy gap is reduced, and the band gap and energy level spacing are narrowed, which leads to the red shift of the optical absorption band and absorption edge caused by the transition of electrons from low energy level to high energy level and semiconductor electrons from valence band to conduction band.

The peak position of each light absorption band of nano-materials is determined by the combined action of blue shift and red shift factors. When the blue shift factor is greater than the red shift factor, the light absorption band will be blue shifted. On the contrary, red shift. For example, in the wavelength range of 200–1400 nm, single crystal NiO shows eight light absorption bands, and their peak blue shifts are 3.52, 3.25, 2.95, 2.75, 2.15, 1.95, 1.75 and 1.1 3eV, respectively, while nano-NiO (particle size in the range of 54–84 nm) does not show 3.53 eV. The peak positions of the other seven bands are 3.30, 2.93, 2.78, 2.25, 1.92, 1.72 and 1.07 eV, respectively. The first four optical absorption bands are blue-shifted and the last three optical absorption bands are red-shifted.

3.4.2 Optical Absorption Characteristics of Nanomaterials

When light passes through a substance, the spectrum produced by the absorption of certain wavelengths of light by the substance is called absorption spectrum. When an atom emits light, each dark line in the absorption spectrum of various atoms corresponds to a bright line in the emission spectrum of the atom. Whether it is nano-semiconductor material or metal material, as long as the photon energy given by the outside world meets the conditions of its energy level transition, photon transition can be promoted, so that absorption lines can be detected.

The optical properties of solid materials are closely related to their internal microstructure, especially to electronic state, defect state and energy level structure. Most of the traditional optical theories are based on the crystalline state with translation period. Since 1970s, the study of amorphous optical properties has established a theory to describe optical phenomena of disordered systems. Nanostructured materials are very different from conventional crystalline and amorphous materials in structure. Small quantum size particles, large specific surface area, and large irregularity in atomic arrangement and bond configuration at the interface make the optical properties of nanostructured materials appear some new phenomena different from conventional crystalline and amorphous materials.

3.4.2.1 Brief Introduction of Light Absorption

Generally speaking, in the process of studying the optical properties of solid materials, it is often necessary to introduce the complex form of relative dielectric constant and refractive index:

$$\varepsilon_r = \varepsilon_1 + i\varepsilon_2 \tag{3.12}$$

$$N = n + i\kappa \tag{3.13}$$

where "ε_1" and "ε_2" are the real and imaginary parts of the relative permittivity, respectively, and the imaginary part "κ" of the complex refractive index "N" is called the extinction coefficient; The real part n is commonly referred to as the refractive index. Since the complex refractive index is related to the permittivity "$N = \sqrt{\varepsilon_r}$"), there are:

$$n^2 - \kappa^2 = \varepsilon_1, \quad 2n\alpha = \varepsilon_2 \tag{3.14}$$

People usually use the optical constants "n" and "k" to characterize the optical properties of solids. Experiments show that when light propagates in a solid, its intensity generally attenuates, and the absorption of light is related to its intensity. If the incident light with intensity 10 passes through the solid and is displaced by "x", its intensity will be attenuated to [20]:

$$I = I_0 \exp(-\alpha x) \tag{3.15}$$

In the formula, "α" is called absorption coefficient, which represents the exponential attenuation law of light propagation in solids. Extinction coefficient "κ" also indicates the absorption of substances, and its relationship with absorption coefficient "α" is as follows:

$$\alpha = 2\omega\kappa/c = 4\pi\kappa/\lambda_0 \tag{3.16}$$

where "λ_0" is the wavelength of light in vacuum; "ω" is the frequency of incident light; "c" is the speed of light in vacuum. The reciprocal of the absorption coefficient is called the penetration depth of light in a solid, which is expressed by "d", then:

$$d = \frac{1}{\alpha} = \frac{\lambda_0}{4\pi\kappa} \tag{3.17}$$

The medium with large extinction coefficient has a shallow light penetration depth, which indicates that the absorption of matter is strong, while the penetration depth of long-wave light is greater than that of short-wave light.

When the solid material is irradiated with light with appropriate wavelength, electrons in the solid material can be excited from the valence band to the conduction band, leaving holes in the valence band. This photo-excited electron–hole pair can emit photons in different ways and produce corresponding emission peaks in the spectrum. The spectral details obtained from experiments reflect the information of solid materials.

If we only need to consider the dipole moment interaction between electrons, holes and electromagnetic waves, under the first-order perturbation theory, the recombination probability of electron–hole pairs is proportional to the intensity "$A(\omega)$" of absorption peak (or emission peak). According to the golden rule in quantum mechanics, there are:

$$A(\omega) = \frac{2\pi}{\hbar} \sum_{f} \left| \left\langle f \middle| \vec{e} \cdot \vec{P} \middle| i \right\rangle \right|^2 \delta^2(E_f - E_i - \hbar\omega) \tag{3.18}$$

where "$|i\rangle$" and "$\langle f|$" represent the initial and final states of the system, where all possible end states are summed, for light absorption, "$|i\rangle$" states are empty quantum states, "$\langle f|$" is the single-particle excited state, "$\vec{e} \cdot \vec{P}$"is the dipole moment operator, "$\vec{e}$" is the electromagnetic field vector potential, "$\vec{P}$" is the momentum operator; "$\hbar\omega$" is the photon energy, and the "$-$" sign in the "δ" function corresponds to the absorption photon transition; "$\left\langle f \middle| \vec{e} \cdot \vec{P} \middle| i \right\rangle$" is a dipole moment matrix element. The intensity of light absorption is proportional to the probability of finding an electron and a hole in the same location at the end state [21].

3.4.2.2 Luminescence Mechanism of Nanocrystals

The luminescence of materials in the range from ultraviolet to visible light has always been a hot topic of interest. The so-called photoluminescence refers to the microscopic process in which electrons excited to the high-energy excited state under the irradiation of a certain wavelength of light jump back into the low-energy level and are trapped by holes to emit light [22]. The light emitted only during the excitation process is called fluorescence, and the light that continues to emit for a certain time after the excitation stops is called phosphorescence.

From the analysis of physical mechanisms, electronic transitions can be divided into two categories: non-radiative transitions and radiative transitions. When the energy level spacing is small, the electron transition can emit phonons through the non-radiative decay process (as shown in the dotted arrow in Fig. 3.14), in this case it does not emit light, and only when the energy level spacing is large, it is possible to emit photons, achieve a radiative transition, and produce luminescence. As shown in Fig. 3.15, E0 is the ground state, E_1–E_6 is the excited state, and the electron transition from E_2 to E_1 or E_0 energy levels can emit light. The luminescence phenomena we are going to discuss are all related to the microscopic process of electron radiation transition. Figure 3.15 shows a schematic diagram of absorption emission between quantum wells. From the transmission absorption of different nanoparticle sizes in Fig. 3.16, it can be found that the emission spectrum of nanostructured materials is very different from the conventional state, and new light-emitting bands have been observed in the conventional state.

Small quantum size particles, large specific surface and a large number of defects in the interface may produce many additional energy gaps; Because the translation periodicity is destroyed in nanostructured materials, the selection rule of electronic transition of conventional materials in momentum K space is probably not applicable to nanostructured materials, and the luminescence band generated by nanostate under light excitation is the luminescence that cannot appear in conventional

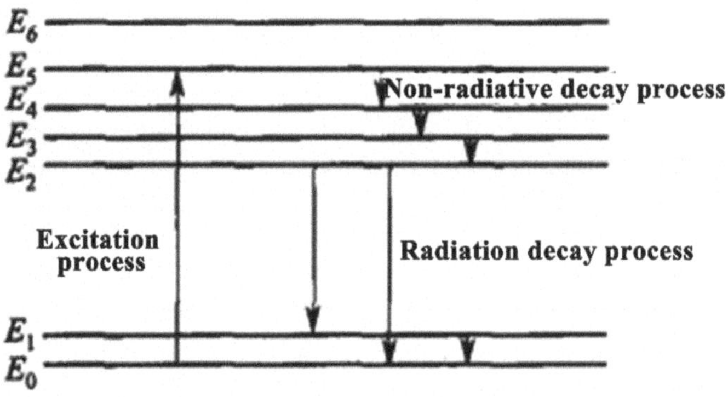

Fig. 3.14 Schematic diagram of the excitation and luminescence process [23]

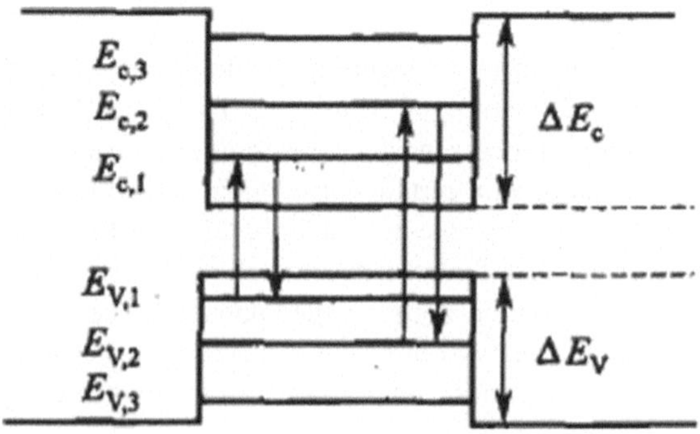

Fig. 3.15 Schematic diagram of absorption emission transitions between quantum wells [23]

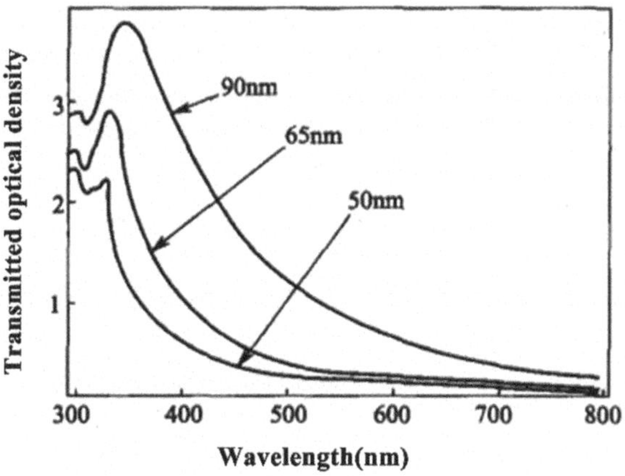

Fig. 3.16 Transmitted optical density (absorption) of different nanoparticle sizes [23]

materials due to the limitation of selection rule. The quantum confinement effect makes the exciton luminescence of nano-materials easy to appear, and the intensity of exciton luminescence band increases with the decrease of particles. The defect energy level makes the contribution of the huge specific surface and dangling bonds and unsaturated bonds of nanostructured materials to luminescence, which is also a new luminescence phenomenon that is rarely observed by conventional materials. Some transition group elements (Fe^{3+}, Cr^{3+}, V^{3+}, Mn^{3+}, Mo^{3+}, Ni^{3+}, Er^{3+}, etc.) will cause some luminescence in disordered systems [24]. The huge interface with low specific surface order in nanocrystalline materials is likely to provide a favorable position for the segregation of transition group impurities, which leads to the formation of impurity energy levels and the generation of impurity luminescence in nanocrystalline materials. Generally speaking, the impurity luminescent band is located at a lower energy position, and the luminescent band is wider.

3.5 Thermal Performance

Nano-materials refer to polycrystalline materials with grain size in the order of nanometer, with a high proportion of internal interfaces (including grain boundaries, phase boundaries, domain boundaries, etc.). Because the vibrational melting, entropy, configuration melting and entropy value of interface atoms are obviously different from those of lattice atoms, nano-materials show a series of thermal characteristics obviously different from ordinary polycrystalline materials, such as the increase of specific heat capacity, the increase of thermal expansion coefficient and the decrease of melting point. Thermal performance of materials is one of the most important physical properties of materials. At present, people's research on the thermal properties of nano-materials mainly focuses on the melting temperature of nano-materials, thermodynamics and mechanics of nanocrystalline-liquid and nanocrystalline-glass transition, growth kinetics of nano-phase or nanocrystalline, heat capacity and thermal expansion of nano-materials and interfacial melting of nano-materials. The thermal properties of nanomaterials are directly related to their grain size.

3.5.1 *Thermal Properties and Size Effect of Nanomaterials*

3.5.1.1 Melting Point and Internal Energy of Nanomaterials

The thermal properties of materials are closely related to the motion behavior of molecules and atoms in materials. When the characteristic dimensions of thermal carriers (electrons, phonons and photons) are equivalent to those of materials (grain size, particle size or film thickness), physical parameters reflecting the thermal properties of materials, such as melting temperature and heat capacity, will show distinct size dependence. In particular, the average free path of thermal carriers will become longer at low temperature, which makes the size effect of thermal properties of materials more obvious.

Figure 3.17 shows the size effect of melting point of several metal nanoparticles. With the decrease of particle size, the melting point decreases. When the size of metal particles is less than 10 nm, the melting point drops sharply, and the point of gold particles around 3 nm is only half of the melting point of its bulk material. By observing the structure of gold nanoparticles with a size of 2 nm with high-power electron microscope, it can be found that the morphology of gold nanoparticles can be continuously transformed between single crystal, polycrystalline and twin. This behavior is completely different from that of traditional materials melting at a fixed melting point. With the decrease of melting point of nano-materials, the latent heat absorption (enthalpy change) of particles per unit mass decreases with the decrease of size. The size effect of melting has also been found in nanocrystals, inert gases and molecular crystals of covalent semiconductors with free surfaces.

According to the basic principle of solid state physics, the fundamental reason for the size effect of thermal properties of materials can be explained. In general, the internal energy u of crystal materials can be estimated under Debye hypothesis according to the wave characteristics of lattice vibration, namely:

$$U = 3\sum_{k} \frac{\hbar\Theta k}{\exp\left(\frac{\hbar\Theta k}{k_B T}\right) - 1} \tag{3.19}$$

where "Θ" is the Debye temperature; "k" is the wave vector; "T" is the thermodynamic temperature; "$\hbar$" is Planck's constant; "k_B" is the Boltzmann constant. The summation is done for all possible k-values. The allowable value of "k" is expressed by its components as [26]:

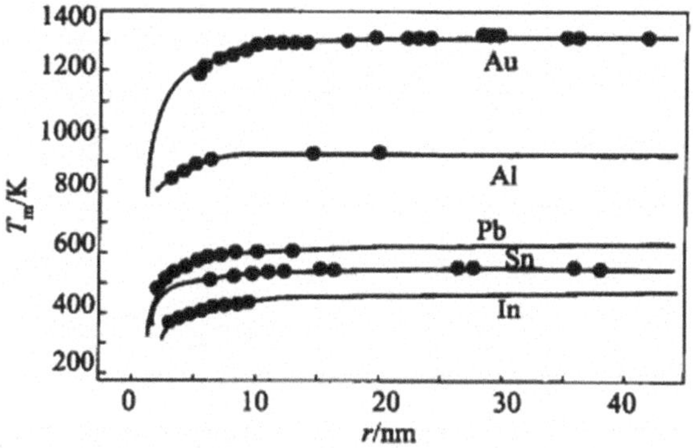

Fig. 3.17 Melting point reduction phenomenon of several metal nanoparticles [25]

$$k_z = 0, \ \pm\frac{2\pi}{L_x}, \ \pm\frac{4\pi}{L_x}, \ldots, \ \pm\frac{N\pi}{L_x}, \ \Delta k_x = \frac{2\pi}{L_x} = \frac{2\pi}{N_x a} \tag{3.20}$$

where "L" is the lattice length; "N" is the degree of state, which "Δk_x" is the difference of continuous wave vectors in a specific direction. Similar relationships exist for components in other directions. In bulk materials, Eq. (3.19) is usually simplified as:

$$u_{bulk} = 9nk_B T\left(\frac{T}{\Theta}\right)^3 \int_0^{x_D} \frac{x^3}{\exp(x) - 1} dx \tag{3.21}$$

where "u_{bulk}" is the internal energy per unit volume of bulk material; "n" is the atomic number density, which "x_D" is the integral limit corresponding to Debye temperature. The above statement only gives the contribution of phonon modes from bulk materials, while the contribution of surface phonons is ignored. In bulk materials, the contribution of surface phonons can indeed be ignored; However, when at least one dimension of the material is greatly reduced to nanometer level, this simplification is not correct, that is, it is necessary to consider the size effect for nano-materials.

As the scale of the material decreases, the method of calculating the internal energy and heat capacity with Eq. (3.21) is no longer effective, and the original summation expression (3.19) should be used directly. If the number of atoms in at least one direction of the material decreases significantly, the change of k in this direction is no longer negligible compared with all allowable k values, so k in this direction will increase by a considerable distance length in the range of $2\pi/L$, so that the method of replacing the distance length with the integral approximation of Eq. (3.21) cannot be applied, resulting in two effects: (1) the exact number of points in space differs from the value of the solid material; (2) The k space volume Ω must be calculated by discrete summation. Thus, the exact number of points in a certain region of k-space must be determined by discrete summation. From this, the internal energy of the tiny volume lattice can be derived:

$$u_{micro} = 3\frac{\varepsilon\hbar c}{(2\pi)^3} \times \sum_{kz}\sum_{ky}\sum_{kz} \frac{\sqrt{k_x^2 + k_y^2 + k_z^2}}{\exp\left(\frac{\hbar c\sqrt{k_x^2+k_y^2+k_z^2}}{k_B T}\right) - 1} \Delta k_x \Delta k_y \Delta k_z \tag{3.22}$$

Among them:

$$\varepsilon = \frac{3}{N_x^3} \sum_{i=0}^{N_x/2} \left(N_x^2 - 4i^2\right) \tag{3.23}$$

Here "u_{micro}" represents the internal energy obtained by addition, that is the internal energy of the tiny lattice volume. It can be seen that the dependence of the basic thermal properties of the material on the crystal size will inevitably be caused by the size effect in the lattice.

3.5.1.2 Heat Capacity and Characteristic Temperature of Nanocrystals

Heat capacity refers to the rate of change of energy "Q" of thermal motion of molecules or atoms of materials with temperature "T", and the expression of heat capacity "C" of materials at temperature "T" is:

$$C = \left(\frac{\partial Q}{\partial T}\right)_T \tag{3.24}$$

If the volume of the material is constant during heating, the measured heat capacity is constant volume heat capacity (C_V); If the pressure of the material is constant during the heating process, the heat capacity at constant pressure (C_p) is measured. Namely:

$$C_V = \left(\frac{\partial Q}{\partial T}\right)_V = \left(\frac{\partial U}{\partial T}\right)_V \tag{3.25}$$

$$C_p = \left(\frac{\partial Q}{\partial T}\right)_p = \left(\frac{\partial U}{\partial T}\right)_p \tag{3.26}$$

Substituting formula (3.22) into formula (3.25) and formula (3.26), the heat capacity of nanocrystals can be calculated.

Figure 3.18 shows the calculated relationship between the ratio of constant volume heat capacity C_{nano} of several nano-film materials to the corresponding block heat capacity C_{bulk} and the number of atomic layers n. It can be seen that the thermal capacity of nano-film is less than that of bulk, but it is equivalent for thicker film. It is worth pointing out that in the above calculation, it is assumed that the crystal lattice structure is still intact when the size of nanocrystals is extremely small, and the surface phonon softening effect is ignored, so the calculated heat capacity value will be smaller than the actual value.

Table 3.3 lists the changes of constant pressure heat capacity of several nanocrystalline materials prepared by amorphous crystallization, high-energy ball milling and inert gas condensation methods relative to that of coarse-grained materials:

$$\Delta C_p^{nc}\left[\Delta C_p^{nc} = \left(C_p^{nc} - C_p^{c}\right)/C_p^{c}\right] \tag{3.27}$$

From the measurement results, it can be seen that the excess heat capacity "ΔC_p^{nc}" of nanocrystalline materials prepared by inert gas condensation method and high-energy ball milling method is very large, such as the "ΔC_p^{nc}" of nanocrystalline Pd prepared by inert gas condensation method is as high as 48%; The "ΔC_p^{nc}"

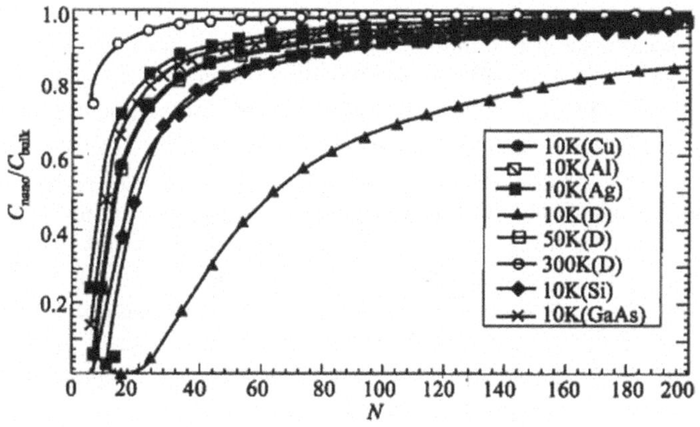

Fig. 3.18 The ratio of C_{nano} and C_{bulk} (The symbol D in the figure represents diamond) [27]

Table 3.3 Excess heat capacity of nanocrystalline materials prepared by different methods

Nanocrystals	Preparation method	Average grainsize (nm)	ΔC_p^{nc} (%)
Cr	High-energy ball mill	9	10
Cu	Inert gas condensation	8	8.3
Hf	High-energy ball mill	13	9
Ru	High-energy ball mill	11	20
Pd	Inert gas condensation	8	48
Zr	High-energy ball mill	13	20
AlRu	High-energy ball mill	11	15
Ni	Electrolytic deposition	20	2.5–5
Se	Amorphous crystallization	10	1.7
Ni80P20	Amorphous crystallization	6	0.9

of nanocrystalline materials prepared by amorphous crystallization and electrolytic deposition is small, usually less than 5%.

The reason for this difference lies in the different defect densities introduced by different preparation methods. For nano-materials prepared by inert gas condensation and high-energy ball milling, there are a lot of micropores, impurities and structural defects in the materials, which make the materials have great differences of the "ΔC_p^{nc}". This great difference cannot represent the intrinsic heat capacity difference of nano-materials [28]. For nanocrystals prepared by amorphous crystallization and electrolytic deposition, the material is formed under the condition of near equilibrium, so its internal structural defects are few, and there are few micropores and impurities, and its heat capacity is not increased much compared with coarse crystals. Especially, the amorphous method is equivalent to annealing the material once. The interface and grains in the nanocrystal are in a relaxed state, and the micro-strain inside the nanocrystal is extremely small (one order of magnitude smaller than the stress inside the nanocrystal obtained by other methods), so that the excess heat capacity of the amorphous nanomaterial is minimal, and it can also be concluded that the excess heat capacity of grain boundary components is very small.

The heat capacity of a material is closely related to its structure, or vibration entropy and configuration entropy, which are strongly influenced by the configuration of the nearest neighbor atom. In nano-materials, a large number of atoms are located on the grain boundary, and the nearest neighbor configuration of interface atoms is significantly different from that of grain atoms, or the grain boundary has a certain excess volume relative to the complete lattice. Thermodynamic calculation shows that the heat capacity of nanocrystals increases with the increase of excess volume of grain boundaries, and therefore increases with the increase of grain boundary energy. The lower the excess heat capacity of grain boundary components, the lower the corresponding excess grain boundary volume and interfacial energy will be. Due to the contribution of high proportion of grain boundary components, the specific heat capacity of nano-materials will be higher than that of their corresponding coarse-grained materials.

According to the theory of solid physics, Debye characteristic temperature is defined as:

$$\Theta = \frac{\hbar\omega_m}{k_B} \tag{3.28}$$

where "ω_m" characterizes the highest frequency of lattice vibrations "k_B" as the Boltzmann constant. Therefore, the Debye characteristic temperature is related to the lattice vibration of the material, and also reflects the strength of the interatomic bonding force.

Table 3.4 shows the rate of change of the characteristic temperature "Θ" of nanocrystalline materials prepared by different methods relative to the coarse crystal value. In the table "$\Delta\Theta_{nc} = (\Theta_{nc} - \Theta_c)/\Theta_c$", where "$\Theta_{nc}, \Theta_c$" is the characteristic temperatures of nanocrystals and coarse crystals. It can be seen that the characteristic temperature of the nanocrystals prepared by each method is less than the value of the

Table 3.4 Characteristic temperature and coefficient of change of thermal expansion of nanocrystalline materials

Sample	Average grain size (nm)	Preparation method	$\Delta\Theta_{nc}$ (%)	Δa_1^{nc} (%)
Al	–	Magnetron sputtering	–	0
Cu	8	Inert gas condensation	–	94
Cu	21	Magnetron sputtering	–	0
Pd	8.3	Inert gas condensation	−5	0
Ni–P	7.5	Amorphous crystallization	–	51
Se	13	Amorphous crystallization	−12	61
Ni	20	Electrolytic deposition	–	−2.6
Ni	152	Severe plastic deformation	−22	180
Fe	8	High-energy ball mill	−17	130
Cr	11	Inert gas condensation	−22	–
Sn	7	Inert gas condensation	−17	–
FeF_2	8	High-energy ball mill	−71	–
Au	10	Evaporative deposition of electron beams	−15	0
Ag	15	Evaporative deposition of electron beams	−25	–
Pb	14.4	Chemical precipitation	−13	–

coarse crystals, and the range of reduction is 5–71%. In addition, the characteristic temperatures of ultrafine powders Ni and Pd also showed a decreasing trend. It is generally believed that the decrease in the characteristic temperature of nanocrystalline materials is caused by the weakening of the harmonic effect of atomic vibration due to their structural defects (such as lattice static distortion, grain boundaries, etc.), but there is no quantitative explanation.

3.5.1.3 Thermal Expansion of Nanocrystals

Thermal expansion refers to the phenomenon that the length or volume of a material increases with the increase of temperature without pressure. The essence of thermal expansion of solid materials lies in the anharmonic vibration of material lattice. Because the force between adjacent particles in lattice vibration is nonlinear and the lattice energy curve is asymmetric, the material expands during heating. Generally speaking, the crystal with dense structure has larger thermal expansion coefficient than the material with loose structure. Table 3.4 also shows the changes of thermal expansion coefficient of nanocrystalline materials prepared by different methods relative to coarse crystals [29].

To date, the thermal expansion behavior of nanocrystalline materials has been poorly studied, and only a few cases have been reported (see Table 3.4). Birringer reported that the linear expansion coefficient (a_1^{nc}) of nanocrystalline Cu (8 nm) prepared by inert gas condensation method was 1.94 times that of coarse-grained Cu. Eastman used in situ X-ray diffraction to find that the nanocrystals Pd (8.3 nm) prepared by inert gas condensation showed no significant change compared with "a_1^{nc}" in the temperature range of 16–300 K compared with the coarse crystals. The expansion coefficients of nano-Ni–P and Se prepared by amorphous crystallization increased by 51% and 61%, respectively, compared with the coarse crystals. The "a_1^{nc}" of nanocrystalline Ni (20 nm) prepared by SPD method increased by 1.8 times compared with amorphous Ni, while the a_1^{nc} of non-porous nanocrystals Ni (20 nm) prepared by electrolytic deposition was lower than that of coarse-grained Ni (100 μm) between 205 and 500 K, and $a_1^{nc} = -2.6\%$ at 500 K; The coefficient of expansion of Cu films deposited by magnetron sputtering is also the same as that of Cu of coarse grains. In addition, it was found that the coefficient of thermal expansion of ultra-powdered nano-Au and Pt evaporated by gas was the same as that of coarse crystals. Obviously, "a_1^{nc}"is closely related to the preparation method and structure of nanosamples, especially micropores.

3.5.2 *Grain Growth of Nanocrystals*

The structural instability of nanocrystalline materials includes grain growth, phase separation and second phase precipitation. Because of these changes, the microstructure changes, especially the shape and quantity of grain boundaries will inevitably affect the properties of nanocrystalline materials, which may make nanocrystalline materials lose their excellent mechanical or physical and chemical properties. Therefore, it is of great practical significance to study the thermal stability of nanocrystalline materials.

3.5.2.1 Long Dynamics Characterization of Nanocrystals

Nanocrystalline materials are in a thermodynamic metastable state, but they are often dynamically stable at room temperature and normal pressure, and their structural transformation process often needs to overcome certain activation energy, so it is necessary to study the thermal stability of nanocrystalline materials from the perspective of kinetics.

Kinetics research is usually divided into two aspects: one is to express the relationship between grain size and annealing temperature or time by kinetic formula, and the other is to obtain some characteristic parameters in the process of instability by monitoring the changes of physical properties of nanocrystalline materials, so as to study its dynamic process.

The process of grain growth in polycrystalline materials can usually be expressed as:

$$d^N - d_0^N = k_T t \quad (3.29)$$

where "d_0" is the initial grain size; "d" is the grain size after annealing in the "t" time period, and "N" is the grain growth index; "k_T" is a kinetic constant. This formula accurately reflects the grain growth law in metal materials at lower temperatures. According to the classical grain growth mechanism, different "N" values represent different grain growth mechanisms, and the N value is usually between 2 and 4. The kinetic constant "k_T" has the following relationship with temperature "T".

$$k_T = k_{T_\diamond} \exp\left(-\frac{Q}{RT}\right) \quad (3.30)$$

where "$k_{T_\diamond}$" is the factor; "R" is the gas constant; "Q" is the activation energy for grain growth. Activation energy during grain growth is another important parameter for grain dimensional stability, which represents the energy barrier that needs to be overcome during the diffusion process corresponding to grain growth. In the process of studying grain growth, the mechanism of grain growth is usually judged by calculating the grain growth index "N" and the grain growth activation energy "Q", and then comparing the experimental value with the theoretical prediction value.

Some dynamic parameters of the thermal stability of nanocrystalline materials can also be obtained by monitoring the changes of other physical parameters. For example, the activation energy of the grain growth process can be inferred by measuring the changes of the temperature rising rate by using differential thermal analysis or resistance analysis, which is the commonly used Kissinger equation:

$$\ln\left(\frac{B}{T^2}\right) = -\frac{Q}{RT} + C \quad (3.31)$$

where "B" is the heating rate; "C" is constant and "Q" is activation energy; "T" is the characteristic temperature of a process (e.g. the starting temperature "T_{on}" or the peak temperature "T_P")

Table 3.5 collects some data about activation energy of nano-grain growth. Generally speaking, the greater the activation energy during grain growth, the better the stability of grain size. The experimental results show that the activation energy of grain growth of alloys and compounds is often high, which is close to the activation energy of bulk diffusion of corresponding elements. However, the activation energy of elemental nanocrystalline growth is low, which is close to the activation energy of grain boundary diffusion, which shows that the process of nanocrystalline growth can not be simply described by classical grain growth theory, and there are some essential influencing factors of nanocrystalline structure, which are not fully understood.

The recent research results show that the thermal stability and internal grain growth mechanism of nanocrystalline materials are not only related to kinetics, but also closely related to the microstructure, chemical composition and grain morphology of grains. At present, many studies on the thermal stability of nanocrystalline materials are carried out by cold pressing samples with ultrafine powder, which generally

Table 3.5 Nanograins growth activation energy

Sample	Average grain size (nm)	Preparation method	$\Delta\Theta_{nc}$ (%)	Δa_{l}^{nc} (%)
Al	–	Magnetron sputtering	–	0
Cu	8	Inert gas condensation	–	94
Cu	21	Magnetron sputtering	–	0
Pd	8.3	Inert gas condensation	−5	0
Ni–P	7.5	Amorphous crystallization	–	51
Se	13	Amorphous crystallization	−12	61
Ni	20	Electrolytic deposition	–	−2.6
Ni	152	Severe plastic deformation	−22	180
Fe	8	High-energy ball mill	−17	130
Cr	11	Inert gas condensation	−22	–
Sn	7	Inert gas condensation	−17	–
FeF_2	8	High-energy ball mill	−71	–
Au	10	Evaporative deposition of electron beams	−15	0
Ag	15	Evaporative deposition of electron beams	−25	–
Pb	14.4	Chemical precipitation	−13	–

contain a lot of pores, pollution, micro-strain and defects. For example, when nano-pure Ag is heated at a speed of 10 K min^{-1}, the grain growth process starts at 423 K with obvious hardness decrease, and the activation energy of grain growth is 108 K min^{-1}, which is equivalent to the activation energy of grain boundary diffusion of Ag. If 7% oxygen is added to this nano-Ag sample, the initial temperature of grain growth is increased by about 80 K, and the activation energy of grain growth is increased to 209 kJ mol^{-1}, which is equivalent to the activation energy of bulk diffusion of Ag, and the stability of grain size is improved. In addition, the micro-pores in nanocrystalline materials will also increase their thermal stability by preventing grain boundary movement *x*.

3.5.2.2 Interface Energy of Grain Growth

The growth process of nano-grains is often accompanied by the release of some excess energy. It is assumed that:

1) The corresponding thermal effect in grain growth is the release of interface energy due to the decrease of interface;
2) The structure of grain boundaries remains unchanged before and after grain growth;
3) The energy state of the grain does not change with grain size. Since the grain boundary fraction of a nanocrystalline material is usually inversely proportional

to its grain size, for a sample with a volume “V”, the excess enthalpy stored at the interface can be obtained:

$$H_0 = g\gamma_H V/r_0 \tag{3.32}$$

where “γ_H” is the excess energy of the interface; “g” is a numerical factor that depends on grain shape and its size distribution, with “H_0” and “r_0” representing the initial excess enthalpy and grain radius, respectively. When the grain radius increases to “γ_H”, the total interfacial excess energy becomes:

$$H(t) = g\gamma_H V/r(t) = H_0 r_0/r(t) \tag{3.33}$$

During this period, the energy changes of the system are as follows [30]:

$$\Delta H = H(t) - H_0 = H_0 r_0 \left(\frac{1}{r} - \frac{1}{r_0}\right) = g\gamma_H \frac{W}{D}\left(\frac{1}{r} - \frac{1}{r_0}\right) \tag{3.34}$$

where “W” and “D” are the mass and density of the sample in question, respectively.

The thermal effect can usually be measured by differential scanning calorimetry (DSC). Chen and his collaborators have developed a complete set of theories to discuss the application of DSC measurement method in the study of grain growth. According to Eq. (3.32), the average intensity of DSC signal in time can be expressed “$\delta_H = \Delta H/\delta_t$”.

Taking pure Cu samples as an example, simply consider the grain shape factor as constant, D = 8.91 g cm^{-3}, and take W = 5 mg, $\delta_t = 50$ s, $\gamma_H = 01$ J m^{-2}, the thermal effect of grain growth with different initial grain sizes can be obtained, as shown in Fig. 3.19. The horizontal dashed line in the figure is the current energy accuracy limit of DSC. It can be seen that the intensity of the DSC signal is always changing as the grain grows, and for smaller grain sizes, “δ_t” increases faster with “d”, and for larger grain sizes, the curve becomes flattened. As far as the sensitivity of the current DSC equipment is concerned, the measurement accuracy can reach 0.01 mJ s^{-1}.

According to this precision, when the initial grain size is small ($d_0 < 10$ nm), the grain growth process can be easily measured. For samples with large initial grain size, such as $d_0 = 30$ nm, DSC can only detect its thermal effect when the grain size is about 37 nm long. Therefore, DSC measurement often has a certain lag effect.

Figure 3.20 shows the thermal effect of the grain growth process of nanocrystalline Cu samples with small angle grain boundaries, which increases the sample mass to 30 mg. Other parameters are the same as those in Fig. 3.18, and the horizontal dotted line is the energy accuracy limit of DSC at present. For nanocrystalline Cu used in this experiment, the temperature range of 125–175 °C is the fastest growing temperature range, so the thermal effect in this temperature range is the most concentrated. When the grain size grows from 37 nm (125 °C) to 70 nm (175 °C), its heat release per unit time is about 0.0143 mW. Therefore, the thermal effect of grain growth of nanocrystalline Cu is too weak to be accurately measured by current DSC testing

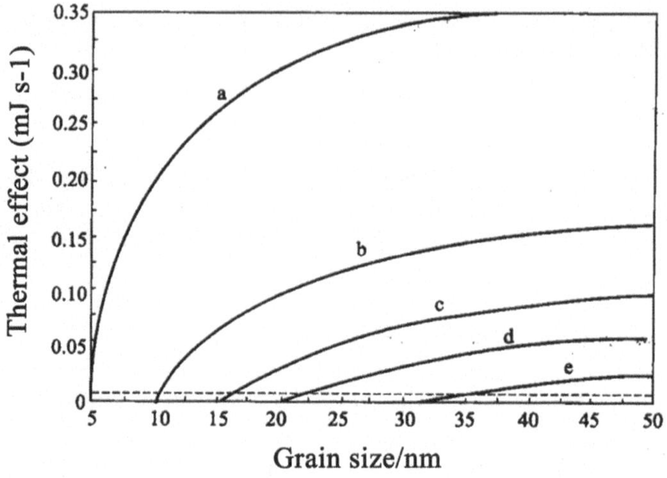

Fig. 3.19 Thermal effect a—5 nm when the large-angle grain boundary nanocrystalline Cu corresponds to different starting grain sizes; b—10 nm; c—15 nm; d—20 nm; e—30 nm [31]

equipment. In addition, the existence of a large number of growth twins and stacking faults in the sample will also affect the thermal effect of grain growth to some extent.

It can be seen that according to the accuracy of DSC measurement method and the technical conditions of equipment, not all the grain growth processes of nanocrystalline materials can be detected by DSC [32].

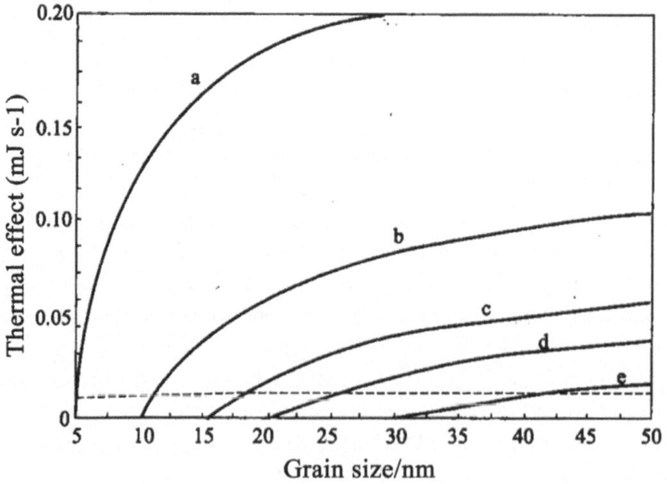

Fig. 3.20 Thermal effect a—5 nm when the small-angle grain boundary nanocrystalline Cu corresponds to different starting grain sizes; b—10 nm; c—15 nm: d—20 nm; e—30 nm [31]

Generally speaking, there is no single measurement method that can reflect all the structural and energy changes during the grain growth of nano-materials. It is difficult to determine the parameters of some changes by conventional analysis methods, and sometimes the corresponding structural changes can only be inferred by monitoring the changes of physical properties of samples. Therefore, it is very important to establish the corresponding relationship between physical properties and microstructure. When studying the thermal stability of nanocrystalline materials, we should not only consider the microscopic defects of samples, but also use a variety of measurement methods and compare them to reveal the essence of thermal stability of nanocrystalline materials.

3.5.3 Lattice Thermodynamic Properties of Nanocrystals

For a long time, people attributed the unique properties of nano-materials to the contribution of grain boundaries, while ignoring the study of grain parts. Until recently, grain structure has become the object of attention. The experimental results show that the microstructure of nano-grains is quite different from that of complete lattice. For example, in the Ni–P system and Fe–Mo–Si–B system nanocrystalline alloys prepared by amorphous crystallization method, the lattice distortion of Ni_3P and Fe_2B was found, respectively. The A-axis became longer, the C-axis became shorter, and the cell volume ($V = a^2c$) expanded, and the smaller the crystal grains, the more obvious the distortion was. Lattice distortion is also found in nanocrystalline elements such as Se, Si, Co, Pd and Ni, and the non-monotonicity of unit cell volume with grain size is consistent with the law of mechanical properties with grain size.

Whether the atomic structure of nano-crystalline materials is complete or not is an important issue that people have been concerned about. Gleiter once pointed out that the mismatch between the grains of nanocrystalline materials will produce a stress field from the grain boundary to the inside of the grain, which will change the atomic structure within the grain. However, in early research reports, people often assume that the crystal grain has an ideal crystal structure. Only recently have people discovered that the crystal grain of nanocrystalline materials has obvious structural defects, such as the change of lattice parameters, lattice distortion, lattice static distortion and so on.

The essential reason for the change of lattice parameters of nanocrystalline materials is not clear at present, but it confirms the stress effect of grain boundaries predicted by Gleiter earlier. Because the grain boundaries of nanocrystalline materials have large excess energy and excess volume, the grain boundaries will act on the grains to reduce their own energy.

Ke Lu et al. used classical thermodynamic theory to calculate the change of lattice parameters of nanocrystalline materials. Compared with infinite crystals, the increase of Gibbs free energy of nanocrystalline materials composed of spherical grains with size "D" is:

$$\Delta G(T, D) = \frac{4\Omega\gamma}{D} \quad (3.35)$$

where "Ω" is the atomic volume of the grain; "γ" is the interface energy and "T" is the thermodynamic temperature. The resulting increase in solid solubility is:

$$\Delta C^B(T, D) = \frac{4\Omega\gamma C_O^B}{k_B TD} \quad (3.36)$$

where "C_O^B" is the equilibrium solid solubility of the coarse grain; "k_B" is Boltzmann's constant. From Eq. (3.36), it can be seen that the nanograins have a supersaturated solid solution of a solute. For nanoparticles and intermetallic compounds, vacancies or other point defects can be regarded as their special "solute atoms", and the defect concentration increases with the decrease of grain, resulting in changes in lattice parameters.

The lattice parameters of nano-Si films deposited by PVD are studied. It is found that the lattice parameters change from 0 to 1.0% when the grain size is reduced from 100 to 3, and 3 nm is the minimum grain size of nano-Si films. Thermodynamic calculation shows that when the lattice parameter of Si expands to 1.0%, its free energy is equal to that of amorphous Si, which shows that the lattice parameter change of nanocrystalline materials has an upper limit, and the structural instability caused by the lattice parameter change is a factor of the lower limit of grain size.

3.5.4 Interface Thermodynamics of Nanocrystals

The thermal properties of nanomaterials are directly related to the interfacial thermodynamic properties in nanocrystals. However, because micropores are often introduced into the samples in the traditional preparation process of nanocrystals, it is difficult to test the interface structure and properties, so there are few reports on the interface thermodynamics of nanocrystals. In recent years, the research group led by Ke Lu has obtained samples of nano-crystals without micropores by using the developed technology of amorphous preparation of nano-crystals, and studied the interface thermodynamics of nano-crystals experimentally, and obtained a series of important interface thermodynamic parameters, which have had great influence in the world.

3.5.4.1 Thermodynamics of Amorphous Crystallization

Amorphous crystallization means that ultrafine polycrystalline materials with nano-sized grains are produced by crystallization of amorphous alloys. Because the crystal grains and the internal interface are naturally formed from the amorphous phase matrix during the phase transition, there is no external pressure to make the crystals combine to form the interface, so this method can easily control the grain size of nanocrystals, and the sample does not contain micropores and impurity pollution. At the same time, because amorphous crystallization is a phase transition process from amorphous to nanocrystalline, it provides excellent conditions for studying the formation process, interfacial thermodynamic characteristics, grain nucleation and growth, metastable phase transition kinetics and thermodynamics in nanocrystals.

Crystallization of amorphous alloy is a solid-state phase transition process. If the crystallization product is general polycrystalline (general crystallization process), there are few internal interfaces formed, and the corresponding thermodynamic parameters can be calculated from the basic thermodynamic relations.

$$\Delta H^{c\to a}(T) = -\Delta H_m + \int_T^{T_m} \left(C_p^a - C_p^c\right) dT \tag{3.37}$$

$$\Delta S^{c\to a}(T) = -\Delta S_m + \int_T^{T_m} \left(C_p^a - C_p^c\right) d(\ln T) \tag{3.38}$$

$$\Delta G^{c\to a}(T) = -\Delta H^{c\to a}(T) - T\Delta S^{c\to a}(T) \tag{3.39}$$

where "ΔH_m" and "ΔS_m" are the material at the melting point "T_m", respectively; "C_p^a" and "C_p^c" are the molar heat capacities of amorphous alloys and crystals, respectively.

For the phase transition from amorphous alloy to nanocrystal, a similar expression can be obtained:

$$\Delta H^{nc\to a}(T) = -\Delta H^{nc\to a}(T_x) + \int_T^{T_x} \left(C_p^a - C_p^{cn}\right) dT \tag{3.40}$$

$$\Delta S^{nc\to a}(T) = -\Delta S_c^{nc\to a} + \int_0^{T} \left(C_p^{nc} - C_p^a\right) d(\ln T) \tag{3.41}$$

$$\Delta G^{nc\to a}(T) = \Delta H^{nc\to a}(T) - T\Delta S^{nc\to a} \tag{3.42}$$

where "$\Delta H^{nc\to a}(T_x)$" is the enthalpy change of the amorphous state to nanocrystals at crystallization temperature "$\Delta\ T_x$", "$\Delta S^{nc\to a}$" is the difference between the nanocrystals and non-0 K temperatures; "C_p^{nc}" is the heat capacity of the nanocrystal.

It can be seen from the above that in order to obtain the three basic thermodynamic parameters of the phase transition process from amorphous to nanocrystals, it is necessary to measure "C_p^{nc}","C_p^a","$\Delta H^{nc\to a}(T_x)$" and "$\Delta S^{nc\to a}$", the first three parameters can be accurately determined by the DSC method. The entropy value of nanocrystals at absolute zero (0 K) can be approximated as the configuration entropy

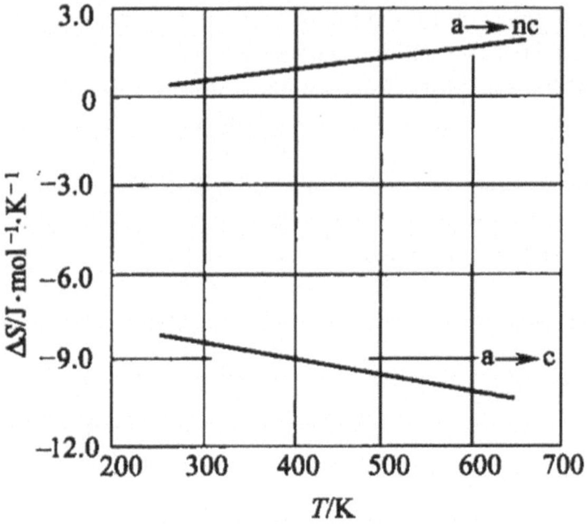

Fig. 3.21 The entropy change of the two transitions calculated from the real result. (a → nc is the transition from non-fret to nanocrystalline a → c is the transition from amorphous to crystalline) [33]

of the grain boundary part (ignoring the grain part and thermal vibration entropy), and the structure of the grain boundary part is disordered, similar to amorphous alloys, and its configuration entropy can be approximated by the following formula:

$$S = -k_B \sum_i x_i \ln x_i \tag{3.43}$$

where "x_i" is the concentration of class "i" components (atoms, atomic clusters, vacancies, or free volumes). The free volume (vacancy) of the interface between amorphous and nanocrystals can be measured by density. Based on the density measurements of nanocrystals, "$\Delta S_c^{nc \to a} = -0.04\ k_B$" was calculated.

Using Eqs. (3.35) to (3.38), thermodynamic parameters of Ni–P amorphous crystals and amorphous nanocrystals were calculated, as shown in Figs. 3.21 and 3.22.

It can be seen from the figure that the thermodynamic parameters of the two phase transitions are very different: in the transition from amorphous to crystal, "ΔS" is negative, and with the increase of temperature, the value of "ΔS" decreases, while the value of "ΔG" increases; For the phase transition from amorphous to nanocrystal, the entropy change "ΔS" is positive, indicating that the nanocrystals are larger than the amorphous state, and the values of "ΔS" and "ΔH" increase with the increase of temperature, while the phase transition free energy "ΔG" decreases. The obvious difference between these two phases transition thermodynamic parameters indicates that the interface of nanocrystals plays an important role in amorphous crystallization phase transition.

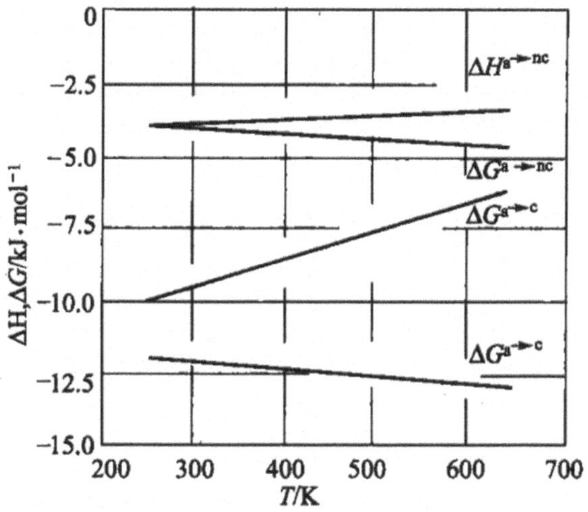

Fig. 3.22 Calculates the enthalpy change and Gibbs free energy change of the two transitions based on the experimental results [34]

3.5.4.2 Interfacial Thermodynamics of Nanocrystals

The grain boundary structure of nanomaterials can be directly reflected by its interfacial thermodynamic properties, such as excess enthalpy, excess entropy and excess Gibbs free energy. At present, people mainly use DSC method to measure the enthalpy released during the grain growth of nano-materials to get the excess enthalpy of grain boundaries. However, as mentioned above, in many cases, the grain boundary enthalpy cannot be accurately obtained by DSC when measuring grain growth, and the measurement results are easily affected by micro-strain, texture and grain size distribution.

Considering the internal interface of nanomaterials as an independent "phase", the phase transition from amorphous alloy to nanocrystal can be understood as a decomposition reaction:

$$\text{Amorphous solid phase} \rightarrow \text{nanoscale grain} + \text{inner interface} \tag{3.44}$$

If the mole fraction (atomic fraction) of the internal interface is "x_{in}", the total enthalpy of phase transition becomes:

$$\Delta H^{a\rightarrow nc}(T) = (1 - x_{in})\Delta H^{a\rightarrow c}(T) + x_{in}\Delta H^{a\rightarrow in}(T) \tag{3.45}$$

It can be considered that the transition of amorphous to grain is the same as the phase transition of amorphous to general crystal, then the enthalpy change "$\Delta H^{a\rightarrow in}(T)$" of the amorphous transition to the inner interface can be expressed as:

$$\Delta H^{a\rightarrow nc}(T) = \Delta H^{in}(T) - \Delta H^{a}(T) \tag{3.46}$$

where "$\Delta H^{in}(T)$" and "$\Delta H^{a}(T)$" are the excess enthalpy of the amorphous solid phase at the inner interface (as opposed to the complete crystal), respectively. Among them, from the thermodynamics of the solid crystallization process, it can be seen that the excess enthalpy "ΔH^{a}" of the amorphous solid is:

$$\Delta H^{a} = \Delta H_{f} - \int_{T_m}^{T} \left(C_p^l - C_p^l\right) dT \tag{3.47}$$

where "ΔH_f" is the melting enthalpy change at the melting point "T_m", and "C_p^l" is the molar heat capacities, respectively. Then there are:

$$x_{in}\Delta E_{in} = \Delta H^{anc}(T) - \Delta H^{ac}(T) \tag{3.48}$$

where "ΔE_{in}" is the molar excess energy of the relatively intact crystal at the inner interface. Since the enthalpy change "$\Delta H^{anc}(T)$" of the amorphous to nanocrystalline transition and the enthalpy change "ΔH^{ac}" of the transition to the crystalline state can be measured, the interfacial excess energy "$\Delta E_{in}(T)$" of different grain samples can be measured using "x_{in}" at different grain sizes.

Figure 3.23 is the results of "ΔH^{anc}" and "ΔH^{ac}" obtained by measuring the crystallization of amorphous Se at different annealing temperatures, where "T_g" is the glass transition temperature; "T_m" is the melting point of Se. It can be seen that with the decrease of annealing temperature, the difference between the two increases, reaching a maximum value at 373 K, about 1.80 kJ mol^{-1}. Using Eq. (3.47), "ΔE_{in}" at different annealing temperatures (different grain sizes) can be found.

Similarly, in the thermal analysis experiment of amorphous Ni–P alloy into nanocrystals, it was found that the interfacial excess enthalpy of nanocrystalline Ni–P alloy decreased with the decrease of grain size. At the same time, the density measurement results show that the interfacial excess volume of nanocrystalline Ni–P

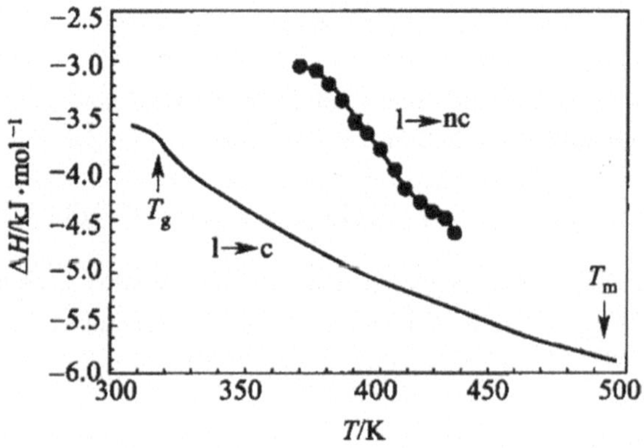

Fig. 3.23 Amorphous SE enthalpy of crystallization and nanocrystallization exothermic at different crystallization temperatures [35]

Table 3.6 Grain boundary enthalpy of several nanomaterials

Sample	Preparation method	Average grain size (d/nm)	Enthalpy of grain boundaries ($\Delta H_{gb}/J\ m^{-2}$)
Ni–P	Amorphous crystallization	6.5	0.16
		60	0.47
TiO_2	Amorphous crystallization	34	1.28
		76	1.55
Se	Amorphous crystallization	8.6	0.27
		22.4	0.32

alloy also decreases with the decrease of grain size, which is confirmed by positron annihilation experiment.

Similarly, by measuring the thermal effect of grain growth of TiO_2 nanocrystals with different grain sizes, it is also found that the excess enthalpy of interface in TiO_2 nanocrystals also decreases with the decrease of grain size.

Table 3.6 lists the grain sizes and corresponding interfacial enthalpies of several nanocrystalline alloys with different grain sizes. They all show that the enthalpy of grain boundary or phase boundary decreases with the decrease of grain size.

It is generally believed that the excess volume of the interface (relatively complete lattice) "$\Delta V = V/V_0 - 1$" (where "V_0" and "V" are the volumes of complete single crystals and grain boundaries respectively) is the most reasonable parameter to describe the energy state of grain boundaries, which determines the thermodynamic properties of the interface, such as interface entropy, interface enthalpy, and interface Gibbs free energy. The thermodynamic calculation can be used to obtain the relationship between interface enthalpy, interface entropy, interface Gibbs free energy and interface excess volume. Figures 3.24, 3.25, and 3.26 is the calculation result of nanocrystalline pure Ni. It can be seen that the interfacial enthalpy, entropy, and Gibbs free energy all decrease with the decrease of excess volume at the interface.

Therefore, it can be concluded that with the decrease of grain size of nanomaterials, the excess volume, interfacial enthalpy, interfacial entropy and interfacial Gibbs free energy in the unit area of grain boundaries or phase boundaries of nanomaterials also decrease.

3.6 Mechanical Properties of Nanocrystalline Materials

Due to the unique and attractive properties of nano-materials, in recent decades, people's research on their preparation, structure, properties and their relationship has formed a worldwide upsurge. Nanocrystalline materials are single-phase or multi-phase polycrystalline materials, and their grain size is in the order of nanometer (1×10^{-9}–250×10^{-9} m). Nanocrystalline materials are structurally composed of grains of nanometer size and a lot of volume fraction boundaries between grains. Compared

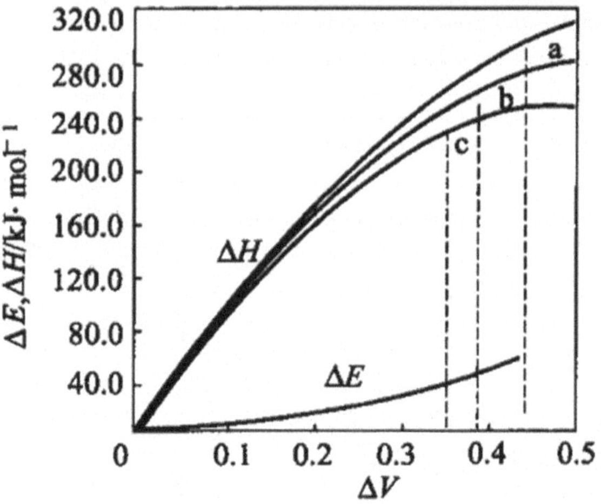

Fig. 3.24 The relationship between interface excess energy and excess enthalpy and interface excess volume change at different temperatures a—T = 300 K, b—T = 800 K, c—T = 1300 K [35]

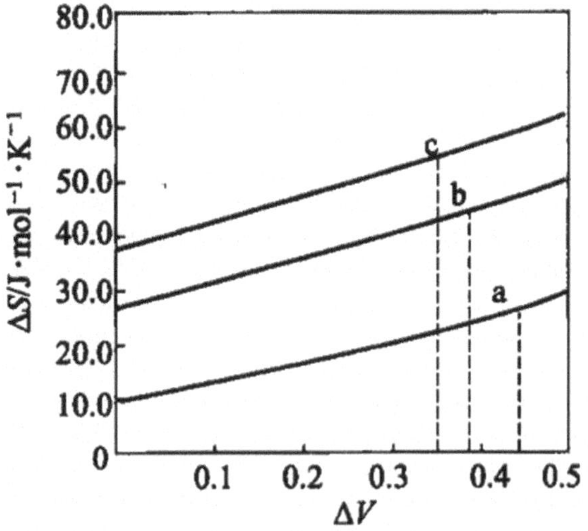

Fig. 3.25 Relationship between interfacial excess enthalpy and interfacial excess volume change a—T = 300 K at different temperatures; b—T = 800 K; c—T = 1300 K [36]

with traditional coarse-grained materials (grain size is 10–300 μm), their mechanical, physical and chemical properties have been obviously improved to some extent. Nanocrystalline materials have high strength/hardness, good toughness, low elastic modulus and ductility, high diffusivity, high specific heat capacity, high thermal expansion coefficient and excellent soft magnetic properties [37].

Nanocrystalline materials not only provide people with an opportunity to understand the microscopic characteristics of the interface of materials, but also open up a new way to explore the changes in the relationship between the structure and properties of materials after the grain size is as small as nanometer order. At the same time, the unique properties of nanocrystalline materials have laid a solid foundation for

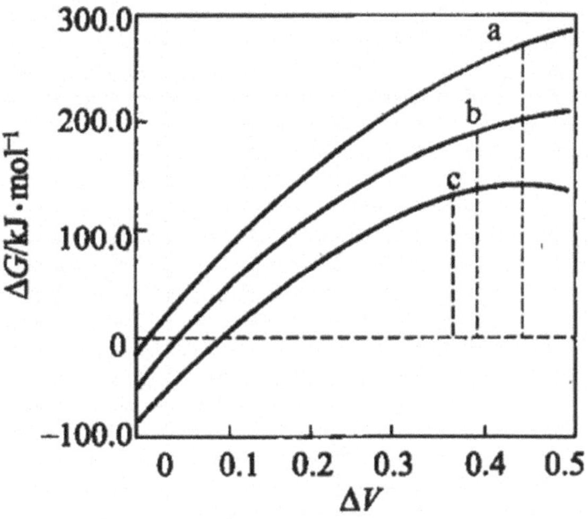

Fig. 3.26 Relationship between interface excess Gibbs free energy and interface excess volume change at different temperatures a—T = 300 K, b—T = 800 K; c—T = 1300 K [36]

the development of high technology. This chapter will focus on the understanding of mechanical properties of nanocrystalline materials so far, including yield stress, ductility, strain hardening, strain rate sensitivity and dynamic response, creep, fatigue and fracture of nanocrystalline metals and alloys, ceramics and composites.

3.6.1 Mechanical Properties of Nanocrystalline Metals and Alloys

Before the beginning of this section, the influence of porosity on the properties of nanocrystalline materials caused by the preparation method is briefly introduced. Pore is a kind of preparation/processing defect, and its existence may cover up the real properties of nanocrystalline materials or distort them, which is generally considered to have an adverse effect on the properties. Figure 3.27 shows the functional relationship between elastic modulus and porosity of Pd and Cu nanocrystals.

It can be seen that the elastic modulus decreases with the increase of porosity. This has been confirmed by many mechanical simulation and calculation results. Wachtman proposed an expression of the relationship between porosity and elastic modulus:

$$E = E_0\left(1 - f_1 p + f_2 p^2\right) \tag{3.49}$$

where "p" is porosity; "f_1" and "f_2" are equal to 1.9 and 0.9, respectively. For relatively low porosity, "p^2" is negligible, so an approximation is given "$E/E_0 = 1 - 1.9$ p". Yield stress and tensile ductility are also affected by porosity, and Fig. 3.28 shows the yield strength versus compaction density in Cu and Pd metals. It is clear that the

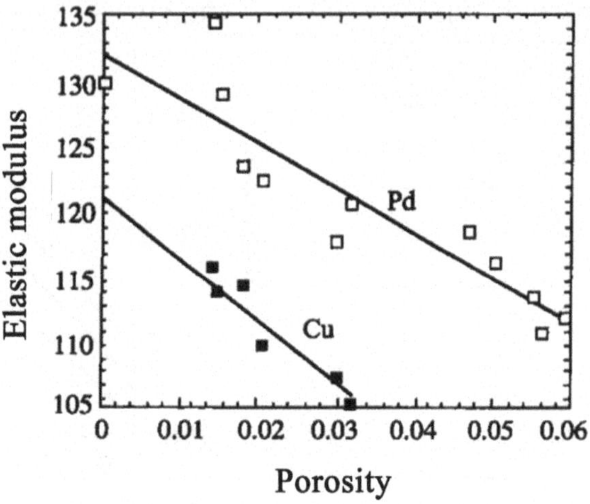

Fig. 3.27 Elastic modulus of Pd and Cu nanocrystals as a function of porosity [38]

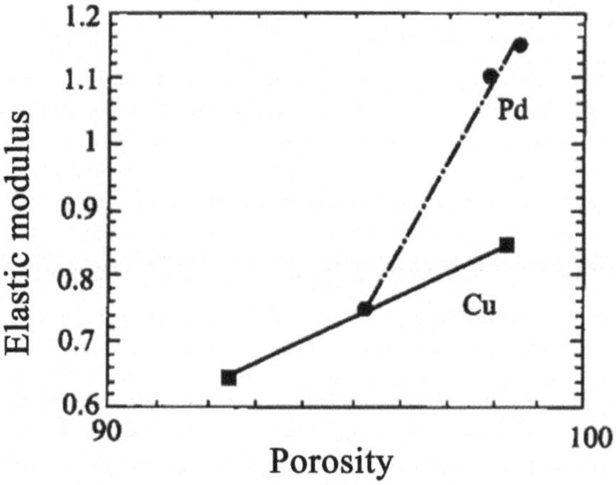

Fig. 3.28 Compressive yield strength of Cu and Pd as a function of compaction density [38]

yield strength is strongly affected by the compaction density. This may be that the existing pores provide the initial location for the propagation of the crack and even the fracture failure of the material.

3.6.1.1 Yield Strength

Grain size has obvious influence on the mechanical properties of materials, especially the yield strength. The relationship between yield stress and grain size of materials has been very clear in traditional polycrystalline materials, and they follow the Hall–Petch (hereinafter referred to as H–P) relationship, that is

$$\sigma_y = \sigma_0 + kd^{-1/2} = \sigma_0 + kd^{-1/2} \tag{3.50}$$

Here "σ_y" is the yield stress of the material, "σ_0" is the frictional stress, and "k" is a constant. This is only an approximation, and a more general form is that d has an exponent of $-n(0.3 \le n \le 0.7)$. The mechanical properties of nanocrystalline materials with face-centered cubic structure were measured by equiaxed tension/compression and micro/nano indentation experiments. In order to avoid the adverse effects of defects such as holes on mechanical properties testing, micron-scale samples are usually used.

Table 3.7 shows the compressive yield strength of nanocrystalline Cu and Pd synthesized by inert gas condensation (IGC), and the corresponding curve is shown in Fig. 3.28. It can be seen that the yield strength of nanocrystalline Cu and Pd is obviously higher than that of the corresponding coarse-grained materials, and they are strongly dependent on the density of the materials. The compressive strength of the strongest nanocrystalline Cu sample studied by Suryanarayana et al. can reach 500 MPa. Table 3.7 shows the 1/3 Vickers hardness of nanocrystalline materials, which is basically consistent with the compressive yield strength, which indicates that the work hardening phenomenon of the materials is not very significant in the experiment. The results also show that the hardness of Cu and Pd nanocrystals prepared by IGC method is much higher than that of the corresponding annealed materials. However, it is not known whether these improvements are entirely caused by the reduction of grain size and are not affected by defects.

The H–P relation listed in Eq. 3.50 predicts that the stress will increase with the increase of the reciprocal of the square root of grain size. However, some research results show that the H–P relationship obtained in coarse-grained materials cannot be well extrapolated to the case that the grain size is smaller than lum. Many data on the relationship between yield strength and grain size of Cu with grain size ranging from micro to nano are collected in the literature. It is found that the curve trend changes with the grain size decreasing to 25 nm $d^{-1/2} = 0.2$. Some results predict a stable platform; Some decrease the yield strength with the further decrease of grain size. It is an example of the latter phenomenon that nanocrystalline pure Ni and Ni–P alloys with different sizes are obtained by amorphous crystallization. It is worth noting that amorphous is the limit state in which the grain size continues to decrease. It is

Table 3.7 Compression yield strength of nanocrystalline Pd and Cu synthesized by IGC method

Test samples	Compaction temperature/ °C	Density (theoretical)/ %	Test samples/ nm	Yield strength/ GPa	(Strength/3)/ GPa
Pd1	335	98.5	54	1.15	1.0
Pd2	183	97.9	38	1.10–1.13	1.1
Pd3	RT	95.3	24	0.75	0.75
Cu1	106	92.5	19	0.65	0.77
Cu2	106	98.4	20	0.85	0.87

considered that this trend change of H–P relationship may be attributed to different deformation mechanisms, that is, when the grain size is reduced below a certain critical grain size, the new unknown deformation mechanism becomes an important dominant factor.

Chokshi et al. reported the phenomenon of anti-H–P relationship for the first time by studying the properties of Cu and Pd nanocrystals prepared by IG method. Both metals they measured showed negative slopes, as shown in Fig. 3.29a. They believe that this negative slope is caused by the diffusion creep of nanocrystalline samples at room temperature, which is similar to the grain boundary sliding of traditional crude samples at high temperature. In addition, other researchers have reported similar anti-H–P relations. Figure 3.29b is the result of hardness test by Weertman et al. It can be seen that there is no anti-H–P relation phenomenon, but this phenomenon appears in tensile test. They attribute the decrease of yield strength to the existence of defects at a smaller grain size. Weertman thinks that the negative slope of the H–P relationship of nano-crystalline obtained by Chokshi et al. may be due to the use of the same sample, which has been repeatedly annealed and changed the grain size, so it is artificially caused.

Chokshi and others believe that the negative slope of nanocrystalline copper is caused by the rapid diffusion creep at room temperature. Coble creep is considered as the main deformation mechanism, namely:

$$\xi = \frac{150\Omega\delta D_{gb}\sigma}{\pi kTd^3} \tag{3.51}$$

where "ξ" is the strain rate, "Ω" is the atomic volume, "δ" is the grain boundary width, "D_{gb}" is the grain boundary diffusion coefficient, "σ" is the stress, "k" is the Boltzmann constant, and "T" is the thermodynamic temperature. Chokshi assumes: $\Omega = 1.3 \times 10^{-29}$ m^3, $\delta = 1$nm, $D_{gb} = 3 \times 10^{-9}\exp[-62000/(RT)]$ m^2 s^{-1}

Figure 3.30 shows the relationship between strain rate and grain size at 300 K when the stress is 100 and 1000 MPa respectively. It can be seen from this curve that when the grain size is less than about 20 nm, the grain boundary diffusion process becomes more and more important. In addition, it is worth noting that Chokshi and others neglected the role of plastic deformation in their analysis, and plastic deformation is necessary for grains to slide and jump over each other.

Figure 3.31 shows the changing trend of H–P relationship of four different metals Cu, Ni, Fe and Ti when the grain size changes from micron to nanometer. It can be seen that within the scale range of traditional coarse-grained materials,

Most data points are coincident. However, in the range of nanocrystals, the distribution of data points is very discrete, and the H–P relationship obtained from this obviously deviates from the law of microcrystalline materials, and the slope is also reduced [41]. So after the grain size is less than about 10–15 nm, what is the relationship between yield stress and grain size? The understanding is not very uniform. This is still a subject worthy of further study. Especially, the debate about the existence of anti-H–P relationship is very fierce. Another point of view is that there may be

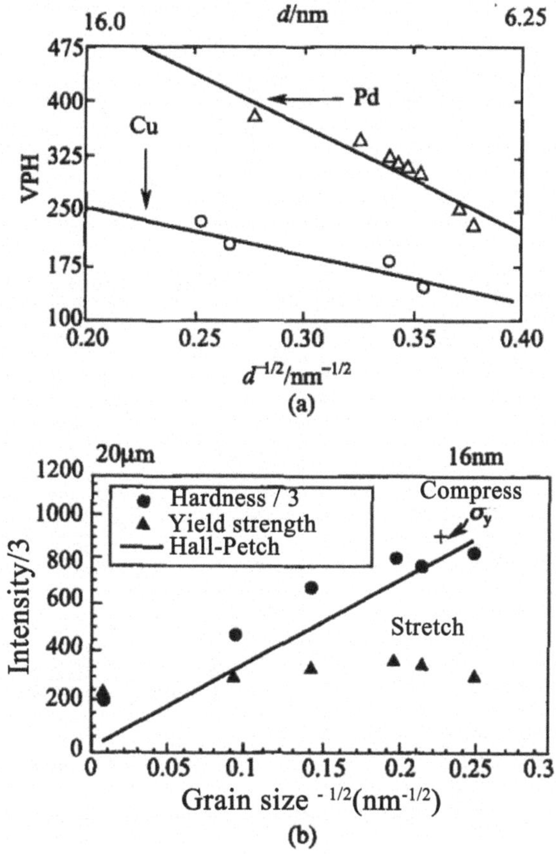

Fig. 3.29 The anti-H–P relationship (**a**) and compression experiments observed by Chokshi et al. in Cu and Pd nanocrystals have higher positive HP relationship slope values than tensile experiments (**b**) [39]

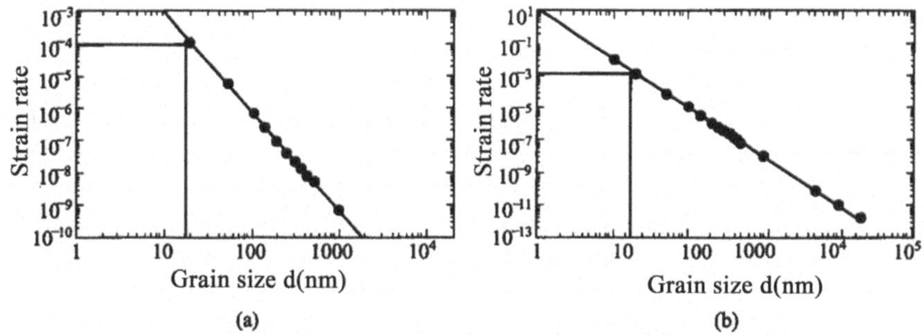

Fig. 3.30 Relationship between strain rate and grain size in Coble creep **a** 100 MPa, **b** 1000 MPa [40]

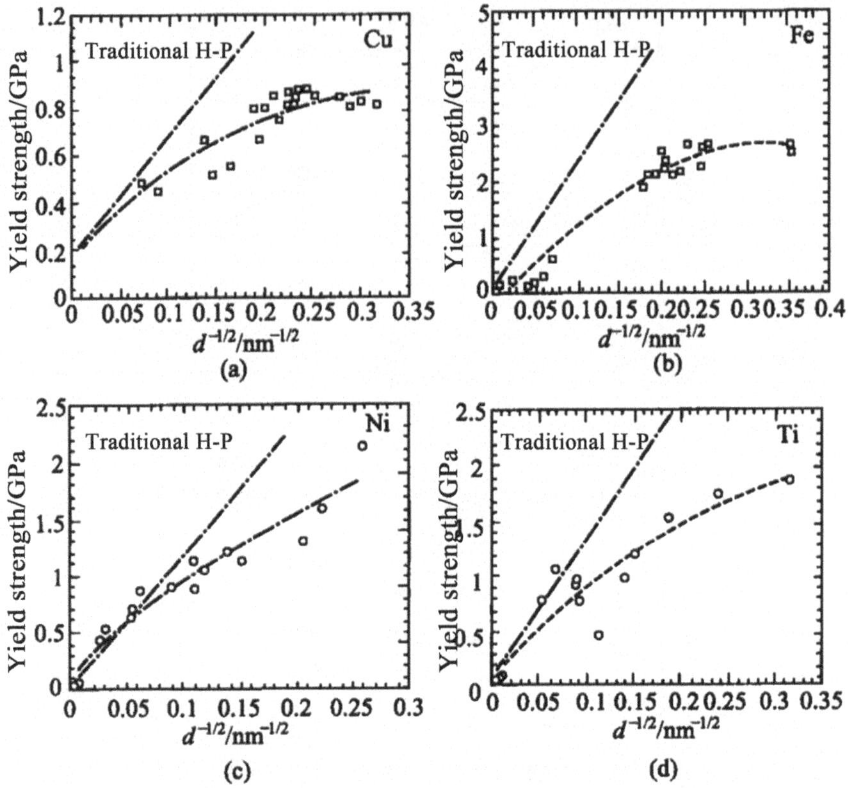

Fig. 3.31 Relationship between yield stress of different metals and grain size [42]

a platform of yield strength below a certain critical grain size, which may change according to different materials.

3.6.1.2 Ductility

In the traditional research of coarse-grained materials, it is thought that the smaller the grain size, the higher the ductility of the material. Therefore, it is expected that the ductility may be greatly improved when the grain size is reduced to nanometer order. But in fact, the existing research shows that when the grain size of metals is less than or equal to ≤25 nm, their ductility is very poor, only 40–60% of the ductility of these metals in the traditional grain size range. Koch thinks that there are three main reasons for the limited ductility of nanocrystalline materials:

1) The introduction of defects (such as pores) during processing;
2) Tensile instability;

3) Crack nucleation or shear instability. Under the current processing conditions, it is very difficult not to have artificially introduced defects. These defects seriously conceal or distort the intrinsic mechanical properties of nanocrystalline materials. Fortunately, the appearance of molecular dynamics simulation provides an important auxiliary means for people to understand the deformation mechanism of nanocrystalline materials. The simulation work at atomic level has been carried out to describe different plastic deformation mechanisms as a function of grain size. The results are basically consistent with the current understanding, namely:
1) The grain size d > 1 μm, and the dislocation and work hardening in the material control the plastic deformation;
2) At the smallest grain size d < 10 nm, in which the number and activity of intragranular dislocations are limited, grain boundary shear is considered as the main deformation mechanism;
3) In the middle grain size range (10 nm–1 μm), people still have little understanding, and it is these unknown mechanisms that strongly affect the ductility of materials. This will be discussed in detail in the deformation mechanism of nanocrystalline materials later.

Figure 3.32a shows the relationship curve between yield strength and elongation of metal materials in the nanometer scale, and the ductility decreases obviously with the increase of strength. Comparatively speaking, ultrafine grained materials UFG (100–500 nm) have good ductility while improving the yield strength (Fig. 3.32b).

Zhang et al. obtained the mixed microstructure of Zn nanocrystals and ultrafine grains by changing the ball milling time. At the temperature of liquid nitrogen, with the extension of ball milling time, the hardness has a very obvious mutation. The strength and toughness of the sample milled at low temperature for 4 h are the best.

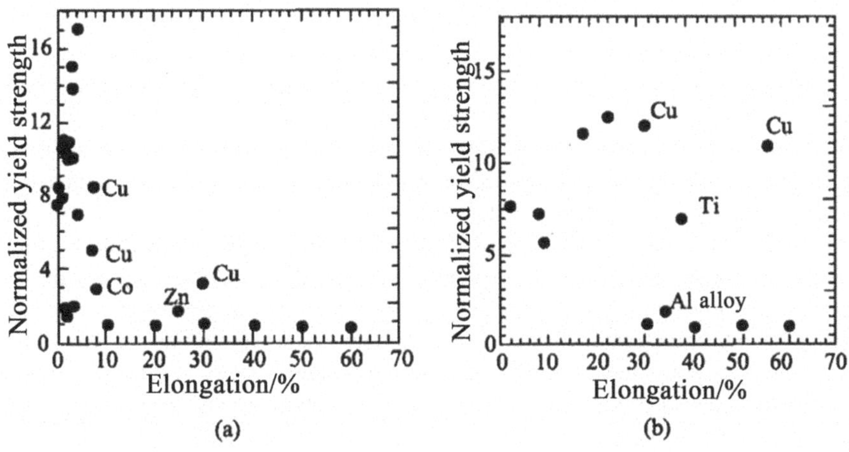

Fig. 3.32 Curve of yield strength and elongation, showing that nanocrystalline materials have low ductility (**a**) and curves of yield strength and elongation of different ultrafine grain metals (**b**) [42]

In this sample, 30% of the grains are larger than 50 nm in size, and the others are nano-grains. Compared with the microstructure obtained by ball milling for a longer or shorter time, the microstructure of nano-crystal and ultra-fine crystal has the best strain hardening effect, mainly because they combine the strengthening of small grain size with the strain hardening caused by dislocation activity of large grain size, and this strain hardening makes the material have better ductility. Therefore, the bimodal distribution of grain size may be an effective means to improve the ductility of materials.

Non-equilibrium grain boundaries are also considered as an important factor to improve the ductility of materials. It is considered that this grain boundary provides more dislocations for slip, and even makes the grain slip or rotate at room temperature, which leads to an obvious increase in strain hardening index. Another way to improve ductility is to reduce the strain rate so that the specimen can maintain more plastic strain before necking.

Figure 3.32a shows the relationship curve between ductility and strength of metal in the traditional coarse grain size region. It can be seen that with the increase of strength, ductility decreases, and these data points basically fall in the gray area of the diagram. But above this area, there are five data points that are inconsistent with this law, that is, they maintain good ductility under high strength. It is considered that three factors determine the ductility of materials: work hardening, strain rate sensitivity and thermal softening. In some cases, materials show high ductility because shear localization is prevented. The strain rate sensitivity m can be expressed as follows:

$$m = \frac{3^{1/2}kT}{V\sigma_y}. \tag{3.52}$$

where "V" is the volume fraction of plastic deformation activation (it is directly related to the physical mechanism of deformation, and "T" is the temperature and "σ_y" is the yield/flow stress. Lu et al. think that high strain rate sensitivity indicates that the activated volume fraction of the material is small, so it is closely related to the change of nano-structure (such as the appearance of twins). Therefore, the ductility, strain rate sensitivity and deformation mechanism of materials are interrelated, and the ductility of materials can be improved by manipulating nanostructures. Zhu and Liao improved the ductility of the metal nanocrystals they studied by greatly increasing the density (annealing) of the growth twins.

Figure 3.32b shows the mechanical properties of zinc samples with different grain sizes. As the grain size changes from 238 to 23 nm, the ductility decreases obviously. Zhang et al. think that the decrease of elongation with the decrease of grain size may be an inherent characteristic of nano-materials. These nano-crystalline materials they studied have no pores and are well bonded during synthesis. Early results have shown that the mechanical properties of nanocrystalline materials may be misinterpreted because of the lack of understanding of its internal structure details. At the same time, the pollution and porosity caused by the preparation process are also considered to be very harmful to ductility (Fig. 3.33).

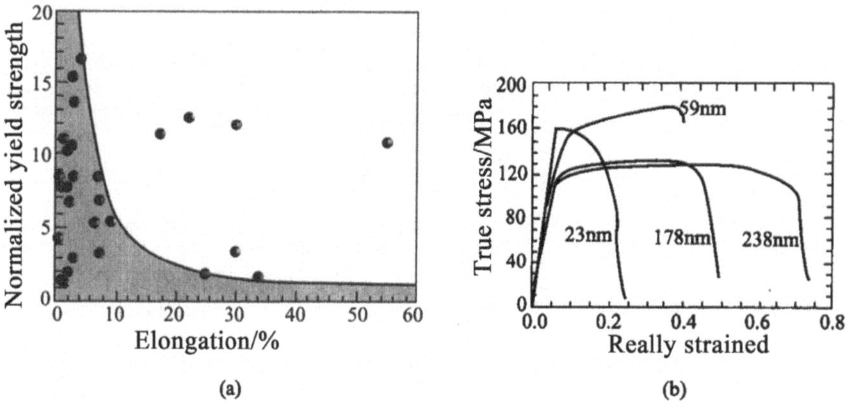

Fig. 3.33 The data point on the right-hand side of the figure illustrates that the ductile material has high ductility (**a**) at the nanocrystalline area and at room temperature, constant strain rate 10^{-4}–10^{-3} s^{-1}, the ductile ductility of the ball milled zinc specimen decreases with the decrease of grain size (**b**) [42]

3.6.1.3 Strain Hardening

Nanocrystalline and ultrafine grained materials generally cannot maintain uniform elongation. It has been reported that after the initial rapid strain hardening of these materials in a small strain range (about 1–3%), there will be no strain hardening phenomenon, which is different from the behavior of coarse-grained materials.

The dislocation density in nanocrystalline samples is saturated, which is mainly caused by dynamic recovery or dislocation annihilation at grain boundaries. This ultimately leads to a low strain hardening rate. Therefore, only in the process of large additional strain can work hardening be observed. Dynamic recovery is considered to occur only during severe plastic deformation. Due to the increase of temperature, the deformation structure is transformed into ultrafine grains during the recovery process, which has both low-angle and high-angle grain boundaries. Low strain hardening has been observed in samples prepared by equal channel angular pressing (ECAP) and powder consolidation. Figure 3.34 shows the compressive and tensile stress–strain curves of UFG copper prepared by ECAP process (8 passes). There is basically no work hardening when compressed. During tension, the specimen necked at the yield stress, showing low tensile ductility.

During tensile deformation, the stress–strain curve of nanocrystalline copper first appears a peak quickly, and then appears softening phenomenon, which may be caused by necking. The disappearance of strain hardening ($d\sigma/d\varepsilon = 0$) causes local deformation and leads to low ductility. Flat compression curves were observed for other nanocrystalline materials, including Fe(bcc) and Ti(hcp). Necking mostly occurs under the condition of severe instability, and shear bands are also observed in compacted Fe samples.

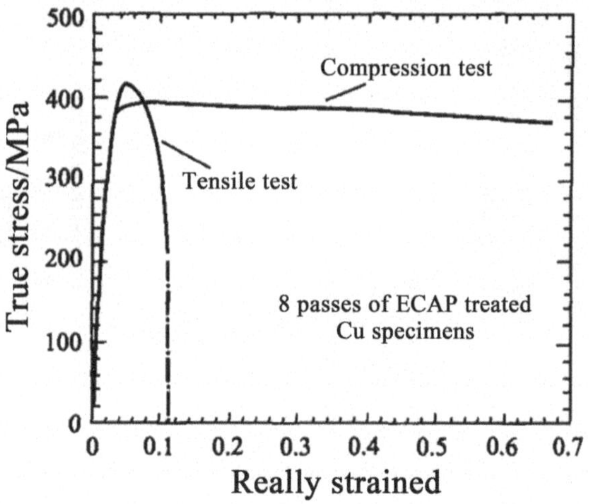

Fig. 3.34 Copper compression and tensile stress–strain curves obtained by the ECAP method [42]

It is very common that dynamic recovery occurs in nanocrystalline samples at room temperature. The competition between dislocation generation and dislocation annihilation during plastic deformation determines the density of steady-state dislocation. Although there are still some controversies, it is still believed that dislocations are generated at the grain boundary and disappear there at the same time. In addition, when the grain size of the material is in the range of 10–15 nm, other deformation mechanisms may occur.

3.6.1.4 Strain Rate Sensitivity

There have been many studies on the sensitivity of metal materials with the decrease of grain size. Iron is generally sensitive to strain rates, which are defined as m (defined as "$m = \frac{\partial \ln \sigma}{\partial \ln \varepsilon}$"/"$\frac{1}{\sigma_y} \cdot \frac{\partial \sigma}{\partial \ln \varepsilon}$"), which is generally on the order of 0.04 and will decrease to 0.004 when the grain size is reduced to 80 nm. Malow et al. prepared nanocrystalline iron using ball milling and compaction and found that at $d \approx 20$ nm, the $m \approx 0.006$.

Gray et al. found an opposite phenomenon in fcc metals of UFG prepared by ECAP method, such as Cu, Ni and Al–4Cu–0.5Zr, whose mechanical properties depend on the strain rate (0.001–4000 s^{-1}), as shown in Fig. 3.35. Within the above strain rate range, the strain rate sensitivities m of Cu, Ni and Al–4Cu–0.5Zr alloys are 0.015, 0.006 and 0.005, respectively [43]. They believe that this relatively high strain rate in Cu and Ni of UFG, and the yield work hardening rate which is almost zero, are mainly caused by the pre-existing high dislocation density.

Higher m values also appeared in nanocrystalline Ni, nanocrystalline Au and electrodeposited Cu. However, these data are obtained from creep tests, in which case the grain boundary deformation is dominant. In general, quasi-static tensile tests, these nanocrystalline metals do not show high strain rate sensitivity. It is impossible

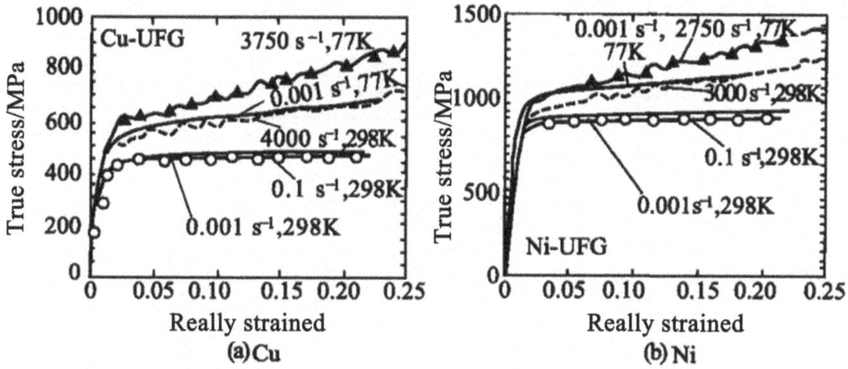

Fig. 3.35 Stress–strain curve of ultrafine grain material [44]

for a small value of m to keep the material deformed all the time without local shear during deformation, which is similar to the behavior of amorphous alloys.

Wei et al.'s research shows that the strain rate sensitivity will be improved when the grain size is below a certain critical value, and this result is shown in Fig. 3.36. This phenomenon has also been confirmed in the nano-indentation hardness test experiment. The research by Goken and his collaborators (Fig. 3.37) shows that the observed strain rate sensitivity is almost the same when the grain size d is as small as about 60 nm in pulse electrodeposition of nickel (Fig. 3.37b). Yes, the sample with d = 20 nm shows a higher slope. A similar phenomenon appears in Al nanocrystals prepared by ECAP method. The slope of ultrafine grained aluminum is about m = 0.027, while the strain rate sensitivity of traditional polycrystalline aluminum is less than 1/3 of this value, which is about 0.007 (Fig. 3.37a). This higher strain rate sensitivity is directly related to the strain rate change that controls the deformation mechanism. It can be seen from formula (3.52) that m is proportional to V^{-1}, that is, m is inversely proportional to the activation volume. Traditional fcc metals have a large activation volume V:

$$V \propto 10^2 - 10^3 b^3$$

This is related to the dislocation of cutting forest dislocation. On the other hand, the activation volume of grain boundary diffusion process is lower, which is in the order of atomic volume:

$$V \propto (1 \sim 10) b^3$$

This is a lower range. The reduction of v itself proves the enhancement of the exchange effect of atomic size order of magnitude. The m observed by Hoppel et al. in ECAP has the following values: 2 passes ($\gamma \approx 2$)m = 0.06; 16 passes ($\gamma \approx 2$)m = 0.14.

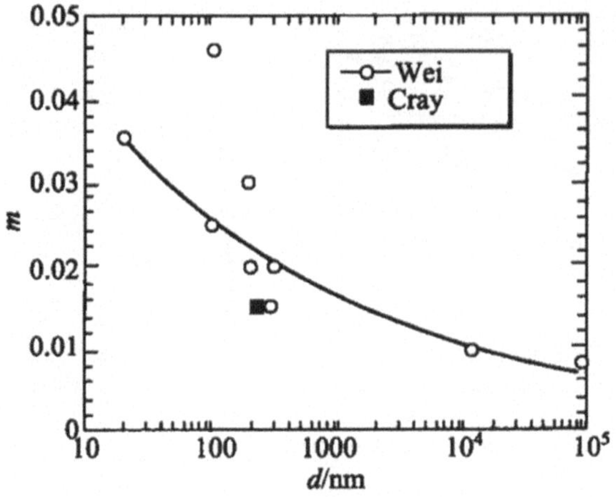

Fig. 3.36 Strain rate sensitivity of Cu as a function of grain size [43]

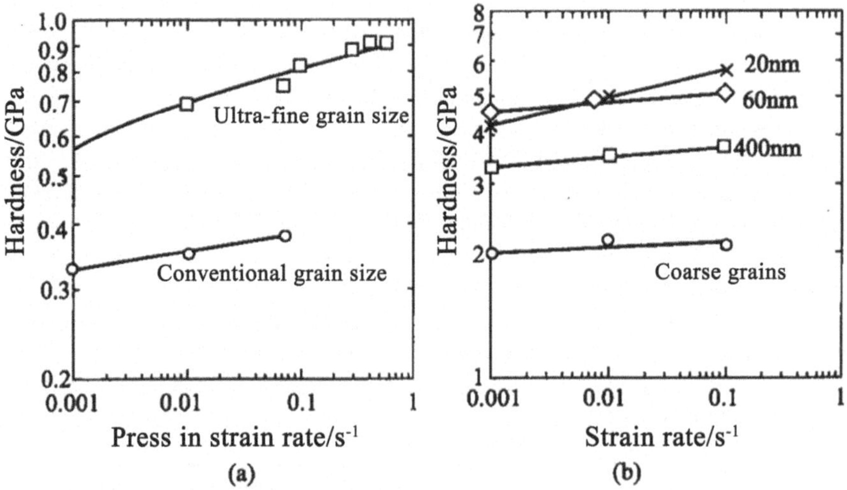

Fig. 3.37 Comparison of strain rate sensitivity of coarse-grained and ultrafine grained aluminum measured by hardness measurement (**a**) and pulse electrodeposition of hardness measurement Strain rate sensitivity of Ni (**b**) [43]

With the increase of passes, the strain rate sensitivity is obviously improved, which is not attributed to the change of deformation mechanism to grain boundary sliding, but to the annihilation of thermal activation dislocation at grain boundary. This conclusion is reliable, because under the condition of ultra-fine grain, because of the size effect, forest dislocation cutting is impossible to work.

3.6.1.5 Creep of Nanocrystalline Materials

Creep of coarse-grained materials has been extensively and deeply studied for about a century. A more accurate model has been established, and the deformation characteristics and the mechanism involved have been clarified. Due to the complexity of creep of nanocrystalline materials, research has only begun in recent years. Firstly, there are many difficulties in preparing nanocrystalline bulk materials with uniform size without defects (pores and impurities), so it is impossible to provide reliable data to explain the creep deformation process of nanocrystalline materials. Secondly, grain boundaries and intergranular defects, such as the obvious increase of the volume fraction of trident lines and quadruped nodes, make the creep mechanism more complicated, which brings great challenges to the establishment of models to explain the deformation mechanism. Thirdly, compared with coarse-grained metal materials, the growth of nanocrystals appears in a lower temperature range, which limits the testing temperature.

Because the volume fraction of grain boundary is very high, very obvious diffusion creep will occur. The high temperature deformation of crystal materials is given by the following Bird-Dorn-Mukherjee equation:

$$\varepsilon = \frac{\mathrm{ADGb}}{\mathrm{kT}}\left(\frac{\mathrm{b}}{\mathrm{d}}\right)^{\mathrm{p}}\left(\frac{\sigma}{\mathrm{G}}\right)^{\mathrm{n}} \tag{3.53}$$

where "ε" is strain rate, A is dimensionless constant, "G" is shear modulus, "B" is Burgers vector, "K" is Boltzmann constant, "T" is thermodynamic temperature, "P" is the reciprocal index of grain size, and "N" is stress index. For coarse-grained materials, the established diffusion creep mechanisms are Nabarro-Herring creep and Coble creep, which involve vacancy flow through the lattice and along the grain boundary respectively. Nabarro-Herring creep equation is

$$\varepsilon_{\mathrm{NH}} = \frac{\mathrm{A_{NH}D_LGb}}{\mathrm{kT}}\left(\frac{\mathrm{b}}{\mathrm{d}}\right)^{2}\left(\frac{\sigma}{\mathrm{G}}\right) \tag{3.54}$$

where "D_L" is the lattice diffusion coefficient, with exponent $p = 2$ and $n = 1$. The dimensionless constant $A_{NH} = 28$. Coble creep equation is:

$$\varepsilon_{\mathrm{CO}} = \frac{\mathrm{A_{CO}D_{gb}Gb}}{\mathrm{kT}}\left(\frac{\mathrm{b}}{\mathrm{d}}\right)^{3}\left(\frac{\sigma}{\mathrm{G}}\right) \tag{3.55}$$

Here "D_{gb}" is the grain boundary diffusion coefficient, with exponent $p = 3$ and $n = 1$. The dimensionless constant $Aco = 33$.

Polumbo et al. estimated the total grain boundary fraction on the assumption that the grain shape is regular tetrahedron, and found that this value increased from about 0.3% when the grain size was 1 μm to 87.5% (the maximum) when the grain size was 2 nm. It is worth noting that the volume fraction of trident grain boundaries shows greater sensitivity to grain size than grain boundaries. Wang et al. revised the standard diffusion creep equation to meet the conditions of diffusion along the trident boundary. The following is the expression of diffusion creep along the trident boundary:

$$\varepsilon_{TL} = \frac{K_{TL} D_{TL} \Omega \delta^2 \sigma}{kTd^4} \tag{3.56}$$

Here "K_{TL}" is the constant of the geometry and boundary conditions of the grain boundary, "D_{TL}" is the diffusion coefficient of the trigeminal grain boundary, "Ω" is the atomic volume, and "δ" is the grain boundary width. Chokshi et al. modified the Bird-Mukherjee-Dorn equation to describe changes from diffusion mechanism to intergranular dislocation power-law creep [45].

$$\varepsilon_{pL} = \frac{AD_L Gb}{kT}\left(\frac{\sigma}{G}\right) \tag{3.57}$$

where $n \geq 3$, $p = 0$, $D = D_L$.

Early reports on the creep of nanocrystalline materials include Wang et al. research on 28 nm Ni–P alloy (Fig. 3.38a), Deng et al. research on Ni–P alloy and Nieman et al.' s research on nanocrystalline Pd. As can be seen from Fig. 3.38a, the slope of the curve n = 1, which supports the creep mechanism of Coble and Nabarro-Herring [Eqs. (3.54) and (3.55)]. Wang et al. attributed this creep to grain boundary diffusion, and in their subsequent work, they concluded that grain boundary diffusion is the main operating mechanism in the process of nanocrystalline creep, while in coarse-grained materials, the composite mechanism of dislocation creep and grain boundary diffusion may be dominant.

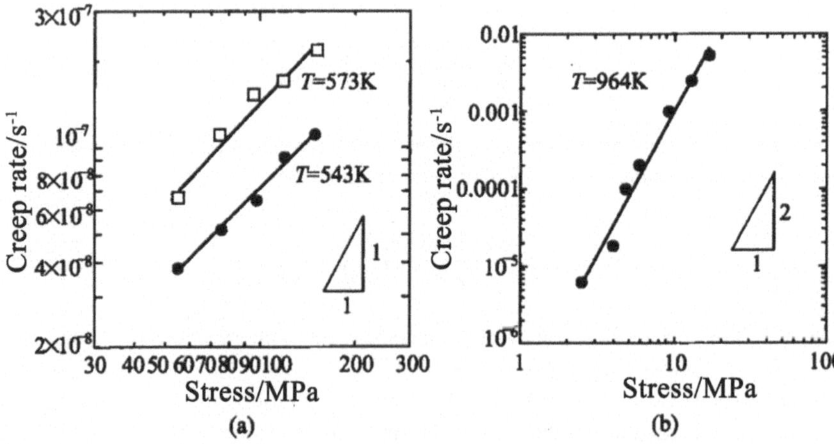

Fig. 3.38 Stress rate curves for nanocrystalline Ni–P(d = 28 nm) (**a**) and nanocrystalline TiO_2 (**b**) [44]

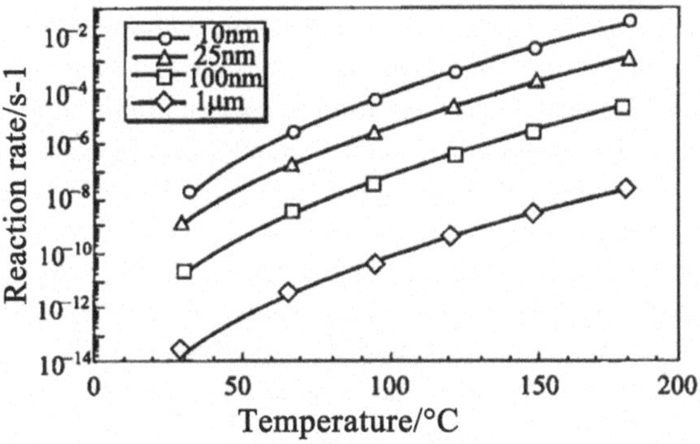

Fig. 3.39 Creep curve calculated based on grain boundary diffusion mechanism under different grain size conditions (note that the strain rate increases by 6 orders of magnitude when the grain size decreases from 1 μm to 10 nm) [44]

Hahn et al. conducted a compression creep test on TiO_2, and the result is shown in Fig. 3.38b, where n = 2. Nieman et al. considered that nanocrystalline Pd did not show obvious room temperature creep when the load was greater than the yield stress of coarse-grained Pd samples. Therefore, it is not feasible to directly evaluate the mechanical properties of nanocrystalline Pd at room temperature by grain boundary diffusion creep. Sanders et al. conducted creep tests on nanocrystalline Cu, Pd and Al–Zr alloys condensed and compacted by inert gas in a wide temperature range (0.24–0.64 TM). The creep rate observed in the experiment is 2–4 orders of magnitude smaller than that predicted by Coble creep equation. The predicted creep rate as a function of temperature for different grain sizes is shown in Fig. 3.39. This figure shows the creep curves calculated on the basis of grain boundary diffusion mechanism under different grain sizes (note that when the grain size decreases from 1 to 10 nm, the strain rate increases by about 6 orders of magnitude). Sander and others concluded that the existence of a large number of low-energy grain boundaries and the inhibition of dislocation activity caused by small grain size are the main reasons for low strain rate and high creep resistance.

Ogino compared the lower strain rate observed experimentally with the value calculated theoretically by Coble creep equation. He attributed this result to the increase of grain boundary area caused by grain deformation in the early stage of creep. Wang et al. studied the effect of grain size on the steady-state creep rate of pure nickel prepared by electrodeposition. They believe that at high stress level, grain boundary sliding becomes the main deformation mechanism at room temperature. For small grain size, the contribution of diffusion creep mechanism is very obvious in the intergranular region, so it is suggested that dynamic end change should be considered when analyzing the stress–strain curve of nanocrystalline materials. Cai et al. studied the tensile creep of pure copper nanocrystals with an average grain size of 30 nm prepared by electrodeposition at 20–50 °C ($0.22\ T_m \sim 0.24\ T_m$), and found that the steady-state creep rate was proportional to the effective stress "$\sigma_e = \sigma - \sigma_0$"; Here "$\sigma$" is the applied stress and "σ_0" is the threshold stress. Using "σ_e", it can be found that the resulting creep rate is of the same order of magnitude as the value calculated by the Coble equation [Eq. (3.55)]. The presence of sill stress means that grain boundaries are not emitting sources and traps for atoms or vacancies. This creep mechanism is defined as "interface-controlled diffusion creep". Li et al. believe that high minimum creep rate values are caused by early fracture, and they disagree with Cai et al. Grab et al. used nanocrystalline Cu, Ni and Cu–Al_2O_3 composites as examples to study the creep characteristics of materials prepared by violent plastic deformation at temperature T < 0.3 T_m. They concluded that the steady-state rate of change of nano-Cu and Ni can be well described by an exponential law, where the index n is about 5.5 for copper and about 8 for nickel. They hypothesized that during the deformation of fine-grained materials, local deformation macroscopic and mesoscopic zones appeared. They also estimated the apparent creep activation energy of nanocrystalline Cu and Ni at a temperature interval of 0.2–03 T_m, which was 1/3.5 of the coarse-grained material. They experimentally found that this difference was due to the outstanding contribution of grain boundary sliding controlled by grain boundary diffusion.

In addition, there are some examples of creep research of nano-materials close to the real situation. Estrin et al. studied the diffusion-controlled creep of nanocrystalline materials under the condition of co-growth of grains. The creep mechanism of Nabarro-Herrin and Coble is modified to explain the influence of vacancy on creep.

Based on molecular dynamics simulation, researchers have proposed several creep models of nanocrystalline materials. Yamakov et al. simulated the microstructure of three-dimensional and face-centered cubic nanocrystalline metals in detail to study the grain boundary (GB) diffusion creep. It simulates the situation at higher temperature, so that we can understand the different behaviors of GB diffusion more clearly. In order to prevent synergistic grain growth and then maintain steady-state diffusion creep, the microstructure is optimized to have uniform grain shape (nano-size) and only contain high-energy grain boundaries. The results show that under relatively high tensile stress, these microstructures show uniform steady-state diffusion creep, and their strain rates are consistent with those given by Coble creep equation. When the grain diameter and grain boundary thickness are in the same order of magnitude, the grain size parameter to measure creep is reduced from d^{-3} (typical Coble creep) to d^{-2} (typical Nabarro Herring creep). Grain boundary sliding, known as Lifshitz sliding, is Coble creep mechanism. In their molecular dynamics simulation, Hasla et al. explained the effect of synergistic grain growth on grain boundary diffusion creep and grain boundary sliding of simulated structure of nanocrystalline Pd during high temperature deformation. Before the obvious grain growth appeared, the deformation was carried out by Coble mechanism, accompanied by grain boundary sliding. Because grain growth increases the average grain size, it is generally believed that the creep rate is reduced [45]. However, in this study, the results show that the creep rate increases when the grain grows and the grain boundary migrates rapidly. The high creep rate comes from the topological change in the initial phase growth process, which improves the grain boundary diffusion flux and grain boundary sliding caused by stress. In the vicinity of some trident grain boundaries, the dislocation caused by grain merging and grain boundary decomposition caused by grain rotation also contributes to deformation.

The recent research work shows that the grain size for compensating diffusion creep rate measured in the experiment is basically the same for cubic, tetragonal and monoclinic Zr, which shows that the absolute size of grain boundary diffusion coefficient of different crystal structures is very close. The grain growth of monoclinic cubic Zr is very slow due to obvious grain boundary segregation.

The creep test results of Yin et al. show that the minimum creep rate and creep strain decrease obviously with the increase of sulfur content (or boron doping in nickel). When the temperature changes from room temperature to 100 °C and 200 °C, the stress index n in Coble creep expression increases from 2 to about 5. They explained the change of stress index with a grain boundary sliding model considering grain boundary dislocation and drag stress. The calculated drag stress shows that the interstitial atoms at the grain boundary effectively prevent the grain boundary from sliding.

3.6.1.6 The Fatigue of Nanomaterials

There are not many reports about the fatigue properties of nanocrystalline materials. Whitney et al. first reported the pull–pull cycle of nanocrystalline copper prepared by inert gas condensation. The maximum stress is 50–80% of the yield stress, and the minimum stress is 10 MPa. After several hundred thousand cycles, it is found that the grain size grows and grows (about 30%). After extending the fatigue test time, the sample has a slight elongation. The strain is similar to the creep strain at room temperature previously observed in nanocrystalline copper, in which the experimental constant stress is equivalent to the maximum cyclic stress. Cyclic deformation is elastic. However, the measured elastic modulus is 1/2 of that of ordinary copper.

Yan et al. tested the tensile fatigue of Zr (Y stabilized treatment) with an average grain size of 100 $\pm$ 20 nm. At the same time, the microcrystalline samples with the same chemical composition were also tested for comparison. It is proved by AFM imaging that on the surface at or near the fatigue fracture, some grains with the size of 100 nm are found to have local superplasticity, while on the microcrystal sample, the grains retain their original equiaxed morphology. It is believed that the microscopic mechanism of this phenomenon is basically controlled by grain boundary diffusion. At the same time, the climbing and proliferation of dislocations may also play a role.

Vinogradov et al. studied the cyclic behavior of UFG material of industrial pure titanium obtained by severe plastic deformation of ECAP. It is found that the fatigue life and fatigue limit of fine-grained Ti processed by ECAP are greatly improved compared with those of coarse-grained Ti under constant test load. The S–N curve of ECAP titanium is shown in Fig. 3.40. Contrary to Cu and A1 alloys, which show obvious degradation in strain-controlled cyclic performance, the fatigue performance of Ti has not been reduced under the condition of constant plastic strain cyclic testing, which is almost the same as the Coffin-Manson behavior under the condition of fine-coarse grain size. It is considered that grain refinement and work hardening caused by the increase of dislocation density play a very important role in the properties of materials prepared by severe plastic deformation.

Patlan et al. studied the fatigue response of 5056 Al–Mg alloy, and thought that some fatigue properties of nanocrystalline materials were improved in high stress

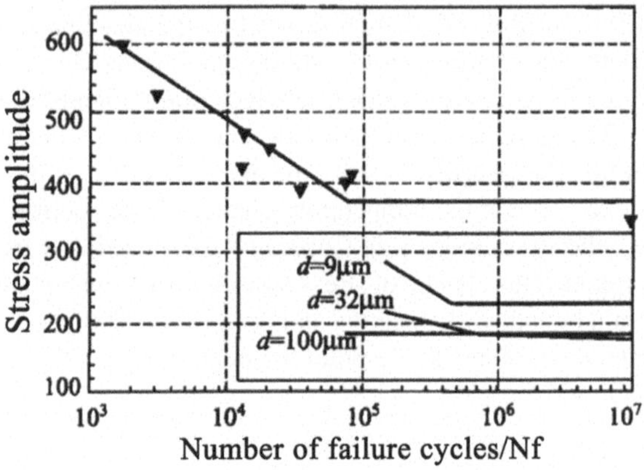

Fig. 3.40 ECAP treated S–N curve (the box inserts the S–N curve of conventional polycrystalline titanium with grain sizes d = 9, 32 and 100 μm, respectively) [42]

region compared with traditional fully annealed materials, but this phenomenon did not appear in low stress cycle region. They pointed out that the easy strain localization of ECAP materials may be a major factor, which limits the tensile and fatigue ductility of materials and determines the poor fatigue performance compared with traditional materials. Short-time annealing at appropriate temperature can greatly change the ductility of Al alloy of ECAP, which is mainly due to the partial recovery of grain boundaries. They think that annealing at 150 °C for 15 min and constant plastic strain amplitude can greatly improve the low cycle fatigue life.

Chung et al. studied the fatigue behavior and mechanical properties of 6061Al alloy after ECAP treatment. They found that compared with the industrial 661A1 alloy treated by T6, the fatigue life was obviously improved, which was about 9 times higher. This phenomenon appears in the low and high cycle areas of single-pass ECAP processing materials. However, after further ECAP deformation, this performance improvement disappeared, especially in the high cycle area. After one pass of ECAP, the typical fine-grained microstructure with low grain boundary dislocation can improve the fatigue life of 6061Al alloy. At the same time, it is pointed out that if the fatigue life of materials is the main index in engineering application, then more attention should be paid to single pass rather than multi-pass treatment technology.

Kim et al. studied the fatigue properties of UFG low carbon steel (carbon mass fraction 0.15%) obtained by ECAP, including cyclic softening and crack growth rate. ECAP steel shows cyclic softening characteristics. After the first cycle, the peak stresses in tension and compression gradually decrease with the increase of the number of cycles. With the increase of R, the steel treated by ECAP shows a higher crack growth rate and a lower crack growth rate. This is mainly due to fewer tortuous crack channels in the material. Kim et al. also studied the effect of grain size (ranging from 1 to 47 μm) on the fatigue behavior of AISI304 stainless steel. The results show that the fatigue properties of stainless steel are improved with the decrease of grain size.

Chape et al. studied the fatigue strength of two kinds of UFG steels (grain size 0.8 and 1 μm respectively). They found that its fatigue strength was improved compared with that of coarse-grained steel. The relationship between fatigue limit and grain size of materials follows the Hall–Petch relationship.

Chape et al. found that the fatigue crack growth threshold of UFG steel with grain size of 1 μm is relatively low, which is consistent with previous conclusions. However, if the same average stress is applied in the middle and high stress intensity range, the ultrafine grained steel has a lower fatigue crack growth rate than SM90 steel or HT80 steel. Lukas et al. studied the fatigue notch sensitivity of UFG copper with 99.9% purity processed by ECAP and compared it with that of traditional copper. It is found that the fatigue notch sensitivity of UFG copper prepared by ECAP technology is higher than that of polycrystalline copper [45]. The tendency of grain coarsening of ECAP copper with purity of 99.9% is small during the cycle. The grain structure in the plastic deformation zone around the crack is quite different from that outside the plastic zone, and the grains are obviously elongated, but their sizes have not changed.

Mughrabi et al. studied the cyclic deformation and fatigue behavior of UFG metal materials prepared by ECAP to determine cyclic softening, micro-shear band and fatigue life. The tensile micro-yielding, unloading and reloading tests of high purity UFG and industrial pure UFGAl were carried out at low cycle (strain amplitude is about 10^{-3}–10^{-2}) and high cycle (strain amplitude is about 10^{-4}). For example, the unloading and reloading stress–strain curves of super grained copper specimens are nonlinear in the ECAP processing state, while the coarse grain material exhibits the unloading/re-truncated stress–strain behavior of linear elasticity. Therefore, it is believed that the nonlinear unloading/reloading behavior is a characteristic of UFG materials, indicating that UFG materials form greater internal reaction force (back stress) than coarse grain materials during the strain process. Low cycling fatigue behavior is worse than coarse-grained materials, while high cycling fatigue behavior is significantly improved than coarse-grain specimens. The optimized annealing process results in a bimodal distribution of grain size, resulting in a high low cycle fatigue life for UFG copper and, in fact, better performance than coarse grain materials. Figure 3.41 shows the fatigue life of UFG copper as represented by the Woehler curve S–N (stress amplitude "$\Delta\sigma/2$" corresponding to "$\lg N_f$"). Ding et al. see UFG as a composite material with soft grain and hard grain boundaries, and believe that this microstructure has a hindering effect on fatigue crack propagation. An important aspect of this model is the introduction of a suitable cyclic stress–strain curve, i.e. the "saturation" stress ("$\Delta\varepsilon_{pl}/2$"). This model works well with experimental data.

Kim et al. studied the fatigue properties of ultra-fine grained AZ31 magnesium alloy processed by ECAP, and found that the fatigue life was reduced by more than 1/10 in low and high cycle fatigue regions compared with the samples without ECAP processing. Different from the earlier research results on fatigue life of nanocrystalline materials, AZ magnesium alloy with fine grain size shows higher crack threshold and lower crack growth rate. It is found that the ductility of AZ31 magnesium alloy treated by ECAP is improved due to grain refinement, which leads to the increase of crack growth resistance, because the material has better ability to coordinate plastic strain during cycling.

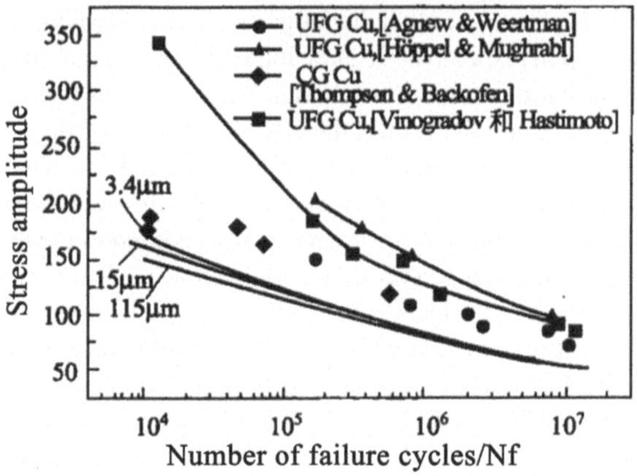

Fig. 3.41 S–N (Woehler) fatigue life curves for UFG copper, ECAP and coarse grains [45]

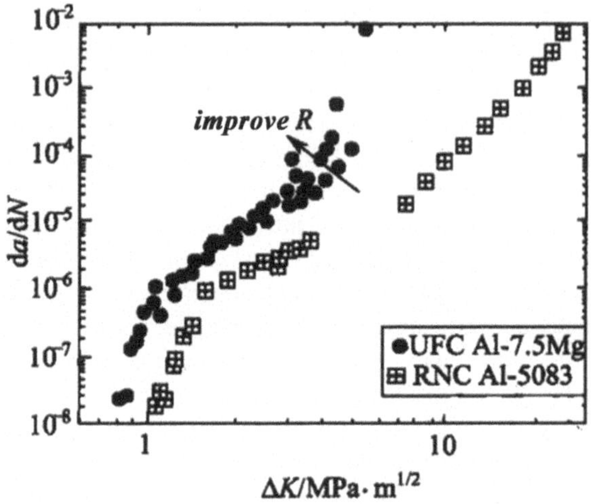

Fig. 3.42 Under the conditions of room temperature and fatigue rate of 10Hz R = 0.1–0.5, the fatigue crack cracking growth rate da/dN of low-temperature ball-milled Al-7.5Mg alloy varies with the stress intensity factor ΔK [45]

Hanlon et al. studied the fatigue response of electrodeposited nanocrystalline Ni and Al–Mg alloy of UFG ball milled at low temperature. They found that in strain-controlled fatigue, grain refinement has a great influence on the whole fatigue life and fatigue crack growth. Figure 3.42 shows the da/dN − ΔK curve of ultra-fine grain Al–Mg alloy in the range of R = 0.1–0.5, and compares it with the crack growth data of 5083 aluminum alloy used in industry under the condition of R = 0.33. Compared with traditional microcrystals, fully dense nanocrystalline nickel and UFG nickel prepared by electrodeposition show higher fatigue resistance. However, the fatigue growth rate shows that the grain refinement within the scope of nanocrystals has an adverse effect on the subcritical fatigue fracture resistance. This conclusion has been confirmed in A1-7.5Mg alloy ball milled at low temperature.

References

1. K. E. Drexler, *Nanotechnology.* **1991**, *2*,113.
2. F. C. Simmel, B. Yurke, H. R. Singh, *Chem. Rev.* **2019**, *119*, 6326–6369.
3. R. D. Mcmichael, R. D. Shull, L. J. Swartzendruber, L. H. Bennett, R. E. Watson, *J. Magn. Magn. Mater.* **1992**, *111*, 29–33.
4. A. D. Ellington, J. W. Szostak, *Nature.* **1990**, *346*, 818–822.
5. D. K. Kim, Y. Zhang, W. Voit, K. V. Rao, M. Muhammed, *J. Magn. Magn. Mater.* **2001**, *71*, 133–137.
6. X. J. Lu, W. K. Olson, *Nat. Protoc.* **2008**, *3*, 1213–1227.
7. A. Irrera, P. Artoni, F. Iacona, E. F. Pecora, G. Franzò, M. Galli, B. Fazio, S. Boninelli, F. Priolo, *Nanotechnology.* **2012**, *23*, 075204.
8. C. F. Monson, A. T. Woolley, *NanoLett.* **2003**, *3*, 359–363.
9. R. M. Dirks, N. A. Pierce, *Proc. Natl. Acad. Sci. USA* **2004**, *101*, 15275–15278.
10. X. L. Zuo, Y. Xiao, K. W. Plaxco, *J. Am. Chem. Soc.* **2009**, *131*, 6944–6945.
11. T. Gong, D. T. Wu, D. W. M. Marr, *Langmuir.* **2003**, *19*, 5967–5970.
12. Y. Fang, B. M. Phillips, K. Askar, B. Choi, P. Jiang, B. Jiang, *J. Mater. Chem. C.* **2013**, *1*, 2113–2136.
13. Y. Huang, J. Zhou, B. Su, L. Shi, L. Jiang, *J. Am. Chem. Soc.* **2012**, *134*, 17053.
14. S. Wu, J. Nan, Y. Wu, *ACS. Appl. Mater. Interfaces* **2022**, *14*, 141–164.
15. G. Fu, X. Zhang, X. Chu, Y. Zhou, Y. Song, *DyesPigm.* **2021**, *195*, 109747.
16. Y. Wang, D. Maspoch, S. Zou, G. C. Schatz, R. E. Smalley, C. A. Mirkin, *Proc. Natl. Acad. Sci. U. S. A.* **2006**, *103*, 2026–2031.
17. J. H. Seol, I. Jo, A. L. Moore, L. Lindsay, Z. H. Aitken, M. T. Pettes, X. Li, Z. Yao, R. Huang, D. Broido, *Science* **2010**, *328*, 213–216.
18. L. M. Jiang, Z. F. Liu, J. Tang, L. Zhang, K. Shi, Z. Q. Tian, P. K. Liu, L. N. Sun, Z. W. Tian, *J. Electroanal. Chem.* **2005**, *581*, 153–158
19. Michael, D, Austin, Haixiong, Ge, Wei, Wu, Mingtao, Li, Zhaoning, *Appl. Phys. Lett.* **2004**, *84*, 5299–5299.
20. J. J. L. Morton, D. R. Mccamey, M. A. Eriksson, S. A. Lyon, *Nature.* **2011**, *479*, 345–353.
21. L. Postler, S. Heuβen, I. Pogorelov, M. Rispler, T. Feldker, M. Meth, C. D. Marciniak, R. Stricker, M. Ringbauer, R. Blatt, *Nature.* **2022**, *605*, 5422–5438
22. Y. Jiang, B. Li, X. Chen, A. D. Ellington, *Molecules.* **2012**, *17*, 13211–13220.
23. J. Fern, R. Schulman, *Nat. Commun.* **2018**, *9*, 3766.
24. F. He, H. Liu, X. Guo, B. C. Yin, B. C. Ye, *Anal. Chem.* **2017**, *89*, 12968.
25. W. Zhou, R. Saran, J. Liu, *Chem. Rev.* **2017**, *117*, 8272.
26. A. A. Fokina, D. A. Stetsenko, J. C. François, *ExpertOpin. Biol. Ther.* **2015**, *15*, 689–711.
27. D. H. Bunka, P. G. Stockley, *Nat. Rev. Microbiol.* **2006**, 4, 588–596.
28. S. D. Mendonsa, M. T. Bowser, *J. Am. Chem. Soc.* **2004**, *126*, 20–21.
29. R. Stoltenburg, C. Reinemann, B. Strehlitz, *Anal. Bioanal. Chem.* **2005**, *383*, 83–91.
30. K. Sefah, D. Shangguan, X. Xiong, M. B. O'Donoghue, W. Tan, *Nat. Protoc.* **2010**, *5*, 1169.
31. L. Guerrini, F. Mckenzie, A. W. Wark, K. Faulds, D. Graham, *Chem. Sci.* **2012**, *3*, 2262–2269.
32. J. Zheng, Y. Hu, J. Bai, C. Ma, J. Li, Y. Li, M. Shi, W. Tan, R. Yang, *Anal. Chem.* **2014**, *86*, 2205–2212.
33. S. M. Douglas, H. Dietz, T. Liedl, B. Hogberg, F. Graf, W. M. Shih, *Nature.* **2009**, *459*, 414–418.
34. E. G. Gwinn, P. O'Neill, A. J. Guerrero, D. Bouwmeester, D. K. Fygenson, *Adv. Mater.* **2008**, *20*, 279–283.
35. M. Zhang, Y. Ou, D. Xin, X. Li, H. Huang, Y. Wen, X. Zhang, *J. Porous Mater.* **2017**, *24*, 13–20.
36. J. E. Park, Y. Jung, M. Kim, J. M. Nam, *ACS Cent. Sci.* **2018**, *4*, 1303–1314.
37. P. Ghosh, G. Han, M. De, C. K. Kim, V. M. Rotello, *Adv. Drug Delivery Rev.* **2008**, *60*, 1307–1315.
38. Z. Chen, C. Liu, F. Cao, J. Ren, X. Qu, *Chem. Soc. Rev.* **2018**, *47*, 4017–4072.

39. J. T. Petty, Z. Jie, N. V. Hud, R. M. Dickson, *J. Am. Chem. Soc.* **2004**, *126*, 5207–5212.
40. S. Bosi, T. D. Ros, G. Spalluto, M. Prato, *Chem Inform.* **2004**, *38*, 913–923
41. S. R. Bhattarai, E. Muthuswamy, A. Wani, M. Brichacek, A. L. Castaneda, S. L. Brock, D. Oupicky, *Pharm. Res.* **2010**, *27*, 2556–2568.
42. M. Uchida, S. Kang, C. Reichhardt, K. Harlen, T. Douglas, Biochim. Biophys. *Acta Gen. Subj.* **2010**, 1800, 834–845.
43. M. Cordey, U. Gundimeda, R. Gopalakrishna, C. J. Pike, *J. Neuro chem.* **2003**, *84*, 1340–1348.
44. J. Zhao, M. Liu, Y. Zhang, H. Li, Y. Lin, S. Yao, *Anal. Chim. Acta.* **2013**, *759*, 53–60.
45. W. H. Chen, W. C. Liao, Y. S. Sohn, M. Fadeev, A. Cecconello, R. Nechushtai, I. Willner, *Adv. Funct. Mater.* **2018**, *28*, 1705137.

Chapter 4
Chemical Properties of Nanomaterials

Physical properties and chemical properties complement each other, which jointly determine the properties, application fields and prospects of nano-materials. After studying the physical properties in the previous chapter, we will learn the chemical properties of nano-materials. Nano-materials are usually regarded as a reconstruction of existing bulk materials, and their chemical properties will change predictably with different material scales. Various synthesis methods can control the strict scale, shape and surface properties of nano-materials, while self-assembly technology can combine these structural units into structures with special functions according to different needs. In this chapter, the chemical characteristics of nano-materials will be discussed in detail.

4.1 Assembly of Nano-materials

The term "self-assembly" itself shows its meaning, which is a collection of some entities or a collection that is not affected by any external influence. The term "self-assembly" has been scientifically recognized in the past 3–4 years, and this kind of assembly has been recognized in biology, chemistry or physics. Centuries ago, some philosophers, such as Canard, Democritus and Descartes, assumed that everything in the world, from small objects around us to solar energy systems, galaxies and the universe, was a tiny and inseparable unit (or atom). Demokert first used the word atoms to express the inseparability of these particles. In the case of chaos, various forms of matter still follow some or some orderly natural laws. Today we know that atoms can actually be further crushed into smaller particles with very high energy, namely electrons, protons and neutrons. Protons and neutrons are also composed of elementary particles such as quarks. However, the concept that matter consists of tiny units such as atoms or molecules is still very useful for understanding most phenomena around us [1]. People may tend to think that "self-assembly" can be used

S. Peng and P. Li, *Nanomaterials and Nanotechnology*,
https://doi.org/10.1007/978-981-97-8433-2_4

for any substance with atoms or molecules stuck together, such as solids. However, this is not the case. As will be discussed in this section, self-assembly is suitable for spontaneously formed, reversible, locally ordered and thermodynamically stable assemblies. Self-assembly is very sensitive and can be turned back to disorder. In the disordered state, some "building blocks" or "motifs" with uniform shapes and sizes spontaneously assemble into ordered structures through some weak interactions. A "building block" itself is a collection of strongly interacting particles (atoms or molecules).

Self-assembly was originally considered to be limited to the biological world in the twentieth century. The origin of peacock feathers, butterfly wings and many beautiful colors of birds or insects was understood as an orderly structure. Microscopic analysis shows that these usually involve micro or nano structures. Self-assembly occurs in many length scales, not only in tiny objects, but also in giant celestial bodies such as stars, planets, galaxies and the whole universe. In the natural biological and abiotic worlds, the observation of self-assembly enables scientists to understand what factors lead to the orderly or random assembly of some smaller entities, blocks, patterns or units.

We know that the main bonds in inorganic solids are ionic bonds, covalent bonds or metal bonds. They have considerable formation (or dissociation) energy, usually from about 0.5 eV to several electron volts. Self-assembly is spontaneously formed by phase interaction (such as $\pi - \pi$ phase force, van der Waals force, colloid force, electricity, light, shear force or capillary force). In nanotechnology, self-assembly plays an important role. The transformation of dense organic molecules and nanoparticles can be used for new devices. Recognizing the importance and origin of self-assembly, scientists intentionally try to make self-assembled organic, inorganic or composite materials to obtain unprecedented electrical, mechanical, magnetic or optical properties. Langmuir–Blodgett (L–B) films, micelles, liquid crystals, composite layers obtained by dip-pen lithography, deposition of materials in the gaps of self-assembled spheres and anodic oxidation of alumina templates with ordered pores can all realize the self-assembly of required materials. This kind of structure can be used to create photonic band gap materials, new sensors, lasers, Bragg mirrors, electroluminescent devices, photovoltaic solar cells and so on. In addition, nano-fabrication technology based on DNA double helix self-assembly structure will also play a potential role in nano-electronics, nano-mechanical equipment and computers [2].

4.1.1 Self-assembly Characteristics

From the broadest point of view, the most typical chemical characteristics of nanomaterials are the preparation of nano-scale structural units with different sizes and shapes, different compositions and surface structures, and different charges and functions by synthetic chemical means. As shown in Fig. 4.1, these structural units or nanomaterials have their own independent applications, which can build assembly structures with fine functions and specific purposes. Such an assembly structure

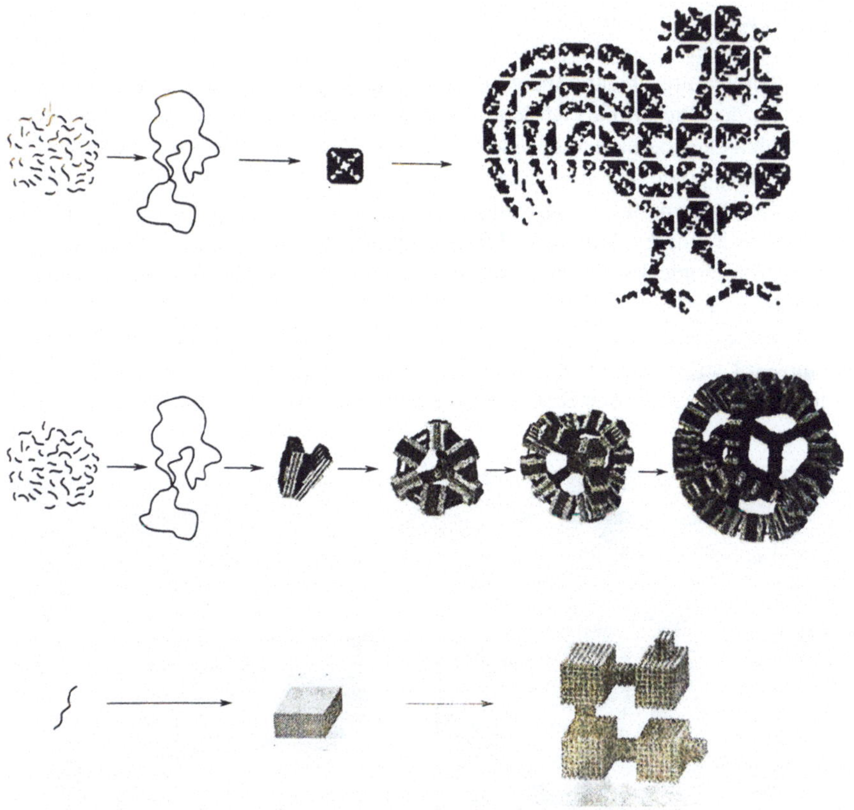

Fig. 4.1 Self-assembly characteristics of nanomaterials [3]

can be prepared by spontaneous or template-guided self-assembly methods, or by chemical synthesis and printing.

There are various driving forces of molecular self-assembly, which can be ionic interaction, covalent interaction, hydrogen bonding, noncovalent interaction or metal–ligand bonding. These driving forces may promote the formation of structures and properties that cannot be observed in a single component. The self-assembly chemistry of nano-materials goes beyond molecular assembly. It is a unique solid material chemistry, and its structural additions and assemblies are not limited by scale range, nor are they limited by chemical bonds. On the "full" scale, the self-assembly of materials mainly involves various binding forces between structural units at different scales. The forces that drive the self-assembly of materials at a scale larger than molecules include capillary action, van der Waals force, elastic force, electrostatic force, acting force and shear force. In a self-assembly system with nanocrystals, nanorods or nanosheets as building blocks, the size and shape of each structural unit and various forces between them can make it spontaneously form a specific assembly configuration. This assembly system will evolve towards a state of lower free energy and higher structural stability [4].

Another feature of nano-materials self-assembly is that they form multi-level structures, and the primary structural units form more complex secondary structures, which can continue to be assembled into more advanced assembly structures. The assembly process continues until a higher-level structure is formed. As shown in Fig. 4.2, these multilevel structures may show unique properties that cannot be observed in a single component. Multi-level structure is a characteristic of many assembled biological structures, and it is also a sign of multi-scale self-assembled materials.

Self-assembly of nanomaterials can also be guided by templates. In the guided assembly of structural units, in addition to the structural units themselves, some molecules and organic substances are often used as structural guiding components,

Fig. 4.2 Multistage structure of nanomaterials [5]

so guided self-assembly is different from spontaneous self-assembly. In the process of self-assembly of structural units, templates can be used to fill space, balance charges and guide the formation of specific structures. For example, in the synthesis of microporous aluminosilicate, a single organic molecule is used as the template of micropores, while in the synthesis of mesoporous silica, the lyotropic liquid crystal formed by block copolymer micelles or surfactants is used as the template. From this definition, template-guided self-assembly has the same meaning as co-assembly, but it is different from spontaneous self-assembly. Template-guided self-assembly can be carried out on a flat or curved substrate with patterns, and the patterns of the substrate are obtained by soft printing or other means. The unit and the substrate can be assembled to form a preset configuration through designed hydrophilic-hydrophobic interaction, electrostatic interaction, hydrogen bonding interaction, metal–ligand interaction or acid–base interaction. For example, nonpolar polymers are assembled into microlens arrays with hydrophobic patterned surfaces. Relief patterns prepared by soft printing on the surface of the substrate can be used to guide the assembly of structural units, such as preparing nanorod arrays in nano-holes on the surface of the substrate. The assembly process can also use nano-scale patterned porous templates, which can be nano-porous films of polymers, silicon and alumina.

4.1.2 Self-assembly Mechanism

To understand the mechanism of self-assembly, the most important thing is to realize that self-assembly involves one, two or three dimensions, from weak to strong interaction between nano-materials and nano-technology, and structural evolution from micro to macro. As mentioned in the previous section, self-assembly may be the result of very weak forces, such as van der Waals interaction, ammonia bond, electrostatic charge, magnetic interaction and so on. The driving force behind self-assembly has been proved to be the principle of minimum energy. The ability of the system to enter an ordered low-energy state depends on the availability of units with the same size and shape. Molecules with definite shape, atomic number and size that are already in low energy state are good candidates for self-assembly. It is worth noting that the atoms in these molecules or the self-assembled building blocks themselves are bound together by stronger forces.

When "building blocks" (one, two or more) can be self-assembled, they can spontaneously reach the state of minimum energy without any external force. However, self-assembly may also occur in the presence of external driving forces (such as temperature, pressure magnetic field, etc.). These two types are called static self-assembly and dynamic self-assembly respectively (Fig. 4.3). Only the system can be minimized.

Static self-assembly can only be realized when the energy state is maintained. This state can only be destroyed if the system is subjected to a strong external force. On the other hand, dynamic self-assembly involves the continuous influence of external forces from the environment. If the absorption of energy from the environment stops,

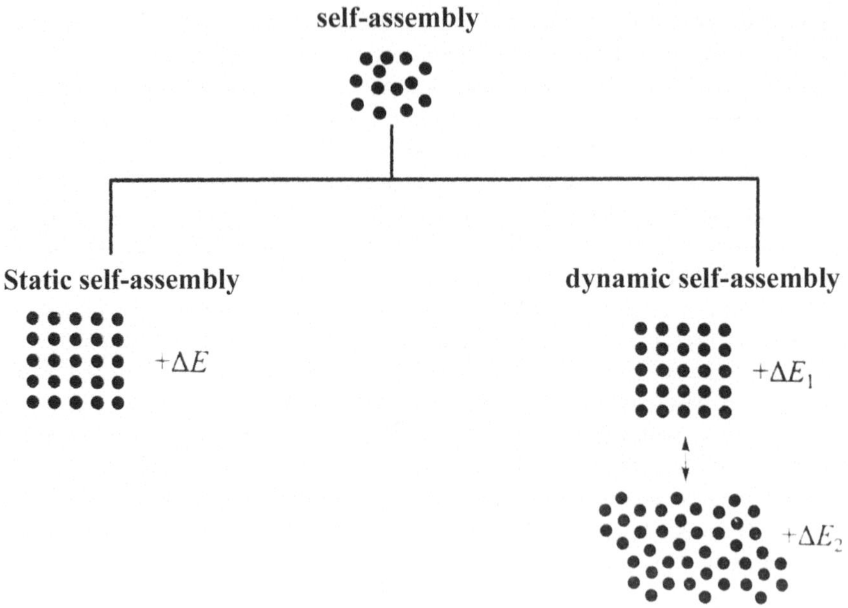

Fig. 4.3 Two types of self-assembly schematic diagram [6]

self-assembly can leave the organized structural state and disintegrate. The formation of an ordered crystal structure from a melt can be considered as an example of static self-assembly. Thin films formed by L–B technology or dip-pen nanolithography technology are also examples of static self-assembly. Living animals or plants are examples of dynamic self-assembly. Once the food supply, proper temperature and air pressure are interrupted, animals and plants will decompose. Static and dynamic self-assembly can be further divided into hierarchical self-assembly, directional self-assembly and collaborative self-assembly, as shown in Fig. 4.4.

Hierarchical self-assembly is characterized by the interaction of one type of building block in a small, medium or even long range. Directional self-assembly occurs when building blocks occupy pre-designed positions (such as some parts of the photoetching pattern substrate and holes in the memory). Collaborative self-assembly, as its name implies, can be formed by two or more types of blocks, which can cooperate with each other. In most cases, nanoparticles can be placed in a certain position of polymer chains, organic molecules or templates through specific bonding. When a well-structured template (such as DNA or S layer) is used, direct patterning of nanoparticles becomes feasible. These are inorganic substances (nanoparticles), organic molecules and inorganic-biological molecular assemblies. Of course, it can also be an aggregate of pure inorganic particles. The driving forces of such components are usually very different [7]. Below we will discuss various examples of self-assembly.

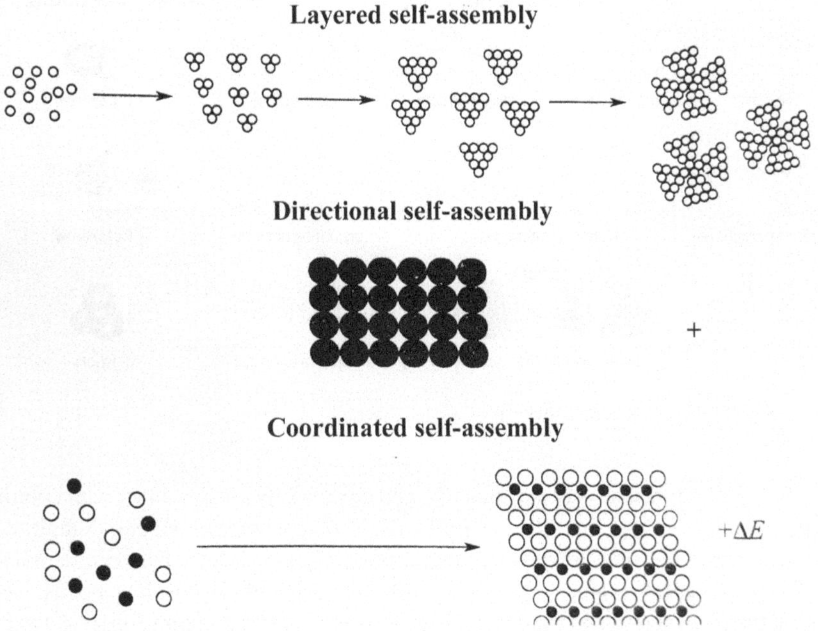

Fig. 4.4 Schematic diagram of hierarchical self-assembly, directional self-assembly and cooperative self-assembly [6]

4.1.3 Self-assembly Example

A very important branch of chemistry is supramolecular chemistry, which is only a form of self-assembly. The word supramolecular chemistry was created by Jean-marie Lane, a Nobel Prize winner, which means chemistry beyond molecules. It is essentially a collection of one or several types of molecules, and aggregates or larger crystals are made by noncovalent interactions. Molecular recognition (like locks and keys) helps to build larger components, such as two intertwined DNA chains. This three-dimensional ordered arrangement of molecules can promote the establishment of "superlattices" or large single crystals of self-assembled molecules. Concepts such as self-assembly and supramolecular structure are closely related to supramolecular chemistry. Supramolecular chemistry, as an important branch of modern chemistry, has developed rapidly in recent decades. At the same time, supramolecular chemistry and nano-materials research are combined, promoted and developed together. Supramolecular chemistry promotes soft chemical preparation of nano-materials and assembly of nano-robots. Moreover, from an aesthetic point of view, some nano-supramolecular structures also give people a pleasant and unfathomable feeling.

In the mid-1960s, research on macrocyclic complexes began in organic chemistry, coordination chemistry and other fields, which is also considered as the symbol of the birth of supramolecular chemistry (Fig. 4.5). In the following years, supramolecular

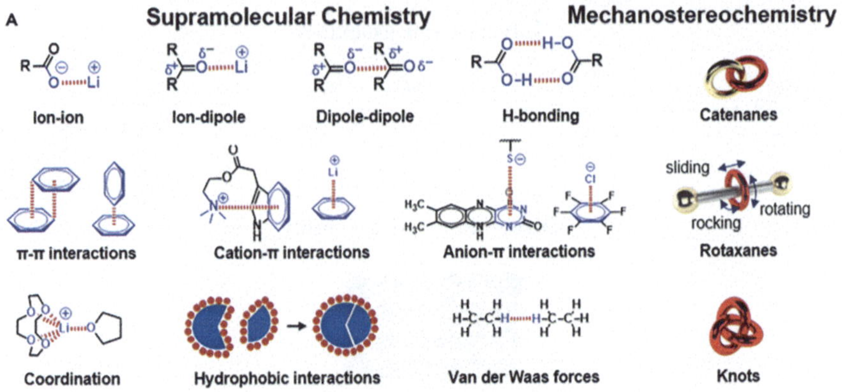

Fig. 4.5 Supramolecular chemistry [9]

chemistry developed slowly and steadily, and always kept close contact with coordination chemistry, but contradictory problems also appeared. In general complexes, the interaction between ligands and metal ions is close, usually between 0.18 and 0.25 nm, which is roughly equivalent to the bond length of covalent bonds. The interaction between ligands and metal ions is strong, and the binding energy between ligands and atoms such as O, N, S and Cl in metal ligands is usually between 139 and 517 kJ/mol, which is within the range of chemical bond energy [8]. Therefore, in this sense, the connotation of supramolecular chemistry should include both supramolecular and intrinsic molecules. For example, the interaction between ligands and metal ions in the above complexes should belong to the same intramolecular interaction.

The concept of self-assembly in the field of science is abstract, and the self-assembly in science should also consider the integration of the components to be assembled, that is, the geometry and interaction of the units to be assembled should also be considered. Another feature of self-assembly in science can be reflected in Fig. 4.6. Its assembly process involves many aspects, such as the construction of toy building blocks and building construction. Therefore, self-assembly in the scientific field can be defined as a process in which basic units (nanostructures such as molecules and atomic clusters) on the molecular or nano scale are assembled and combined with each other by using selective or oriented forces (mainly supramolecular forces), their own geometric shapes and other factors, and form ordered structural aggregates. Figure 4.7 introduces the self-assembly process of inorganic nanoparticles, which is also related to the concept of superlattice, that is, the final assembly product belongs to superlattice structure. The self-assembly process is mainly divided into three stages: Step one, inorganic nanoparticles are prepared by related chemical methods, and nanoparticles with good dispersibility can be obtained because stabilizers are added in the preparation process; In the second step, the superlattice structure is formed by the self-assembly process, which requires that the size of nanoparticles should be very uniform, so it is necessary to screen the size of nanoparticles obtained in the first step, which is a key to successful self-assembly. Binary mixed solvents can be used for screening, which can be mixed by polar solvents (such as alcohols) and nonpolar solvents (such as hydrocarbons), and nanoparticles can produce selective precipitation in mixed solvents; In the third step, self-assembly is realized by evaporating the solvent, and finally an ideal superlattice structure is generated.

Iler et al. discovered in 1966 that the surface modified by some cationic surfactants can be combined with a layer of negatively charged silica colloid or polymer latex particles. This negatively charged granular layer can continue to bind cationic colloids, such as boehmite or collagen. In this way, the multilayer film structure can be obtained by repeatedly immersing in the suspension with opposite charges and adsorbing one electrostatic layer at a time. Decher extended this characteristic to the layer-by-layer growth of soluble polyelectrolyte in 1991. This layer-by-layer assembly process is suitable for all systems with different charges. Firstly, a surface can be charged with aminoalkyl chlorosilanes or siloxanes, which can be anchored to hydroxyl groups on the surface of silicon or silicon dioxide, or directly chemically adsorbed onto the surface of gold with aminoalkyl thiols to form a tightly arranged self-assembled monolayer [11].

Under neutral pH condition, the amino functional groups exposed on the surface will be protonated, so that the surface of the substrate will be positively charged in aqueous solution. This surface can be closely combined with water-soluble anionic polyelectrolyte through electrostatic interaction. For molecules with only a small amount of charge, only weak bonds can be formed. The polyelectrolyte layer formed by electrostatic adsorption will overcompensate the alkylammonium cations in the first layer and form the remaining surface negative charges. Although polyelectrolyte is a kind of polymer compound, some of its repeating units are charged enough to compensate all the charges of the previous layer, but some charged repeating units are still exposed, so the deposition process can be repeated in polyelectrolyte with anion and cation. Other factors can also affect the deposition properties of polyelectrolyte, such as pH value, salt concentration or charge density of polyelectrolyte.

In 2021, researchers controllably synthesized two-dimensional polymer materials through self-assembly and supramolecular strategies, and the results are shown in Fig. 4.8. In this study, a two-dimensional nano-flaky polyaniline material was successfully constructed by using 1,4-benzenediboronic acid and aniline (Fig. 4.8b and c). The boric acid group in 1,4-benzenediboronic acid can be added into the polyaniline chain as a dopant, while the hydroxyl group of the boric acid group in 1,4-benzenediboronic acid is crosslinked with the amino group in the polyaniline molecular chain to form hydrogen bonds. Therefore, 1,4-benzenediboronic acid acts as both a crosslinking agent and a dopant in the emulsion polymerization process, and both electrostatic interaction and hydrogen bonding can connect 1,4-benzenediboronic acid molecules with polyaniline chains. This strong interaction can cross-link one-dimensional polyaniline molecular chains into two-dimensional nano-sheets through supramolecular interaction (Fig. 4.8d). In order to clarify the exact self-assembly mechanism, the researchers discussed the growth process at different reaction times, and obtained two-dimensional nanosheets based on forward microemulsion polymerization and supramolecular synergy. When aniline monomer

Fig. 4.6 Self-assembly in the field of science [10]

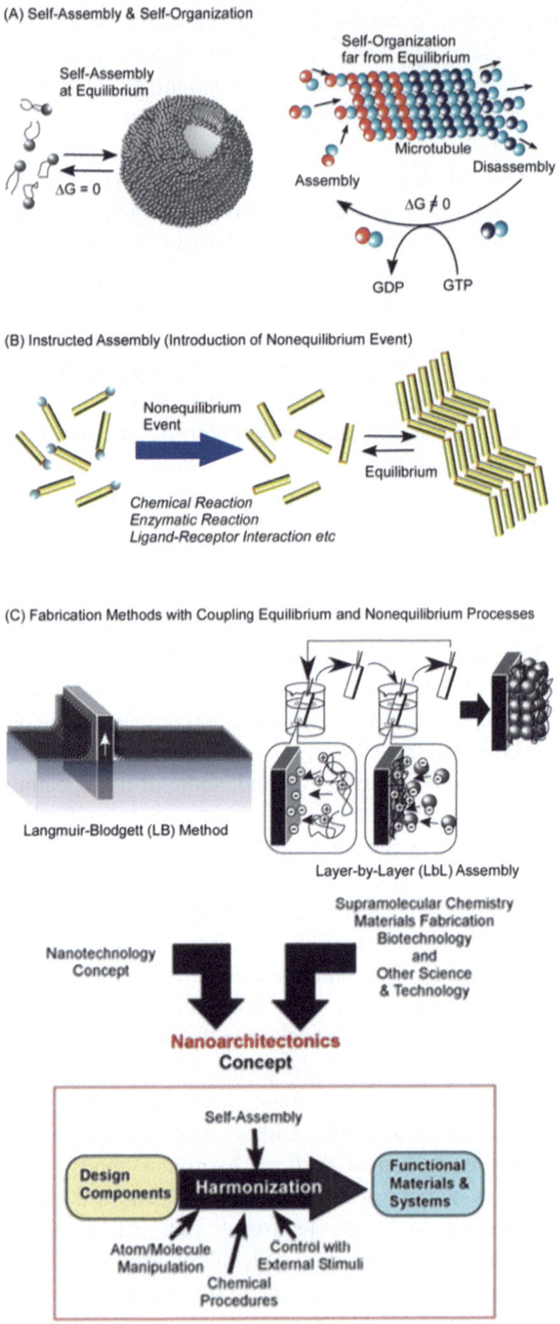

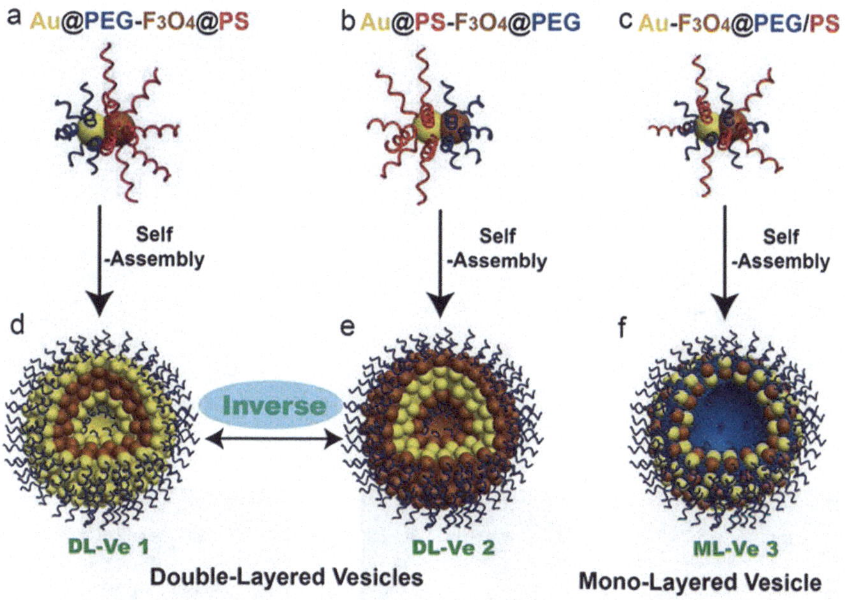

Fig. 4.7 Self-assembly of inorganic nanoparticles [12]

was added, 1,4-benzenediboronic acid quickly adsorbed on aniline. With the increase of assembly time, the surface of aniline was covered with 1,4-benzenediboronic acid, and spherical micelles with 1,4-benzenediboronic acid outside and aniline inside were formed in the solution. The dynamic balance of hydrophilic/hydrophobic effect of 1,4-benzenediboronic acid leads to the gradual decrease of surface tension. Once the interaction between 1,4-benzenediboronic acid and polyaniline is dominant, the above surface tension will be effectively suppressed. This effect destroys the water–oil interface and leads to demulsification of micelles. Due to the continuous consumption of aniline monomer and oxidant, a complete consumption zone is formed in the center of nucleation zone, but the free aniline monomer and initiator can not be replenished in time, resulting in its gradual growth into nanofiber along one-dimensional direction. When the reaction time is prolonged, the material will undergo further crosslinking reaction, and the interaction between hydrogen bonds and $\pi - \pi$ electrons provides the driving force for self-assembly, and finally it will be assembled into two-dimensional nano-sheets. In addition, by controlling the reaction conditions, polyaniline two-dimensional materials with different lamellar thicknesses can be obtained, as shown in the atomic force microscope results of Fig. 4.8e.

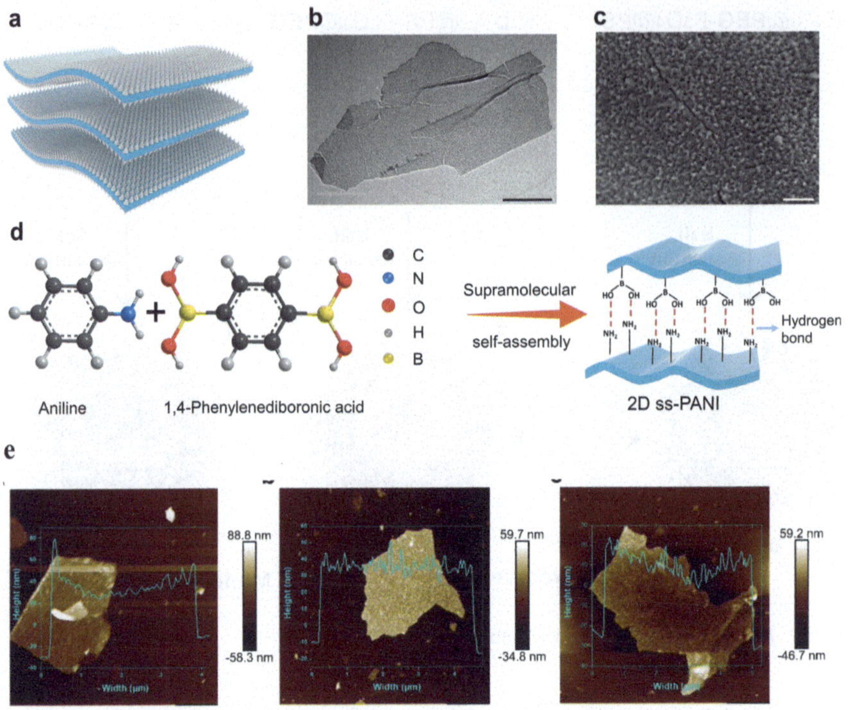

Fig. 4.8 Polyaniline two-dimensional nanosheets synthesized by self-assembly and supramolecular strategy, **a** the structure of polyaniline two-dimensional nanosheets; **b** Transmission electron microscope photos of polyaniline two-dimensional nanosheets; **c** High-power transmission electron microscope photos of polyaniline two-dimensional nanosheets; **d** Supramolecular self-assembly strategy of polyaniline two-dimensional nanosheets; **e** Atomic force microscope photograph of polyaniline two-dimensional nanosheets [13]

4.2 Adsorption Characteristics of Nanomaterials

Adsorption is a combination phenomenon between different phases during the contact of two or more substances. According to the magnitude of adsorption force, adsorption can be divided into physical adsorption and chemical adsorption. In terms of mechanism, when the adsorption between adsorbent and adsorption phase is caused by van der Waals force (that is, weak intermolecular force), it is called physical adsorption. When the adsorption process is accompanied by charge movement interaction or chemical bond force, it is called chemical adsorption. Obviously, the force of chemical adsorption is far greater than the van der Waals force of physical adsorption. Nano-materials have larger specific surface area than bulk materials, which leads to insufficient coordination of atoms on the surface of nano-materials and shows stronger adsorption than bulk materials. The surface structure of nanomaterials is incomplete, and only by adsorbing other substances can the materials be stabilized. For example, it is confirmed by mass spectrometry that all kinds of transition metal nanoparticles have good hydrogen storage capacity. For example, nano-metal powder can store hydrogen at low pressure, which greatly reduces the risk of hydrogen explosion. The adsorption properties of nano-materials are related to the properties of adsorbed substances, solvents and solutions.

4.2.1 Non-electrolyte Adsorption

Molecules whose charges are electrically neutral are collectively called non-electrolytes, and the driving force for such molecules to adsorb on the surface of particles mainly comes from a series of weak interaction forces. However, hydrogen bonding is the main role in the adsorption process of non-electrolyte. For example, in the process of metal oxide nanoparticles (such as SiO_2) adsorbing some organic solvent molecules, only silanol layer of SiO_2 nanoparticles interacts with the organic solvent, and hydroxyl groups in silanol layer will form hydrogen bonds with oxygen atoms or nitrogen atoms in the organic solvent, resulting in the adsorption behavior of metal oxide nanoparticles with these organic solvents. However, only one hydrogen bond can be formed between hydroxyl groups on the surface and alcohol molecules, so the binding force is relatively weak, which belongs to physical adsorption and easy desorption. For high molecular polymers, such as polyvinyl alcohol (PVA), the adsorption on the surface of SiO_2 is through hydrogen bonds formed between more oxygen atoms and hydrogen atoms, so this chemical adsorption has stronger adsorption force and is more difficult to desorb. Non-electrolyte adsorption process is not only affected by the surface properties of nanoparticles, but also by the properties of adsorption phase. If different solvents are selected for adsorption, even the same adsorption phase, the final adsorption amount may vary widely. For example, linear fatty acids are adsorbed in benzene and n-hexane, respectively. Because the number of hydrogen bonds formed between fatty acids and n-hexane is much larger than that formed with benzene, the adsorption capacity in n-hexane is higher.

4.2.2 Electrolyte Adsorption

Electrolyte usually dissociates in the solution in the form of ions, and these ions show different adsorption capacities according to different coulomb force, while the adsorption of nanoparticles in electrolyte solution mostly belongs to physical adsorption. Generally speaking, nanoparticles often produce some unsaturated bonds because of their larger specific surface area. It is these unsaturated bonds that make the surface of nanoparticles spontaneously attractive to ions with opposite charges in electrolyte solution. This adsorption based on Coulomb interaction can be explained by Fig. 4.9.

For example, clay nanoparticles are negatively charged, so it is easy to adsorb positively charged Ca^{2+} on the particle surface in alkali metal electrolyte. In this

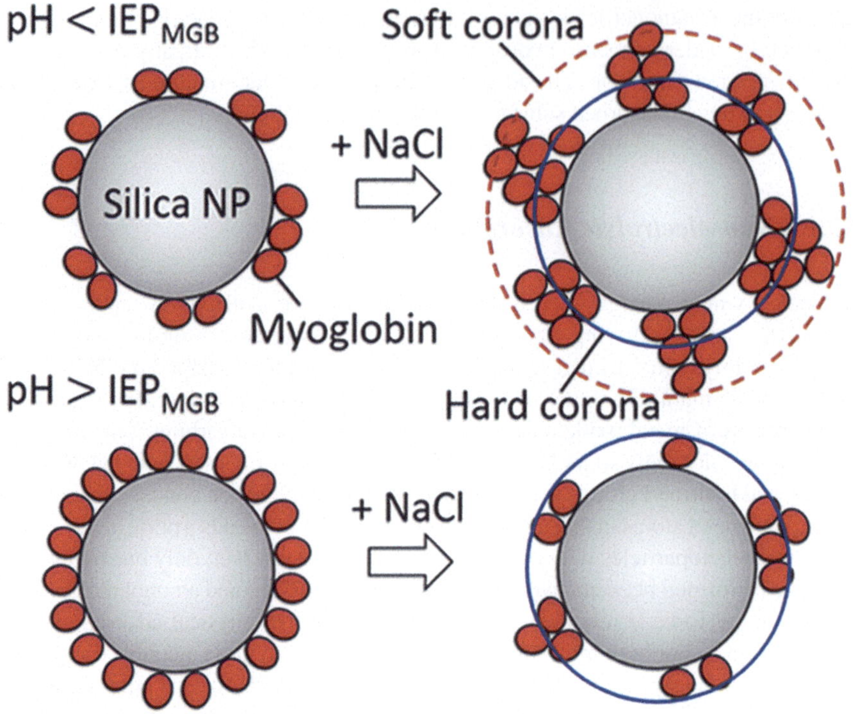

Fig. 4.9 Schematic diagram of electrolyte adsorption caused by Coulomb interaction of nanomaterials [14]

adsorption process, the adsorbed electrolyte ions are called heteroelectric ions. At the same time, the physical adsorption process also has a certain level. The electrical properties of different adsorption layers are also obviously different. Generally speaking, the layer near the surface of nanoparticles belongs to strong physical adsorption, which is called compact layer, and its function is to balance the electrical properties of nanoparticles. Electrolyte ions slightly away from the surface of nanoparticles form a weak adsorption layer, which is called dispersion layer. The dense layer and the dispersed layer will form an electric double layer adsorption, which is embodied in that once the potential in the strong adsorption layer drops sharply, the potential in the weak adsorption layer will slowly decrease, and as a result, a potential drop gradient will be generated in the whole adsorption layer.

4.3 Catalytic Characteristics of Nanomaterials

Catalysts play an indelible role in chemical manufacturing, industrial synthesis and other fields. The catalyst can not only control the reaction time, but also significantly improve the reaction rate and efficiency. However, the preparation of traditional catalyst materials is often based on experience, which leads to huge waste of raw materials, makes it difficult to improve economic benefits, and also causes serious pollution to the environment. Nanomaterials have the advantages of higher specific surface area and more surface-active centers, which provide the necessary conditions for it to replace traditional catalysts. Generally speaking, compared with traditional catalyst materials, nano-catalyst materials can improve the catalytic efficiency by 10–15 times.

4.3.1 Nano-metal Catalysis

The research and application of nano-metal catalysts began before the appearance of nano-materials science. Metals are the active components of catalysts, such as platinum black component of platinum black electrode and porous alumina supported palladium catalyst for olefin hydrogenation in industry. In this section, metal heterogeneous catalysts are briefly introduced as an example. The performance of metal catalysts is mainly determined by the following structural characteristics.

1. The metal atoms on the catalyst surface have periodic end points and one or more unsaturated coordination sites (dangling bonds), which can effectively promote the chemical reaction of reactant molecules.
2. The atomic position on the metal surface is basically fixed, and it is in metastable state in energy. This shows that metal catalysts have strong ability to activate reactant molecules, but poor selectivity.
3. The chemical bonds between metal atoms are non-localized, and there is cohesion between atoms on the metal surface. This requires that metal catalysts have very strict reaction conditions, and are often structure-sensitive catalysts.
4. When metal atoms show catalytic activity, they are always in a considerable group, that is, in the form of "phase". Such as metal single crystal catalyst, the catalytic activity of different crystal faces is obviously different. But its adaptability is also easy to predict.

4.3.1.1 Nanometer Precious Metal Catalyst

Precious metal catalyst is the most excellent metal catalyst. At present, researchers have successfully introduced nano-noble metal catalysts in selective hydrogenation. For example, Rh nanoparticles are good catalysts for selective hydrogenation of olefins, and large groups are often attached to the double bonds of olefins, which

Table 4.1 Nano noble metal catalyst and its application

Metal element	Application
Ag	Selective hydrogenation to prepare mono-olefin; Selective oxidation of ethylene to prepare ethylene oxide; Alkylation of aromatic hydrocarbons; Selective oxidation of methanol to formaldehyde
Au	Low temperature oxidation of carbon dioxide; Combustion of hydrocarbons; Selective oxidation of hydrocarbons
Pd	Catalytic oxidation of hydrocarbons; Selective hydrogenation of unsaturated nitro compounds; Methanol synthesis; Dehydrogenation of cycloalkane and cycloolefin; Selective hydrogenation of olefins, aromatics, aldehydes and ketones
Pt	Selective hydrogenation of olefins, dienes and alkynes; Catalytic purification treatment of automobile exhaust; Dehydrogenation of cycloalkane and cycloolefin; Catalytic oxidation of silicon dioxide; Deep oxidation and combustion of hydrocarbons; Decarboxylation of aldehydes and ketones
Rh	Selective hydrogenation of olefins; Catalytic purification treatment of automobile exhaust; Hydroformylation reaction
Ru	Selective oxidation of ethylene-to-ethylene oxide; Selective hydrogenation of organic carboxylic acids; Hydrocarbon catalytic reforming reaction; Alcohol production

makes it difficult to open the double bonds. If nano-sized Rh particles are used, the opening of double bonds will be easier and the hydrogenation activity will be significantly improved. At present, the main uses of nano noble metal catalysts are shown in Table 4.1.

4.3.1.2 Nanometer Transition Metal Catalyst

Catalysts based on transition metal elements play an important role in modern chemical industry. For example, Fe-based catalysts are used in synthetic ammonia, Ni-based catalysts are used in light hydrocarbon gasification, Co-based catalysts are used in hydrofining fuel oil products, and Cu-based catalysts are used to prepare methanol. The rapid development of nano-material preparation technology promotes the iterative development of traditional transition metal catalysts. However, there are still some technical problems to be solved, such as poor structural regularity, insufficient performance index and high preparation cost.

In the aspect of nano transition metal catalyst, how to solve the stability of its particles is very important. Jairton DuPont used imidazole-based ionic liquid to stabilize the Ir nanoparticle catalyst for selective hydrogenation of olefins. It was found that the stabilized Ir nanoparticle catalyst showed excellent olefin hydrogenation performance. In addition to single-component nano-transition metal catalysts such as Ir, multi-component nano-particle catalysts have also attracted a lot of attention. The nanoparticles of these multicomponent catalysts are mostly amorphous, so they have extremely high specific surface area, and the atoms on the surface of the nanoparticles also have high degree of unsaturated coordination. At the same time, for nano-sized catalyst particles, the anisotropy of multiphase surface structure will transition to isotropy. It is precisely because of these structural characteristics that promote catalytic reaction that multicomponent transition metal catalysts have also attracted extensive attention from researchers.

Ni–B and Ni–P ultrafine alloys are very representative systems. Okamoto found that the 3d electron density of nickel component will change under the induction of another component B or P, which is the key to its excellent hydrogenation performance. However, Chen thinks that after adding metalloid components B and P into Ni metal to form an alloy, there will be different electron transfer between Ni component and this kind of metal component, which makes their catalytic hydrogenation performance show great differences. At this time, if the B/P ratio can be well controlled, the highest catalytic hydrogenation efficiency can be achieved by synthesizing multicomponent nano-transition metal catalysts, which shows that nano-sized particles with narrow particle size distribution and high surface area will be more efficient alloy catalysts.

4.3.2 Photocatalysis of Semiconductor Nanoparticles

Since the photocatalytic effect of semiconductor nanoparticles was discovered, it has been widely used in environmental protection fields such as water quality treatment and waste degradation. Photocatalytic effect means that once the material is excited by light, the electrons in the valence band will jump to the conduction band, and the holes left by the jumping electrons in the valence band can excite oxygen molecules and water molecules around the material into highly active free radicals ·OH and O^{2-}. These free radicals are highly oxidizing, which can almost oxidize and decompose most organic and inorganic harmful substances, and complete the efficient degradation of garbage waste. For example, TiO_2 photocatalytic reaction (Figure 4.10). Oxides such as TiO_2, ZnO, Nb_2O_5, WO_3, SnO_3, ZrO_2 and sulfides such as CdS and ZnS are typical photocatalysts, among which TiO_2 is the most widely used because of its outstanding redox property, high chemical stability and non-toxicity to human body and environment. In the actual use process, it is only necessary to process TiO_2 into hollow pellets and put them on the water surface containing organic wastes, and the organic wastes can be degraded by sunlight without other steps. In some developed countries, the pollution caused by offshore oil spill accidents is usually treated in this way. If nano-TiO_2 powder is added to the glaze during the ceramic sintering process, it can also be sintered to obtain appliances with cleaning and sterilization functions.

Nano-TiO_2 also has a good ultraviolet shielding function. If a certain amount of nano-TiO_2 (0.5–1% by mass) is added to cosmetics, the ultraviolet shielding ability of cosmetics can be effectively improved. At present, some cosmetics with TiO_2 nanoparticles have been put on the market.

Ultraviolet rays can not only oxidize meat food to change its color, but also destroy vitamins and aromatic compounds in food from the inside and reduce its

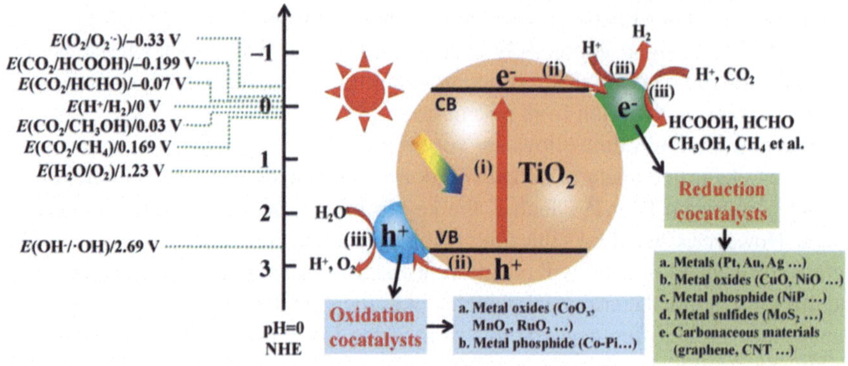

Fig. 4.10 Schematic diagram of photocatalytic characteristics of nanomaterials [15]

nutritional value. When adding 0.1–0.5% mass fraction of TiO_2 nanoparticles into food packaging materials, the protection ability of packaging bags against ultraviolet rays is significantly improved, which can effectively improve the shelf life of food. At the same time, if the surface of TiO_2 nanoparticles is further modified with copper ions and silver ions, the antibacterial ability of the materials can be further improved. Such modified materials have been widely used in medical devices and household appliances.

4.3.3 Thermal Catalysis of Nano-metal and Semiconductor Particles

Metal nanoparticles have high chemical properties, so they usually act as combustion improvers in fuels, as shown in Fig. 4.11. In order to further improve the combustion-supporting efficiency, nano-metal and semiconductor particles can also be co-doped into the fuel to improve the combustion efficiency. This kind of enhanced combustion-supporting agent has been applied to rocket boosters and coal combustion. Besides being used in fuel, nano-metal particles can also be used as dopants and detonators to improve the explosion efficiency of explosives.

4.4 Crystal Characteristics of Nanomaterials

Crystallography plays an important role in the study of nanomaterials, because many nanomaterials have certain crystallization properties. The crystal chemistry of nanomaterials is based on general crystallography. Crystal defects are an important problem in the structure of nano-materials. The defects of nano-materials mainly include vacancies, adsorbed atoms, dentate structures and dislocations. This section

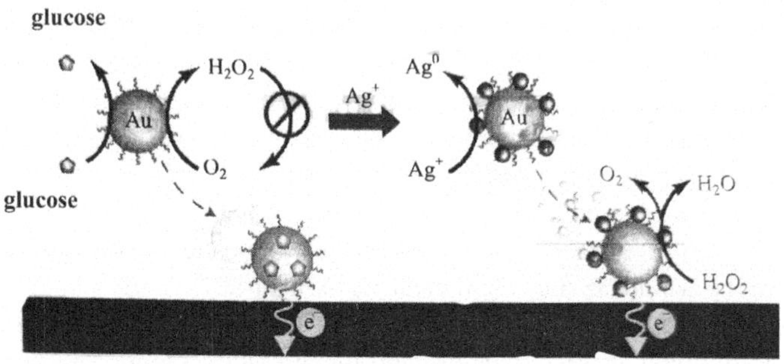

Fig. 4.11 Thermal catalytic characteristics of nanoparticles [16]

will take ZnO hexagonal crystal as an example to introduce the crystal chemistry of nano-materials.

Among nano-ZnO structures, the most common crystal structure is hexagonal wurtzite structure. As shown in Fig. 4.12, the unit cell of ZnO is equivalent to hexagonal crystal form. The unit cell parameters of hexagonal crystal are a = 0.325 nm and c = 0.52121 nm. In the crystal structure diagram, the A-axis and the B-axis are located on the regular hexagonal plane, and the C-axis is perpendicular to this plane. The ZnO hexagonal cell belongs to the P63mc space group. Besides ZnO, BeO, ZnS, AlN and other crystals also have similar structures.

According to crystallography, some crystals (single crystals) can have polar crystal faces. Taking ZnO as an example, if the upper surface of hexagonal ZnO cell is (001) plane, the corresponding lower surface of the cell must be $(00\bar{1})$ plane. When $(00\bar{1})$ surface is used as a surface, the crystal face is actually composed of O atoms, but the coordination between O atoms and Zn atoms is different from that inside the crystal.

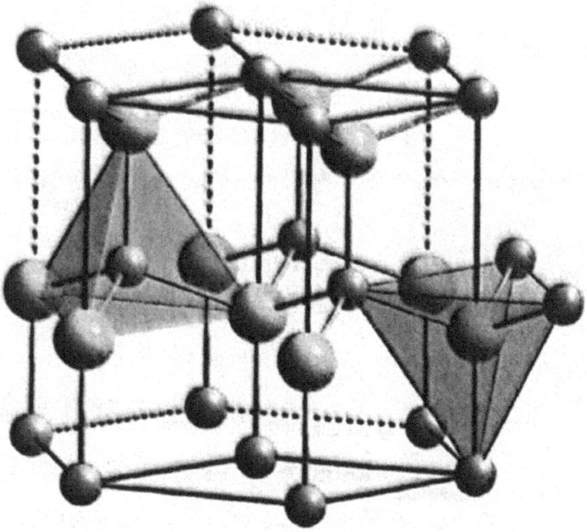

Fig. 4.12 Hexagonal crystal structure of ZnO [17]

Inside the crystal, the tetrahedron structure is formed by the coordination of four Zn atoms and one O atom. On $(00\bar{1})$ surface, only three Zn atoms coordinate with one O atom, forming a coordination vacancy, which leads to the surplus negative charge of the O atom. On the other hand, when the (001) plane composed of Zn atoms is used as the surface, the coordination vacancy of Zn atoms will lead to the surplus positive charge of Zn atoms. It can be seen that (001) and $(00\bar{1})$ two crystal planes are polar when they are used as surfaces, and the charges are opposite, but the absolute values are equal, so the whole crystal is still electrically neutral.

In most cases, many nano-metal oxides are easy to form surface hydroxyl structure, which is an important characteristic of nano-metal oxides. Figure 4.13 analyzes the relevant mechanism and causes.

Theoretically, when ZnO crystal $(00\bar{1})$ is used as the surface, the bonding ability between O atom and H atom on the surface is strong, because the O atom on the surface is rich in negative charge and tends to bond with H atom with positive charge (such as H^+). At the same time, the O atom is in the convex position on the surface, so the steric hindrance during the reaction combination is small and the reaction resistance is reduced. In contrast, when the $(\bar{1}00)$ surface is used as the surface, the reaction of this ZnO crystal without surface defects is uncharacteristic.

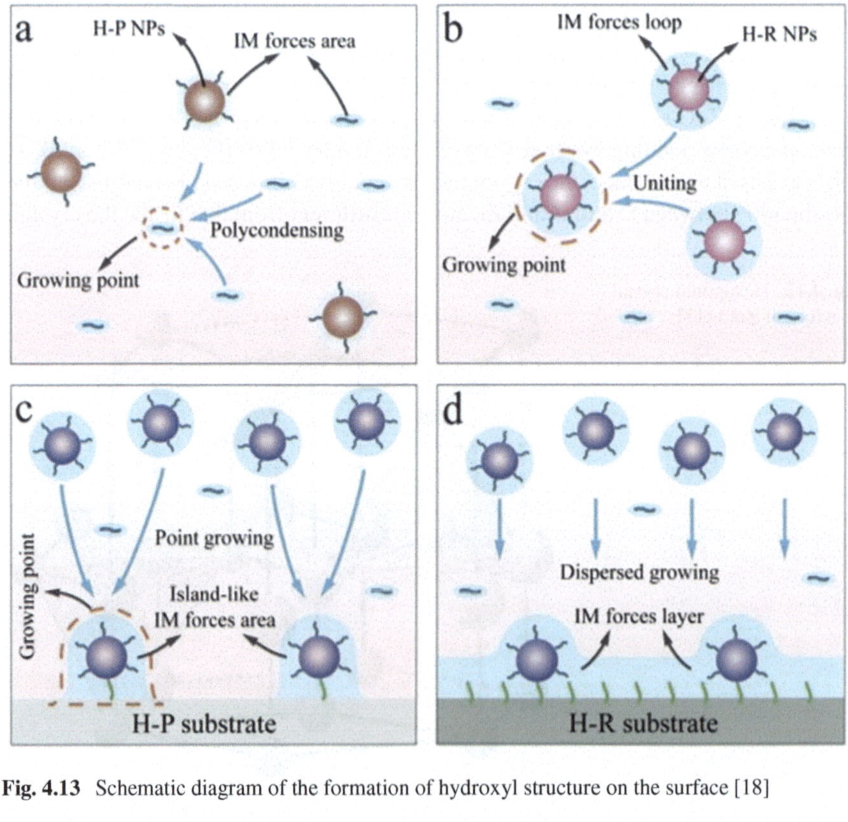

Fig. 4.13 Schematic diagram of the formation of hydroxyl structure on the surface [18]

Fig. 4.14 ZnO crystals with different morphologies [19]

Researchers have confirmed that on different surfaces of the same crystal, different activity differences can be shown for a specific chemical reaction.

The crystallography of nanomaterials is not only applied in chemical reactivity, but also plays an important role in guiding the growth of nanocrystals. Figure 4.14 shows two nano-ZnO patterns with different morphologies. As can be seen from the figure, the growth orientations of the two ZnO nanocrystals are significantly different [19].

4.5 Surface Defect Characteristics of Nanometer Materials

One of the main problems in the use of nanomaterials in basic research and industrial conditions is their tendency to aggregate. Nanomaterials can be dispersed in solvents by ultrasound, but the strong van der Waals force between nanomaterials leads to their rapid precipitation. In order to solve this problem, there are many different methods to chemically modify nano-materials to disperse them in a specific matrix. For example, oxidation with nitric acid or other strong acids can open the end of carbon nanotubes and introduce carboxyl groups, so that they can be dispersed moderately in polar solvents. Nano-materials can be functionalized by chemical reactions involving some molecules and polymers. However, these chemical reactions usually cause atomic defects in the structure (Fig. 4.15), which have a certain impact on its electrical and mechanical properties. Non-covalent modification may be more advantageous to

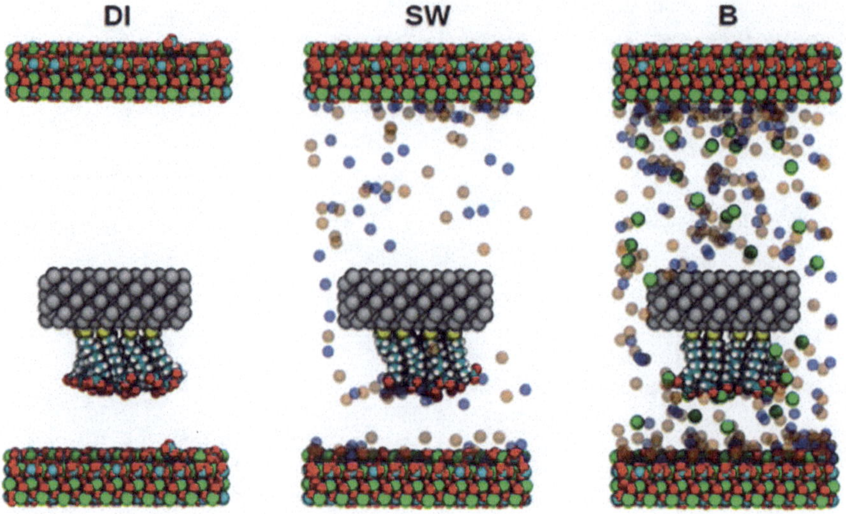

Fig. 4.15 Surface defect characteristics of nanomaterials [20]

functionalize nano-materials without affecting their properties as much as possible. For example, using starch to treat carbon nanotubes, on the one hand, starch is a rich renewable natural resource, on the other hand, when carbon nanotubes are immersed in starch solution, this biomacromolecule is wrapped around carbon nanotubes by synergistic van der Waals interaction, which prolongs their time in dissolution. It is also very easy for dissolved carbon nanotubes to precipitate again. Simply add amylase and decompose starch to peel off the protective shell of nanotubes.

With the help of the internal and external structures of carbon nanotubes, nanowires and nanotubes of other materials can be replicated for electronic and optical fields. For example, open-ended carbon nanotubes with carboxylate or biotin end functionalization are beneficial to site selective recognition and molecular resolution atomic force microscopy imaging. There are various chemicals that can be anchored on the carbon nanotube probe, which lays a good foundation for ultra-high precision chemical force microscope.

The surface defects of nano-materials greatly affect the physical and chemical properties of materials. This influence includes both favorable and unfavorable factors. The reason can be attributed to that the formation of surface defects changes the surface electronic structure and surface chemical environment of nano-materials. On the one hand, the formation of surface defects can increase the specific surface area and surface energy, which may promote some fields (such as ion adsorption, chemical and electrochemical behavior, catalysis, etc.). On the other hand, the formation of defects will cause problems such as electron scattering and photon recombination, which will inevitably affect the properties of electronic devices and photoelectric devices. In addition, the intrinsic physical and chemical properties of nano-materials also have a great influence on the electrochemical properties of materials. How to reasonably regulate the surface structure and electronic state of nano-materials will ultimately determine the electrochemical properties of materials.

Defect construction of nano-materials has been proved to significantly improve the related properties of materials. These surface defects can not only be used as active sites of the reaction to promote catalytic performance, but also improve the specific surface area and promote adsorption. Of course, there is still a serious agglomeration problem in the process of defect construction of nano-materials. If it cannot be effectively solved, the surface defect enhancement effect will be masked by this problem. At present, nano-diamonds, fullerenes, carbon nanotubes, graphene and carbon dots have attracted more and more attention because of their unique electrical, mechanical, optical, thermal and chemical properties, which paved the way for the development of energy, sensing and environmental remediation. Among them, it is very important for the development of modern science and technology to customize carbon nanomaterials with different sizes with ideal characteristics. Molecular shearing method not only successfully realizes the controllable construction of carbon nanomaterials with different dimensions at the molecular level, but also inhibits the agglomeration characteristics of carbon nanomaterials [21]. The molecular shearing method proposed by researchers (Fig. 4.16a) has realized the controllable construction of micro-defects and superstructures of carbon nanomaterials with different dimensions. The carbon nanomaterials obtained by shearing have high specific surface area (1685 m^2/g) and suitable pore size distribution. This technology is different from the surface modification technology widely studied at present. It uses the active shear characteristics of metal vapor and carbon dioxide molecules to act on the surface interface of carbon nanomaterials with different dimensions to realize the cutting on the molecular scale. Moreover, by adjusting the shear activity of molecular scissors to act on different carbon materials, such as carbon nanotubes and activated carbon, the controllable cutting of different carbon nano-materials has been successfully realized, and the results are shown in Fig. 4.16b–d. Further exploring the defect construction mechanism, zinc vapor and magnesium vapor will enter the interior of carbon materials, and once they encounter carbon dioxide, they will form a molecular front knife, thus realizing the controllable cutting of carbon nanomaterials. By adjusting the active temperature and cutting time of molecular scissors, the controllable construction of carbon nanomaterials with different dimensions can be realized.

As we all know, defect engineering is an indispensable basic means to regulate the surface-interface characteristics of carbon nanomaterials. It is very important to realize the multi-scale controllable construction of carbon nano-materials with rich vacancy defects for the improvement of ion transmission rate. Molecular pre-embedding technology and hierarchical double vacancy defect construction platform effectively regulate the defect degree of carbon nano-materials, and realize the controllable preparation of highly interconnected three-dimensional porous carbon fibers. This technology is different from the chemical method to manufacture defects, which is widely studied at present. It makes full use of the molecular pre-embedding technology to rivet active molecules in the material. In the process of high temperature reaction, a large number of vacancy defects are formed by the internal and external

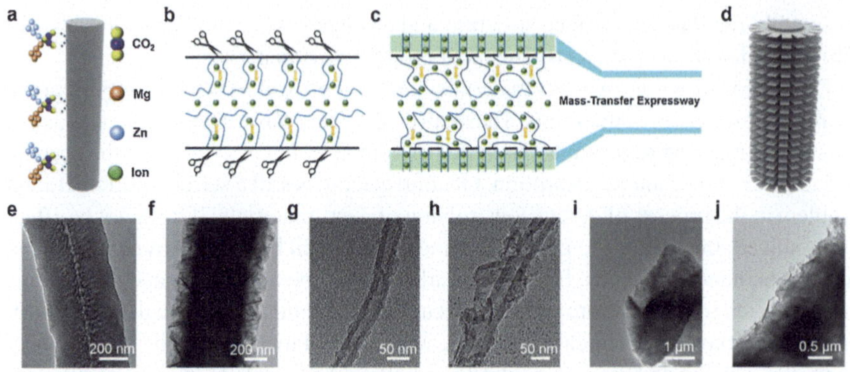

Fig. 4.16 Molecular scissors reasonably customize carbon nanomaterials. **a**, **b** Molecular scissors acting on the surface of carbon nanomaterials. The highly interconnected microstructure of **c**, **d** nano-carbon tubular superstructure (TSNCs) promotes a large number of ion storage and rapid mass transfer channels. **e** Transmission electron microscope images of layered carbon nanotube nanostructures. **f** TEM images of TSNC obtained by scissors engineering. **g** Transmission electron microscope images of multi-walled carbon nanotubes. **h** TEM images of MWCNT superstructure obtained through scissors engineering. **i** Transmission electron microscope image of activated carbon. **j** Transmission electron microscope images of the upper structure of the air conditioner obtained through scissors engineering [22]

synergy between active molecules and carbon dioxide molecules. The test shows that the three-dimensional porous carbon fiber constructed with double vacancy defects not only has suitable pore size distribution, but also effectively solves the storage space and transmission rate of ions. To sum up, surface defect construction is not only expected to become a new wind direction in the field of carbon nano-materials, but also provides reference for energy storage, sensing and environmental protection.

On the other hand, electronic devices (such as diodes and transistors) should try to avoid the formation of surface defects. For example, the two-dimensional material-based electronic devices developed in recent years put forward new requirements for the control of nano-material surface defects [23].

Graphene, a two-dimensional material that can exist at normal temperature and pressure, was first discovered in experiments, and its discovery broke people's understanding of two-dimensional materials. Graphene has become a complex research field involving many fields and dimensions because researchers have been deeply studying graphene. In addition, it has played a great role in the field of 2DLMs. Although graphene has many extraordinary characteristics, which makes it a unique platform for exploring low-dimensional physics and can be used to build a new generation of electronic devices with atomic thickness and perfect combination of speed and flexibility, it still has many shortcomings to be overcome. Other 2DLMs have wider properties, including conductors, semiconductors with different band gaps (such as black phosphorus MoS_2 and WSe_2) and insulators (such as boron nitride). The 2DLMs library with optional material properties increases the possibility of heterogeneous integration at atomic scale and creating a hybrid structure with new physical meanings and unique functions.

This kind of 2DLMs relies on Van der Waals force to form a two-dimensional material heterojunction. Because of the low-dimensional nature of two-dimensional materials, the interface between materials can be acted by the best Van der Waals force in close contact, so 2DLMs is more suitable for Van der Waals integration than rigid bodies. More importantly, each layer of atoms in two-dimensional materials is formed by covalent bond interaction, and there is no problem of a large number of dangling bonds and surface reconstruction on the surface of the materials we discussed earlier, that is to say, there are no defects and trap States on the surface of two-dimensional materials, which will not induce electron scattering and photon capture annihilation. Van der Waals heterojunction has a well-defined interface and mutual diffusion at zero atomic level, which opens up a new dimension of engineering regulation of electronic and optical properties at atomic level. This makes it possible to fabricate a series of ultra-thin, flexible and transparent electronic and optoelectronic devices, including tunnel transistors, vertical field effect transistors, photodetectors, solar cells, light emitting diodes (LEDs) and so on [24].

In fact, it is very necessary to strictly avoid the formation of defects in the development of two-dimensional material-based optoelectronic devices. There is a universal structure in these photoelectric devices-metal–semiconductor junction. Schottky barrier will be formed in metal–semiconductor junction due to the difference of work function and energy band structure between metal and semiconductor materials. Schottky barrier has an important influence on regulating the operation of carrier transport. It can be said that it fundamentally determines the charge transfer efficiency, thus essentially determining the performance of the device. Theoretically, Schottky barrier obeys Schottky–Mott law, but the experimental results seriously deviate from this ideal behavior. In fact, the interface loss between semiconductor layer and metal has always been an important problem that puzzles high-performance new photoelectric devices, and the conventional energy band calculation often seriously underestimates the adverse effects of the interface Schottky barrier.

Defect control has always been considered as the most effective means to reduce interface loss and improve device performance. As shown in Fig. 4.17, in 2018, researchers from UCLA and China Hunan University took a different approach. By transferring metal resistance at atomic level, the phenomenon of chemical disorder and interface Fermi level being pinned is avoided, and the extreme height of Schottky–Mott limit barrier is approached perfectly, thus greatly improving the electrical properties of the metal–semiconductor junction of gold-molybdenum disulfide, which is quite different from the traditional vapor-deposited metal electrode, which easily leads to the phenomenon of serious Fermi level being pinned. However, by transferring metal electrode, The Schottky barrier at the interface only depends on the corresponding metal work function and strictly obeys the Schottky–Mott law calculated by theory. Based on this, researchers used molybdenum disulfide as the research system to prepare electronic and optoelectronic devices, such as electrons at room temperature, field effect transistors with hole mobility as high as 260 and

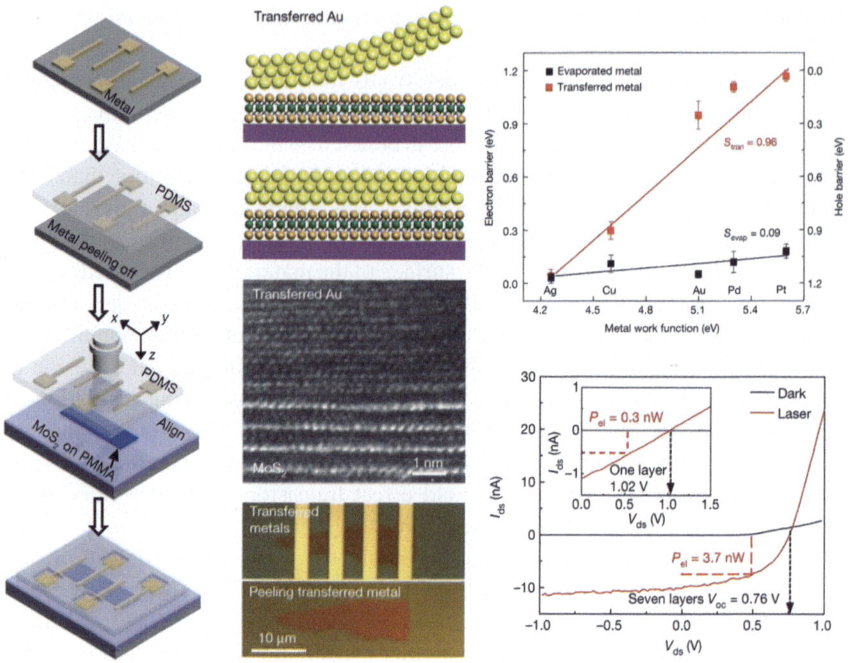

Fig. 4.17 Schematic diagram of Van der Waals metal–semiconductor junction and related test results [25]

175 cm^2/Vs, and single-layer molybdenum disulfide photodiode with open circuit voltage as high as 1.02 V [26].

In recent years, the semiconductor industry has always been shrouded in the unsustainable haze of Moore's Law. These new materials, unique device development and surface defect control all provide more possibilities for opening up the way for advanced electronics and optoelectronics applications in the future.

4.6 Agglomeration Characteristics of Nanomaterials

4.6.1 Brief Introduction of Agglomeration Characteristics

The agglomeration of nano-materials refers to the phenomenon that the original nano-powder particles are interconnected or larger in the process of preparation, separation, treatment and storage. The agglomeration of nano-materials has a serious impact on their properties. Agglomeration not only reduces the chemical activity of nanoparticles, but also affects other properties of nanoparticles. Secondly, the agglomeration of nano-materials brings great inconvenience to the mixing, homogenization and packaging of nano-materials, which makes it very difficult in actual production and application.

The agglomeration of nano-materials is directly affected by the surface effect and small size effect of nano-particles. Specifically, the agglomeration of nanoparticles is mainly caused by the following four reasons: (1) Electrostatic attraction on the surface of nanoparticles. In the process of nanocrystallization of materials, a large number of positive and negative charges gather on the surface of the new nanoparticles, which is extremely unstable. For stability, they attract each other, holding the particles together. This process is mainly driven by electrostatic coulomb forces. (2) The surface energy of nanoparticles is high. In order to reduce the surface energy, particles tend to cluster together to achieve a stable state, which causes particle agglomeration. (3) Van der Waals forces between nanoparticles. When materials are nano-sized to a certain size, the distance between particles is short, the van der Waals forces between particles are much greater than the gravity of the particles themselves, and particles tend to attract each other. (4) Van der Waals forces between nanoparticles. When materials are nano-sized to a certain size, the distance between particles is short, the van der Waals forces between particles are much greater than the gravity of the particles themselves, and particles tend to attract each other [27]. Figure 4.18 shows the diagram of nanomaterials prepared by sol–gel method.

The agglomeration of nanoparticles is generally divided into soft agglomeration and hard agglomeration. Figure 4.19 shows the agglomeration characteristics of different nanoparticles. Soft agglomeration is mainly caused by electrostatic force and van der Waals force between particles, which can be eliminated by chemical

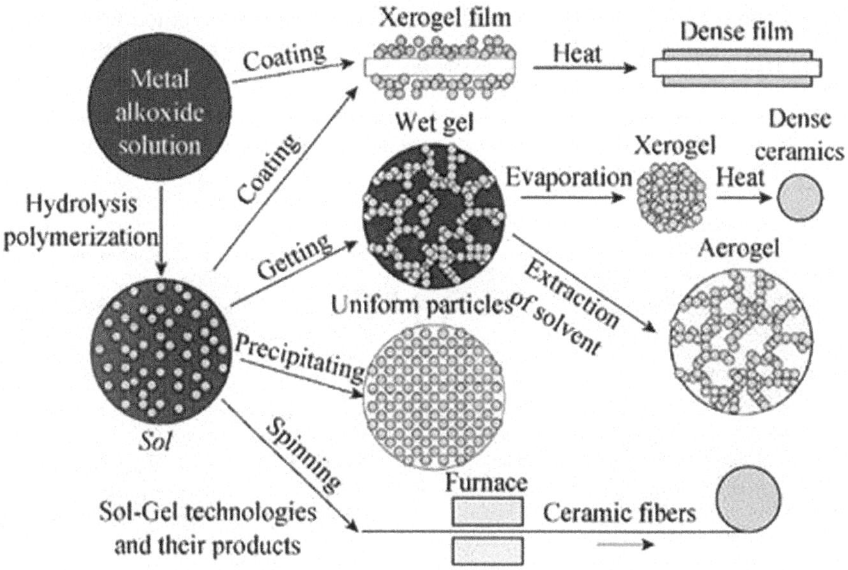

Fig. 4.18 Schematic diagram of preparation of nano-materials by solution-gel method [28]

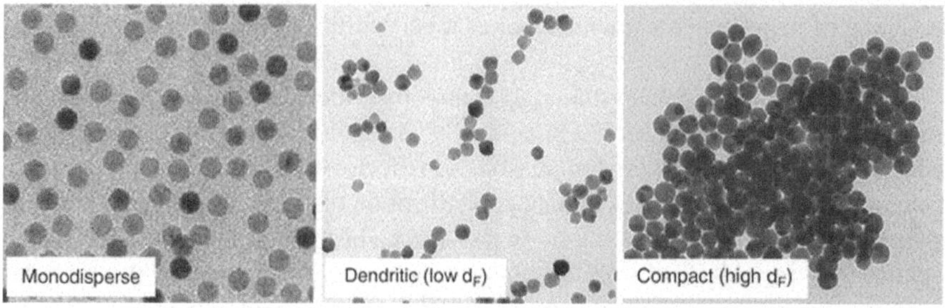

Fig. 4.19 Agglomeration characteristics of different nanoparticles [29]

action or mechanical energy because of its weak force. Besides electrostatic force and van der Waals force, strong chemical bond force also exists. Therefore, hard agglomeration is not easy to destroy and needs some special methods to control it.

As shown in Fig. 4.20, the particles in the liquid phase have Van der Waals attraction, overlapping of electric double layers, overlapping of liquid bridge and solvation layer, and solid bridge and sintering neck exist in the solid phase. The interaction between particles is very complicated in liquid medium.

Besides van der Waals force and coulomb force, there are solvation force, capillary force, hydrophobic force and hydrodynamics, which are related to liquid medium and directly affect the degree of agglomeration. Particles in the liquid medium will form a solvation layer because they attract a layer of polar substances. When the particles are close to each other, the solvated layers overlap and generate repulsive force or solvated layer force. If the display particles are well wetted by the medium and the two particles are close enough, their necks will form a liquid bridge. The liquid bridge has a certain pressure difference, which makes particles attract each other and then become a capillary force between particles. Hydrodynamic force generally exists in suspension with high solid phase. When two particles approach each other, it will generate shear stress between liquid and prevent the particles from

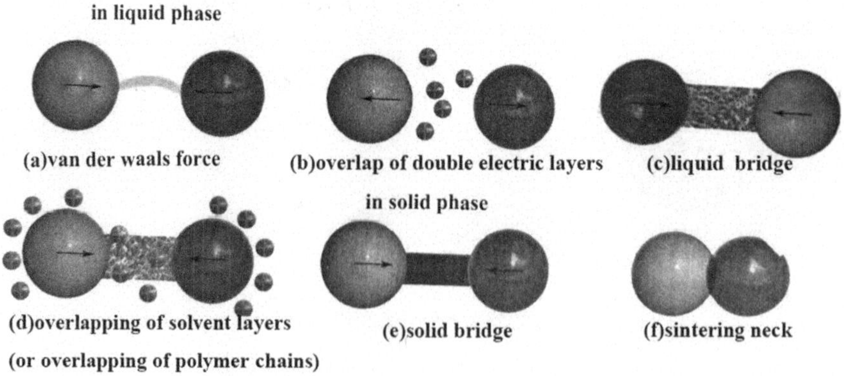

Fig. 4.20 Interaction between particles in various phase states [30]

approaching each other. When two particles are separated, it plays an attractive role, which is quite complicated. In solid phase, agglomeration is mainly caused by solid bridge and sintering neck. If the gel particles are in close contact, it is easier to form hard agglomerates [31].

The agglomeration of nanoparticles in liquid medium is the result of adsorption and repulsion. The adsorption of nanoparticles in liquid media includes the following aspects: quantum tunneling effect, charge transfer and interfacial atom coupling adsorption; Intermolecular forces, hydrogen bonds and electrostatic adsorption; The larger specific surface area between nanoparticles accelerates the growth rate of particles and makes them easy to adsorb. In the presence of adsorption, there is also repulsion between nanoparticles in liquid medium, which mainly includes the formation of solvation film on particle surface, electrostatic effect of electric double layer and spatial protection of polymer adsorption layer. The sum of these effects tends to disperse nanoparticles. If the adsorption is greater than the repulsion, the nanoparticles are agglomerated; If the adsorption is less than the repulsion, the nanoparticles are dispersed.

At present, there is no unified theory about the agglomeration mechanism of nanoparticles in liquid medium. Deryagin and Landau, Verwey and Overbeek respectively put forward the calculation method of the interaction energy between morphological particles and the repulsive energy of the electric double layer, which is called DLVO theory, that is, $V_t = V_a + V_r$, where V_t represents the total interaction energy; V_a represents van der Waals action energy; V_r represents the action energy of the electric double layer.

The theory holds that the agglomeration and dispersion of particles depend on the relative relationship between van der Waals interaction energy between particles and electric double layer interaction energy. When $V_a > V_r$, the particles spontaneously approach each other and eventually form agglomeration; When $V_a < V_r$, the particles repel each other to form a dispersed state [32]. Figure 4.21 shows the relationship between the existing state of particles and repulsive potential energy and gravitational potential energy. Because there are other effects between particles besides the above two forces, DLVO theory can not fully explain the agglomeration between particles. Considering that the interaction between particles is influenced by environmental medium properties, particle surface properties and the composition, coverage and adsorption strength of the adsorption layer on the particle surface, the total potential energy between particles can be expressed as follows: $V_t = V_a + V_b + V_s + V_{st}$. ($V_a$ is the van der Waals action energy; V_b is the action energy of electric double layer; V_s is the solvation energy; V_{st} is the spatial repulsion energy).

Nanoparticles are easy to agglomerate in gas, which brings inconvenience to the processing and storage of powder. The main reasons for the agglomeration of nanoparticles in gas are as follows: intermolecular force (van der Waals force), electrostatic force between particles, adhesion of particles in wet gas phase and wettability of particle surface. It is the result of the above action that leads to the agglomeration of nanoparticles in gaseous medium. Although the magnitude of Van der Waals force is inversely proportional to the 7th power of molecular spacing, its attraction is obvious because of the small distance between nanoparticles, which is the

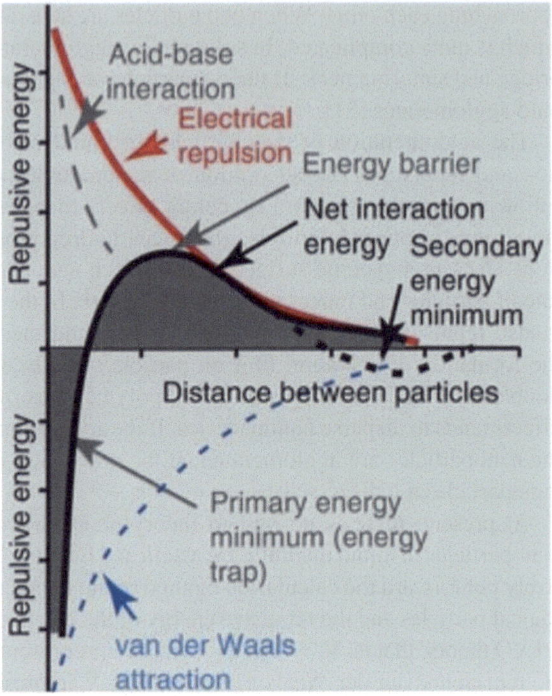

Fig. 4.21 Repulsive potential energy, gravitational potential energy and total potential energy curve [29]

fundamental reason for the agglomeration of nanoparticles. In air, most particles are naturally charged, and electrostatic attraction is inevitable. This force is an important factor in the agglomeration of nanoparticles. The reason for the adhesion of particles in the air is that when the relative humidity of the air is too high, water vapor gathers on the surface of particles and between particles, which increases the adhesion between particles. The wettability of particle surface has a significant effect on the adhesion between particles, thus affecting the agglomeration of particles [33].

Compared with the clean surface, the surface of oxides, metals, carbides, nitrides and other substances in the atmospheric environment has a layer of hydroxyl structure (ROH), which is different from the internal composition and structure. This is the result of the reaction of suspended bonds on the surface with O_2 and water in the air. Therefore, the agglomeration and dispersion mechanism of powder is the same. The newly prepared porous silicon nanoparticles have a complete hydroxyl (–OH) "passivation layer" on their surfaces, which has an important influence on luminescence and catalytic performance [34].

On the one hand, the formation of hydroxyl base changes the surface structure, and the electrostatic repulsion is reduced due to relaxation phenomenon, so the powder becomes attractive and agglomeration is inevitable. With the increase of the density, quantity and activity of hydroxyl groups, the agglomeration is intensified. Surface hydroxyl activity is related to powder structure, cationic polarizability, quantum size effect and electronic structure. Because one end of hydroxyl base is high metal ion

Rn^+ and the other end is low charge hydrogen ion, the structure is unbalanced, and the surface excess energy is still very large, so that hydroxyl groups continue to adsorb water and other polar substances in the form of physical adsorption. With the polarization of hydroxyl groups, the nature of adsorbed water becomes active (for example, the concentration of H^+ and OH^- is much higher than that of free water), which is beneficial to the thickening of adsorbed water layer. When the adsorbed water layer reaches a certain thickness, a water film will be formed on the surface of nano-powder, which will produce another great attraction, namely the surface tension of the water film. Because of the small radius of curvature of nano-powders, the capillary surface force in contact with water film becomes very large (>100 MPa), and such a large force will inevitably lead to aggregation and condensation between nano-powders. Polarization and anti-polarization will promote the dissociation or hydration of surface ions and form a new substance-solid bridge. The further reduction of activation energy makes it possible to form new connecting phases between powders, forming primary and secondary agglomeration. Soft agglomeration is called soft agglomeration, which is caused by van der Waals force, hydrogen bond and capillary force between hydroxyl bases (including adsorption layer) on the surface of agglomerated nano-powder. The chemical adsorption or chemical reaction between hydroxyl bases on the surface of nano-powder has great mixing force with the powder, which is called hard agglomeration. Nano-powders with hydroxyl structure are composed of van der Waals force (including hydrogen bond), capillary force and solid bridge (connection between hard aggregates). The electrostatic repulsion between powders is greatly reduced, the force between powders is greater than 0, and the powders automatically reunite. The highly active hydroxyl structure on the surface is the source of nano-powder agglomeration, which leads to the formation of hydrogen bonds and capillary forces. The chemical reaction between hydroxyl structures is the driving force for the source and powerful nano-powder agglomeration. A lot of work has been done to eliminate the agglomeration of nano-powders. The nano-powder is dynamically dispersed by intense stirring, because the dispersion of nano-particles in the solvent inevitably leads to the increase of system energy, and the work is completed on the system by stirring, thus providing the energy needed to increase the surface area [35]. The kinetic energy of nanoparticles is enough to overcome the attractive force between particles without aggregation. But when the stirring was stopped, the nanoparticles reunited. Soft agglomeration can be redispersed after stirring again, while hard agglomeration is difficult to redisperse.

For nano-powder in gas medium, general chemical or mechanical action can eliminate soft agglomeration. High power ultrasound or ball milling can weaken hard agglomeration. Generally speaking, the dispersion methods of nano-powder in gas can be divided into physical dispersion method and chemical dispersion method (according to whether there are foreign substances in chemical reaction). Physical dispersion includes mechanical dispersion method, electrostatic dispersion method, high vacuum method and inert gas protection method. Chemical dispersion method is to change the structure, change the chemical composition and electrochemical characteristics of the surface of nanoparticles through the chemical reaction between modifiers and nanoparticles, so as to achieve the purpose of surface modification and promote the dispersion of nanoparticles.

4.6.2 Dynamic Model of Agglomeration

Colloidal stability theory considers collision frequency (β) and collision efficiency (α). Among them, the collision efficiency has been solved by the theory put forward by Smoluchowski, and Fuchs gives the basic theory to evaluate the collision efficiency. In order to use Fuchs theory, the interaction energy as a function of the distance between interacting particles must be solved by classical DLVO theory. Von-Smoluchowski's population equilibrium equation can be used to describe the nonaggregation dynamics of particles:

$$\frac{dn_k}{dt} = \frac{1}{2}\sum_{i+k=k} \alpha(r_i, r_j)\beta(r_i, r_j)n_i n_j - n_k \sum_{i=1} \alpha(r_i, r_j)\beta(r_i, r_j)n_i \tag{4.1}$$

In the formula, it is the number of agglomerations of many particles composed of k; And $\alpha(r_i, r_j)$ and $\beta(r_i, r_j)$ represent collision efficiency and collision frequency functions of class i and class j particles respectively; It represents the particle radius. Taking van der Waals force and fluid mechanics into consideration, the collision frequency can be expressed by the following formula:

$$\beta(i, i) = \frac{8kt}{3\mu}\left\{2\int_0^\infty \lambda(u)\frac{\exp\left[U_{iwi}^{vdw}(h)/(kT)\right]}{(2+u)^2}du\right\}^{-1} \tag{4.2}$$

where $U_{|w|}^{vwd}$ (h) represents the total energy between two interacting particles; $u = h/r$ and $\lambda(u)$ the correlation factor representing the diffusion efficiency, and the relationship between it and the distance is as follows:

$$\lambda(u) = \frac{6u^2 + 13u + 2}{6u^2 + 4u} \tag{4.3}$$

For micro-scale flocculation of heterogeneous particles with diameters of d_i and d_j, the collision frequency can also be expressed as:

$$\beta = \left(\frac{2k_B T}{3u}\right)\left(\frac{1}{d_i} + \frac{1}{d_j}\right)(d_i + d_j) \tag{4.4}$$

where k_B is the reaction rate constant.

When monodisperse nanoparticles have the same particle size, the above formula can be simplified to $\beta = \frac{8k_B T}{3\mu}$, , where the viscosity of the solution is (1×10^{-3} Pa·s).

Therefore, the absolute agglomeration rate can be simplified as the following formula:

$$\frac{dn_k}{dt} = -\frac{4}{3}\alpha\frac{k_BT}{\mu}n_i^2 \tag{4.5}$$

The early aggregation kinetics can be described by the formation rate of two peaks, which is obviously dominant compared with the formation of other high-order aggregates. The loss of primary particles in the early stage of agglomeration can be expressed as a second-order rate range:

$$\left(\frac{dn_1}{dt}\right)_{t\to 0} = -k_{11}n_0^2 \tag{4.6}$$

where is the function of initial particle concentration and time; K_{11} is the absolute agglomeration rate constant $\left(k_{11} = \frac{4}{3}\alpha\frac{k_BT}{\mu}\right)$ and the initial concentration of particles. In the Rayleigh-Gans-Debye (RGD) approximation, which is effective for relatively small primary particles compared with the incident wavelength, the absolute aggregation rate constant k_{11} can be determined by time-resolved DLS measurement at a fixed scattering angle:

$$\frac{1}{r_H(0, q)}\left(\frac{dr_H(t, q)}{dt}\right)_{t\to 0} = k_{11}n_0\left[1 + \frac{\sin(2aq)}{2aq}\right]\left(1 - \frac{1}{\delta}\right) \tag{4.7}$$

where q is the scattering vector defined by $4\pi n/[\lambda \sin(\theta/2)]$, where n is the refractive medium coefficient, λ is the wavelength of incident light, and θ is the scattering angle; r_H (t, q) represents the function of hydrodynamic radius; δ is the relative fluid radius of the doublet, which is about 1.38; A is the particle radius. Generally, DLS operates at a fixed angle, and the only independent variable in Eq. (4.7) is that t changes with time. In order to obtain the value of k_{11}, Chen et al. proposed a linear least square method for regression analysis. The k_{11} value at the initial stage of nanoparticle agglomeration can be expressed by the following formula:

$$\left[\frac{dr_H(t)}{dt}\right]_{t\to 0} \propto k_{11}n_0 \tag{4.8}$$

Agglomeration adsorption efficiency α, also known as inverse stability ratio, is used to express the relative aggregation rate, and it is also used to divide the aggregation process into the aforementioned diffusion-limited aggregation model (DLCA) and reaction-limited cluster aggregation model (RLCA). By quickly determining the diffusion-limited polymerization rate constant k_{11} under favorable polymerization conditions, the measured k_{11} is normalized to calculate the agglomeration adsorption efficiency:

$$\alpha = \frac{k_{11}}{(k_{11})_{fast}} \tag{4.9}$$

The fast stands for fast calculation.

The agglomeration adsorption efficiency a is the reciprocal of the stability ratio W, which can be defined by the following formula:

$$\alpha = \frac{1}{W} = \left\{ \int_0^{\infty} \lambda(u) \frac{\exp\left[U_{iwi}^{DLVO}(h)/(k_B T)\right]}{(2+u)^2} du \right\} \left\{ \int_0^{\infty} \lambda(u) \frac{U_{iwi}^{DLVO}(h)/(k_B T)}{(2+u)^2} du \right\}^{-1} \tag{4.10}$$

where U_{iwi}^{DLVO} represents the total energy between two interacting particles. Formula (4.10) has been widely used to study the kinetic system of agglomeration of various nanoparticles. However, the stability ratio calculated according to the classical DLVO theory is inconsistent with the value obtained based on experimental observation, which also proves that the classical DLVO theory has certain limitations. Although Formula (4.10) is widely used in colloidal adhesion system, it still has some problems when applied to nano-agglomeration system. For example, the calculated stability ratio is more unreasonable than the experimental value. This difference may come from the assumption that van der Waals gravity is the only driving force for particle agglomeration, which is feasible in colloidal particles, but may not be suitable in nano-particle system. Because for small-sized nanoparticles, the interaction force will be greatly reduced, and random kinetic energy will play a leading role in the transportation process. On the contrary, for colloidal particles, the interfacial interaction plays a major role. According to the Stokes–Einstein equation, the stochastic dynamics (i.e., diffusion) is much lower than that of nanoparticles. On the other hand, the good fitting between 1/W and experimental A is usually achieved by changing Hamaker constant as fitting parameter. However, Hamaker constant ($A_{H,123}$) reflects the inherent characteristics of the materials of two interacting particles, which reflects the strength of long-range mutual attraction between two small-volume materials. In this sense, the Hamaker constant should be a fixed value of the pair interaction of nanoparticles, or at least there should be no significant change. In fact, the Hamaker constant ($A_{H,123}$) between particles can be calculated by van Oss method. In addition, according to the assumption of the derivation of formula (4.8), this experimental method should only be used to determine k_{11} or α in early aggregation kinetics (such as the hydrodynamic size of aggregates is less than 30% of the size of primary nanoparticles). Therefore, more experiments and theoretical work are needed to clarify the basic mechanism of nano-scale aggregation kinetics [36].

Although colloid science has confirmed that several interaction forces with different properties and origins control the dynamics and degree of particle aggregation, there are certain interactions among these different factors, which challenge the two main theories in colloid science (DLVO theory and EDLVO theory). This challenge comes from the specific role of nano-particle morphology (size or shape), surface characteristics (such as organic surface coating), composition or crystallinity, and solution chemistry. For example, pH, ionic strength, valence state of electrolyte ions and NOM (natural organic matter), as well as environmental parameters (such as temperature and light) will all have different effects on agglomeration kinetics.

4.7 Rheological Properties of Nanomaterials

When the shear stress τ of a fluid is proportional to the shear speed $\dot{\gamma}$, that is, $\tau = \eta\dot{\gamma}$, and the viscosity η is constant, this fluid is called Newtonian fluid, but some fluids do not follow the above relationship, and their viscosity η changes with τ and $\dot{\gamma}$. Figure 4.22 shows the relationship curve between τ and $\dot{\gamma}$ of various fluids, the fluid that follows curve a is Newtonian fluid, and the fluid that follows curves b and c is called non-Newtonian fluid.

When particles are dispersed in a dispersant (Newtonian fluid) to form a sol, the sol is a non-Newtonian fluid.

When the viscosities of solution and solvent are expressed by η and η_0 respectively, the relative viscosity $\eta_{rel} = \frac{\eta}{\eta_0}$ the increase rate of solution viscosity relative to solvent viscosity η_{sp} is called specific viscosity, $\eta_{sp} = (\eta - \eta_0)/\eta_0 = \eta_{sp} - 1$. The specific viscosity of spherical particle dispersion system formed by solid particles dispersed in liquid obeys Einstein viscosity expression $\eta_{sp} = 2.5\varphi$, φ is the volume fraction of particles. This viscosity formula is suitable for Au colloid, latex and other particle dispersion systems, but the experimental value deviates from this formula when the concentration is high, and the viscosity increase rate at unit concentration. $\eta_{recl} = \eta_{sp}/C$, where C is the solute concentration and η_{recl} is called reduced viscosity.

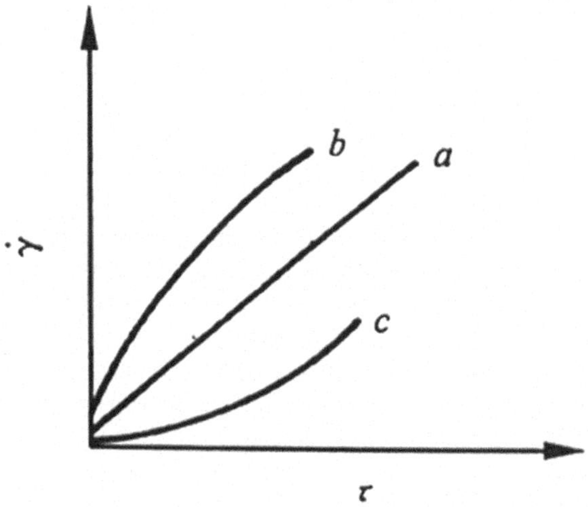

Fig. 4.22 Fluid behavior: a—Newtonian fluid; b, c—non-newtonian fluid [37]

4.7.1 Viscosity of Typical Colloidal Suspension

Generally, spherical dispersed particles (0.1 μm) of various synthetic resin latexes made by emulsion polymerization are regarded as typical colloidal particles, and this colloidal particle dispersion system is studied. Saunders studied the influence of the concentration of monodisperse polystyrene latex on viscosity. The results showed that when the latex concentration (volume fraction) was lower than 0.25, the latex dispersion system was a Newtonian fluid, while when the latex concentration was higher than 0.25, the latex dispersion system was a non-Newtonian fluid. When the latex concentration increases, the reduced viscosity η_{recl} increases. Even if the latex concentration is the same, the viscosity increases with the decrease of latex particle size. The relationship between latex concentration and relative viscosity can be expressed by Mooney formula, that is

$$\eta_{recl} = \exp \frac{(\alpha_0 \varphi)}{(1 - K\varphi)} \tag{4.11}$$

φ is the concentration (volume fraction) of latex, α_0 is the shape factor of particles, equal to 2.5, and k is the electrostatic attraction constant ($\approx$1.35). With the decrease of latex particle size, the increase of viscosity is due to the smaller particle size, the larger specific surface of latex, the larger electrostatic attraction between latex and the larger K in Mooney formula.

4.7.2 Viscosity of Nano-Al_2O_3 Suspension

Figure 4.23 shows the curves of viscosity of aqueous suspension of Al_2O_3 particles with different concentrations with different particle sizes and shear rate ($\dot{\gamma}$). It can be seen that this kind of suspension shows a shear thinning behavior that viscosity decreases with the increase of shear speed.

In the past, it was generally believed that the shear thinning behavior was caused by the coagulation of particles, but the results of electrodynamics experiments showed that the particles in suspension were very dispersed. Therefore, Yeh and Sacks pointed out that the shear thinning behavior can not be attributed to the aggregation of particles, but is caused by Brownian motion and electric viscosity effect. Krieger et al. have investigated the influence of Brownian motion on viscosity of "neutral stable" suspension of monodisperse latex particles, and observed shear thinning behavior, high shear limit viscosity and low shear limit viscosity. The high shear limit viscosity of the suspension with a concentration of 50 vol% and a particle diameter of 150 nm is twice as high as the low shear limit viscosity. With the decrease of concentration and the increase of particle diameter, the difference between the two limit values decreases rapidly. However, Fig. 4.24 shows that the high shear viscosity of Al_2O_3 suspension with a concentration of 38vol% and a particle size of about 100 nm is three

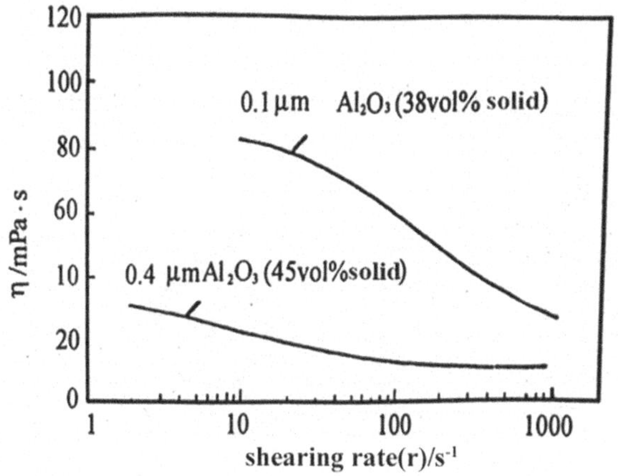

Fig. 4.23 Relationship between viscosity and shear rate of suspension of Al_2O_3 particles with different concentrations [38]

times that of the low shear viscosity, which contradicts Krieger's result. Because the suspension investigated by Krieger is electrically neutral, but the Al_2O_3 suspension is not, Yeh and Sacks believe that the difference between the behavior of Al_2O_3 suspension and that investigated by Krieger is due to the electrical viscosity effect in the Al_2O_3 suspension. Especially, when the surface charge density and Zeta potential of particles increase and the ionic strength and particle size decrease, the effect of electro-viscosity becomes very important, and its influence on Bublens motion is much greater, which may lead to several orders of magnitude of viscosity change at high and low shear rates.

4.7.3 *Viscosity of Magnetic Liquid*

(1) Basic knowledge of magnetic fluid: In magnetic materials, whether it is single crystal or polycrystalline, metallic or oxide, crystalline or amorphous, blocky or film-like, it is a solid state, but this section discusses liquid magnetic materials, hereinafter referred to as magnetic fluid.

Up to now, no material with Curie temperature higher than melting point has been found, so the real liquid ferromagnetic material is still being explored. The magnetic liquid discussed in this section is a stable colloidal system composed of magnetic particles highly dispersed in the carrier liquid through surfactants. It has both strong magnetism and fluidity, and can exist stably for a long time under the action of gravity and electromagnetic force without precipitation and stratification.

Magnetic fluid is composed of magnetic particles, surfactant and carrier fluid, and the relationship among them is shown in Fig. 4.24.

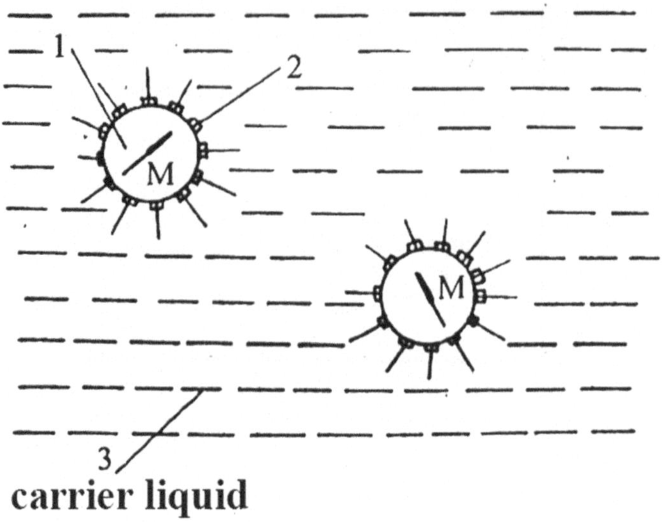

Fig. 4.24 Composition of magnetic fluid: 1—sexual particles; 2—surfactant; 3—carrier liquid [39]

The selection of surfactants is mainly to make the corresponding magnetic particles stably suspended in the carrier liquid, which is very important for the preparation of magnetic liquid. It is related to whether the magnetic liquid can be made and whether its stability meets the requirements. See Table 4.2 for the commonly available surfactants.

Table 4.2 Examples of suitable surfactants

Name of carrier liquid	Examples of surfactants suitable for the carrier liquid
Water	Unsaturated fatty acids, such as oleic acid, linoleic acid, linolenic acid and their derivatives. Salt and soap, dodecanoic acid, dioctyl sodium sulfosuccinate, etc.
Esters and diesters	Oleic acid, linoleic acid, linolenic acid, or corresponding ester acid, such as di (2-ethylhexyl) phosphate and other nonionic surfactants
Hydrocarbon group	Oleic acid, linoleic acid, linolenic acid and other nonionic surfactants
Fluorocarbon group	Fluoroether acid, fluoroether sulfonic acid, and their corresponding derivatives, perfluoroisopropyl ether, etc.
Silicone oil group	Silane coupling agent, carboxyl polydimethylsiloxane, hydroxyl polydimethylsiloxane, mercaptopolydimethylsiloxane, amino polydimethylsiloxane, carboxyl polyphenylmethylsiloxane, amino polydimethylsiloxane, carboxyl polyphenylmethylsiloxane, hydroxyl polyphenylmethylsiloxane, mercaptopolyphenylmethylsiloxane, amino polyphenylmethylsiloxane
Polyphenyl ether	Phenoxy undecanoic acid

Magnetic particles can be: Fe_3O_4, γ-Fe_2O_3, single or composite ferrite, pure iron powder, pure cobalt powder, iron-cobalt alloy powder, rare earth permanent magnet powder, etc. At present, Fe_3O_4powder is commonly used.

The selection of carrier fluid depends on the characteristics and uses of the prepared magnetic fluid. Generally speaking, when selecting magnetic fluid, the type of carrier fluid should be considered first, followed by other indicators of magnetic fluid. For users, when selecting magnetic fluid specifically, please refer to Table 4.3.

Generally speaking, the main performance indexes of magnetic fluid are: magnetization M_s(A/m) at high field ($H_0 \geq 650$ A/m), viscosity η (Pa·s), service temperature range, and sometimes vapor pressure and other physical and chemical parameters, such as flow point (K), boiling point (K), flash point (K) and density (kg/m^3).

(2) Viscosity of magnetic fluid: The viscosity of magnetic fluid is an important parameter to measure magnetic fluid. Nanoparticles have good fluidity in magnetic liquid, but the viscosity of magnetic liquid is low. On the contrary, the viscosity of magnetic liquid is high, which is influenced by many factors, the most important of which are the volume percentage of particles in magnetic liquid, the viscosity of carrier liquid and the properties of surfactant.

When there are many magnetic particles in the magnetic liquid, the relationship between viscosity and concentration can be expressed by the following formula:

$$\frac{\eta_s - \eta_0}{\eta_s} = 2.5\phi - \frac{2.5\phi - 1}{\phi_c^2}\phi^2. \tag{4.12}$$

Table 4.3 Characteristics and uses of magnetic fluids with different carrier fluids

Carrier liquid type	Characteristics and uses of the prepared magnetic liquid
Water	The pH value can be changed in a wide range, the price is low, and it is suitable for medical treatment and magnetic separation Refer to magnetic tape and bubble inspection
Esters and diesters	Low vapor pressure, suitable for vacuum and high-speed sealing, magnetic liquid with good lubricity, especially suitable For devices with low friction and damping devices, others are used for speakers and stepping motors
Silicate ester	Good cold resistance, suitable for low temperature occasions
Hydrocarbon	Low viscosity is suitable for high-speed sealing, and magnetic fluids with different hydrocarbon-based carrier fluids can be mixed with each other
Fluorocarbon	Non-flammable, wide temperature, insoluble in other liquids, suitable for active environment, such as odor Application in oxygen, chlorine and other environments
Polyphenyl ether	Low vapor pressure, low viscosity, suitable for high vacuum and strong radiation occasions
Mercury	It can be used as carrier liquid for cobalt and iron-cobalt particles, with high saturation magnetization and good thermal conductivity

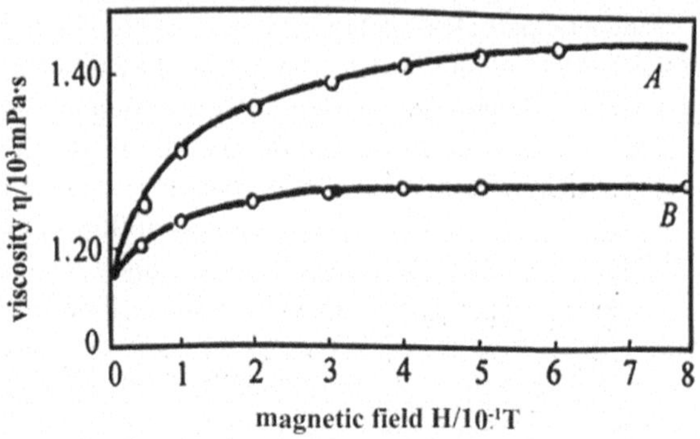

Fig. 4.25 A—The variation of viscosity with magnetic field: the magnetic field is parallel to the rheological direction of the magnetic fluid; B—the magnetic field is perpendicular to the rheological direction of the magnetic fluid [40]

where η_s is the viscosity of the magnetic liquid; η_0 is the viscosity of the carrier liquid; φ is the particle volume percentage (including the thickness of the surface adsorption layer); φ_c is the critical concentration when the liquid loses its viscosity.

The applied magnetic field has obvious influence on the viscosity of the magnetic fluid. When the applied magnetic field is parallel to the rheological direction of the magnetic fluid, the viscosity of the magnetic fluid increases rapidly while the applied magnetic field is perpendicular to the rheological direction of the magnetic fluid, and the viscosity of the magnetic fluid also increases, but it is not as obvious as the former, as shown in Fig. 4.25.

References

1. A. Jana, A. Meena, S. A. Patil, Y. Jo, S. Cho, Y. Park, V. G. Sree, H. Kim, H. Im, R. A. Taylor, *Prog. Mater. Sci.* **2022**, *129*, 12–18.
2. M. Pagliaro, *J. Chem. Educ.* **2005**, *21*, 43–48.
3. Dr., Douglas, Philp, Prof., J., Fraser, Stoddart, *Angew. Chem. Int. Ed.* **1996**, *35*, 1154–1196.
4. L. R. Khanal, J. A. Sundararajan, Y. Qiang, *Energy. Tech.* **2019**, *8*, 1901070.
5. S. Tao, X. Li, X. Wang, Y. Wei, Z. J. Tian, *Angew. Chem.* **2019**, *132*, 223428.
6. Y. Zhao, G. Chen, G. Chen, *Books.* **2015**, *17*, 32–38.
7. A. A. Lahcen, A. Amine, *Electroanalysis.* **2019**, *31*, 188–201.
8. J. Sun, H. Sun, Z. Liang, *Chem. Electro. Chem.* **2017**, *4*, 1651–1662.
9. T. W. Kwon, J. W. Choi, A. Coskun, *Joule.* **2019**, *3*, 662–682.
10. K. Ariga, X. Jia, J. Song, J. P. Hill, D. T. Leong, Y. Jia, J. Li, *Angew. Chem. Int. Ed.* **2020**, *6*, 132545.
11. G. R. Patzke, Y. Zhou, R. Kontic, F. Conrad, *Angew. Chem. Int. Ed.* **2011**, *50*, 826–859.
12. J. Song, B. Wu, Z. Zhou, G. Zhu, Y. Liu, Z. Yang, L. Lin, G. Yu, F. Zhang, G. Zhang, *Angew. Chem. Int. Ed.* **2017**, *16*, 11358.
13. Y. Wang, X. Chu, Z. Zhu, D. Xiong, H. Zhang, W. Yang, *Chem. Eng. J.* **2021**, *423*, 130203.
14. J. G. Lee, K. Lannigan, W. A. Shelton, J. Meissner, B. Bharti, *Langmuir.* **2020**, *36*, 14157–14165.
15. A. Meng, L. Zhang, B. Cheng, J. Yu, *Adv. Mater.* **2019**, *31*, 1807660.
16. C. Gao, F. Lyu, Y. Yin, *Chem. Rev.* **2020**, *8*, 31–35.
17. Y. Lu, Y. Lu, Z. Niu, J. Chen, *Adv. Energy. Mater.* **2018**, *8*, 1702469.
18. S. Deng, S. Jia, X. Deng, Y. Qing, S. Luo, Y. Wu, *Chem. Eng. J.* **2021**, *416*, 111–118.

19. H. Wang, J. Xie, K. Yan, M. Duan, *J. Mater. Sci. Technol.* **2011**, *27*, 153–158.
20. H. Chen, S. L. Eichmann, N. A. Burnham, *J. Phys. Chem. C.* **2020**, *12*, 34–38.
21. H. Inaba, K. Matsuura, *Chem. Rec.* **2019**, *19*, 843–858.
22. Q. Wang, Y. Zhou, X. Zhao, K. Chen, G. Bingni, T. Yang, H. Zhang, W. Yang, J. Chen, *Nano Today.* **2021**, *36*, 112–114.
23. C. Dai, Y. Pan, B. Liu, *Adv. Energy. Mater.* **2020**, *10*, 2002474.
24. K. Ariga, T. Mori, J. P. Hill, *Adv. Mater.* **2012**, *24*, 158–176.
25. Y. Liu, J. Guo, E. Zhu, L. Liao, S. J. Lee, M. Ding, I. Shakir, V. Gambin, Y. Huang, X. Duan, *Nature* **2018**, *557*, 696–700.
26. J. Qiu, X. J. Liu, B. A. You, N. Ren, H. Liu, *Small.* **2023**, *19*, 2206487.
27. X. Kang, X. Sun, B. Han, *Adv. Mater.* **2016**, *28*, 1011–1030.
28. F. Taghizadeh, *Optics & Pho. Journal.* **2016**, *06*, 62–68.
29. O. N. Oliveira, L. Caseli, K. Ariga, *Chem. Rev.* **2022**, *122*, 6459–6513.
30. J. Ni, Y. Li, *Adv. Energy. Mater.* **2016**, *6*, 1600278.
31. C. Tan, Z. Lai, H. Zhang, *Adv. Mater.* **2017**, *29*, 1701392.
32. L. B. Huang, W. Xu, J. Hao, *Small* **2017**, *13*, 1701820.
33. H. Kim, S. Beack, S. Han, M. Shin, T. Lee, Y. Park, K. S. Kim, A. K. Yetisen, S. H. Yun, W. Kwon, S. K. Hahn, *Adv. Mater.* **2018**, *30*, 1701460.
34. P. Karthick Kannan, P. Shankar, C. Blackman, C. H. Chung, *Adv. Mater.* **2019**, *31*, 1803432.
35. C. Hu, R. Chen, N. Zheng, *Adv. Mater.* **2021**, *33*, 2006159.
36. Z. A. VanOrman, L. Nienhaus, *Info. Mat.* **2021**, *3*, 962–986.
37. M. H. Nurmawati, P. K. Ajikumar, R. Renu, S. Valiyaveettil, *Adv. Funct. Mater.* **2008**, *18*, 3213–3218.
38. K. Zhang, Y. J. Gao, P. P. Yang, G. B. Qi, J. P. Zhang, L. Wang, H. Wang, *Adv. Healthc. Mater.* **2018**, *7*, 1800344.
39. P. Kumar, *Adv. Mater. Interfaces.* **2019**, *6*, 1901454.
40. S. Wu, R. Xu, M. Lu, R. Ge, J. Iocozzia, C. Han, B. Jiang, Z. Lin, *Adv. Energy. Mater.* **2015**, *5*, 1500400.

Chapter 5
Methods for Analysis and Characterization of Nanomaterials

5.1 SEM Characterization of Nanomaterials

High-resolution scanning electron microscopy (SEM) is excellent for size and shape characterization of nanoparticles due to the relatively quick and simple sample preparation and image acquisition required. Although SEM images are two-dimensional (2D) representations of 3D objects at a certain viewing angle, SEM images contain a certain amount of 3D information, and the shapes of simple structures with subnanometer precision can be reconstructed by model-based measurements [1].

The signals responsible for generating the SEM images are given by secondary electrons (SEs) and/or backscattered electrons (BSE), produced when the sample is bombarded by the primary electron beam (PE). A typical maximum accelerating voltage for a SEM is 30 kV. While SE has a small escape depth in the range of a few nm due to its low energy below 50 eV, BSE has significantly higher kinetic energy (up to the excitation energy) and comes from about half the depth of the interaction volume in the sample, see Fig. 5.1a. Therefore, SE is able to produce high-resolution SEM micrographs with excellent surface topography contrast, while BSE micrographs show excellent compositional contrast but poor spatial resolution. Further, SE electrons can be divided into SE1, that is, the electrons generated at the point where primary electrons collide with the sample; SE2, that is, the energy generated by high-energy electrons inside the sample; and SE3, that is, the energy generated by BSE electrons impacting the surface of the microscope specimen chamber. Detectors installed inside the electron column, for example so-called in-lens or through-the-lens, enable the collection of secondary electrons with very good efficiency at the optimal working (ultimate lens sample) distance and provide the highest spatial resolution.

In recent years, the SEM version of the Scanning Transmission Electron Microscopy (STEM) working mode has gained popularity, especially in the analysis of nanoparticles, due to the excellence of modern high-resolution SEMs with

S. Peng and P. Li, *Nanomaterials and Nanotechnology*,
https://doi.org/10.1007/978-981-97-8433-2_5

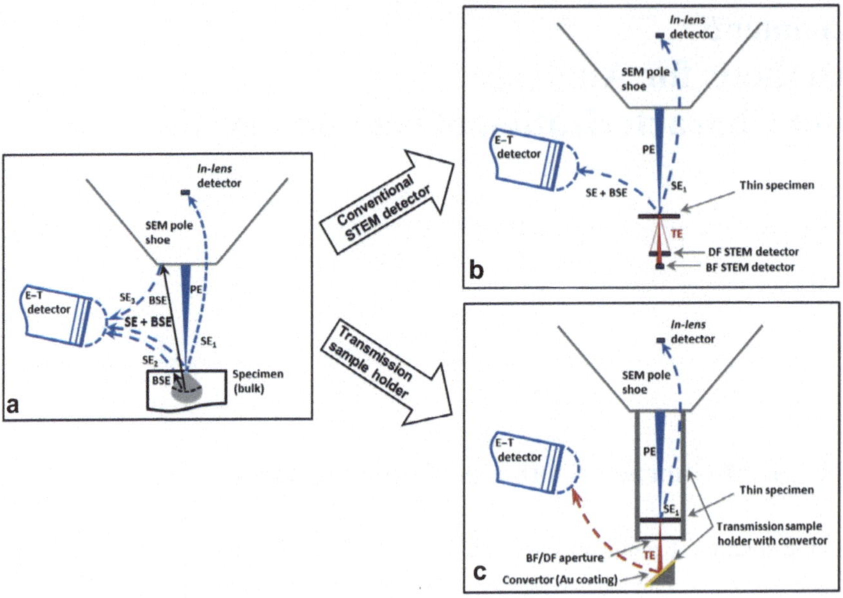

Fig. 5.1 Schematic illustration of the types of electrons generated by the primary electron beam of the SEM and the typical reflected electron detectors, the conventional E–T detector and optional in-lens detector (**a**); and the STEM detection modes, the conventional type (**b**); and transmission sample holder with converter and E–T detector (**c**)

better than 1 nm spatial resolution. The main advantage of operating the SEM in transmission mode is the enhanced mass-thickness contrast, which allows more accurate determination of nanoparticle boundaries and thus more accurate determination of nanoparticle size and size distribution. This STEM mode of operation is STEM-in-SEM (also known as "TSEM" or "Low Voltage TEM"). The operation is similar to conventional TEM, requiring the preparation of electron transparent samples on thin slices (TEM grids). Lower accelerating voltages lead to fewer scattering events in thin samples, which improves nanoparticle contrast. For low atomic number materials with different structures and small differences in electron density, the transmission contrast of SEM is significantly better than that of conventional TEM. Modern SEMs include STEM detectors as part of their standard configuration, as Fig. 5.1b. STEM signals can also be detected by using special sample holders, as shown in Fig. 5.1c. Some of the primary electrons generate transmitted electrons (TE), which are converted into secondary electrons by a thin gold layer. Both STEM versions can be operated in brightfield (BF) mode and darkfield (DF) mode. BF images are produced by inelastically forward scattered electrons, while DF images are produced by elastically scattered electrons, which are more strongly deflected at higher angles. It is worth noting that the relatively simple modeling (physical modeling or via Monte Carlo simulation) of the transmitted signal STEM in-SEM is straightforward and has been recently reported in the literature, for example [2]. This background is critical

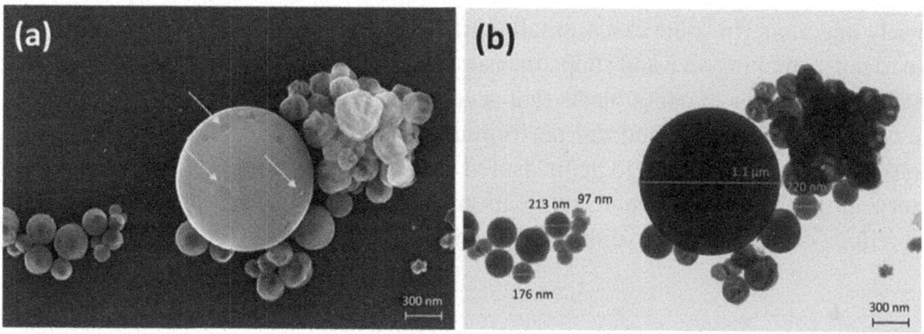

Fig. 5.2 SEM micrographs of SiO_2 particles obtained with the high-resolution in-lens SE detector (**a**) and with converted transmitted electrons detected by the E–T detector (**b**) Note nanometre-scale details on particle surface morphology in the SE image (arrows) and sharp particle boundaries in the STEM image

to the precise measurement of nanoparticle size and makes STEM one of the very few analytical techniques that can provide traceable results on nanoparticle size and size distribution.

Figure 5.2 presents a representative example of complementary analytical top-view SEM and STEM imaging utilizing the same field of view. Thus, nanoscale details of particle surface topography provided by high-resolution in-lens detectors can be complemented with information obtained from transmission micrographs. Further practical examples of nanoparticle analysis utilizing complementary SEM and STEM-in-SEM modes can be found in ref. (3, 4). More elemental composition information can be obtained by energy dispersive x-ray analysis (EDX). The spatial resolution of elemental information itself is poorer than that of SE or STEM images, as it is significantly more informative. However, in the analysis of nanoparticles, compositional information is indispensable.

For simplicity, this chapter only deals with two morphological properties, size and projected shape, for discrete and aggregated or aggregated nanoobjects. Rigorous 3D characterization of nanoparticles will include size, shape, surface structure (e.g. texture), surface and internal material composition, including their chemical state and location in the volume being studied.

Proper sample preparation is key to obtaining high-quality, high-resolution electron microscopy images and excellent nanoparticle characterization. The preferred technique generally varies with the sample material. It is also important to ensure that the SEM itself is suitable for the required uncertainty of the measurement. The size of nanoobjects that can be measured well depends on the sample, the required uncertainty and the performance of the SEM. The resolution of some SEMs may not be sufficient to determine size and shape differences between small particles (eg, nanoparticles smaller than 10 nm). For example, if an electron optical column focuses the primary electron beam with a spot size of 5 nm, the difference between the images of 3-nm and 4-nm sized particles may lie only in the signal intensity at the particle's location, which may make the size measurements of these particles largely uncertain. In some cases, small particles cannot be identified in SEM images due to noise, making size and shape measurements infeasible. To obtain high-quality results, the number of nanoobjects that need to be measured depends on the sample, the required uncertainty and the performance of the SEM. Typically, hundreds or thousands of particles need to be measured for statistical sound size and shape characterization. Optimization and automation of image acquisition and data analysis can reduce costs and improve the quality of results.

5.1.1 Sample Preparation

The quality of SEM nanoparticle size and shape measurements largely depends on sample preparation. For all nanoparticles, there is no single best preparation method, but rather a more or less excellent method is determined by the properties of the nanoparticles and the analyte, measurand, or nature of the measurand. Ensuring that a test or analysis sample is representative and relevant to the measurement is essential. Tools used to prepare and handle nanoparticles must be kept clean and stored dry in a clean enclosure or bench, preferably in high-efficiency particulate air (HEPA) filtered air for sample preparation. Information on the health effects of various nanoparticles is incomplete. Some nanoparticles are known to be harmful to health, so proper personal protective equipment including disposable gloves, safety glasses, lab coat, and filter respirator must always be worn and appropriate precautions taken when handling nanomaterials.

Nanoparticle samples are usually powders or suspensions and often require dilution to image discrete particles or minimize particle packing. If the presence of small grains on top of large grains can be seen with a high-resolution SEM, the small grains hidden beyond the large grains cannot be detected. Measuring exposed particles is possible, but it may require operator assistance to separate the particles to avoid unacceptably high measurement uncertainties. The goal is to isolate a small number of nanoparticles from the original nanoparticle material and deposit on the substrate surface in order to prepare a homogeneous and representative sample for testing or analysis. The contrast between the particles and the substrate should be high and the particles should not be too far apart so that a large number of nanoparticles can be measured in relatively few images. Charge-based trapping of nanoparticles on functionalized substrate surfaces has shown better results compared to simple dispersion [3]. All TEM/STEM samples are suitable for SEM imaging, so published methods of nanoparticle preparation for TEM/STEM imaging and measurement are applicable to samples prepared for SEM, even with non-electronically transparent substrates, such as silicon wafers. For SEM and STEM imaging, powder samples are typically dispersed in a liquid and then deposited on a substrate (eg, a 10 mm $\times$ 10 mm or smaller conductive silicon wafer) or a 3 mm electronically transparent grid [4].

Many nanoparticles can be in the liquid colloidal phase, that is, suspended in another substance, and electrostatic or steric stabilization methods can prevent agglomeration. In order to minimize the number of nanoparticles on the contact

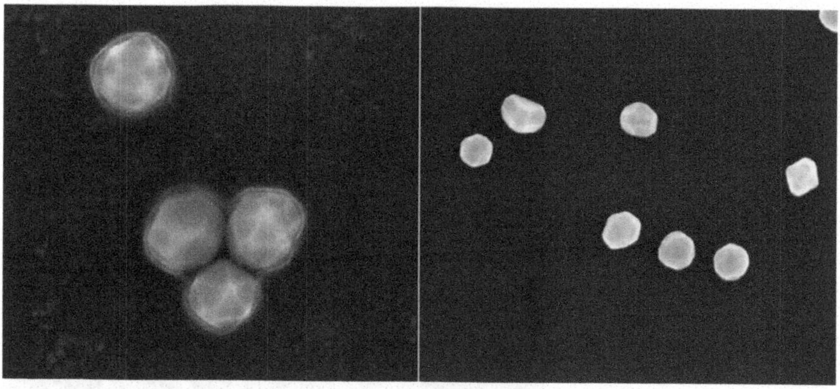

Fig. 5.3 SE images of colloidal Au particles deposited on Si chips with thin SiO_2 on top; 80-nm nominal size, deposited after dilution only, left; and 60-nm nominal size, diluted, poly-L-lysine captured and rinsed particles, right. HFW 429 and 600 nm, respectively

measurement substrate or grid, these nanoparticles must be diluted. Charge-based trapping of nanoparticles and careful flushing out of the suspending medium yielded better results than simple deposition (Fig. 5.3). Notice on the right that there is no residual material, the background is cleaner, and the contrast is better, allowing for easier and more accurate measurement of size and shape.

There are commercially available TEM grids for nanoparticle sample preparation. They may have surface coatings or functionalized surfaces designed to ensure good particle dispersion and capture. Adding diluents and surfactants to the stock nanoparticle material may help prepare useful test samples. Depending on the imaging parameters, both of these factors may alter the apparent size of the nanoparticles, so appropriate measures must be taken and, if necessary, the consequences of these effects must be compensated for. Depending on the viewpoint, non-spherical particles can appear very different in size and shape. Quasi-spherical and non-spherical nanoparticles can be deposited preferentially and on their largest surface area. Some nanoparticles with elongated shape, such as TiO_2 nanorods, may not always adhere to the sample surface at the same angle, so their apparent size and shape vary greatly (Fig. 5.4). Pairs of images taken from two angles help determine whether the nanorods were indeed deposited parallel to the planar substrate. When determining their size and shape, it is essential to evaluate and use information about how 3D particles are deposited on the surface of a 2D sample in order to minimize measurement errors and resulting uncertainties in the results. For this purpose, operators are preferred to aid in the determination of particle boundaries.

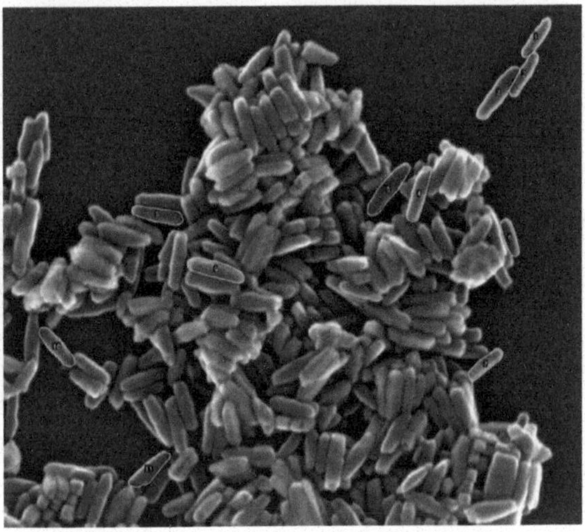

Fig. 5.4 TiO_2 nanorod sample prepared on Si chip. The boundaries of 11 horizontally deposited particles were determined manually. 635-nm HFW

5.1.2 SEM Image Acquisition

The first thing to do before starting to acquire images for nanoparticle characterization is to make sure the SEM is working properly. SEMs are easy to operate and provide fast results, but a number of well-known issues prevent optimal performance of SEMs and lead to a lack of excellent reproducibility in imaging and measurements. The most troublesome ones are unintended motion, drift of the sample stage and the main electron beam, contamination by the electron beam, image blur (lack of sharp electron beam focus), geometric distortion, wrong scale, charging and noise. Quantification of these basic performance parameters is necessary; it also aids in the calculation of measurement uncertainties as well as in maintenance and repair. Contamination from the electron beam can easily make measurements of nanoparticles erroneous or outright impossible. Contamination can come from carbonaceous (mainly alkanes) molecules covering the surface of the sample or the instrument, or both. These molecules have high mobility under vacuum conditions, so they may flow in from a distance, but deposit when secondary electrons cut their bonds. In some cases, the energetic electrons of the main beam may overwhelm the deposit, and the center of the image may be cleaner than the periphery of the image frame. Figure 5.5 shows a pair of SE and bright-field STEM images of a TiO_2 sample, where the deposited contamination makes it impossible to identify the particles in SE mode, while the STEM image better reveals the particles. Contamination makes dimensional measurements impossible with SEM and prone to errors with transmitted images. Cleaning with low-energy oxygen or hydrogen plasma or ultraviolet light in vacuum allows the source molecules to be volatilized and pumped out, enabling essentially pollution-free operation [5].

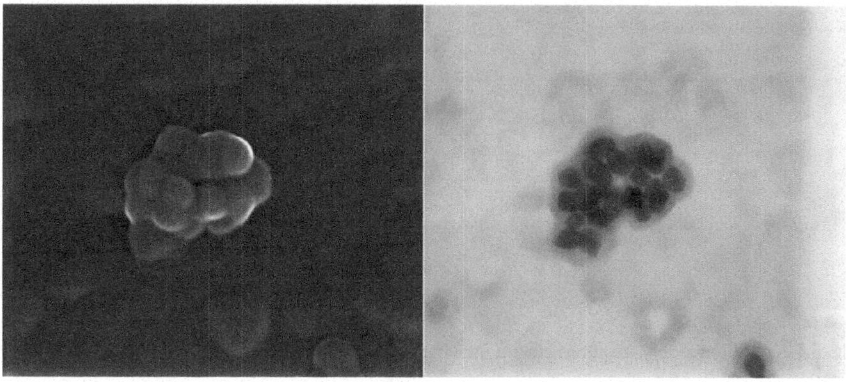

Fig. 5.5 A severe case of contamination: secondary electron (left) and dark-field STEM (right) images of TiO_2 nanoparticles. 15-keV landing energy, 635-nm HFW

For high-quality SEM characterization of nanoparticles, obtaining high-quality images is essential. To do this, in addition to ensuring that the SEM meets its performance specifications, once focus, astigmatism is set, most gray levels need to be used without over- or under-saturation of the image by setting contrast and brightness appropriately, an example is shown in Fig. 5.6.

In SEM, both reflection SE and BSE as well as transmission electron imaging modes are possible, see Fig. 5.7. Simultaneous acquisition of multiple signal images has its advantages as they provide additional useful information for nanoparticle characterization with no time burden, only additional file storage, which is a negligible additional cost. Simultaneous acquisition also removes distortion differences between image pairs. In scanning transmission mode of operation, the landing energy

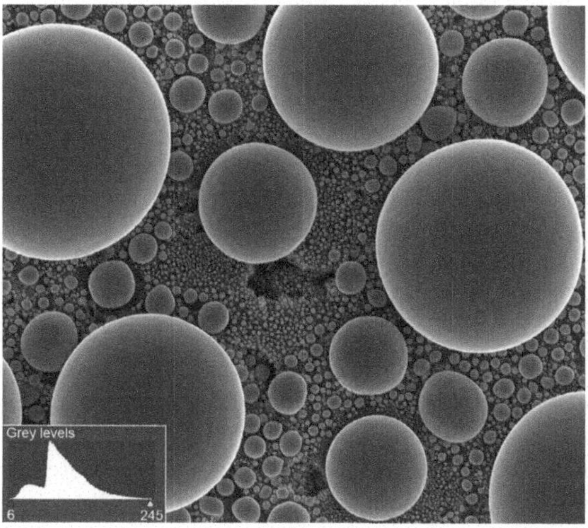

Fig. 5.6 Properly set contrast and brightness, with no under- or oversaturation, as shown in the histogram at the lower left corner. SE image of evaporated tin spheres, varying in diameter from several micrometers to a few nanometers

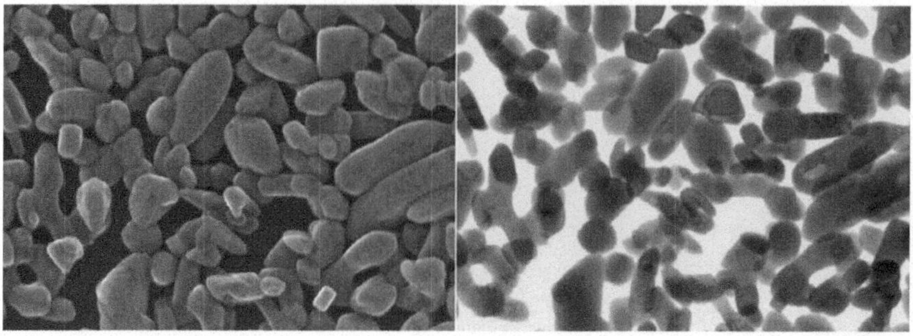

Fig. 5.7 Simultaneously acquired SE image (left) and bright-field STEM image (right) of TiO_2 nanoparticles at 30-keV landing energy. 508-nm HFW

of SEM is much lower than that of transmission electron microscope (TEM), which is advantageous in most cases, namely contrast enhancement for near-Z elements. Measurements of discrete nanoparticle samples are often easier to perform with automated image acquisition and particle analysis. Overlapping, aggregated or agglomerated particles and highly variable particle/background samples may require operator assistance in image acquisition and analysis.

5.1.3 Particle Size Measurement Based on SEM Images

Size measurements determine a set of parameters related to the size or size distribution of 3D particles, for example, the circular equivalent diameter in a projected view image. The first step is to obtain a high-quality image of the nanoparticles; the second is to use suitable processing, criteria, and algorithms to generate a binary (black and white) version of the grayscale image in which the particles are isolated, i.e. determine their boundaries from the background and each other, and finally use the binary image for size measurements, as shown in Fig. 5.8. Depending on the resolving power of the SEM, the size and size distribution of particles below 10 nm can be measured. For the highest resolution SEMs with a resolution of 0.4 nm or better, the smallest measurable nanoparticles start at 1 nm. Other instruments, especially those with conventional tungsten-filament electron sources, have a spatial resolution of only a few nanometers and may not be able to image the smallest nanoparticles. The actual minimum measurable nanoparticle size depends on the instrument, environment and sample. Particle counting remains feasible due to the lack of high enough resolution for reliable size determination of nanoparticles. For this, detecting the presence of nanoparticles may be sufficient. Using a suitable sample, the limits of characterizing very small-sized nanoparticles with a given SEM can be determined by comparing high-resolution and more sensitive other measurements. Measurement uncertainty also depends on the resolving power of the SEM. In general, a sharper, more focused electron beam results in images showing more detail and what appear to be smaller

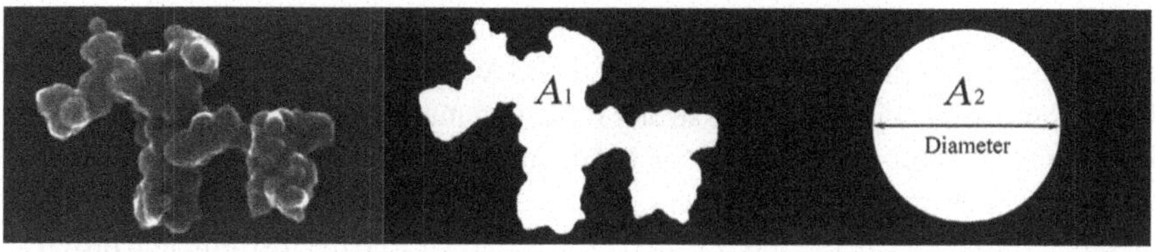

Fig. 5.8 Determination of circular equivalent diameter: SEM SE image of a carbon black aggregate nanoparticle (left), its binary version (middle), and its area-equivalent circle, A1 ¼ A2. 424 nm HFW

particles, while a lack of focus or resolution makes them appear larger but with less detail and clarity. This does not necessarily mean that high-quality size and shape distribution measurements cannot be made. Proper calibration using relevant reference samples of known particle size [6] can help to significantly reduce measurement uncertainty.

5.2 Transmission Electron Microscopy Characterization of Nanomaterials: Measurement Procedures

Transmission electron microscopy (TEM) is a versatile technique that can analyze the size, morphology, crystal structure, and chemical composition of various nanomaterials (NMs). It can be considered the gold standard for nanoscale characterization for several reasons [7]: (i) TEM analysis is one of the few methods that can provide reliable spatial resolution covering the entire nanometer size range from 1 to 100 nm. (ii) TEM readily produces projected 2D images of NM. (iii) TEM imaging combined with image analysis can quantitatively determine the physical properties (size, shape, and surface morphology) of individual nanoobjects based on their characteristics in 2D projections. As required by regulations and guidelines, multiple properties of each individual particle can be measured simultaneously, from which descriptive statistics and corresponding number-based distributions can be determined [8]. (iv) TEM can assess the agglomeration and/or aggregate state of the material and, to some extent, can identify primary particles within agglomerates/agglomerates. (v) Spectroscopic methods (EDS and EELS) can be incorporated into TEM for elemental analysis of nanoobjects and examination of chemical bonds, allowing the characterization of nanoobject subpopulations in mixtures and nanoobjects in complex matrices. (vi) Selected area electron diffraction (SAED) allows the study of the crystal structure of nano-objects.

Despite the wealth of information available, the applicability of TEM to characterize nanoobjects is currently limited by the following drawbacks: (i) The reliability of TEM analysis depends on transferring a representative portion of the sample containing a sufficiently large number of particles to the sample support (TEM grid). This is affected by sample preparation (including purification and concentration steps) and TEM sample preparation. (ii) The spatial resolution of TEM depends on the accelerating voltage of primary electrons. Therefore, traditional TEM methods usually use accelerating voltages of 100–300 kV to obtain high spatial resolution. While compatible with metals or strong inorganic materials (so-called hard matter), beam damage limits the analysis of beam-sensitive "soft matter" such as certain zeolites, porous materials, polymers, hybrid materials, and selected carbon-based nanomaterials. Ke et al. [9] and Borrnert et al. [10] reviewed the instrumental progress using aberration correction techniques to overcome the aberrations caused by the imperfection of the electromagnetic lens when the electron microscope operates at low accelerating voltages, thereby improving high-resolution imaging at low accelerating voltages. (iii) Conventional TEM applications are further limited by artifacts generated under high vacuum conditions. Technological advances, such as liquid-cell electron microscopy (EM) [11], ambient EM [12] and cryo-EM-based applications [13], can help overcome this limitation, at least in part. (iv) TEM methods generally estimate the two dimensions of particles perpendicular to the electron beam, but not the dimension parallel to the electron beam. Therefore, measurements of particles with a preferential orientation on the EM grid, such as platelets, may be biased. Operational modes such as energy-filtered TEM (thickness mapping) and transmission electron tomography can be used to acquire 3D information of particles [14].

Clearly, recent technological innovations can overcome the limitations of conventional TEM and become an important tool for understanding the relationship between the physical properties, structure, and composition of nanomaterials down to the atomic scale. However, since many of the advanced methods pointed out earlier are still very expensive and labor-intensive, dedicated infrastructure is still limited to highly specialized research institutions. Initiatives to standardize and validate these methods are just beginning. In this regard, this chapter describes methods for the TEM characterization of nanomaterials using widely available equipment. The proposed measurement procedure is intended to be applicable to NMs that are currently produced and sold in large quantities. The method requires that the fraction of nanoobjects deposited on the EM grid represent the entire material and that individual nanoobjects can be detected. To be suitable for TEM analysis, samples and EM samples need to be carefully prepared, for example, following the guidelines described in Ref. [15]. The focus of this chapter is on the description and measurement of TEM-accessible physicochemical properties, which are crucial in defining a material as an NM [16] or assessing its safety and toxicological potential [17] in a legislative and regulatory context.

5.2.1 Measuring Principle of Transmission Electron Microscope

A transmission electron microscope uses an electron source to generate a beam of primary electrons, which is accelerated by an electric potential and projected onto a thin sample through a set of lenses and apertures. Figure 5.9 shows how an electron beam passes through an electron microscope in TEM mode under the influence of lens and aperture settings. Part of the electron beam can pass through the sample without interacting with it. Other electrons will be scattered by atoms in the sample. The information contained in the electron waves that exit the sample is the means by which an image is passed through the objective lens system. Portions of the sample can be selected or excluded by placing an aperture (i.e., the objective aperture) in its back focal plane. In imaging mode (that is, when the image is projected onto a screen), when the aperture is positioned so that only transmitted electrons can pass through, a so-called bright-field image is formed.

When the aperture selects only diffracted electrons, a dark field image is formed. The imaging plane of the objective is projected onto the viewing screen through a series of additional lenses, which also ensure further magnification and correct positioning of the electron wave distribution. Usually, the observation screen is a fluorescent screen, and a CCD camera is used to record the image. At low magnification, TEM image contrast arises from the absorption and scattering of electrons in the sample due to the thickness and composition of the material (i.e., mass-thickness contrast), and crystallographic orientation (i.e., diffraction contrast). Combining bright-field TEM imaging with image analysis based on the gray-scale values of a

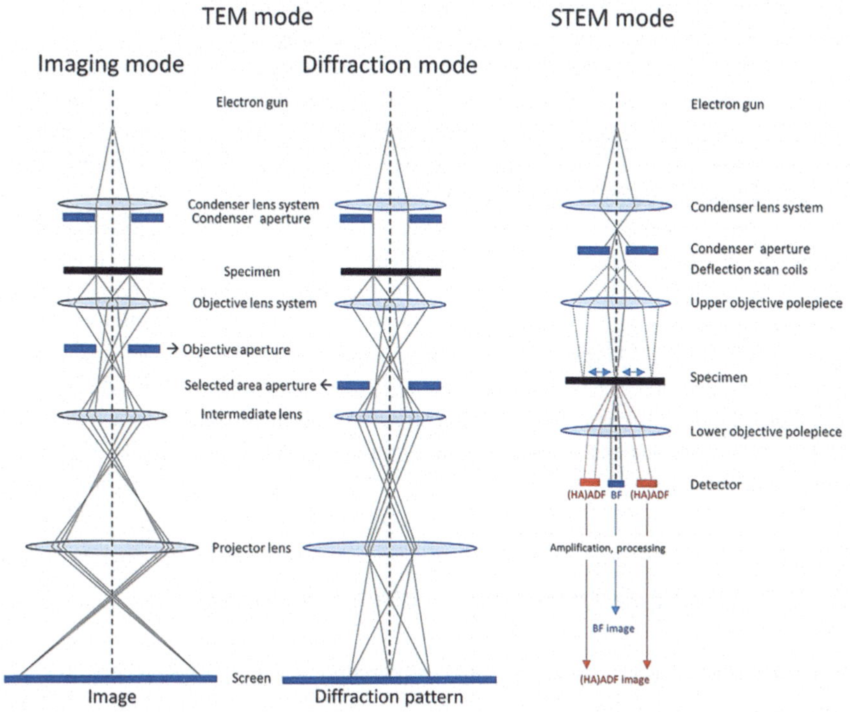

Fig. 5.9 Illustration of how an electron beam passes through the electron microscope in TEM mode (imaging and diffraction) and in STEM mode, under the influence of the lens and aperture settings

large number of nanoobjects yields a number-based distribution of projected nanoobject size and shape properties, and the sample can be modeled as an object that changes the phase of the incident wave. In crystalline nanoobjects, elastic coherent scattering causes an incident electron beam to diffract at regularly arranged atoms. Whether the interference of diffracted electron waves with undiffracted transmitted waves is constructive or destructive depends on the defocus and phase shift of the electron waves in the sample. The interference of transmitted and diffracted electrons produces so-called phase contrast. Thus, the electrostatic potentials of the atomic columns can be visualized and related to the crystal structure of the nanoobjects, although they are not a direct projection of the atomic positions. In order to provide information about atomic arrangements, it is often necessary to compare HR-TEM images with computer simulation images based on atomic models. In addition to providing information on crystal structures, HR-TEM can also be applied to study defects and measure the thickness of nano-objects based on focal sequences [18]. Diffraction by three-dimensional periodic structures, such as crystals, is the result of interference between waves reflected from different crystal planes. For crystalline materials, the electron beam undergoes Bragg diffraction. This can be used to obtain crystal structure information of crystalline nanoobjects [19]. The electron diffraction pattern of larger regions containing many crystalline nanoparticles consisted of a ring pattern similar to X-ray powder diffraction. This ring pattern originates from the different orientations of the particles (which can be single or polycrystalline) relative to the electron beam, and can be used to identify the texture in addition to obtaining crystallographic information. The recorded diffraction patterns should be indexed using a diffraction database such as AtomWork [20]. An aperture can be inserted in the imaging plane of the objective to obtain a diffraction pattern from only this selected area (SAED) of the specimen.

More complex behavior can also be observed in diffractive planes. For example, in convergent beam electron diffraction (CBED), the interaction of the convergent beam with the specimen can provide information beyond structural data, such as determining the space group or determining specimen thickness. A relatively new application to study NM is precession electron diffraction (PED), in which an electron beam is tilted and precessed on a cone that has a common axis with the TEM optical axis. PED allows solving crystalline nanostructures ab initio, performs artifact-free 3D BF image reconstruction, and creates nanometer-resolution EBSD-TEM-like orientation/phase maps.

5.2.2 *Scanning Transmission Electron Microscopy*

Similar to SEMs, TEMs can be operated in scanning mode (STEM), where an electron beam is focused into a (sub) nanometer-sized probe and scanned across the sample. Spatial resolution is mainly controlled by the nanoscale illuminated electron probe, which can reach 0.1 nm in high-resolution STEM mode when using a probe-corrected TEM. Various signals resulting from electron scattering can be detected and displayed as a function of probe position. Interaction with the specimen gives precise local physical and chemical information. Typically, images are created using small on-axis detectors (bright field) or large annular detectors with fixed dimensions (dark field). Similar to HR-TEM, BF-STEM can be used to visualize atomic columns. If a ring detector is used and the inner collection half angle is set to collect only the diffracted beam, then a diffraction contrast dark field image can be obtained. This is typically used for samples composed only of light elements. High Angle Annular Dark Field (HAADF) imaging refers to the use of geometrically larger annular detectors, placed outside the optical far field of the sample. Due to the detector geometry, only electrons scattered at high angles are detected. Electrons scattered at low angles are mostly coherent (and like this conventional brightfield and darkfield images can have contrast inversion with changes in sample thickness, orientation, or defocus), while electrons scattered at high angles are mostly incoherent. The incoherence mainly comes from the detector geometry and thermal diffuse scattering. Since high-angle incoherent scattering is related to scattering from atomic nuclei, it gives an intensity more or less proportional to the square of the atomic number, as would be expected from Rutherford scattering. Therefore, this imaging method is also called z-contrast imaging. One of the advantages of HAADF-STEM imaging is the ability to visually distinguish materials with different chemical compositions. Furthermore, if the elements contained in the sample are known, the HAADF-STEM image directly gives the qualitative two-dimensional distribution of different materials in the sample. Since the ring detector is used to rule out Bragg scattering and eliminate phase problems, the contrast inversion of HAADF-STEM images caused by defocus and thickness variation is largely suppressed.

5.2.3 Transmission Electron Microscopy Analysis

The elemental composition of a sample can be obtained by analyzing the energy of emitted x-rays (energy dispersive x-ray analysis, EDS) or the energy loss of transmitted electrons (electron energy loss spectroscopy, EELS). Both of these microanalytical techniques can be incorporated into TEM systems, and EDS can also be incorporated into most SEM systems. Inelastic scattering occurs when energy is exchanged between the primary electrons from the electron beam and the sample. If enough energy is transferred, an electron is ejected from the inner atomic shell of the sample material, creating a vacancy that puts the atom in an excited state. To stabilize the atom, an electron in the outer atomic shell (higher energy level) then fills the vacancy in the inner atomic shell (lower energy level). The energy difference between the two levels will be released during the transition as X-rays or Auger electrons. Spectrometers measure the energy of x-rays emitted from a sample. Since x-ray energy is characteristic of the electronic structure of atoms, it can be used for qualitative and qualitative analysis. Elements with atomic numbers ranging from beryllium to uranium can be detected. The minimum detection limit varies from about 0.1 to a few atomic percent, depending not only on the elemental sample but also on the type of EDS detector used. Furthermore, quantitative results can be obtained by extracting the intensities of the characteristic peaks of the constituent elements after correcting for signal intensities related to the background (bremsstrahlung), according to calculated or experimentally determined sensitivity factors. In TEM, EDS is usually performed in STEM mode to obtain high spatial resolution. EDS elemental mapping combines the spatial information obtained in STEM mode with the x-ray signal to construct the elemental distribution of each detected element within a predefined area. Recent high-throughput EDS systems record the entire EDS spectrum at each position of the main electron beam, allowing the extraction of virtually any elemental profile from the encoded data cube. Elemental maps can be overlaid with STEM images. More details and examples on EDS vs. SEM operation. EELS (see Fig. 5.10) measures the energy lost by inelastically scattered primary electrons to obtain information about the scattering atoms in the sample. The energy required to extract an electron from the inner shell of an atom in a sample is element-specific. Therefore, the primary electron energy lost by exciting an atom in a sample is also element-specific and can be used to obtain qualitative and quantitative information on the elemental composition of the sample [21]. Quantitative analysis in EELS typically uses computed cross-sections that have undergone background removal and deconvolution to exclude multiple scattering effects.

Thus, both EELS and EDS are related to the same phenomenon (ejection of electrons from atoms' inner shells), but use different methods to probe the process. EELS is often considered to complement EDS. While spectrometers are good at identifying the elemental composition of materials, are easy to use, and are particularly sensitive to heavier elements, spectrometers tend to be more effective for elements with relatively low atomic numbers, where the excitation (core loss) edges are sharper and more defined. EELS-based elemental distribution maps can be obtained by sequentially acquiring EELS spectra from each position of the (focused) main electron beam (STEM EELS) while scanning the electron beam on the sample, or by collecting a series of images obtained with increasing energy loss around the excitation edge of the element of interest (parallel illumination).

5.3 X-ray Diffraction of Nanomaterials

The X-ray diffraction (XRD) technique was first proposed by German physicist Max von Laue in 1912. In late 1895, Wilhelm Röntgen discovered the properties of x-rays, which were unknown at the time; knowledge about the structure of matter was also in its infancy. Thus, von Laue's experiments were revolutionary. Von Laue, with the help of Walter Friedrich and Paul Nipping, directed a beam of x-rays at a copper sulfate crystal, as shown in Fig. 5.11. By analyzing the observed diffraction patterns, they also proved that X-rays are electromagnetic waves whose wavelength is very short compared with visible light, and that crystals are composed of ordered atoms periodically distributed in space, with characteristic distances in the same order. These important discoveries earned von Laue the Nobel Prize in Physics in 1914. In

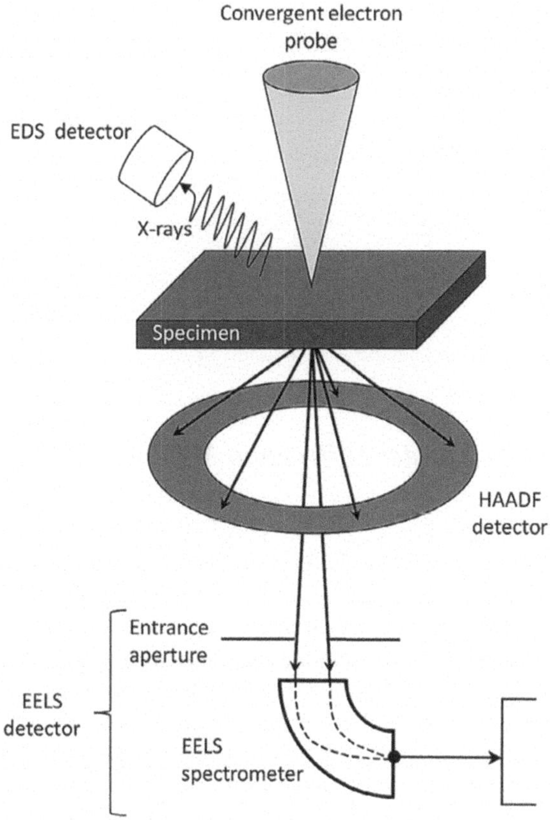

Fig. 5.10 Illustration of how the electron beam passes through the electron microscope in STEM mode and where the EDS and EELS detectors are positioned

1913, William Henry Bragg and William Lawrence Bragg (father and son), winners of the 1915 Nobel Prize in Physics, determined the first crystal structure using x-rays. They studied and characterized the atomic order of sodium chloride and other simple compounds. A few years later, researchers began applying the technique to determine the crystal structures of more complex inorganic compounds, starting the field of science now known as "crystallography".

The importance of XRD is obvious, and it clears the way to determine the atomic or molecular structure of all types of materials, which is a prerequisite for understanding their properties. Currently, this technique covers a wide range of applications, from relatively simple but very important materials science research to the structure determination of complex biological macromolecules, including proteins, nucleic acids, and ribosomes.

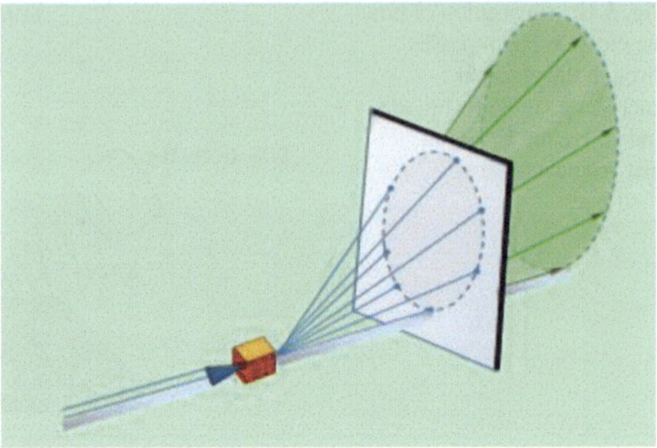

Fig. 5.11 Schematic view of the experiment conducted by Max von Laue in 1912, which resulted in the development of X-ray diffraction and modern crystallography

5.3.1 Powder Diffraction Method

XRD applied to crystalline powders: X-ray powder diffraction (XPD) is a variant of the single crystal XRD technique in which the sample is a polycrystalline material with a large number of randomly oriented small crystals. This method is also applicable to powders formed from crystals with preferred orientation. Typically, powders of polycrystalline materials with a crystal size of a few micrometers are investigated. In this case, it is assumed that for any crystal plane with an interplane distance, there will always be a considerable portion of correctly oriented crystals satisfying Bragg's law, so all reflections satisfying the criterion $|Fhkl| \neq 0$ will be observed in the experiment. Note that XPD experiments cannot produce diffracted beams that follow a well-defined single direction. Instead, there is a set of directions in which the diffracted beam forms the so-called "Bragg diffraction cone", as shown in Fig. 5.12. The reason for this phenomenon is that for an ideal powder formed from a large number of randomly oriented crystals, there will always be some crystals whose crystal planes form an angle θ with respect to the incident beam, which diffract the x-ray beam in different possible directions around the incident beam. Therefore, for each interplanar distance of a given crystal powder, a diffraction cone with an opening of 4θ is obtained, so that the diffraction intensities corresponding to all these cones can be recorded simultaneously.

Conventional laboratory diffractometers typically use step-and-scan procedures and x-ray detectors with small sensitive areas, or, one-dimensional position-sensitive detectors, which in both cases allow one to measure the intensity associated with only a very small fraction of the diffraction cone, resulting in the loss of most of the diffracted beam. However, two-dimensional detectors allow the simultaneous collection of diffracted intensities corresponding to the entire diffraction cone. When the incident x-ray beam is generated by a synchrotron light source, and the photon energy E is relatively high, that is, the wavelength is relatively short, a two-dimensional planar position-sensitive detector [$\lambda(\text{Å}) = 12.39/E(\text{keV})$] is generally used. In this case, all Bragg angles are smaller than those observed at low energies, allowing one

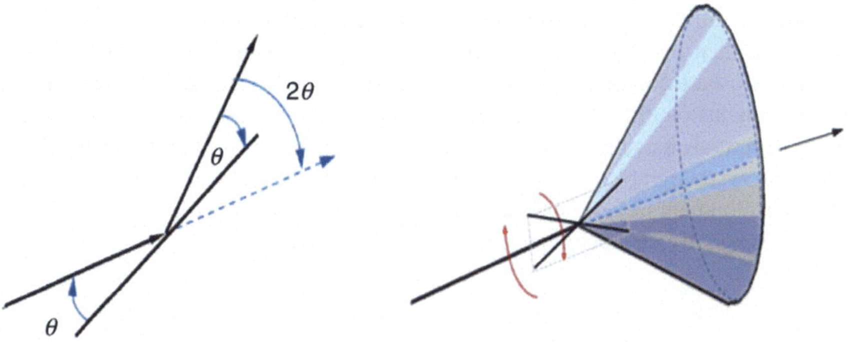

Fig. 5.12 The origin of the diffraction cones observed using the XPD method. XPD, X-ray powder diffraction

to measure the intensity associated with more cones. This type of measurement is faster, typically exhibits better photon counting statistics, and is often used to study in situ phase transitions of materials in real time.

5.3.2 Modern Methods for XPD Data Analysis of Nanomaterials

So far, classical methods for analyzing Bragg peak spectra have been proposed and discussed, providing a relatively simple method for obtaining information on average grain size and microstrain. These methods are very useful in improving synthetic processes and understanding the physicochemical properties of nanostructured materials. Currently, new methods are available to determine structural parameters from complete powder diffraction patterns rather than from individual diffraction peak profiles. In particular, the rietveld method allows one to characterize crystallographic and microstructural aspects in a single analysis, using subroutines to determine crystal shape, strain direction, etc. With the development of computers, the traditional methods discussed in the previous sections have led to more complete procedures called full pattern profile fitting (FPPF) and full pattern profile modeling (FPPM), based on the simultaneous analysis of the entire diffraction pattern, thus solving the problems associated with the superposition of diffraction peaks. In FPPM, the diffraction pattern across the sample is modeled by convolving different effects with well-defined physical meanings, such as those due to the distribution of crystal size, dislocations, and twins. A different, more modern approach to studying nanocrystalline materials from XPD modes—using high-energy (i.e., small-wavelength) x-rays—is based on the pair distribution function (PDF). PDFs contain information about the relative distances between atoms, which can be used to determine the degree of local order in a material, the distances between neighboring atoms, coordination numbers, etc. Compared to processing conventional diffraction data, the

PDF method allows the determination of the local atomic order of a material. The results can then be correlated with X-ray absorption spectroscopy data, the most commonly used technique for structural analysis of the local atomic order of materials. In conclusion, the XPD technique provides complete and valuable information for the structural characterization of nanomaterials, including information not strictly related to crystallography, including crystal morphology and local atomic order. The recent development of new analytical strategies, as mentioned earlier, clearly shows that XPD experimental techniques are still evolving.

5.4 X-ray Photoelectron Spectroscopy

5.4.1 Introduction to X-ray Photoelectron Spectroscopy

The few atoms in the outermost layer of any object are critical to many important functional properties, including wettability, electrical conductivity, friction and chemical reactivity. Where interfacial properties are critical, techniques sensitive to only the uppermost 10 nm are used and find applications in fields as diverse as semiconductor manufacturing, biomaterials, corrosion, coatings, and packaging. A common finding of surface chemical analysis is that the top few atomic layers of a material bear little resemblance to the bulk unless the environment is carefully controlled and rigorous surface cleaning procedures are followed. For colloidal nanomaterials, surface atoms can constitute a substantial fraction of the total number of atoms in the material. However, measuring the surface properties of nanomaterials is still not an easy task. A spherical particle 10 nm in size has 15 percent atoms on its surface, and it is a challenge to distinguish these atoms from other atoms using conventional spectroscopy. To identify and measure surface atoms, it is appropriate to use surface-sensitive methods that selectively collect information from the outermost layers of the material so that bulk signals do not dominate the data. A related example is citric acid "gold nanoparticles". If these are carefully prepared for analysis by first removing all salts from the suspending medium and then drying, the surface of the particles typically consists of carbonaceous material 1 nm thick. However, XPS showed that most of the carbon does not come from citrate ions, but mainly from hydrocarbons [22]. This is not surprising since almost all surfaces have such an indeterminate layer of hydrocarbons. These findings may be important because, if the surface of the gold particle needs to be coated, for example, with a thiol-bound ligand, it is important to confirm that the coating has replaced the non-immobilized material [23]. If you don't, you run the risk that your particles won't work or reproduce. In general, processes that rely on particles interacting through their surfaces need to be verified using surface chemical analysis.

Electron spectroscopy encompasses a range of techniques in which electrons emitted from materials are used to determine their chemical and structural properties. When they are applied to solids and liquids, they provide surface-sensitive information because electrons can travel very short distances in matter, typically just a few nanometers. Such scale lengths are ideal for the detection of nanostructured materials, and both XPS and AES are routinely used for this purpose [24]. These techniques should be distinguished from x-ray emission techniques such as EDX, which generate information from hundreds of nanometers of material and cannot be described as surface sensitive. For decades, XPS and AES have made significant contributions to the understanding of heterogeneous catalysts. Such materials typically consist of a few nanometer-sized metal particles dispersed on a solid support. Thus, there is a long history and extensive literature on electron spectroscopy of nanoparticles. As the fields of application of nanomaterials continue to expand, so does the need for the chemical measurements provided by electron spectroscopy. Over the past decade, the technique has been applied more frequently to various forms of nanoparticles, and the number of publications using XPS to study nanoparticles has increased tenfold since the early 2000s.

5.4.2 Analysis Usage

Since each element has a set of characteristic peaks in the XPS spectrum, the main use of this technique is to detect the presence of elements on the surface of a sample. If more than one element is present, it becomes important to know whether there is a chemical interaction between some of these elements. In these cases, it is instructive to obtain detailed spectra of the peaks themselves. These spectra are usually acquired at higher resolution, often referred to as "narrow scans". The shape and location of peaks can provide valuable information. The main effect observed is caused by electrostatic charges on the atoms. If the atoms are bound to a more electronegative substance, they become positively charged relative to the element, which increases the energy required to remove an electron. As a result, the electron binding energy increases, and XPS peaks are observed at lower kinetic energies. The opposite happens if it is combined with an electropositive element. This difference in energy from the element is known as the chemical shift. In many cases, spectra can be interpreted in this simple manner, and binding energies of elements in different chemical states can be found in textbooks, research papers, and web resources. The chemical shift of core shell electrons does not strongly affect the cross section for photoionization, so if more than one chemical environment can be distinguished in a material, the proportion of atoms in each environment can be deduced from the relative peak areas.

This simple form of interpretation is useful in many situations, but physical processes that can occur in XPS spectra complicate the shapes and positions of elemental peaks. These processes arise from the interaction of multiple electrons during photoionization. The binding energy of an electron is the energy difference between the initial and final state of an atom. During photoionization, the electrons around the atoms can rearrange to accommodate the positive nuclear holes. This relaxation process is strongly influenced by chemical bonds. For carbon in organic compounds, this is a negligible effect, easily explained by the binding energy. For example, each bond to an oxygen atom increases the binding energy of C1s by about 1.5 eV. For silver, it has a significant effect, with only a small change in the Ag three-dimensional binding energy of silver oxide, which is counterintuitively slightly lower than that of metallic silver. The appearance of additional peaks in a narrow scan of an element does not necessarily mean that the element has more than one chemical state. For some elements in specific chemical environments, photoionization events are likely to involve the excitation of multiple electrons. These events are evidenced by additional peaks on the high binding energy (low kinetic energy) side of the main photoelectron peak. These moons are known as oscillating moons, and they're usually weaker than the main peak, but not always. These characteristics, if known, are very useful in determining the specific chemical environment of an element. Specific examples include cerium and copper in oxidation state IV, and plasmonic losses in metals such as aluminum, silicon, and germanium. Other features can arise from the interaction of unpaired core electrons with unpaired electrons in the valence band. This is known as multiple dissociation and is most commonly observed in transition and rare earth metal compounds. The most typical manifestation is broadening of XPS peaks, but multiple splitting can also appear as clear and distinct peaks. Figure 5.13 summarizes these effects. For quantitative XPS analysis, the most important thing is the intensity of the peaks in the spectrum. Since XPS peaks vary in width and shape, peak intensities are usually measured in terms of peak area. The peak is accompanied by a tail to reduce the kinetic energy generated by electrons that lose energy during inelastic scattering. Careful analysis of this background can provide useful information about the depth distribution of elements, even extending to cases where the peak itself is no longer visible. Separating the peak from the inelastic background is a necessary step in measuring peak area; however, it is not trivial to draw the line between the two. This can only be achieved with accurate knowledge of the electron energy loss function, which can be obtained by direct REELS measurements. However, in most analytical situations this is not possible to achieve in a timely and cost-effective manner; therefore, an approximate description of the background close to the peak is used.

Figure 5.14 shows the high intensity Au 4f peak of pure gold. The three most commonly used backgrounds are applied to peaks using the same fixed endpoint. Linear backgrounds are self-explanatory and convenient, but have no physical basis. The intensity of the Shirley background varies proportionally to the area above the background, so it must be defined iteratively. Likewise, the Shirley background has no clear physical basis, but in cases where the background rise is larger than the peak, it provides more accurate results than the linear background, which becomes highly sensitive to the choice of endpoint. However, linear backgrounds are still useful for insulating materials and components that exist as thin surface layers. The Tougaard background is based on the physics of electron energy loss, which essentially changes the slope as the intensity above the background changes, so no iterative process is required. To obtain accurate results in the Tugard context, it is necessary to use an appropriate energy loss function. In general, Tugard backgrounds are not as sensitive to the choice of endpoints as Linear or Shirley backgrounds. There are other functional forms of the background. In analytical work, variance represents

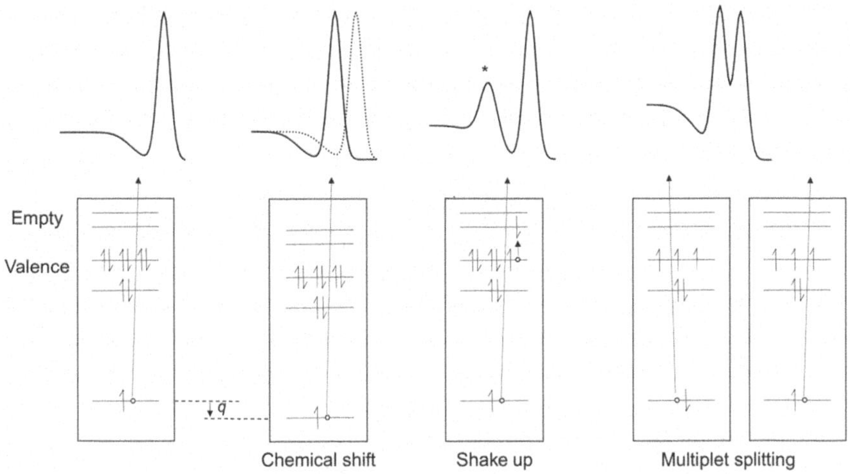

Fig. 5.13 Overview of processes that affect the position and shape of peaks in an XPS spectrum

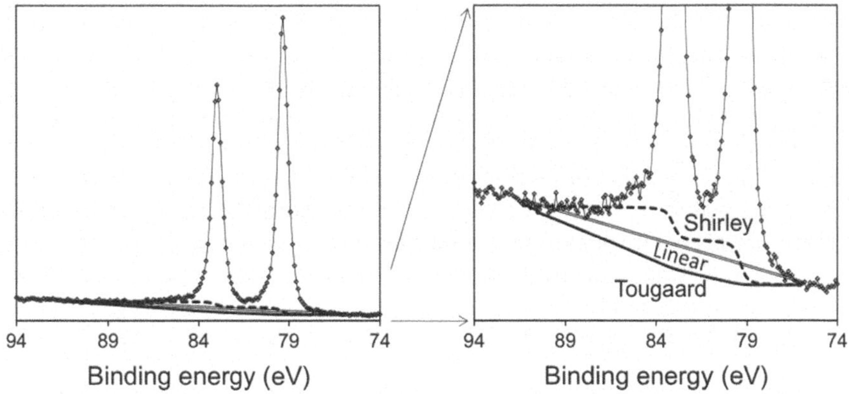

Fig. 5.14 Au 4f peaks from clean gold with three backgrounds applied. Linear background is a double line, the Shirley background is dashed and the Tougaard is a solid line

relative uncertainty in peak area measurements. When comparing sample data, it is important to use consistent background shapes and background endpoints.

5.4.3 UV–Vis Absorption Spectrum

Ultraviolet–visible (UV–vis) absorption spectroscopy is a versatile technique that can be used for the characterization and quantification of different types of organic, inorganic and biological materials. The studies by Skoog et al. and Petrozzi and

Goudie demonstrate a conceptual approach using UV–Vis absorption techniques. In particular, UV–vis is often applied to the synthesis of new materials and the study of nanoparticles in the field of nanotechnology by using colorimetric sensors in catalysis research. Furthermore, this technique is widely used to monitor the growth of nanostructured thin films grown by various techniques.

The synthesis of new materials, mainly carbon-based materials, has become an area of intense research with the aim of creating materials with new properties and applications. Carbon nanotubes and graphene are two of the most thoroughly studied materials in recent years, with their mechanical, thermal and electrical properties studied. Figure 5.15a shows the UV–vis absorption spectra of multiwalled carbon nanotubes (MWCNTs) dispersed in water, MWCNTs of different lengths. Figure 5.15b shows the changes in the UV–vis absorption spectra of large MWCNTs at different oxidation degrees, denoted by the nomenclature in parentheses. Figure 5.15c shows the deconvolution analysis of the UV–Vis absorption spectrum. The analysis breaks down the spectrum into three main bands, the second being the main band in both intensity and breadth. By analyzing the position and intensity of this energy band in different MWCNTs samples, the authors were able to conclude that its intensity is related to the degree of oxidation and the variation of MWCNTs length. Based on resolved spectra, MWCNT samples from different sources can be identified and compared. Another area of research in nanotechnology is the synthesis, characterization and application of colloidal nanoparticles, such as organic materials, including polymers and ceramics with magnetic properties, especially metallic nanoparticles. UV–Vis absorption is the main spectroscopic tool for studying these materials. Xue and Mirkin reported a photochemical synthesis route that can be used to control the growth of Ag nanoprisms. Their approach involves adjusting the pH of the solution and exciting it at specific wavelengths, which allows the synthesis of nanoprisms with uniformly controllable wavelength bands ranging from visible to near-infrared absorption. Figure 5.16 is the extinction spectrum of the nanoprism obtained at pH = 11.2 and the excitation wavelength is 488 ~ 633 nm.

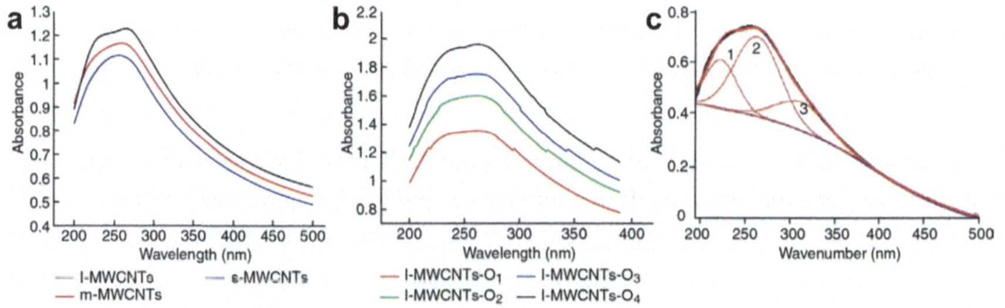

Fig. 5.15 UV–vis absorption spectra of MWCNTs **a** with different lengths and **b** with different oxidation degrees; **c** A UV–vis absorption spectrum of MWCNTs resolved into three bands. MWCNT, Multiwalled carbon nanotube; UV–vis, ultraviolet–visible

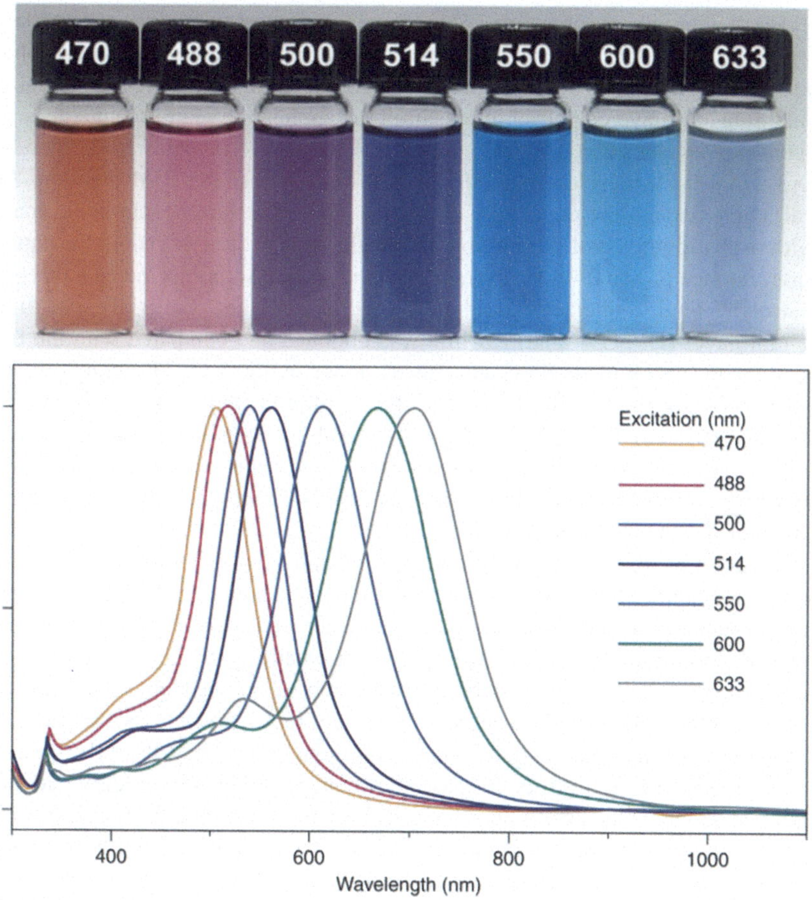

Fig. 5.16 Ag nanoprism solutions with different edge lengths and the corresponding UV–vis absorption spectra. The numbers on the flask and spectra labels correspond to the irradiation wavelength used to prepare the nanostructures. UV–vis, Ultraviolet–visible

5.4.4 Fourier Transform Infrared Spectroscopy

One of the most promising areas of development in materials science in recent decades has been techniques for manipulating materials at the nanoscale. This technique makes it possible to revisit materials with interesting macroscopic properties to examine their nanoscale properties. For thin films, manipulations at the nanoscale can be achieved by immobilizing different materials on the thin film using different techniques. Fourier transform infrared spectroscopy is widely used for the identification of thin film materials and provides information on the mechanism of interaction of an analyte of interest with molecular organization. For example, phospholipids are used as simple analog systems to study cell membranes. Langmuir membranes and LB membranes are one of the main technologies used to produce ordered layers of phospholipids that exist as monolayer or bilayer structures in aqueous phases and solid substrates. However, difficulties encountered in producing LB membranes containing multilayered phospholipids limit the application of this technique, which depends on the sensitivity of the experimental analysis to be performed. Therefore, there is a need to develop methods for multilayer phospholipid deposition. Aoki et al. demonstrated the deposition of the LB multilamellar phospholipid dipalmitoylphosphatidylglycerol (DPPG) by adding the polyelectrolyte Pah to the aqueous phase. Surface pressure versus average molecular area isotherms indicate that polyelectrolytes tend to interact with phospholipid monolayers. The FTIR spectrum of the multilayer (DPPG + Pah) LB film was obtained and compared with the FTIR spectra of the cast film of DPPG and Pah, as shown in Fig. 5.17. The main difference in the spectra of the LB film lies in the shifts of the PO^{-4} groups of DPPG (1221, 1094, and 1048 cm^{-1}) and the NH^{3+} groups of Pah (1551 and 1607 cm^{-1}) compared with the cast film. These differences strongly suggest the existence of electrostatic interactions between DPPG $(PO)^{-4}$ and Pah (NH) $^{3+}$, which is responsible for the growth of (DPPG + Pah) LB multilayers.

Centurion et al. reported the production of thin films with active elements for the development of humidity sensors. Using the LbL technique, nanofilms of cobalt tetrasulfonated phthalocyanine (CoTsPc) were alternately deposited with polyamine-amine dendrimers (PaMaM) and polycyclic aromatic hydrocarbons (Pah). FTIR measurements revealed electrostatic interactions between the CoTsPc sulfonic acid groups and the polyelectrolyte amine groups, and these interactions are critical for film growth. Stronger interactions, such as covalent bonds, may also be responsible for the growth of LbL films. In this case, the changes in the FTIR spectrum are more intense and may include shifts and changes in relative band intensities and the disappearance and/or appearance of new bands. In the study of hu and Ji, LbL multilayer films were deposited by alternating covalent bonds, with the aim of applying the films to the controlled release of therapeutic agents.

5.4.5 Raman Scattering Spectrum

Raman spectroscopy can also be used to characterize different nanostructures made from the same compound, for example, zinc sulfide (ZnS). Different ZnS nanostructures such as nanowires, nanocombs, and nanoribbons have been studied by Raman spectroscopy. SEM images of the nanostructures are shown in Fig. 5.18a–c. Nanowires are shown in Fig. 5.18a, nanocombs are shown in Fig. 5.18b, and nanoribbons are shown in Fig. 5.18c. Raman spectra of nanostructures are shown in Fig. 5.18d for nanowires, Fig. 5.18e for nanocombs, and Fig. 5.18f for nanoribbons. The nanowires, nanoribbons, and nanocombs were irradiated with two excitation sources with energies of 1.96 eV and 2.71 eV (632.8 and 457.9 nm, respectively). Raman spectroscopy analysis shows that under the irradiation of 2.71 eV, the signal-to-noise ratio of the nanowires is low, and there are more bands between 200 and

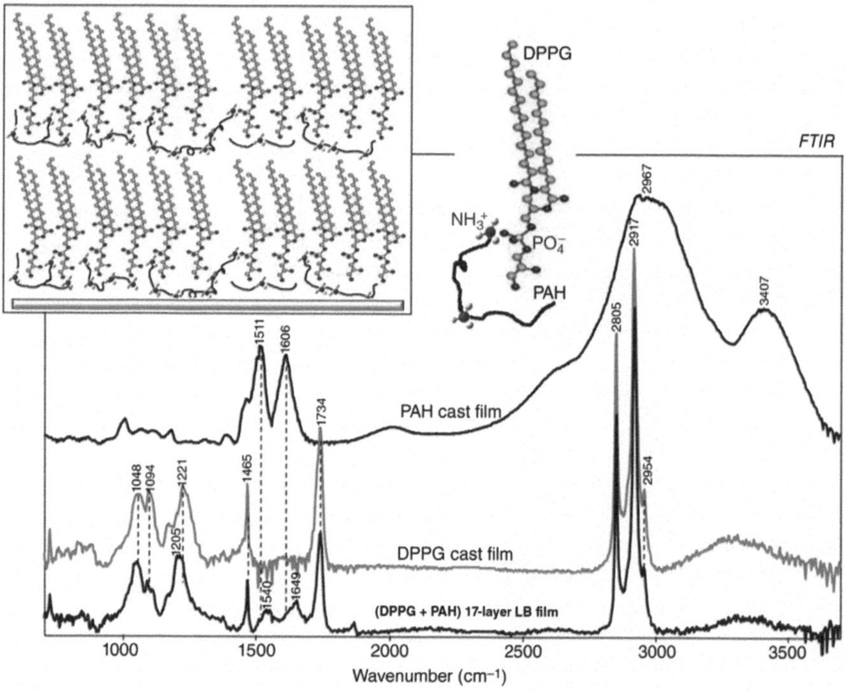

Fig. 5.17 FTIR spectra of 17 LB layers of (DPPG + PAH) deposited on ZnSe. The FTIR spectra of the cast films of PAH and DPPG are given as reference. The inset shows the molecular architecture proposed for the LB films containing (DPPG + PAH). DPPG, Dipalmitoylphosphatidylglycerol; FTIR, Fourier transform infrared; LB, Langmuir–Blodgett; LbL, layer-by-layer; PAH, poly (allylamine hydrochloride)

300 cm^{-1}. The peak at 521 cm^{-1} is caused by the Si substrate, while the disappearance of this peak in other spectra is due to the dense growth of nanocombs and nanoribbons. During the nanoribbon formation, the change in wavenumber from 350.2 to 349.2 cm^{-1} indicates the generation of elastic tension.

The presence of surface defects in the nanomaterials allows them to adsorb O_2, and Raman spectroscopy can be used to monitor the presence of this gas. Wu et al. characterized cerium oxide nanocrystals of different shapes: nanorods, nanocubes, and nanooctahedrons. Transmission electron microscopy (TEM) images (high resolution) of nanorods are shown in Figs. 5.19a, b, nanocubes are shown in Figs. 5.19c, d, and nanooctahedrons are shown in Figs. 5.19e, f. In the Raman spectrum characterization, two excitation sources of 532 and 325 nm were used. The excitation is better at 325 nm due to the resonance Raman effect. CeO_2 has a main band at 462 cm^{-1} and three weak bands at 258, 595 and 1179 cm^{-1}. In the nanorods, band broadening occurs at 462 cm^{-1} due to the size distribution of the nanostructures. Characterization of surface defects is done using the presence of adsorbed oxygen. The results showed that no Raman bands were observed in the calcined (heated)

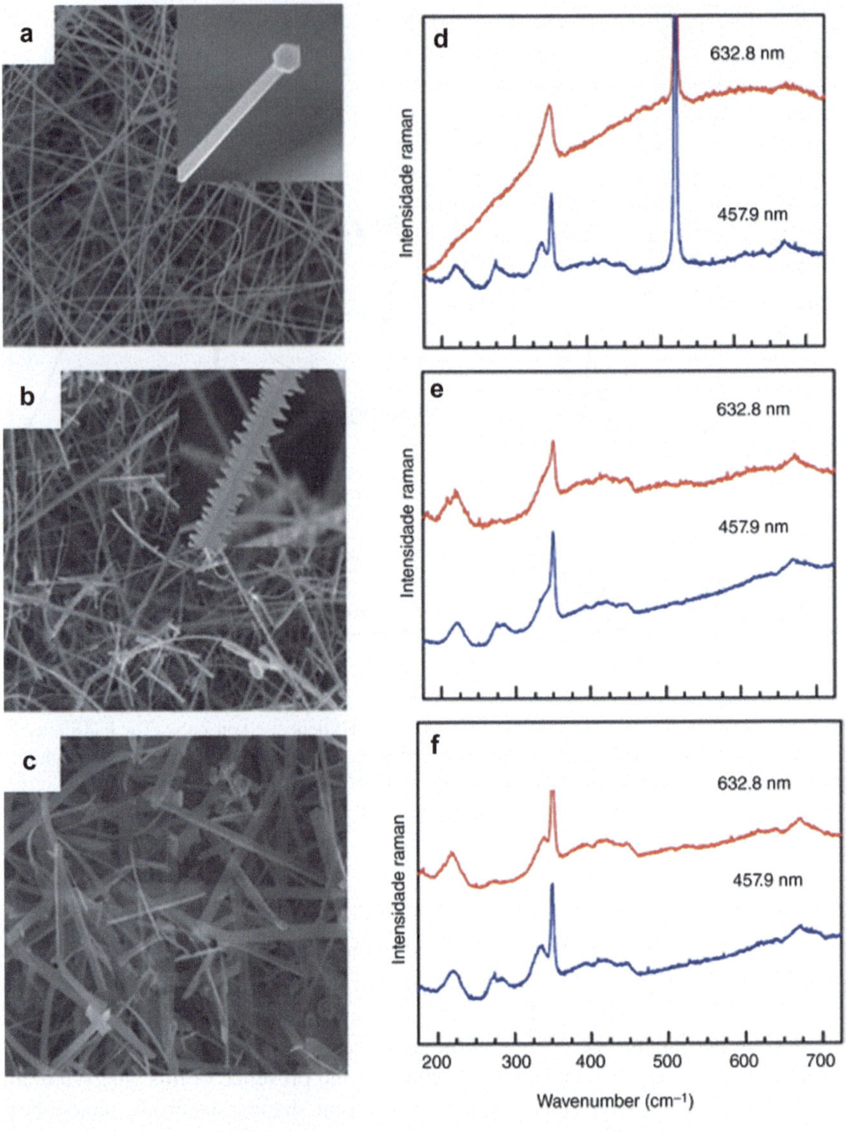

Fig. 5.18 SEM images of **a** nanowires, **b** nanocombs, and **c** nanobelts. Raman spectra of **d** nanowires, **e** nanocombs, and **f** nanobelts excited with 1.96 and 2.71 eV. SEM, Scanning electron microscopy. Adapted from Kim et al.: Raman spectroscopy of ZnS nanostructures, J Raman Spectrosc 43(7):906–910, 2012

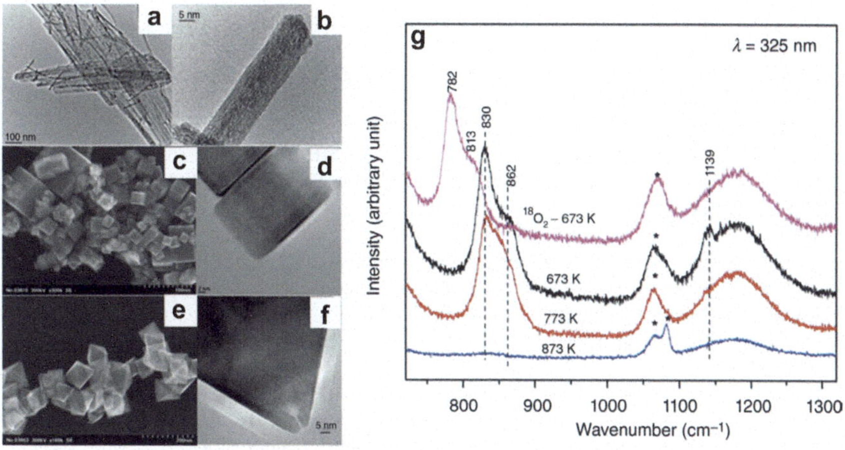

Fig. 5.19 TEM images of nanorods (**a, b**), nanocubes (**c, d**), and nanooctahedra (**e, f**). **g** Raman spectra of nanorods reduced with thermal treatment and followed by O_2 adsorption. The asterisks indicate the presence of nitrate/phosphate. TEM, transmission electron microscopy

samples due to the presence of adsorbed O_2. This shows that the oxidized sample does not adsorb O_2, but the reduced sample does adsorb O_2. Figure 5.19g shows the Raman spectra of reduced and heated nanorods at 673, 773, and 873 K. The Raman spectrum of the sample after heating and reduction at 673 K has three bands at 1139, 862 (shoulder), and 830 cm^{-1}, which are attributed to the surface-adsorbed O_2. In the Raman spectrum of the reduced sample after heating at 773 K, the band at 1139 cm^{-1} is hardly observed, and the shoulder band at 862 cm^{-1} is stronger than the main band at 830 cm^{-1}. No adsorbed oxygen bands were observed in the samples treated at 873 K. When $18O_2$ is used instead of $16O_2$, the 830 and 862 cm^{-1} bands are shifted to 782 and 813 cm^{-1}. Raman spectroscopy is an effective technique for characterizing surface defects in nanomaterials. Roro et al. reported MWCNTs chemically functionalized and mixed with NiO solution. Raman spectra of pure MWCNTs, f-MWCNTs and NiO nanocomposites (f-MWCNTs/NiO) were obtained. The results showed that the nanocomposite had broader D and G band peaks and a larger I_d/I_g ratio compared to f-MWCNTs, indicating an increased number of defects in the nanocomposite.

References

1. M. Mousazadeh, B. Paital, Z. Naghdali, Z. Mortezania, M. Hashemi, E. Karamati Niaragh, M. Aghababaei, M. Ghorbankhani, E. Lichtfouse, M. Sillanpää, K. S. Hashim, M. M. Emamjomeh, *Environ. Dev. Sustain.* **2021**, *23*, 12738–12760.
2. M. L. Liu Jizhen, Wang Qinghua, Fang Fang, Zhu Yankai, *Int. J. Hydrogen Energy* **2021**, *23*, 149–159.

3. E. Asghari, M. I. Abdullah, F. Foroughi, J. J. Lamb, B. G. Pollet, *Curr. Opin. Electroche.* **2022**, *31*, 100879.
4. M. Mogensen, S. Primdahl, M. J. Jørgensen, C. Bagger, *J. Electroceram.* **2000**, *5*, 141-152.
5. D. Zhang, J. Wang, Y. Lin, Y. Si, C. Huang, J. Yang, B. Huang, W. Li, *Renew. Sust. Energ. Rev.* **2017**, *76*, 865-871.
6. Z. Li, P. Guo, R. Han, H. Sun, *Energ. Explor. Exploit.* **2018**, *37*, 5-25.
7. S. Anwar, F. Khan, Y. Zhang, A. Djire, *Int. J. Hydrogen Energy* **2021**, *46*, 32284-32317.
8. Y. Liu, X. Liang, H. Chen, R. Gao, L. Shi, L. Yang, X. Zou, *Chinese J. Catal.* **2021**, *42*, 1054-1077.
9. X. Ding, S. Chen, L. Wang, P. Zhang, *Electrochim. Acta* **2022**, *426*, 140837.
10. A. A. H. Tajuddin, G. Elumalai, Z. Xi, K. Hu, S. Jeong, K. Nagasawa, J.-i. Fujita, Y. Sone, Y. Ito, *Int. J. Hydrogen Energy* **2021**, *46*, 38603-38611.
11. O. Farrok, M. R. Islam, K. M. Muttaqi, D. Sutanto, J. Zhu, *IEEE T. Ind. Electron.* **2020**, *67*, 3409-3418.
12. I. Vincent, D. Bessarabov, *Renew. Sust. Energ. Rev.* **2018**, *81*, 1690-1704.
13. R. S. El-Emam, H. Ozcan, I. Dincer, *Int. J. Hydrogen Energy* **2015**, *40*, 11168-11177.
14. I. Khamis, U. D. Malshe, *Int. J. Hydrogen Energy* **2010**, *35*, 8398-8406.
15. X. Vitart, P. Carles, P. Anzieu, *Prog. Nucl. Energy* **2008**, *50*, 402-410.
16. I. I. Bulyk, *Mater. Sci.* **2019**, *54*, 761-775.
17. J. C. Amphlett, K. A. M. Creber, J. M. Davis, R. F. Mann, B. A. Peppley, D. M. Stokes, *Int. J. Hydrogen Energy* **1994**, *19*, 131-137.
18. O. Sanz, I. Velasco, I. Pérez-Miqueo, R. Poyato, J. A. Odriozola, M. Montes, *Int. J. Hydrogen Energy* **2016**, *41*, 5250-5259.
19. R. Accorsi, E. Bergmann, *J. Electrochem. Soc.* **1980**, *11*, 804-811.
20. M. Mrad, C. Gennequin, A. Aboukaïs, E. Abi-Aad, *Catal. Today* **2011**, *176*, 88-92.
21. Y. Matsumura, H. Ishibe, *J. Catal.* **2009**, *268*, 282-289.
22. Y. Matsumura, H. Ishibe, *Appl. Catal. B Environ.* **2009**, *86*, 114-120.
23. R.-C. Wu, C.-W. Tang, H.-H. Huang, C.-C. Wang, M.-B. Chang, C.-B. Wang, *Int. J. Hydrogen Energy* **2019**, *44*, 14279-14289.
24. A. A. Lytkina, N. V. Orekhova, M. M. Ermilova, S. V. Belenov, V. E. Guterman, M. N. Efimov, A. B. Yaroslavtsev, *Catal. Today* **2016**, *268*, 60-67.

Chapter 6
Preparation Technology of Nanomaterials

6.1 Preparation of One-Dimensional Nanomaterials

With the rapid development of nanoscience and technology, the research on some mesoscopic physical phenomena, such as nanoscale structure, light absorption, luminescence, and quantum size effects related to low dimensions, has become more and more in-depth, which puts forward higher requirements for new functional materials and the miniaturization of component devices. Therefore, while zero-dimensional materials have made great progress, the preparation and research of one-dimensional nanomaterials have made great progress. Since the discovery of carbon nanotubes in 1991 by Iijima of Japan's NEC Corporation, it has attracted great attention from many scientists. Later, with the development of technology and demand, on the basis of the original carbon nanotubes, material experts devoted a lot of attention and exploration to the nanomaterials that were further grown in the one-dimensional direction, and prepared nanowires and nanorods.

Carbon nanotubes (CNTs) are tubular nano-scale graphite crystals, which are seamless nano-scale carbon tubes formed by curling single-layer or multi-layer graphite sheets around the central axis at a certain helical angle. Each layer of nanotubes is a cylindrical surface composed of hexagonal planes formed by carbon atoms fully bonded to the surrounding 3 carbon atoms through sp^2 hybridization. The side length of the plane hexagonal unit cell is 0.246 nm, and the shortest C–C bond length is 0.142 nm. According to different preparation methods and formation conditions, NTs exist in two forms: multi-walled carbon nanotubes (MWCNTs) and single-walled carbon nanotubes (SWCNTs) [1].

One-dimensional nanomaterials have important development prospects in the field of mesoscopic and nano-devices research and application. It can be used as the needle tip of scanning tunneling microscope (STM), connection wires in nano-devices and ultra-large scale integrated circuits (ULSIC), optical fibers, micro-drills in microelectronics, and reinforcing agents for composite materials [2].

S. Peng and P. Li, *Nanomaterials and Nanotechnology*,
https://doi.org/10.1007/978-981-97-8433-2_6

6.1.1 *Single-Walled Carbon Nanotubes*

Although single-walled carbon nanotubes generally yield only a few grams per unit, it is still relatively difficult to control single-walled carbon nanotubes with the desired structure. There are many methods used to prepare single-walled carbon nanotubes, mainly arc method, laser method, organic catalytic pyrolysis method, solar method, etc. These methods have their own characteristics, which will be introduced separately below.

6.1.1.1 Arc Method

Single-wall carbon nanotubes are prepared by arc method, and single-wall carbon nanotubes are synthesized in the reaction area of the device. The experimental conditions are: two electrodes are located in the center of the reaction chamber, and the anode and cathode are separated by a certain distance to facilitate the formation of arcs. The anode is a graphite carbon rod and the cathode is a graphite carbon rod with a shallow groove for a small amount of iron powder. The mixed gas of methane and hydrogen filled in the reaction chamber makes the carbon rod arc discharge by applying a direct current between the two electrodes. Under the action of the arc, the iron in the shallow groove first melts to form small droplets, then evaporates, and finally cools on the cathode and condenses into iron-carbon compounds. The cathode product was placed under an electron microscope, and it was observed that most of the product was assembled into bundles, and each bundle was composed of several single-layer tubes with a diameter of only about 1 nm, but single single-walled carbon nanotubes could also be observed [3].

In a similar way, single-walled carbon nanotubes can also be obtained by changing the position and form of the supporting metal. Also a graphite rod is used as the anode, on which a small hole with a diameter of a few millimeters is drilled and filled with a mixture of pure metals (Fe, Co, Ni) and carbon. These filled anodes form an arc discharge in a nitrogen atmosphere under the action of a current of about 100 A. As a result, spider web-like deposits are formed in the space of the reaction chamber, and soot-like deposits are formed on the walls of the reaction chamber. Electron microscope observation revealed that the obtained carbon tubes were composed of a layer of graphite with a uniform diameter, mostly within 1 nm. The characteristic of this method is that the metals contained in the electrodes, such as iron, cobalt, silver, etc., act as catalysts.

At the beginning of the discovery of single-walled carbon nanotubes, the products prepared by the arc method contained a large amount of impurities such as amorphous carbon and metal catalyst particles, while the content of carbon nanotubes was very low. Using Ni/Y as a catalyst can obtain single-walled carbon nanotubes with a purity as high as 70–90%.

In order to further improve the yield and quality of single-walled carbon nanotubes, a semi-continuous hydrogen arc discharge method was used. Different from the traditional arc method, this method uses a large anode and a small cathode, and the cathode and anode form an oblique angle. The cathode is a graphite rod, and the anode consists of finely mixed graphite powder and a catalyst. The positions of the anode and the cathode are both adjustable. When the anode reactant is consumed to a certain extent, the position of the anode can be adjusted to continue the synthesis, thus realizing the semi-continuous production process. Replacing nitrogen with hydrogen as the medium gas can not only reduce the cost, but also effectively improve the quality and yield of SWCNTs, because hydrogen can etch impurities such as amorphous carbon produced in the reaction and promote the evaporation of the catalyst.

In contrast, this device has the following advantages. (1) The large-diameter anode disc provides sufficient reaction raw materials, which is conducive to the mass preparation of single-walled carbon nanotubes. (2) The cathode rod is opposite to the upper surface of the anode disc at an oblique angle. Under the action of the arc force, a plasma flow can be formed in the reaction chamber. Such a relative position is conducive to timely carrying the single-wall carbon nanotubes produced by the reaction out of the high-temperature reaction area, avoiding the product sintering phenomenon that may occur in the traditional arc method; the product concentration in the reaction area can be kept low, which is also conducive to the continued generation of carbon tubes. (3) The orientation and angle of the cathode and anode can be adjusted. After some raw materials have been reacted, the electrode position can be adjusted to facilitate the raw materials in other positions to continue to react, which can realize the semi-continuous production process [4].

The products prepared by such a method are basically composed of uniformly distributed and intertwined single-wall carbon nanometer tube bundles, and the content of impurities is low. Observation of the obtained product with an electron microscope showed that the bundles of single-walled carbon nanotubes were intertwined with each other, arranged randomly and had a high density; no broken ends of the bundles of carbon nanotubes were found, indicating that the bundles were relatively long. In the higher magnification scanning electron microscope observation, it can be seen that the diameter of single-walled carbon nanotube bundles is mostly between 10 and 50 nm.

6.1.1.2 Laser Evaporation Method

Single-wall carbon nanotubes are prepared by laser evaporation. A graphite target made of a metal catalyst/graphite mixture is placed in the middle of a long quartz tube. The tube is placed in a heating furnace with a working temperature of 1200 °C. The heated tube is filled with an inert gas and the laser beam is focused on a graphite target. The graphite target generates gaseous carbon under laser irradiation, and these gaseous carbon and catalyst particles are carried by the gas flow from the high temperature area to the low temperature area, and grow into single-walled carbon nanotubes under the action of the catalyst [5].

Laser evaporation of graphite electrodes is one of the basic methods for studying carbon clusters. A certain amount of catalyst particles can be added to the electrodes to obtain single-walled carbon nanotubes. Using a 50 ns double-pulse laser to irradiate a graphite target containing Ni/Co catalyst particles, high-quality single-wall carbon nanotube tube bundles can be obtained. Each tube bundle is composed of several single-wall carbon nanotubes arranged in the axial direction, and a two-dimensional triangular lattice formed by weak van der Waals interaction between single-wall carbon nanotubes. The content of single-walled carbon nanotubes in the product is greater than 70%, and the diameter is about 1.38 nm.

In the process of preparing carbon nanotubes, with the evaporation of graphite, metal enrichment is formed on the surface of the metal/graphite target, which will reduce the yield of single-walled carbon nanotubes. Change the metal-graphite mixed target to two targets of pure transition metal or its alloy and pure graphite, and place the two targets facing each other to be irradiated by laser light at the same time, which can eliminate the production decline caused by the metal enrichment on the surface of the graphite target due to graphite volatilization [6].

In the above preparation process, with the change of the catalyst, the output of carbon nanotubes will also change greatly. It was found that Ni/C was used as catalyst. In the case of alloys, the output of single-walled carbon nanotubes is 10–100 times that of pure metals as catalysts. Co/Pt alloys and Ni/Pl alloys can also be used as catalysts to obtain high-yield single-walled carbon nanotubes, but when Pt is used as a catalyst, the content of single-walled carbon nanotubes in the product is low. Co/Cu alloys are used as catalysts to obtain only a small amount of single-walled carbon nanotubes.

6.1.1.3 Catalytic Pyrolysis

Compared with the arc method and laser evaporation method, the catalytic pyrolysis method of organic matter has a lower reaction temperature and simpler equipment, and can be used for semi-continuous or full-continuous preparation of single-walled carbon nanotubes. The purity of the product obtained by preparing single-walled carbon nanotubes by the catalytic pyrolysis method is also high, so that it is possible to realize low-cost, large-scale synthesis of single-walled carbon nanotubes. Single-wall carbon nanotubes were prepared by catalytic pyrolysis. The experimental device used a quartz tube as a reactor and a silicon carbide heating element as a heating source. The catalyst precursor is placed directly at the front of the reactor. Adjusting the location of the catalyst precursor can control the volatilization speed and concentration of the catalyst. A pyrolytic carbon substrate is placed in the middle of the reaction zone, which is used to change the flow of gas and collect single-walled carbon nanotubes. A thermocouple is placed in the reaction zone to directly measure the temperature of the reaction zone.

When the reaction zone is heated to a certain temperature, the mixed gas of hydrogen, benzene (methane) and growth promoter is introduced, and when the mixed gas passes through the front of the reactor, the volatilized ferrocene is brought into the reactor together. After the reaction has been carried out for a certain period of time, it can be observed that a film-like product is formed in the reaction zone, floats out of the reaction zone under the action of the airflow, and adheres to the placed substrate, thermocouple sleeve and the inner wall of the low-temperature end of the reactor. In the slightly higher temperature zone of the reactor, rope-shaped products are produced along the direction of gas flow. The appearance time of film and rope-like products changed with the change of gas flow and the amount of growth promoter [7].

After the reactor is cooled to room temperature, the quartz tube is taken out, and the product in the form of net-like, black translucent, very light and thin film-like deposits can be obtained from the inner wall of the reactor near the lower temperature of the gas outlet and the thermocouple sleeve. In the intermediate position between the reaction zone and the film-like product, a rope-like product-single-walled carbon nanotube rope arranged along the gas flow direction is obtained. This kind of single-walled carbon nanotubes is in the shape of strips, which can be peeled off with the naked eye or under an optical microscope with a mount, with a length of several centimeters and a diameter of several microns to tens of microns. A black, lumpy product can be obtained at the high temperature near the reaction zone.

The preparation of single-walled carbon nanotubes by catalytic pyrolysis generally uses hydrocarbons as the carbon source for the growth of single-walled carbon nanotubes. However, hydrocarbons are easily decomposed at 600–700 °C (the temperature required for the growth of single-walled carbon nanotubes) to form graphitic carbon deposits or amorphous carbon. Therefore, using catalytic pyrolysis, the surface of single-walled carbon nanotubes grown from hydrocarbons is often covered by a certain amount of amorphous carbon. To remove these amorphous carbons, it needs to be purified in the later stage, which complicates the method of mass-preparing single-walled carbon nanotubes with hydrocarbons as carbon sources. RE Smalley et al. used the disproportionation reaction of CO to simplify this problem and synthesized a large number of high-purity single-walled carbon nanotubes. This method uses CO as a carbon source, ferrocene as a catalyst, and controls the reaction conditions at 800–1200 °C to synthesize high-purity single-walled carbon nanotubes with a diameter of 0.7 nm. The method is a continuous flow process, so it can realize relatively large amount of preparation of single-walled carbon nanotubes [8].

6.1.1.4 Solar Law

The solar method is to focus sunlight on graphite samples to evaporate carbon atoms, and these sublimated carbon atoms are blown to the low temperature area by oxygen or nitrogen, and deposited in the low temperature area of the reactor (where no sunlight is focused), and finally form carbon nanotubes. The sunlight is collected by a flat lens and reflected to the parabolic mirror, so that the parabolic mirror directly focuses the sunlight on the graphite target, and its maximum temperature can reach 3000 K. The device has a graphite boat made of graphite powder and metal catalyst powder as a target, and the graphite target is placed in the center of the reactor. There is a graphite tube heated by sunlight, which transports the carbon vapor and acts as a heat shield, forming an annealing zone. The graphite tube equipped with the target is placed in the center of the reactor and is evacuated, flushed with inert gas, and then adjusted to the position where the solar energy is focused. The following process is similar to the laser evaporation method.

The experimental results show that the yield of single-walled carbon nanotubes is only 1–2% when the reaction temperature is 2850 K, and the yield increases to 20% when the reaction temperature is 3000 K. For this reason, Alvarei successfully improved a solar furnace with a power of 1000 kW. Its characteristics are: the reaction chamber is a stainless steel chamber, which is connected with a hemispherical silicon glass, which is used to reflect excess energy and is supported by a water-cooled device; it is composed of a graphite rod with a diameter of 60 mm and a length of 80 mm; during the experiment, hydrogen gas was passed into the reaction chamber (pressure 300 mbar, 1 bar = 10^5 Pa, flow rate 20 L/min). After improvement, the reaction temperature can reach 3400 K, and the yield of single-walled carbon nanotubes can be significantly increased [9].

6.1.2 Preparation of Double-Walled Carbon Nanotubes

The research on the preparation technology of double-walled carbon nanotubes is helpful to fully understand the structural parameters controlling carbon nanotubes, and has a guiding role in obtaining high-quality multi-walled carbon nanotubes. The methods for preparing double-walled carbon nanotubes mainly include fullerene filling treatment method, arc discharge method, flow catalysis method, etc., which will be introduced separately below.

6.1.2.1 Fullerene Filling Treatment Method

Double-walled carbon nanotubes can be obtained by filling the hollow tubes of single-walled carbon nanotubes with human C_{60} by gas phase reaction. Through the transmission electron microscope and Raman spectrum analysis of the obtained samples, it is proved that the obtained product is a double-walled carbon nanotube, because the C_{60} molecules enter the hollow tube of the single-walled carbon nanotube through a gas phase reaction, and the arrangement is very regular, like a one-dimensional crystal arranged in an independent tube. The temperature of the gas phase reaction has a very important impact on the formation of carbon nanotubes. The test results show that: when the sample is kept at 800 °C for 14 h, most of the C_{60} remains unchanged. When the temperature exceeds 800 °C, more C_{60} begins to fuse. As the temperature rises, the fusion accelerates; when the temperature rises to 1000 °C, the mutually fused C_{60} molecules lead to a structural change, from C_{60} molecules to double-walled carbon nanotubes. The temperature continued to rise to 1200 °C, so that the single-armed carbon nanotubes that had been filled with C_{60} turned into double-walled carbon nanotubes [10].

6.1.2.2 Arc Discharge Method

Double-walled nanotubes can also be prepared by the arc method. The experimental principle and method are similar to the method of preparing single-walled carbon nanotubes by hydrogen arc method. The difference is: the poles are perpendicular to each other, and the distance is 2 mm; the composition of the pole is 2.6% Ni, 0.7% Co, 1.45% Fe, 0.75% S, and the rest is graphite powder. The reaction chamber is filled with a mixed gas of nitrogen and hydrogen (volume ratio 1:1:1), the voltage difference between the two electrodes is 26–28 V during arc discharge, the current is 75–80 A, the arc discharge evaporation reaction time is 40 min, and the obtained product is double-walled carbon nanotubes. The interlayer spacing of the double-walled tube is 0.39 nm, which is larger than the common multi-walled tube interlayer spacing of 0.34 nm and the interlayer spacing of single crystal graphite 0.335 nm [11].

The transmission electron micrograph of the sample shows that the obtained double-walled carbon nanotubes are similar to the single-walled carbon nanotubes, and also form a tube bundle structure, and there are discontinuous tubes around the tube bundle. The carbon nanotubes in the tube bundle are not uniform, and most of the tubes are composed of two concentric cylindrical graphite layers, that is, double-walled carbon nanotubes. In addition to these bundles of double-walled tubes, individual double-walled tubes were also observed.

6.1.2.3 Flow Catalytic Method

The flow catalytic method has the characteristics of easy control of the preparation process, high preparation efficiency and possibility of continuous production. The device used by this method is exactly the same as that used for preparing single-wall carbon nanotubes. The carbon source is methane, ferrocene is used as the precursor of the catalyst, and diphenhydramine is used as a growth accelerator. The resulting product adheres to the thermocouple ceramic sheath and appears as a translucent mesh. Observation of the product by high-resolution transmission electron microscopy shows that it contains more than 70% double-walled carbon nanotubes, the rest are catalyst particles and amorphous carbon, and a small amount of single-walled carbon nanotubes and triple-walled carbon nanotubes. The cross-section of the double-walled carbon nanotubes has a concentric circle structure. The double-walled carbon nanotubes formed by this method mostly exist in bundles under the action of van der Waals force. The double-walled carbon nanotubes in the bundle form a triangular lattice arrangement under the action of van der Waals force. The lattice constant is 2.47 nm, and the tube spacing is about 0.324 nm. The double-walled carbon nanotubes with larger diameters are close to the adjacent double-walled carbon nanotubes with smaller diameters under the van der Waals force to form an elliptical structure. It may be the competition between elastic force and van der Waals force that affects the structure of double-walled carbon nanotubes in bundles. The number of carbon tubes in bundles is less than that of single-wall carbon nanotube bundles, which may be caused by the decrease of van der Waals force due to the influence of inner tubes.

Through the precise measurement of the diameter of a group of double-walled carbon nanotubes, more than 90% of the inner and outer diameters of the double-walled carbon nanotubes are distributed between 1–2.0 nm and 1.6–2.8 nm respectively [12].

Geometrically, double-walled carbon nanotubes can be regarded as composed of two coaxial single-walled carbon nanotubes with a distance of only a few tenths of a nanometer. Theoretical calculations show that the interlayer interaction failed to open the energy band, so although the energy dispersion relationship is slightly affected by the interlayer interaction, both the inner and outer layers maintain the basic electronic structure of the graphite monolayer tube. It can be considered that the interaction between single-layer carbon nanotubes is similar to single-walled carbon nanotube bundles, and the weak van der Waals force between double-walled carbon nanotubes does not affect the vibration and electronic states enough to affect the physical properties of single-walled carbon nanotubes.

6.1.3 Preparation of Multi-walled Carbon Nanotubes

Compared with the synthesis of single-wall carbon nanotubes, the preparation of multi-wall carbon nanotubes has fewer influencing factors and is easier to control. The preparation technology of multi-wall carbon nanotubes is relatively mature, and the output can reach kilograms per hour, and the diameter and orientation of the product can be controlled. At present, the relatively mature methods for preparing multi-walled carbon nanotubes mainly include arc method and catalytic pyrolysis method (also known as catalytic chemical vapor deposition method). The catalytic pyrolysis method uses easily decomposed organic matter as the carbon source. In the range of 500–1200 °C, under the action of a transition metal element catalyst, the carbon source is decomposed to generate carbon atoms. In comparison, the arc method has complex preparation equipment, but the process parameters are easier to control; while the catalytic pyrolysis method is considered to be the most likely method to achieve mass production of high-quality multi-walled carbon nanotubes.

6.1.3.1 Arc Method

The arc method vaporizes the solid carbon source and rearranges its structure at a high temperature of 3000–4000 °C. Under an ammonia atmosphere, by optimizing the reaction conditions such as the type of inert gas, the pressure, the nature of the current (AC or DC), the voltage and the relative size of the electrodes, carbon nanotubes can be prepared in large quantities. The experimental results show that the nitrogen atmosphere pressure is the main factor affecting the yield of carbon nanotubes. When the air pressure increases to 13.3 kPa, a small amount of multi-wall carbon nanotubes can be formed. When the pressure reaches 6.65 kPa, more carbon nanotubes with high purity can be obtained [13].

In 1999, Ishigami et al. improved the arc method and related experimental equipment to continuously prepare multi-walled carbon nanotubes. The graphite anode is inserted into the reaction chamber containing liquid nitrogen, and after forming an arc with the cathode made of shorter graphite rods already installed in the reaction chamber, the carbon nanotubes generated in the arc area are allowed to fall and deposit on the bottom of the funnel-shaped reaction chamber. This bottom is sealed by a valve that opens automatically at regular intervals to allow the carbon nanotubes to flow out of the reaction chamber. In this setup, an anode graphite rod is continuously supplied with a carbon source through a slidable electronic contact. The content of liquid nitrogen in the reaction chamber is automatically adjusted through sensors and liquid nitrogen feed pipes. The characteristics of the method and device are firstly that no expensive and complicated vacuum sealing is required; secondly, the protective atmosphere and buffer gas source are provided by liquid nitrogen, and when an arc is generated, the liquid nitrogen can provide the temperature gradient required for preparation; the product is very easy to store and transport under an inert atmosphere.

The method and device can prepare products rich in carbon nanotubes, the yield of which is related to the area of the cathode and the duration of the electric arc, the quality of the obtained product is good, the number of layers is 4–8 layers, each layer is arranged in parallel, and the surface is very pure.

6.1.3.2 Catalytic Pyrolysis

The catalytic pyrolysis method has low cost, high yield, and easy control of experimental conditions. By controlling the mode of the catalyst, multi-walled carbon nanotubes in an oriented array can be obtained. Therefore, it is the most promising method for mass production of high-quality multi-walled carbon nanotubes. Catalysts used in catalytic pyrolysis are generally transition metal elements Fe, Co, Ni or their combinations, and sometimes other elements and compounds such as rare earths are added. The preparation of carbon nanotubes by catalytic pyrolysis can be mainly divided into three methods according to the way in which the catalyst is added or present: matrix method, spray method and flow catalytic method (or floating method).

Preparation of carbon nanotubes by matrix method: The so-called matrix method initially uses graphite or ceramics as a matrix, attaches catalysts to the matrix, uses these catalyst particles as "seeds", and passes carbon-containing gas at high temperature to decompose them and form carbon nanotubes on the catalyst particles. The substrate method can prepare carbon nanotubes with high purity, but since carbon nanotubes only grow on the catalyst substrate, the preparation of ultrafine catalyst is very difficult, and it is not easy to spray evenly on the substrate, so the output is not high for some reason, and it is difficult to industrialize production. To achieve large-scale preparation of carbon nanotubes, it is necessary to first solve the problems of continuous feeding of catalysts and timely derivation of catalysts and products. A process called ebullating bed reaction to prepare carbon nanotubes has improved the general matrix method. This method has good gas–solid phase contact, can load a large amount of catalyst, and can make full use of the catalyst surface. In the reactor, the raw gas passes through the gas distribution plate at a certain flow rate, and the activated catalyst is "blown" into a "boiling" state. The catalyst particles are always in motion, and the distance between them is shorter than the distance between the catalyst particles on the fixed substrate, making it easier to grow carbon nanotubes on the surface. Due to the mutual collision of the catalyst particles, the carbon nanotubes are easy to fall off from the surface, which can ensure the formation of the carbon nanotubes with a high opening ratio. When the amount of catalyst in the ebullated bed is large, the raw material gas can still fully contact the surface of the catalyst, which can ensure a high utilization rate of the catalyst. In the closed moving bed catalytic reactor, the reduced nano-scale catalyst is continuously and evenly sprinkled on the moving bed through the nozzle, and the moving bed moves at a certain speed. The flow direction of the raw material gas can be consistent with the movement direction of the bed layer, or it can be opposite, and react on the surface of the catalyst to form carbon nanotubes. The residence time of the catalyst in the constant temperature zone can be adjusted by controlling the moving speed of the moving bed. When the residence time of the catalyst on the moving bed reaches the set value, the catalyst together with the carbon nanotubes formed on it will escape from the moving bed and enter the collector, and the reaction tail gas will be discharged through the exhaust port. The continuous production of carbon nanotubes with designed size can be realized by using the moving bed catalytic cracking reactor, which is expected to greatly reduce the production cost and provide guarantee for the industrial application of carbon nanotubes [14].

Spray method: dissolve the catalyst in the liquid carbon source, use the pump to spray the solution carbon source directly into the reaction furnace, and grow carbon nanotubes at the reaction temperature. The carbon source in the reaction raw material is n-hexane, the catalyst is ferrocene, and phenene is used as an additive to prepare the reaction raw material in a certain proportion. The temperature is raised to above 1000 °C, and the reaction raw material is directly sprayed into the reaction furnace by a pump, and the reaction is carried out in a hydrogen atmosphere. After the reaction is completed, nitrogen is used to cool. The spraying method provides the possibility of preparing a large amount of carbon nanotubes, but because the ratio of catalyst to hydrocarbon is difficult to optimize, the iron particles are unevenly distributed during the spraying process, and the sprayed catalyst particles are difficult to exist in nanoscale form, so the proportion of carbon tubes is small, and a large amount of carbon black is often produced. The reactive carbon sources (benzene, ferrocene, phenophen) required for the experiment were sprayed into the evaporator from a small feeder, and the temperature was controlled at 100 °C. Using this method, several grams of pure multi-walled carbon nanotubes can be prepared within 30 min.

6.1.3.3 Other Methods

There are some other preparation methods for preparing multi-wall carbon nanotubes, such as electrolysis, low-temperature solid pyrolysis, ball milling, diffusion flame and so on. Since the preparation process conditions of these methods are more complicated and difficult to control, and the quality and output of the obtained products are also different, it can be selected according to actual needs and possible conditions.

(1) Electrolysis method

In the electrolysis method, the graphite cathode is first immersed in a dissolved inorganic salt solution, and after electrification, the graphite undergoes a redox reaction under the action of the current to form carbon nanotubes. The specific method is to make graphite into a boat shape, and use it as the cathode of electrolysis, and the anode is a high-purity carbon rod. Put the inorganic salt lithium chloride into the boat, heat it to the melting point (604 °C) under air or hydrogen atmosphere, make it melt into a solution, then immerse the anode in the molten solution, apply current between the two electrodes, and keep the current for several minutes. During this process, the surface of the electrode immersed in the solution begins to corrode, small corrosion pits appear, and carbon nanotubes begin to form. The resulting product contains carbon nanotubes, as well as onion-like structures and wrapped carbon particles. Among them, carbon nanotubes have two shapes: spiral and coiled, with a diameter of 220 nm, composed of 520 layers of coaxial graphite sheets.

(2) Low temperature solid pyrolysis method

The principle of the low-temperature solid pyrolysis method is to pyrolyze the metastable ceramic precursor at a relatively low temperature to obtain carbon nanotubes. The specific method is to place the nanoscale powder in a boron nitride porcelain boat, put it in a graphite furnace, and heat it to above 1200 °C under a nitrogen atmosphere to make it pyrolyze to form multi-walled carbon nanotubes. The formation status and yield of products are closely related to the temperature and state of the system. In a static nitrogen atmosphere at 1400 °C, the yield of carbon nanotubes is the largest, while in a flowing nitrogen atmosphere, the optimum temperature for forming carbon nanotubes is 1850 °C. The diameter of carbon nanotubes formed is 10–25 nm and the length is 0.1–5 m. The biggest advantage of this method is that the experimental steps are simple, but because the carbon nanotubes cover the surface of the raw material, it brings difficulties to separation and purification, and the product quality is not high.

(3) Ball milling method

The ball milling method is to ball mill graphite powder and then anneal to obtain carbon nanotubes. A large number of multi-walled carbon nanotubes can be obtained by ball milling high-purity graphite powder for hundreds of hours in a hydrogen atmosphere, and then heat-treating it at a high temperature of 1200 °C for several hours in a nitrogen or hydrogen atmosphere. The mechanism of the ball milling method is considered to be the nucleation of nanocarbons during the ball milling process, and the heat treatment process is the growth process of carbon nanotubes. During the ball milling process, the trace iron particles that are shed by the stainless steel balls are the catalysts for the growth of multi-walled carbon nanotubes under heat treatment conditions. The method is relatively simple and has industrial prospects [15].

6.1.4 Preparation Method of Nanowire

The preparation of flow nanowires first needs a basis for forming whiskers, and on the basis of whiskers, nanowires grow in one-dimensional direction. In the general gas–liquid-solid (VLS) process, whiskers first form droplets on a solid substrate. A liquid–solid interface is formed between the solution of the droplet and the substrate. Gas atoms are deposited on the droplets due to adsorption on the surface of the solution. When the solution reaches a supersaturated state, crystals begin to precipitate from the solution. According to such a formation mechanism, the formation of nanowires can be divided into physical methods and chemical methods.

6.1.4.1 Physical Method

The preparation of nanowires by physical methods mainly includes evaporation condensation method, laser ablation method, laser deposition method, arc discharge method and other methods.

(1) Evaporation and condensation method

Nanowires can be prepared by a simple physical evaporation method. First, there needs to be a basis for forming whiskers, and then grow to form nanowires in a specific direction on the basis of whiskers. Whiskers first form droplets on a solid substrate, and a liquid–solid interface is formed between the solution of the droplets and the substrate. Gas atoms are deposited on the droplets due to adsorption on the surface of the solution. When the solution reaches a supersaturated state, crystals begin to precipitate from the solution. As the gaseous atoms are continuously adsorbed into the solution, the supersaturation process of the solution continues to appear, causing whiskers to gradually form on the solid substrate [16].

A specific example of preparing silicon nanowires is as follows: the powder material containing 95% Si and 5% Fe is hot-pressed for 8 h to make a target. The target is placed in the evaporation chamber made of quartz tube. Argon, a carrier gas mainly composed of inert gases, is introduced into one end of the evaporation chamber, and is pumped at a constant rate at the other end of the evaporation chamber. The target material is heated, and the entire evaporation system is kept at 1200 °C. After 20 h, Si nanowires with a diameter of 15 nm and a length ranging from tens of microns to hundreds of microns can be collected on the tube wall of the product collection end of the evaporation system. The research results show that the air pressure in the quartz tube has a great influence on the diameter of the nanowires, and the diameter of the nanowires will increase significantly as the air pressure increases. The research results also show that the catalyst is essential for the growth of Si nanowires, and only under the action of the catalyst can the growth of the nanowires be completed and the desired product can be obtained.

(2) Laser deposition method

The principle of laser deposition method (including laser ablation method) is to use high-energy laser to irradiate the target material in a specific atmosphere to evaporate it, and at the same time combine a certain reactive gas to deposit nanowires on the substrate or the wall of the reaction chamber.

In the preparation of nanowires, if there is no catalyst particle by VLS method, generally whiskers cannot be grown. The laser deposition method does not have this problem, and the growth mechanism of the nanowires prepared by the laser deposition method is different from that of the general VLS method. In the laser deposition method, due to the high-power thermal evaporation of the laser, the target material is gasified and separated into plasmas such as single atoms, primary particles, catalyst atoms, and small amorphous particles. These gasified and separated materials flow in the reaction growth chamber made of quartz tubes under the drive of Ar gas, a carrier gas mainly composed of inert gases. Since the thermal movement speed of single atoms and primary particles is much faster than that of amorphous nanoparticles, in the flowing gas, single atoms and primary particles adhere to amorphous particles, that is, these amorphous particles become nucleation sites for nanowire growth. Since the primary particles adsorbed on the amorphous particles have dangling bonds, the atoms constituting the nanowires can be adsorbed on these bonds in the form of chemisorption. Modulating the composition of the nanowires can be achieved by changing the composition of the target or adding other reactive gases. Now, using this technology, Si nanowires, Ge nanowires, Ga-As nanowires and SiO_2 nanowires with diameters of several nanometers have been successfully prepared.

The main factors affecting the preparation of nanowires by laser deposition are laser intensity, gas pressure in the growth chamber, gas flow rate, growth time and growth temperature. The laser intensity affects the state of the excited plasma gas flame, and the gas pressure in the growth chamber determines the density of the internal gas. For some carrier gases with good thermal conductivity, the change of density may change the thermal gradient during the growth of nanowires, and then have corresponding effects on the growth rate and direction of nanowires. The flow rate of the carrier gas can determine the flow rate of the plasma after evaporation, and can also affect the direction and rate of further growth of the nanowires on the nuclei. The reaction of materials and the growth time of nanowires affect the size of nanowires, and if the time is too long, by-products will increase. In the temperature region of reaction and growth, in general, high temperature can reduce the defects generated when forming nanowires.

(3) Preparation of nanowires by arc discharge method

Nanofilaments were prepared using an arc discharge method similar to the preparation of single-walled carbon nanotubes. Using graphite as the anode material, drill a 4 mm-diameter hole with a depth of about 20 mm on a graphite rod with a diameter of 6 mm. Mix C and SiO_2 powder (mass ratio 1:1) thoroughly, fill the hole with acetone, and press it tightly to serve as the anode of the arc discharge; the cathode uses a cooled copper block with a diameter of 25 mm and a thickness of about 5 mm. Put the anode and cathode into a sealed cavity made of quartz glass, and the anode graphite rod is controlled by a motor to move back and forth to facilitate the ignition of the discharge and stabilize the discharge. On the premise of ensuring that the discharge chamber is sealed, first pass in hydrogen gas for flushing for about 1 h to ensure the purity of the gas in the chamber, and then adjust the two vacuum fine-tuning valves that control the gas inlet and outlet to make the hydrogen flow rate about 100 mL/min. The preparation of nanowires adopts a DC power supply, the arc discharge current is about 70 A, and the voltage between electrodes is 25 V. Using an electron microscope, it can be observed that a large number of nano-filaments are distributed in the black soot. The diameter of the nano-filaments is between 30–50 nm, which is a solid one-dimensional nano-material.

6.1.4.2 Chemical Method

The main methods of preparing nanowires (filaments) by flow chemical method include chemical vapor deposition, electrochemical method, polymerization method and template method, among which template method is the most common [17].

(1) Chemical vapor deposition method

The chemical vapor deposition method is to heat the precursor powder under certain gas flow conditions to cause a gas phase reaction with the gas, and then deposit nanowires. Xiaosong Sun et al. used a catalyst-assisted chemical vapor deposition method to prepare ZnO nanowires using a chemical vapor deposition growth device made of quartz tubes. ZnO powder (99.99%) and graphite powder (99%) are uniformly mixed at a ratio of 1:1 (molar ratio) and pressed into a sheet-like reaction source. A 111 tangential silicon wafer was used as a substrate, and a layer of gold with a thickness of about 10 nm was pre-evaporated on this substrate as a reaction catalyst for the growth of ZnO nanowires. Put the ZnO-graphite reaction source sheet in the quartz boat and put it in the chemical vapor deposition growth device. A gold-coated Si substrate was placed 2–10 cm downstream of the gas flow of the ZnO-graphite source sheet. High-purity nitrogen (99.99%) is used as a protective gas and a carrier gas for carrying reactants. This reaction involving gold in catalytic assisted chemical vapor deposition and the formation of ZnO nanowires can be summarized as the following steps: at an appropriate temperature, ZnO in the ZnO-graphite source sheet is reduced to Zn by graphite powder to obtain Zn vapor, while graphite is oxidized to CO, At the same time, the gold coating on the surface of the silicon wafer placed downstream of the gas flow of the ZnO-graphite source sheet and in the lower temperature region forms gold catalyst particles due to surface tension and atomic migration; Zn vapor is carried to the substrate silicon wafer by nitrogen, and reacts with the gold catalyst particles on the substrate surface to form Zn-Au. When the solubility of Zn in the alloy reaches saturation, Zn precipitates from the alloy and reacts with oxygen to form ZnO nanowires.

The growth and preparation conditions of ZnO nanowires are based on maintaining normal pressure in the chemical vapor deposition growth device. In order to fully react the ZnO-graphite source sheet, the source sheet is degassed for 8 h under a nitrogen atmosphere at a temperature of 200 °C with a nitrogen flow rate of 100 mL/min in advance, and then a large number of zinc oxide nanowires can be obtained on the Si substrate. The radius of the product is 50–150 nm, and the length is 2–5 μm.

(2) Electrochemical method

Electrochemical methods include electrochemical deposition, electrochemical hydrolysis and electrochemical polymerization. In the field of preparing nanowires, electrochemical deposition mainly refers to the reduction of metal cathodes to form deposition. This method is suitable for the preparation of metal nanowires by using the nanopores of the template; electrochemical hydrolysis is to make the cathode region alkaline by reducing water, thereby promoting the hydrolysis of metal salts to form deposition. This method can be directly used to prepare metal oxide nanowires; electrochemical polymerization is mainly used for the preparation of conductive polymer nanowires.

The preparation of nanowires by electrochemical deposition needs to first prepare a layer of metal film on the side of the template with nanopores by sputtering or vacuum evaporation as the cathode. By controlling the conditions of voltage, current, temperature and time, the metal is deposited in the nanopores of the template, and then the template is removed to obtain the corresponding nanowire material. Electrochemical deposition methods are generally suitable for the assembly of nanowires and tubes of metal and conductive polymer materials in the pores of alumina templates, for example, the preparation of nanowire and nanotube arrays such as Cu, Pt, Au, Ag, Ni, polypyrrole, polyaniline, and polytrimethylphene. The first step is to prepare a template with nano-holes, and a layer of metal such as gold and silver is deposited on one side of the template as a cathode; the second step is to fix the metal-plated side on a conductive substrate, and the other side is exposed to the electrolyte, and the deposited metal or semiconductor material enters the nano-holes of the template under the action of constant voltage and constant current; the third step is to remove the template by dissolving, so that the nanowires are exposed.

6.1.5 Preparation Method of Nanorods

6.1.5.1 Physical Evaporation Method

According to the conventional preparation method of nano powder, an inert gas is introduced into the evaporation chamber that has been evacuated to a relatively high degree of vacuum, and the metal or oxide is heated to form a vapor that is atomized into an atom mist that collides with the inert atoms, agglomerates to form nano-sized clusters, and finally gathers on a liquid nitrogen cold trap to obtain nanoparticles. During the preparation process, the particle size distribution and morphology of the product particles can be changed by selecting the type of inert gas and adjusting the evaporation rate, gas flow rate and pressure, so as to obtain nanorods.

At present, the generally recognized growth mechanism of nanowires and nanorods with catalyst participation is the VLS (gas–liquid-solid) mechanism, but there are different opinions on the growth mechanism of nanowires and nanorods without catalysts. The latter probably grew by a simple VS. mechanism. Taking the preparation of ZnO nanorods as an example, the specific growth process is as follows: Zn powder is evaporated after heating, transported by vapor, and then deposited on the wall and alumina cover plate to obtain tiny particles. At the same time, the oxygen in the ground smoke makes the surface of Zn particles oxidize and grow to the surroundings, forming ZnO vapor and continuously depositing, resulting in continued growth of ZnO nanowires and nanorods [18].

A catalyst-free high-temperature thermal evaporation method can be used to prepare ZnO nanorods with a good crystal structure and regular shape, and a diameter of about tens of nanometers. Mix 0.3 g of analytically pure Zn powder and ZnO powder in the quartz boat evenly, and use it as the material source for evaporation. The quartz boat is placed in the middle of the alumina tube in the rapidly heating tube electric furnace. At a place about 20 cm away from the evaporation source, which is also the downwind of the alumina tube, put a clean Si sheet as the substrate material for carrying nanorods. Vacuumize the system with a mechanical pump, then start the heating power supply, raise the temperature to 1300 °C within 40 min, keep warm for 1 h and then cool naturally. During the heating and evaporation reaction process, Ar gas was introduced, and the Ar gas flow rate was 40 mL/min, and a certain pressure in the reaction growth chamber was maintained. After cooling, a layer of ZnO rods with hexagonal wurtzite structure can be seen deposited on the Si flakes.

The effects of thermal evaporation reaction temperature and holding time on the formation of ZnO nanocrystals were further studied. The melting point of Zn powder is 420 °C, and a higher temperature is needed to oxidize Zn powder to ZnO. It was found in the experiment that when the temperature is lower than 900 °C, there is no formation of ZnO. When the temperature is higher than 1200 °C, zero-dimensional ZnO particles are formed. Therefore, the suitable temperature ranges for preparing nano-ZnO crystals of various shapes by simple direct physical evaporation method is 900–1200 °C. With the extension of the holding time, nanoparticles and nanowires are connected together and grow into a nanorod tetracone structure.

6.1.5.2 Microemulsion Method

Nanorods are prepared by microemulsion method. The microemulsion system is composed of H_2O/surfactant. Such a microemulsion system has certain self-limiting reaction and self-assembly characteristics. If suitable conditions are selected, this microreactor can be used to synthesize one-dimensional nanomaterials. The size and shape of the nanomaterials synthesized in the microemulsion system are closely related to the diameter of the water core in the system, and the diameter of the water core is determined by the molar ratio of H_2O/surfactant. The change of this ratio will also affect the strength of the interfacial film in the microemulsion at the same time, so that it will eventually affect the collision, coalescence and crystallization process of reactants. Nevertheless, the target product is not necessarily proportional to the diameter of the water core, and different products may have different effects. In addition, the morphology and particle size distribution of the product can also be controlled by properly adjusting the concentration of the reactants.

Heyun Zhao and others successfully prepared SnO_2 nanorods with rutile structure by microemulsion method. The three microemulsions composed of $SnCl_4$, $NaBH_4$ and KCl-NaCl were first mixed and allowed to react to form the SnO_2 precursor. Then the SnO_2 precursor is placed in a molten salt environment for roasting. According to the calcination temperature, calcination time and the influence of molten salt on the SnO_2 nanorods, the calcination temperature range is between 710–900 °C, and the calcination time is from 15 min to 3 h.

6.1.5.3 Hydrothermal Method

Copper nanorods were prepared by hydrothermal method. $CuSO_4$ $5H_2O$ and NaOH were used as raw materials, and the concentration ratio of Cu and OH^- was kept at 1:4, and sorbitol equivalent to Cu^{2+} was added as a reducing agent, and the product copper nanorods were obtained after 20 h of reaction under hydrothermal conditions at 180 °C. The product was determined to be metallic copper by XRD analysis, and the scanning electron microscope (SEM) photos showed that most of the product was a regular hexagon composed of spherical particles, and there was also a small amount of nanorods and nanowires in the product. The diameter of the nanorod is 100–500 nm, and the aspect ratio is above 50, while the nanowire is thinner than the nanorod. The SEM photo shows that it is meandering and has good flexibility.

TiO_2 nanorods are prepared by hydrothermal method, and NaOH solution and TiO_2 powder with anatase structure are stirred and mixed according to a certain ratio to obtain a white suspension. Then add this suspension into the autoclave, heat the autoclave to 130 °C, keep the temperature constant for 50–100 h, and carry out hydrothermal reaction. After the hydrothermal reaction was completed, the white precipitate was repeatedly washed with deionized water until neutral at room temperature, and then dried. The dried powder was put into a diffusion furnace for sintering at 500 °C for 1 h, and the TiO_2 nanorod samples could be obtained after natural cooling with the furnace.

6.1.5.4 Template Method

When the flow template method is used to prepare metal oxide nanorods, the templates used mainly include porous alumina templates (AAM), polycarbonate templates (PC), and carbon nanotubes (CNTs) templates. The sol–gel method is often used to grow and assemble nanorods in AAM or PC.

The essence of the sol–gel preparation of oxide nanorods is the direct filling of sol, that is, the percolation of the target oxide sol in the pores of the substrate. The specific process of using AAM as a template to synthesize MnO_2 nanorods is as follows: first prepare the target oxide sol, and then immerse the AAM template in the sol. After a period of time, the template is taken out and subjected to certain post-processing. The advantage of preparing oxide nanorods by sol–gel template method is that the device is simple, the requirements for reaction conditions are not high, and the preparation process is simple. But the difficulty is that the sol penetrates into the pores of the template through capillary action, so sometimes the filling degree of the mold pores is very low, which directly affects the quality of the obtained nanorods. Zhao Qitao et al. used sol as a template to prepare Au nanorods. They will also use tin nitrate and thioacetamide (TAA) as raw materials, combine the sol–gel method and hydrothermal method under relatively mild conditions, use acetylacetone to control the hydrolysis of butyl titanate, and form network channels in the sol by hydrolysis, and use this as a soft template to synthesize CdS nanorods [19].

Satishkumar et al. used CNTs as templates to prepare nanorods of V_2O_3, WO_3, MoO_3, Sb_2O_5, MoO_2, RuO_2. IrO_2 and other oxides, and most of the nanorods were single crystals. There are two possible mechanisms for the preparation of oxide nanorods using CNTs as templates: one is that when oxide-coated CNTs are heated, CO/CO_2 is generated, O atoms come from metal oxides, and the remaining metals or sub-oxides may be reoxidized and undergo recrystallization, and the grains aggregate into rods; the other possibility is that when CNTs are heated, oxide precursors decompose in situ to produce crystals, and H, O, or CO_2 gases are generated during the decomposition process of the precursors. Driven by the transport of gas, the crystal grains aggregate and grow into nanorods. Taking W_5 as an example, the basic preparation process is to stir H_2WO_3 and an appropriate amount of acid-treated CNTs with magnetic force for 48 h. The obtained samples are filtered and washed, dried at 100 °C, and then burned at 700 °C in an air atmosphere to remove the CNTs, and then WO_3 nanorods are obtained.

6.2 Preparation of Two-Dimensional Nanomaterials

According to the definition of nanomaterials, nanofilms are two-dimensional nanomaterials whose size in the thickness direction is on the order of nanometers or films with special properties of nanostructures. In view of this definition, nano-films can be divided into two categories: films containing substrates such as nanoparticles and atomic clusters; films with a meter-sized thickness, whose thickness is close to the electron free path and Debye length (about 10–100 nm), which can be assembled into new functional devices using their remarkable quantum and statistical properties.

In addition, from the perspective of application, nano-films can also be divided into two categories, namely nano-functional films and nano-structured films. Nano-functional thin films mainly use the optical, electrical, and magnetic properties of nano-materials to make the material as a whole have special nano-functions by forming thin films. Nanostructured films are mainly films with nanostructures such as nanoscale or nanoholes formed by appropriate means, and these structures reflect nanoeffects. Since changes in the composition, performance, and process conditions of nanomaterials have a significant impact on the properties of the film, it is possible to artificially control the formation of the properties of the nano film with a large degree of freedom to obtain materials that meet the needs.

Nano-film is divided into single-layer film and multi-layer film. Single-layer film is relatively simple, while nano-multi-layer film refers to a composite film material with alternating components or structures formed by the alternate deposition of one or several elements or compounds, and the thickness of each layer of material is nanoscale, so it also belongs to nano-structured film materials. The main parameter of the multilayer film is the modulation wavelength A, which refers to the sum of the thicknesses of two adjacent layers of materials in the multilayer film. When the modulation wavelength A is several times or larger than the lattice constant of each layer of thin film single crystal, the multilayer film structure can be called "superlattice" thin film. The nanomaterials that make up the film can be metals, metal semiconductors, insulators, organic polymers and other materials, so there can be many combinations, such as metal/semiconductor, metal/insulator, semiconductor/ polymer materials, etc., and each combination can derive many types of nanostructure films.

The preparation of thin films can be roughly divided into two categories: physical methods and chemical methods. Physical methods mainly include evaporation with different heating methods, DC, high frequency or radio frequency sputtering, ion beam sputtering, molecular beam epitaxy, etc.; chemical methods include various chemical vapor deposition (CVD), sol–gel method (sol–gel) and so on.

This chapter focuses on several main preparation methods of nano-film materials, which are: vacuum evaporation method, sputtering method, microwave method, electrochemical method, sol–gel method, self-assembly and surface modification method, etc.

6.2.1 Vacuum Evaporation Method

The vacuum evaporation method is to vaporize and sublimate the metal, alloy or compound to be prepared by heating the evaporation source in an ultra-high vacuum (10^{-5} Pa) or low-pressure inert gas (Ar) or nitrogen (He) atmosphere (50–1000 Pa), and then condense to form a nano-film material. The vapor deposition method mainly obtains nanostructure films through two ways: controlling the formation of nanostructures during the crystallization process of amorphous films; controlling the formation of nanostructures during the nucleation and growth of films. This is one of the main methods of preparing nano-films by physical methods.

The main process of the material in the heat source being evaporated and condensed by the inert gas is: the vacuum evaporation chamber is filled with a low-pressure inert gas (He or Ar), and the raw material of the target material is heated and evaporated in the evaporation source to generate atomic mist, collide with the inert gas atoms, condense to form nano-sized clusters, and deposit on a relatively cold substrate to form a thin film. The nano-alloy thin film can be obtained by evaporating two or several metal substances at the same time, and can also be obtained by directly evaporating the original alloy material as a raw material. The preparation of the nanometer oxide film can be obtained in the process of evaporation or after the clusters are produced, and then placed in a vacuum chamber to pass through pure oxygen to oxidize it. Due to different evaporation conditions, nanometer films of different thicknesses can be formed, and the atomic arrangement can be crystalline, so it is a thin film crystal; it may also be a disordered amorphous state or a discontinuous film; it may also be different from the corresponding crystalline state and amorphous state, and the transition from close to amorphous state to crystalline state, therefore, its properties are significantly different from crystalline and amorphous states with the same chemical composition.

The characteristic of the vacuum evaporation method is that the state of the nano-film can be regulated by adjusting parameters such as heating temperature, pressure and atmosphere during the preparation process. Vacuum evaporation method is more commonly used in experimental research, especially suitable for the preparation of metal nano-films. Dozens of metal nano-films have been prepared, such as Au, Ag, Cu, Fe, Al, Pd, Be, Bi, Mn, Co, Ni, V, Cr, Cd, Zn, Se. In, Te, etc.; nano-metal oxide films (Fe_2O_3. MnO, NiO, MgO, TiO_2. Al_2O_3), etc.

6.2.2 Sputtering Method

The so-called "sputtering" is very common in people's daily life. When a small stone is shot into the water surface, water droplets will be sputtered on the water surface. Here, it refers to the phenomenon that charged particles bombard the solid surface (target), causing solid atoms (or molecules) to be ejected from the surface. These sputtered particles have higher energy than the aforementioned evaporated particles, and have directionality and film-forming characteristics. The method of applying this phenomenon to deposit these sputtered substances onto the substrate or workpiece surface to form a thin film is called sputtering (coating) method. The sputtering method is a kind of physical vapor deposition. Most of the sputtered particles are in the atomic state, often called sputtered atoms, and there are also sputtered ions. The energetic particles used to bombard the target can be electrons, ions or other neutral particles, because the ions are easy to accelerate and obtain the required kinetic energy under the electric field, so ions are mostly used as the bombarding particles, and the ions are also called incident ions. The sputtering method is widely used in the preparation of various functional nano-films. The sputtering method is divided into DC sputtering, ion sputtering, radio frequency sputtering and magnetron sputtering according to the different ways of exciting sputtering ions and depositing thin films. At present, the latter two are mostly used.

6.2.2.1 RF Sputtering

The sputtering method in which a radio frequency electric field is added to the sputtering target to excite sputtering ions is called radio frequency sputtering. This sputtering deposition method is not only suitable for various metals, but also suitable for the preparation of thin films of non-metallic materials. The high-frequency power frequency commonly used in sputtering by radio-frequency sputtering devices is radio frequency.

The target particles are sputtered by high-power energetic particles. Under the action of high-frequency alternating radio frequency electric field, the particles can oscillate back and forth in the space between the cathode and the anode before being absorbed by the anode, so there are more opportunities to produce impact ionization with gas molecules, so radio frequency sputtering can be carried out at a lower vacuum pressure.

The main feature of RF sputtering is that it can sputter any solid material, including targets made of conductors, semiconductors and insulator materials. This is mainly due to the establishment of a negative bias on the insulating target surface. Since the mobility of electrons is higher than that of ions, when the target electrode is applied with RF voltage through capacitive coupling, the number of electrons arriving on the target will be much larger than the number of ions, so electrons will gradually accumulate on the target, causing the target to have a DC negative potential. In the equilibrium state, the negative potential on the target makes the number of electrons reaching the target equal to the number of ions, so the peak value of the DC voltage will not be equal to the target circuit connected to the external RF power supply through the capacitor. For the target of insulator material, it is precisely because of the negative bias on the target surface that the positive ions can continuously bombard it to make the sputtering continue; and during the positive half cycle of the radio frequency voltage, the bombardment of the electrons on the target surface can neutralize the positive ions accumulated on the target surface and also make the sputtering continue.

RF sputtering does not require a secondary electric field to sustain the discharge. However, when the ion energy is as high as thousands of electron volts, the number of secondary electric field electrons emitted on the insulating target is also quite large, and because the target has a relatively high negative potential, the electrons are accelerated through the dark area, making them high-energy electrons for bombarding the substrate, causing the substrate to heat up, charge and damage the quality of the film. To do this, the substrate needs to be placed in a position that is not directly bombarded by the electrons in the secondary electric field, or the electrons are deflected away from the substrate by using a magnetic field.

6.2.2.2 Magnetron Sputtering

Magnetron sputtering is a high-speed sputtering technology developed rapidly in the 1970s. The orthogonal electromagnetic field is introduced in the magnetron sputtering, which increases the ionization rate of the gas by 5–6%, and the result is that the quiescence rate of the film is increased. For many materials, the sputtering rate reaches the level of electron beam evaporation by means of magnetron sputtering.

The secondary electrons generated by sputtering are accelerated into high-energy electrons in the cathode drop region, but due to the controllable magnetic field, they cannot fly directly to the anode, but move approximately cycloidally under the combined action of electric and magnetic fields. During the movement, high-energy electrons collide with gas molecules continuously, and transfer energy to the latter, making them ionize and become low-energy electrons. These low-energy electrons drift along the magnetic field lines to the auxiliary anode near the cathode and are absorbed, thus avoiding the strong bombardment of high-energy electrons on the workpiece. At the same time, electrons have to fly about hundreds of meters to reach the anode, and the collision frequency is about 10^7 s^{-1}, so the ionization efficiency of magnetron sputtering is high.

6.2.3 Microwave Method

The microwave method uses microwave plasma chemical vapor deposition (MW-CVD) to make thin films. The device is mainly composed of a magnetron microwave source, a matching pin and a plasma coupler, and these components are connected with a waveguide. The power of the microwave source is 0.5–1 kW, and the frequency is 2.45 GHz. The microwave generated by the microwave source is transmitted to the plasma coupling cavity through the waveguide and absorbed by the load. A circulator can also be set up to guide the reflected power into the dark load and be absorbed to protect the magnetron. The tuning pins are used to optimally match the load of the plasma to the total impedance for more efficient use of microwave energy.

Various superhard films (such as diamond film, cubic boron nitride film, SiC film, GaN film, etc.) can be prepared by microwave method. At present, nano-diamond films have been prepared by microwave method, which have very excellent field electron emission characteristics. Chen Guanghua et al. used a metal sheet with a lower work function as a substrate, and used methane and hydrogen as gas sources to deposit thin films in microwave plasma CVD, and obtained cold cathode materials with high current and low threshold voltage. When the voltage of the electron emission cold cathode increases to 600 V, the emission current begins to form. As the field strength increases, the emission current also increases. When the voltage is increased to 1800 V, the emission current reaches 2090 KA, and the emission current density reaches 418 mA/cm^2. The diamond film deposited on the metal sheet by microwave method contains a large number of nano-diamond particles with a size of about 5–10 nm. This film has excellent field emission characteristics.

Zhiming Wang et al. used a MW-CVD system with a microwave frequency of 2.45 GHz and a microwave power of 1000 W to deposit a high-nucleation-density diamond film on an alumina ceramic substrate by applying biased low-voltage nucleation technology and according to the law that the diamond nucleation density increases with the decrease of system pressure. The experiment uses a typical MW-CVD experimental device. The substrate used is an alumina ceramic sheet produced by epitaxy. The alumina substrate is vertically inserted on a substrate frame made of a quartz boat and immersed in the plasma atmosphere in the deposition chamber. The substrate temperature of the alumina substrate was measured by an optical pyrometer. The reaction gas is methane gas diluted with hydrogen, the flows of hydrogen and methane are monitored by a mass flow controller, and the gas pressure during the reaction is characterized by a vacuum gauge. Before deposition, the alumina substrate was ultrasonically cleaned for 30 min in an acetone suspension containing diamond powder, and then rinsed with deionized water. The experiment is divided into two stages: Hydrogen plasma etching stage (to remove impurities on the substrate surface); stage. Different system pressures are used in the nucleation stage, and the interaction between system pressure and nucleation can be studied. The test results of diamond nucleation on alumina substrates under different system pressures show that when the pressure is high, the nucleation density is low, and the particle size and distribution are uneven; when the pressure is low, the nucleation density is high, the diamond particles are dense, and the size and distribution are relatively uniform.

6.2.4 Molecular Beam Epitaxy

Molecular beam epitaxy (MBE) is under ultra-high vacuum conditions, precisely controlling the molecular beam intensity of the raw material, and injecting the molecular beam onto the heated substrate for epitaxial growth. By improving the performance of the evaporation source, monitoring system and analysis system and improving the vacuum environment, extremely high-quality thin film single crystals can be obtained. Therefore, it can be said that MBE is a new thin film growth method based on vacuum evaporation.

Molecular beam epitaxy is an ultra-thin film preparation technology that can precisely control the epitaxial thickness, doping and interface flatness at the atomic scale. The so-called "epitaxy" is to grow a single crystal thin film on some specific single crystal material substrates, extending outward along a certain crystal plane of the substrate. The purity and performance of the epitaxial single crystal thin film may be significantly different or improved compared with the original bulk single crystal material, and it is also possible to manufacture a large area or special material single crystal nano film by using epitaxial technology. The application of this thin film preparation technology has promoted the development of a new generation of semiconductor science and technology based on semiconductor ultra-thin layer structure materials. Molecular beam epitaxy is often used to grow heterojunction compound semiconductor thin films, which can grow high-quality heterojunctions. In addition, the epitaxial single crystal process can unify the material preparation and device manufacturing process, which is conducive to improving the yield and stability of the device, and provides a new way for the development of new nanomaterials and nanoelectronic devices. The semiconductor superlattice nanofilms and quantum well material nanofilms prepared by molecular beam epitaxy have pushed the research and application of nanomaterials to a new stage. In recent years, a variety of epitaxial thin film preparation technologies have been developed, mainly including vapor phase epitaxy, liquid phase epitaxy and molecular beam epitaxy.

The MBE device is mainly composed of a vacuum epitaxy studio, a molecular beam injection source, an ultra-high vacuum system and various corresponding monitoring instruments. The substances and dopants required for the preparation of the film are put into the raw snails of several injection sources in the system, and heated to melt (sublimate) the substances to generate corresponding molecular beams. In MBE, Knudsen jet sources are mostly used at present (that is, the aperture is much smaller than the mean free path of vapor molecules in the container), so that the central part of the molecular beam coming out of each jet source intersects the substrate. By selecting the appropriate furnace temperature and substrate temperature of the jet source, the desired chemical composition and crystalline film grown according to the lattice position can be obtained. A separate baffle is installed between each spray source and the substrate, which can be opened and closed instantly to control the growth of the epitaxial film. In order to control the type and intensity of the molecular beam, a quadrupole mass spectrometer (QMS) can be installed on the path of the molecular beam. In addition to QMS, in-situ detection instruments such as high-energy electron diffractometer (PHEED) and Auger electron spectroscopy (AES) are usually installed, which can be used to observe the changes and components of the mint-adjusted film lattice structure.

6.2.5 Sol–Gel Method

The sol–gel (sol–gel) synthesis method starts from a solution of an organic or inorganic compound of a metal, and through the hydrolysis and polymerization of the compound in the solution, the solution is made into a sol solution in which metal oxide particles are dissolved, and further reaction occurs to gel, and then the gel is heated and dried to form various films. At present, this method is one of the most commonly used methods for preparing nano-films.

When preparing thin films by the sol–gel method, metal alkoxides or other salts are usually dissolved in organic solvents such as alcohols and aldehydes to form a uniform solution. The solution forms a sol through hydrolysis and polycondensation reactions, and further polycondensation reactions undergo sol–gel transformation to form gels. After heat treatment, the remaining organic matter and moisture in the gel are removed, and finally the required film is formed.

There are three main methods for preparing thin films by sol–gel technology: dipping method, spin coating method, and laminar flow method. These three methods have their own characteristics, and different methods can be selected according to the size and shape of the substrate material and the requirements for the thin film. Among them, the dipping method and the spin coating method are currently more commonly used. When these methods are used to prepare nanometer films, the gel film is formed due to the rapid evaporation of the solvent, and the required film material can be obtained by heating the gel film as required.

The sol required for the preparation of thin films by sol–gel method can be divided into organic route and inorganic route according to the method of its formation or the state of existence. The organic approach is to form sols through hydrolysis and polycondensation of organometallic alkoxides. Because this approach involves water and organic matter, the residual stress generated by the evaporation of a large amount of solvents (water, organic matter, etc.) during the drying process of the thin film is likely to cause cracks, which largely limits the possibility of preparing thin films with a certain thickness. In the inorganic way, oxide particles prepared by a suitable method are stably suspended in corresponding organic or inorganic solvents to form a sol. Sometimes it only needs to be dried at room temperature for film formation by inorganic means, so it is easy to make thicker oxide or metal films with more than 10 layers and no cracks. However, the thin film prepared by the inorganic method has poor adhesion to the substrate, and it is difficult to find a suitable solvent that can dissolve multiple oxides at the same time.

6.2.6 Chemical Vapor Deposition

Chemical vapor deposition (CVD) refers to the direct use of gas or various means to change substances into gases, allowing one or several gases to undergo thermal decomposition, reduction or other reactions through the effects of heat, light, electricity, magnetism and chemistry, etc., and precipitate nanoparticles from the gas phase. After cooling, various nano-films such as hydrogen, oxygen, nitrogen, carbon compounds, etc. The gasification of raw materials by the chemical vapor phase reaction method includes various methods such as plasma-enhanced excitation in addition to ordinary resistance electric heating.

Ordinary resistance heating for chemical vapor deposition (CVD) mainly uses resistance heating to vaporize one or several simple substances or compounds containing thin film elements into a gas phase, and then conducts a chemical reaction on the substrate surface to form a thin film. The characteristic of the resistance heating atmospheric pressure chemical vapor deposition (APCVD) reaction device is that the reaction gas is directly sprayed onto the substrate through a nozzle moving at a constant speed, and the size of the crystal grains can be controlled by precisely controlling the reaction temperature and reaction time, so as to obtain nanocomposite thin film materials.

The basic process of film formation by APCVD method includes multiple stages such as gas diffusion, adsorption of reaction gas on the substrate surface, surface reaction, nucleation and growth, gas desorption, diffusion and volatilization. The transport properties in the CVD process (including the transport of heat, mass and momentum), the properties of gas flow formation (including movement speed, pressure distribution, gas heating, activation methods, etc.), substrate type, surface state, temperature distribution, etc. will all affect the composition, structure, morphology and performance of the film. The method can be used to prepare various metal simple substances, metal alloys, and nanocomposite films such as oxides, fluorides, and carbides of non-metals.

Gaorong Han also used this method to successfully prepare Si/SiC nanocomposite film materials. The film is composed of a large number of 5 nm silicon grains and a small amount of silicon carbide grains. The crystalline content is about 50%, of which nano-silicon grains account for 90%. The film presents a good nano-mosaic composite structure.

In addition to ordinary chemical vapor deposition, there is also a technique called plasma chemical vapor deposition (PCVD). Plasma chemical vapor deposition technology is a method of chemically reacting gaseous substances containing the constituent atoms of a film by means of plasma, and then depositing a film on a substrate. It is especially suitable for the synthesis of semiconductor and compound nano-films. It is regarded as the second-generation film deposition technology after CVD.

There are many different forms of PCVD devices, but the basic principle is the same, and the basic structural units are often similar. If divided according to the method of plasma generation, there are several types such as DC glow discharge, radio frequency discharge, and microwave discharge.

6.3 Preparation of 3D Nanomaterials

Nano-solid (three-dimensional) materials refer to bulk materials composed of nanoparticles with a size of 1–100 nm as the main body. From the state of crystal phase, there are three forms of nanoparticles in nano-solids: crystalline state with long-range order, quasi-crystalline state with only orientation order and amorphous state with short-range order. Nano-particles as a unit can be stacked into nano-blocks in three-dimensional space. After artificial control and processing, the nano-particles are arranged in an orderly three-dimensional space to form an array system of different dimensions. There are two main types of preparation methods for nano-bulk materials: one is to prepare nanoparticles first, and then prepare nano-bulk materials through processes such as pressurization and sintering; the other is to directly prepare ordinary amorphous bulk materials into bulk nanomaterials with nanostructures through special processes. Nano-solids can be divided into nano-crystalline materials (also known as nano-crystalline materials), nano-quasicrystalline materials and nano-amorphous materials according to the state of small particle structure. According to the material properties and bond forms of the particles, nanomaterials can be divided into nanometal materials, nanoionic crystal materials, nanosemiconductor materials, and nanoceramic materials. Nanomaterials are solids composed of single-phase particles called nanophase materials. If the nanomaterial itself is composed of two phases (one phase is dispersed in the other phase), the corresponding nanomaterial is called a nanocomposite material. This part will be specially introduced in a later chapter. In this chapter, the preparation method of three-dimensional nanosolids such as single-phase nanometals, nanoalloys and nanoceramics will be introduced.

One of the main goals of the preparation of nano-bulk materials is to obtain large-sized nano-crystal samples, in which the interface is expected to be clean and dense, without micro-porosity, and the grains are fine and uniform. Over the years, material scientists have successively developed a variety of preparation methods related to nano-metal and compound nano-materials, mainly: metal evaporation condensation-in-situ cold pressing method; high-energy grinding and pressing method; amorphous crystallization method; deposition method. The preparation methods of these nano-blocks can be divided into three categories according to their interface formation process: external pressure synthesis, such as ultrafine powder cold pressing method, mechanical grinding method; deposition synthesis, such as various physical and chemical deposition methods; phase change interface formation, such as amorphous crystallization method. Each of these methods has its advantages and disadvantages for different purposes. It is sometimes difficult to achieve this goal by using the above-mentioned several preparation methods. In recent years, on the basis of the original technology and equipment, by improving the heat source and method of metal sublimation (such as induction heating, plasma method, magnetron sputtering method, etc.), the output of ultrafine powder can be greatly increased, creating conditions for the preparation of large-scale nanocrystal samples. In addition, a large number of powder materials with nanocrystalline structure can also be obtained by using mechanical grinding technology. Although there are still some problems such as pollution and oxidation, due to some unique advantages of this method, such as relatively simple process and a large variety of nanomaterials that can be synthesized, especially materials that are difficult to prepare by other methods such as metal compounds (such as Ti-Al, Fe-Cr-Al) and solid solutions (such as Fe-Cu, Fe-Al). Very promising preparation method. The above factors have also prompted the research on the synthesis of bulk samples using mechanically alloyed powders as raw materials. In conclusion, many attempts have been made in recent years to manufacture and use these nanoparticles or powders to synthesize dense bulk nanomaterials, and some new progresses have been made, which will be introduced separately below.

6.3.1 Preparation of Nano-metal and Alloy Materials

6.3.1.1 Inert Gas Evaporation, In-Situ Pressure Preparation Method

Nano-metal materials have a long history of development. As early as the early 1980s, German scientist Professor H. Gleiter first proposed the concept of nano-crystalline materials, and later successfully prepared Fe, Cu, Au, Ag, Mg, Sb, Pd and other nano-crystalline metal bulks and NiAl, TiAl, Fe_5Si_{95}, Si_{25} Pd_{75}, $Pb_{70}Fe_5Si_{25}$, and other nano-alloy materials. Nano-metal and alloy materials in nanostructure materials are crystals or non-crystals that have undergone secondary condensation. The first condensation is usually to form gasified metal atoms into nanoparticles, and then press the nanoparticles together to form a block-like condensed solid under the condition of maintaining a fresh surface. In order to prevent oxidation, the above-mentioned processes are generally carried out in a vacuum, which puts forward very high requirements for the preparation of nano-metal and alloy solids.

The inert gas evaporation and in-situ pressurization method called "one-step method". From the formation process of nano-metal materials, it can be concluded that the steps of preparing nano-metal solids are: preparing nanoparticles; collecting particles; pressing into blocks.

According to the three steps of preparing nano-metal solids, the device is composed of three main parts: the first part is an evaporation source system for preparing nanoparticles by evaporation; the second part is the collection system of nano-particle powder cooled by liquid nitrogen, including the scraping and transportation system of nano-particle powder; the third part is the in-situ pressure forming (sintering) system of nano-particle powder. Wherein the first part and the second part are basically the same as the method for preparing nano metal particles by inert gas evaporation method described in the previous chapters. Here we will focus on the part of preparing nanostructure block by in-situ pressure which is different from the above. After the nano-metal or alloy particles prepared by evaporation of inert gas are formed in a vacuum, they are scraped off from the cold trap by a scraper made of polytetrafluoroethylene, and directly fall into a low-pressure compaction device through a funnel. After being lightly compacted in this device, the particle powder is sent to a high-pressure in-situ press by a robot, and the heating system is hot-pressed into a block, with a pressure of 15 GPa and a temperature of 300–800 K. Since the metal and alloy nanoparticles formed by the evaporation and condensation of inert gases have almost no hard aggregates, even if pressed at room temperature, a block with a relative density higher than 0.9 can be obtained, and the highest relative density can reach 0.97. Therefore, the advantage of this preparation method is that the nanoparticles have a clean surface, rarely agglomerate into coarse aggregates, and the blocks have high purity and high density.

In recent years, on the basis of this device, by improving the heat source and method for evaporating and sublimating metals (such as induction heating, plasma method, electron beam heating method, laser pyrolysis method, magnetic sputtering, etc.) and improving other equipment, several grams to tens of grams of nanocrystal samples can be obtained. Using this method, many studies on nano-supersaturated alloys and nano-composite materials have also been carried out. At present, this method is developing in the direction of multi-component, metering control, multi-pair molds, and ultra-high pressure. The method is characterized by wide application range and clean surface of micropowder, which is helpful to the theoretical research of nanomaterials. However, the process equipment is complex and the output is low, so it is difficult to meet the requirements of performance research and application.

Zhou Yusong et al. improved this method by heating the container containing the nanoparticle powder while pressurizing the nanoparticle powder, which increased the density of the block and successfully obtained a large-sized, nanocrystalline copper and silver bulk material with a clean interface. The specific experimental process and conditions are as follows: the reaction chamber is pre-evacuated to 9X10-6 Pa, and filled with a mixed gas of 0.4–2 kPa high-purity Ar and H_2. The polycrystalline crude copper or silver block with a purity of 99.99% is heated by arc plasma to evaporate, and the evaporated nano-copper or silver particles are transported to the outer wall of the rotating liquid nitrogen cold trap by the flow of inert gas, and quickly condensed into nano-metal particles. At this time, the mold is heated and the original vacuum degree is maintained to desorb the gas adsorbed on the nanoparticles, and the large-size nano-copper and silver blocks with a clean interface are obtained by hot pressing. The nano-copper and silver bulks exhibit high hardness characteristics, and their microhardness is 6 times and 2 times that of polycrystalline coarse-grained copper and silver, respectively. The elastic modulus, yield strength, fracture strength, and elongation of nano-copper are 0.65 times, 1.42 times, 1.82 times, and 0.15 times that of polycrystalline coarse-grained copper, showing its excellent characteristics of high strength, low elastic modulus, and low ductility.

6.3.1.2 High-Energy Milling Combined with Pressurized Agglomeration

High-energy ball milling is a technique used to prepare metal-based or ceramic-based composite powders with controlled microstructures, which can obtain nanoparticles combined with pressure to form nanophase blocks. In the dry ball milling room, under the protection of high vacuum and nitrogen atmosphere, through the collision between the hard steel balls running at high speed and the grinding body during the mechanical grinding process, the original materials are repeatedly broken, sintered, and broken again. This process enables continuous refinement of the material particles to nanometer dimensions, the preparation of which is already known from previous chapters. The nanoparticle powder is pressed by hot extrusion, hot isostatic pressing and other technologies to obtain block nanomaterials. Studies have shown that amorphous, quasicrystalline, nanocrystalline, superconducting materials, rare earth permanent magnet alloys, superplastic alloys, intermetallic compounds, light metal high specific strength alloys, etc. can be synthesized by this method.

The alloy matrix composition of this method is not limited, the cost is low, the output is large, and the process is simple. It shows strong vitality especially in the alloying of refractory metals, the generation of non-equilibrium phases, and the development of special alloys. The problem is that impurities, pollution, oxidation, and stress are easily generated during the grinding process, and it is difficult to obtain a clean nanocrystal interface.

Metal bulk materials with nano-phases can also be obtained by high-pressure, high-temperature solid-phase quenching. This method is called high-pressure, high-temperature solid-phase quenching. Put the raw material into the high-pressure chamber, increase the pressure to several GPa, and then reheat. Since the high pressure suppresses the long-distance diffusion of atoms and the growth rate of crystals, the nanometerization of crystal grains can be realized, and then solid phase quenching at high temperature to retain the influence of high temperature and high pressure on the structure.

6.3.1.3 Amorphous Crystallization Method

The crystallization transition from amorphous structure to crystalline structure will occur under certain conditions. The amorphous crystallization method crystallizes the product into grains by controlling the crystallization kinetics of the amorphous solid, and then obtains a bulk material with a nanophase. By controlling appropriate process conditions, nanocrystalline precipitated phases can be obtained through the crystallization of amorphous materials, and finally nanocrystalline bulk materials or nanocrystalline/non-nanocrystalline composite bulk materials can be produced. The preparation of nano-bulk materials by amorphous crystallization is usually completed by two processes: the acquisition of amorphous solids and the crystallization. According to the mechanism of crystallization, it can be divided into thermal crystallization, electrocrystallization, mechanical crystallization and so on. Amorphous solids can be prepared by melt quenching, high-speed DC sputtering, plasma jet atomization, and solid-state reaction methods. The most commonly used method is single-spoke or double-spoke spin quenching. Since the above methods can only obtain low-dimensional materials such as amorphous powder, silk, and strips, it is necessary to use methods such as hot die compaction, hot extrusion, or high-temperature and high-pressure sintering to synthesize bulk samples, and isothermal annealing methods. In recent years, methods such as hierarchical annealing, pulse annealing, and laser induction have also been developed.

At present, a variety of alloy series nanocrystals based on Ni, Fe, Co, Pd, etc. have been prepared by using the amorphous crystallization method. Metal bond compounds and simple semiconductor nanocrystals can also be prepared, and have developed to a practical stage. Ke Lu et al. took the lead in successfully preparing nanocrystalline Ni-P alloy ribbons by the method of amorphous crystallization. First, the $Ni_{80}P_{20}$ melt was made into amorphous alloy strips by single-spoke quenching method. Then annealing is carried out at different temperatures to crystallize the amorphous strips into strips composed of nanocrystals. When the annealing temperature is lower than 610 K, the particle size of the formed Ni_3P nanocrystals is 7–8 nm, and the grains begin to grow as the temperature rises. The plasticity of the nano-block material prepared by the crystallization method is very sensitive to the grain size, and the plasticity is better only when the grain size is small, otherwise the material will become very brittle. Therefore, for some amorphous alloys with small nucleation activation energy, large grain size and large activation energy, only the amorphous crystallization method can be used to obtain nanocrystalline alloy bulk materials with better plasticity.

Crystallization by heating and annealing is characterized by low cost, large output, clean and dense interface, no micropores in the sample, and easy control of grain size changes, which is helpful for studying the formation mechanism of nanocrystals and testing the possibility of applying the classic nucleation and growth theory under rapid solidification conditions. The limitation of this method is that it is only suitable for alloy systems with strong amorphous forming ability.

The method of deep supercooling can directly crystallize the bulk material. In view of the fact that rapid solidification has a significant effect on grain refinement, deep undercooling is another effective way to achieve rapid solidification of melts and crystallize bulk materials. Deep subcooling and rapid solidification can achieve rapid solidification under large thermodynamic supercooling by avoiding or removing heterogeneous crystal nuclei. Melt growth will not be controlled by external heat dissipation conditions, and grain refinement is only dominated by the special physical mechanism of the melt itself. The use of deep supercooling has become a preferred solution to realize the rapid solidification of three-dimensional large-volume liquid metals to prepare microcrystalline, amorphous and quasicrystalline materials, and some people have successfully obtained nanostructured blocks by rapid cooling and rapid solidification in the process of alloys such as Fe-Ni-Ak Pd-Cu-Si.

6.3.2 Preparation of Nano Ceramics

Nano-ceramics refer to ceramic materials whose grain size, grain boundary width, second phase distribution, pore size, and defect size are all below 100 nm. Nano-ceramics not only has the excellent properties of advanced ceramic materials such as wear resistance, corrosion resistance, high temperature resistance, high pressure resistance, high strength, and not easy to age, but also has the following advantages: low sintering temperature and high sintering rate. This is because the sintering process of nano-ceramics at low temperature is mainly controlled by grain boundary diffusion. The most critical indicator of high-quality nano-ceramics is whether the material is highly dense. At present, material scientists have explored a number of methods for preparing nano-ceramics with high density.

The raw materials for preparing nano-ceramic materials generally use nano-particle powder, so the key step to obtain ideal nano-ceramics is sintering. The sintering of nano-powder into nano-ceramic materials must not only be fully densified, but also maintain the nano-size of the grains, in order to reflect many excellent characteristics of nano-ceramic materials. Compared with conventional powder, nano-powder has stronger particle adsorption, will bring a lot of impurities in the air, agglomeration is more serious, and the bulk density of powder is too low. All of these make nano-sintering face more problems than conventional powder sintering, and the sintering process is more difficult to control. The following preparation methods of nano-ceramics all revolve around how to control the sintering process.

6.3.2.1 Pressureless Sintering

Pressureless sintering (static sintering) does not require special equipment to prepare nano-ceramics, the process is simple, the cost is low, and it is easy to produce industrially. Therefore, it is the most basic and widely used sintering method. The non-pressure sintering process is to mold the non-agglomerated nano powder into a block at room temperature, and then bake it at a certain temperature to make it densified (sintered). Because in the sintering process, the activation energy of particle coarsening, green body densification, and grain growth is different, so the kinetic processes of these three have different dependencies on temperature, that is, particle coarsening, green body densification, and grain growth are carried out in different temperature ranges. Taking advantage of this feature, sintered products with high densification rate and slow grain growth can be obtained by controlling the sintering temperature. Pressureless sintering mainly achieves the densification of the green body under the premise of the least grain growth through the selection of the sintering system. However, during the sintering process, the rapid growth of the grains and the formation of large holes are prone to occur. As a result, the densification of the sample cannot be achieved, which makes the advantages of nano-ceramics lost. The control of the sintering system is mainly to control the temperature rising (falling) speed, holding time and maximum temperature. The most commonly used pressureless sintering is constant velocity sintering.

In pressureless sintering, the densification process is also affected by various factors such as powder properties and green body density. In this process, temperature is the only controllable factor, so the control of material sintering is relatively important and difficult. Taking the process of making nano Y-TZP ceramics by pressureless sintering as an example, we can see the influence of various factors on sintering.

Pressureless sintering is easily affected by other factors in various non-sintering processes, and the sintering temperature of powder is one of the factors affecting sintering. Studies have shown that although the densification process of green bodies made of powders fired at different temperatures is basically similar, there are still some differences in the results. When the sintering temperature is too high (600 °C), due to the large particles, small specific surface energy and low sintering power, the density of the obtained Y-TZP ceramics is low; but the sintering temperature is too low (400 °C), and the densification of the green body will be hindered because the residual groups in the powder are not eliminated. The firing temperature is generally 450 °C, which is more suitable. At this time, the particle size of the powder is smaller, and the residual matrix is also less, so that this kind of green body has a higher density than other green bodies when sintered under the same conditions.

6.3.2.2 Hot Pressing Sintering Method

The non-agglomerated powder is sintered under a certain pressure, which is called stress-assisted sintering (sintering-forging) method, but since this method is accompanied by high-temperature heating during the pressurization process, it should be more accurately called hot-press sintering. Compared with the pressure-free sintering process, the advantage of this process is that many undoped nano-powders can be sintered by stress hot pressing to produce nano-ceramics with higher density, and the grains do not grow significantly. However, this process requires more complex equipment than pressure-free sintering, and the operation is more difficult. Therefore, the press is required to be equipped with a set of molds and heating systems that can be heated and pressurized at the same time, which greatly increases the production cost.

Hot pressing sintering is the purpose of densifying the sample by applying a certain pressure while heating the powder for sintering. In this method, the densification of the sample mainly depends on the migration of substances under the action of external pressure. Hot pressing sintering includes vacuum hot pressing sintering, atmosphere hot pressing sintering, continuous hot pressing sintering and so on. Compared with normal pressure sintering, hot press sintering has much lower sintering temperature, and the sintered body has low porosity and high density. Due to sintering at a lower temperature, the growth of grains is inhibited, and the resulting sintered body has finer grains and higher mechanical strength. Hot pressing sintering is widely used in the preparation of materials that are difficult to densify under normal pressure-free conditions and the preparation of nano-ceramics. The following still takes the preparation of nano-YZP ceramics as an example to analyze the hot-pressing sintering process.

6.3.2.3 Microwave Sintering Method

In order to prevent the crystal grains from growing too much during the sintering process of nano-ceramic materials, rapid heating and rapid cooling must be adopted, and the sintering method of microwave sintering can be used. The heating rate of microwave sintering is fast (500 °C/mm), and the heating time is short, which can effectively solve the problem of abnormal growth of nanocrystals that is inevitable in ordinary sintering methods. Using microwave sintering, the conversion efficiency of microwave energy into heat energy is high, about 80–90%, which is also very beneficial to reduce production costs.

The principle of microwave sintering is to use the inherent dielectric loss of the material in the microwave electromagnetic field to rapidly heat the ceramic material to the sintering temperature to achieve densification. Since microwave heating utilizes the dielectric loss of the ceramic itself to generate heat, the ceramic is both a heating body and a heated body. In the entire prepared device, only the processed ceramic products are in a high temperature state, and the rest are still in a normal temperature state, which is beneficial to maintaining a clean reaction environment. The key to the microwave sintering process is to prevent local overheating and ensure the uniformity of the sintering temperature. This problem can be solved by improving the uniformity of the microwave electromagnetic field, improving the dielectric properties and thermal conductivity of the material, and using insulation materials to protect the sintering. Compared with traditional methods, microwave sintering has the advantages of internal heating, fast heating, fast sintering, material structure refinement, material performance improvement, and high efficiency and energy saving.

Doping is helpful for microwave heating. Because doping causes a large number of holes inside the sample, under the action of a high-frequency electric field, the holes and dopant ions form instantaneous dipoles, and the dipoles vibrate violently, and the friction between them will generate a lot of heat, which can promote sintering. Wang Jiangying et al. carried out microwave sintering on $NiMri_2O_4$ nanopowder. Microwave sintering was used without other doping. The shrinkage of the sintered body was not obvious, and the formation of porcelain was not good. After adding 81% (mole fraction) of A1, sintering under the same conditions, the sintered body shrinks obviously, the sintered density can reach 97% of the theoretical density, and the microstructure is uniform. When selecting the type and quantity of dopants (or additives), the improvement of the microwave coupling performance of the material and the improvement of the performance of the sintered product can be considered together. The selected additives should have multiple functions such as sintering aids, microwave susceptors and modifiers.

References

1. P. V. Snytnikov, S. D. Badmaev, G. G. Volkova, D. I. Potemkin, M. M. Zyryanova, V. D. Belyaev, V. A. Sobyanin, *Int. J. Hydrogen Energy* **2012**, *37*, 16388–16396.
2. T. Fukunaga, N. Ryumon, N. Ichikuni, S. Shimazu, *Catal. Commun.* **2009**, *10*, 1800–1803.
3. L. Selva Roselin, H.-W. Chiu, *J. Saudi. Chem. Soc.* **2018**, *22*, 692–704.
4. P. Clancy, J. P. Breen, J. R. H. Ross, *Catal. Today* **2007**, *127*, 291–294.
5. S. Wongkasemjit, K. Asavaputanapun, T. Chaisuwan, A. Luengnaruemitchai, *Mater. Res. Innovations* **2012**, *16*, 303–309.
6. S. G. Sanches, J. H. Flores, R. R. de Avillez, M. I. Pais da Silva, *Int. J. Hydrogen Energy* **2012**, *37*, 6572–6579.
7. S. Licht, *J. Phys. Chem. B* **2001**, *105*, 6281–6294.
8. L. Bi, E. Fabbri, Z. Sun, E. Traversa, *J. Electrochem. Soc.* **2011**, *158*, B797.
9. M. Enoki, J. Yan, H. Matsumoto, T. Ishihara, *Solid State Ionics* **2006**, *177*, 2053–2057.
10. S. A. Abbas, S.-H. Kim, M. I. Iqbal, S. Muhammad, W.-S. Yoon and K.-D. Jung, *Sci. Rep.* **2018**, 8, 2986.
11. K. A. Stoerzinger, M. Favaro, P. N. Ross, J. Yano, Z. Liu, Z. Hussain and E. J. Crumlin, *J. Phys. Chem. B* **2018**, 122, 864–870.
12. K. F. L. Hagesteijn, S. Jiang and B. P. Ladewig, *J. Mater. Sci.* **2018**, 53, 11131–11150.
13. M. Rashid, M. Al Mesfer, H. Naseem and M. Danish, *Int. J. Eng. Adv. Technol.* **2015**, 4, 2249–8958.
14. O. Farrok, M. R. Islam, K. M. Muttaqi, D. Sutanto, J. Zhu, *IEEE T. Ind. Electron.* **2020**, *67*, 3409–3418.
15. D. Zhang, J. Wang, Y. Lin, Y. Si, C. Huang, J. Yang, B. Huang, W. Li, *Renew. Sust. Energ. Rev.* **2017**, *76*, 865–871.

16. E. Asghari, M. I. Abdullah, F. Foroughi, J. J. Lamb, B. G. Pollet, *Curr. Opin. Electroche.* **2022**, *31*, 100879.
17. N. S. Thomaidis, F. J. Santos-Alamillos, D. Pozo-Vázquez, J. Usaola-García, *Comput. Oper. Res.* **2016**, *66*, 284–291.
18. F. Razi, I. Dincer, *J. Clean. Prod.* **2020**, *264*, 121582.
19. J. K. Nørskov, J. Rossmeisl, A. Logadottir, L. Lindqvist, J. R. Kitchin, T. Bligaard, H. Jónsson, *J. Phys. Chem. B* **2004**, *108*, 17886–17892.

Chapter 7
Surface Modification of Nanomaterials

Due to the small particle size and high proportion of surface atoms, nanomaterials have extremely high specific surface area, surface activity and unique physical and chemical properties, which are the prerequisites for their wide application. But it is also these characteristics that make nanomaterials unstable, have high surface energy, and are easy to interact with each other, leading to agglomeration, thereby reducing the specific surface area of the material and the Gibbs free energy of the system, and also reducing the activity of nanomaterials, making it difficult to give full play to the proper characteristics of nanomaterials. On the other hand, nanomaterials have poor affinity with the matrix with low surface energy, and the two are incompatible when mixed with each other, resulting in voids at the interface and phase separation. To fundamentally solve these problems, the most effective way is to modify the surface of nanomaterials. After the surface modification of the nanomaterials, the dispersibility is greatly improved, and at the same time, the compatibility between the inorganic nanomaterials and the organic matrix is increased, and the interface problem is reduced. Therefore, the surface modification of nanomaterials has become an important content of nanomaterials research.

The so-called surface modification (surface modification or surface treatment) refers to the use of physical and chemical methods to treat the surface of nanomaterials to purposely change the physical and chemical properties of the material surface, such as the structure and functional groups of the surface atomic layer, surface hydrophobicity, electrical properties, chemical adsorption and reaction characteristics, etc. After surface modification, a series of surface properties such as adsorption, wetting, and dispersion of nanomaterials will change, which is beneficial to the storage, transportation and use of nanomaterials. By modifying the surface of nanomaterials, the following objectives can be achieved [1].

(1) Protect nanomaterials and improve their dispersion. Surface-modified nanoparticles have a coating film on their surface, which blocks the surrounding environment, prevents the oxidation of nanoparticles, eliminates the charging effect on

S. Peng and P. Li, *Nanomaterials and Nanotechnology*,
https://doi.org/10.1007/978-981-97-8433-2_7

the particle surface, and prevents agglomeration. At the same time, a potential barrier is formed between the nanoparticles, making it difficult for the nanoparticles to grow up during the synthesis and sintering process.

(2) Improve the wettability of the surface of nanomaterials, enhance the interfacial compatibility between nanomaterials and other substances, make nanoparticles easy to disperse in organic compounds or aqueous media, and improve the application performance of nanopowders. For example, when surfactants are used as modifiers to disperse inorganic nanoparticles in aqueous solution, the non-polar lipophilic groups of the surfactants are adsorbed on the surface of the particles, and the polar hydrophilic groups are compatible with water to achieve the purpose of dispersion in water. Conversely, nanoparticles can be dispersed in oil. The dispersibility of the modified nano-powder is enhanced, and its original unique excellent performance is not affected, so it can better exert its potential in practical applications. The surface state of the modified nanoparticles has changed, which can increase the interfacial binding force with the polymer and improve the performance of the composite material.

(3) Improve the surface activity of nanoparticles. The surface of the modified nanoparticles is covered with the active groups of the surfactant, which greatly improves the reactivity of the nanoparticles with other reagents and creates conditions for the coupling and grafting of the nanoparticles.

(4) Surface modification can also introduce active groups with unique functions on the surface of nanomaterials. Through these groups, they can be recombined with matrix materials, thus endowing materials with special optical, electrical, magnetic and other functional properties.

(5) Selectively linking certain molecules with special functions at specific positions on the surface of nanomaterials has important applications in nanofabrication, self-assembly, nanosensors, bioprobes, drug delivery, coatings, and photocatalysis. After the nanoparticles are modified, an organic coating layer is formed on the surface of the particles. The polar ends of the coating layer are adsorbed on the surface of the particles, and the non-polar long chains point to the solvent. Under certain conditions, the non-polar ends of the organic chains are combined to form a regularly arranged two-dimensional structure. CdTe particles modified by organic molecules can be self-assembled to prepare luminescent CdTe nanowires. In this way, nanomaterials with two-dimensional self-assembled structures such as silver and silver sulfide have also been successfully obtained.

7.1 Reasons for Surface Modification of Nanomaterials

Agglomeration of nanomaterials refers to the phenomenon that nanomaterials are connected to each other during the process of preparation, separation, processing and storage to form agglomeration of multiple nanoparticles. The surface effect, small size effect, surface electronic effect and proximity effect of nanomaterials make them have high surface activity. The specific surface area is large, and the nanoparticles are in a thermodynamically unstable state, which is easy to agglomerate. Nanoparticle agglomeration is a process of free energy reduction, and particle agglomeration proceeds spontaneously. According to different mechanisms, nanomaterial agglomeration can be divided into soft agglomeration and hard agglomeration. Soft agglomeration is mainly due to van der Waals force and Coulomb force between particles or capillary force caused by the presence of liquid in the agglomerate, and the interaction force is small. This agglomeration can be eliminated by chemical methods or by applying mechanical force. In addition to the electrostatic force and van der Waals force, the formation of hard agglomerates also has chemical bonding and a strong combination of liquid bridges or solid bridges between particles. This kind of agglomerates has a strong interaction force and high strength, and it cannot be separated without external energy [2].

Nanomaterials prepared by wet chemical method not only contain a large amount of structurally adsorbed water, but also contain a large amount of physically adsorbed water. Due to hydrogen bonding and the presence of a large number of hydroxyl groups on the particle surface, it is easy to bridge between adjacent particles and combine together. When dehydration occurs, these hydrogen bonds are transformed into stronger bridging oxygen bonds, so that the particles form hard agglomerations; on the other hand, during the water removal process, a huge capillary force will be generated between the precipitate gel network, which will tighten the particles and rearrange them, which is also an important reason for particle agglomeration.

For the preparation of nanomaterials by mechanical ball milling, in fact, the particle size of solid materials is continuously reduced and refined under the action of external force. During the ultrafine process, the particles are subjected to impact friction and particle size reduction, and a large amount of positive or negative charges are accumulated on the surface of the nascent nanoparticles. Because the newborn particles have different shapes and are extremely irregular, the surface charges of the newborn particles tend to concentrate on the corners and protrusions of the particles. Some of these protrusions are positively charged, and some are negatively charged, and these charged particles are extremely unstable. In order to be stable, they attract each other, and the sharp corners contact and connect with each other, which makes the particles agglomerate. The main force in this process is the electrostatic Coulomb force. When the material is ultrafine below a certain particle size, the distance between particles is extremely short, and the van der Waals force between particles is far greater than the gravity of the particle itself. Therefore, such particles tend to agglomerate with each other [3].

For the agglomeration of nano powder materials, simple physical methods can be used to disperse. There are mainly three methods: mechanical stirring dispersion, ultrasonic dispersion and high-energy treatment dispersion. Mechanical stirring and dispersion is to fully disperse nanoparticles in the medium by means of mechanical energy such as external shear force or impact force. Ultrasonic dispersion is an effective method to reduce nanoparticle agglomeration. Ultrasound refers to sound waves in the sound frequency range between 20–5000 kHz. It has the following characteristics: long and short, straight-line propagation, and the propagation characteristics are closely related to the form of the medium. High radiation intensity, easy to concentrate energy. This ultrasonic wave can generate strong vibrations and cavitation of the medium, thereby inducing thermal, optical, electrical, chemical and biological phenomena, and even changing the properties and states of materials. Cavitation is when the expanding circle of ultrasonic waves passes through the liquid medium, the sound waves cause the liquid phase molecules to vibrate violently, forming many tiny bubbles in the liquid, forming a "cavitation effect". The use of local high temperature, high pressure or strong shock waves and micro jets generated by ultrasonic cavitation greatly weakens the interaction between nanoparticles and effectively prevents nanoparticles from agglomerating and dispersing. The high-energy treatment method is to generate active points on the surface of nanoparticles through the action of high-energy particles, increase the surface activity, make it easy to chemically react or attach with other substances, and modify the surface of nanoparticles to achieve the purpose of dispersion. High-energy particles include corona, ultraviolet light, microwaves, plasma rays, etc. Although physical methods can better realize the dispersion of nanoparticles in liquid medium, especially in the process of preparing nanoparticles, combining these methods can achieve better results. But once the external force stops, the particles will gather together again due to the intermolecular force. In order to fundamentally solve the dispersion problem, it is necessary to modify the nanoparticles.

7.2 The Principle of Surface Modification of Nanomaterials

Surface modification is to carry out physical, chemical, mechanical and other deep processing on the surface characteristics of powder, control its internal stress, increase the repulsion between powder particles, and reduce the gravitational force between powder particles. At the same time, the physical and chemical properties of the powder surface, such as crystal structure, surface energy of functional groups, surface wettability, electrical surface adsorption and reaction characteristics, etc. are changed, so as to endow the nano-powder with new functions and improve the physical and chemical properties of the nano-particle surface. There are many methods for surface modification of nanomaterials, and the basic principle is to carry out corresponding physical and chemical treatments on the surface of nanomaterials. According to the interaction between nanoparticles and the surface of the modifier, the surface modification methods of nanomaterials can be divided into surface physical modification methods and surface chemical modification methods.

7.2.1 Physical Modification of the Surface of Nanomaterials

Surface physical modification is to modify the surface of particles by physical means such as adsorption, coating, and coating. The modifier and the surface of nanoparticles are mainly in the form of physical interaction. The surface modification of nanoparticles by ultraviolet rays and plasma rays is also a physical modification.

(1) Surface adsorption is to adsorb heterogeneous materials on the surface of nanoparticles by van der Waals force or electrostatic attraction to prevent the aggregation of nanoparticles, such as modifying nanoparticles with surfactants. Surfactant is a kind of organic compound with hydrophilic and lipophilic structure, which can reduce surface tension and surface energy, and can emulsify, wet, and form a solution. According to the nature of the surface charge of the nanoparticles, the method of adding cationic or anionic surfactants can be used. They form an organic molecular film on the surface of the nanoparticles, which hinders the mutual contact between the particles, increases the distance between the particles, and avoids the formation of bridging hydroxyl groups and chemical bonds. Surfactants can also reduce surface tension and reduce capillary adsorption. Polymer surfactants also have certain steric hindrance. The hydrophilicity, lipophilicity, and surface activity of nanopowders can be controlled by utilizing the adsorption properties of the hydrophilic groups in the surfactant molecules to the surface of nanoparticles, chemical reactivity, and the characteristics of reducing surface tension. Therefore, the role of surfactants: groups combine with surface groups to form new structures, giving new activity to the surface of nanomaterials; reduce the surface energy of nanoparticles to stabilize nanomaterials; the lipophilic groups of active agents form steric hindrance on the surface of particles to prevent re-agglomeration of nanoparticles, thereby improving the dispersion of nanopowders in different media, the surface reactivity and surface structure of nanoparticles, etc. Dongdan Chen et al. used diethanolamine, sodium polyacrylate, sodium dodecylbenzenesulfonate, and polyethylene glycol to modify the surface of nano-TiO_2, which effectively prevented the agglomeration of TiO_2. Yunzhu Ma et al. used N, N-dimethylformamide, hexadecyltrimethyl-trimethylated core (W, Ni, Fe) composite oxides for organic coating, which greatly reduced the capillary action during the drying process, reduced the chance of inter-particle agglomeration, and hindered the transformation of particles into hard agglomerates.

The state of surface adsorption is closely related to the surface charge of nanoparticles. Generally, the properties and strengths of the surface charges of nanomaterials in solutions with different pH values can be used to modify them. For example, for nano-SiO_2 materials, due to the low isoelectric point value of SiO_2 particles, it is between pH = 2–3. When conventional methods are used for surface modification, the pH value of the modified system is often high, making the surface electronegative and easy to combine with cations to obtain modification. But cationic surfactants are expensive and often toxic, which are major disadvantages.

(2) Coating modification is an earlier traditional modification method, which is to use inorganic or organic compounds (water-soluble or oil-soluble polymer compounds and fatty acid soaps, etc.) to coat the surface of nanoparticles to form a heterogeneous coating layer without chemical bonding with the particle surface, which can weaken or shield the agglomeration of nanoparticles. Moreover, due to the presence of coatings, steric repulsion is generated, which makes it very difficult for particles to reunite, so as to achieve the purpose of surface modification. The coating mechanism can be adsorption, adhesion, simple chemical reaction or coating of deposition phenomenon, etc. When preparing nano-TiO_2, hydroxypropyl cellulose modifier is introduced, and the macromolecule of the modifier is adsorbed on the TiO_2 particles to play a role of steric hindrance, effectively preventing the particles from further aggregating and growing, and improving the dispersion and uniformity of TiO_2 hydrated particles. At the same time, these macromolecules are adsorbed on the surface of nano-TiO_2 particles, and the non-bridging hydroxyl groups and adsorbed water between the particles are completely "shielded" to reduce their surface tension and make them less likely to aggregate.

When inorganic substances are used as modifiers, there is no chemical reaction between the inorganic substances and the surface of nanoparticles, and the combination of modifiers and nanoparticles depends on physical methods or van der Waals forces. Generally, inorganic compounds are used to carry out precipitation reaction on the surface of nanoparticles to form a surface coating. After a series of treatments, the coating is fixed on the surface of the particles, which can change the dispersion and stability of nanomaterials in different media, improve their weather resistance, reduce the activity of nanoparticles and prevent their agglomeration. Inorganic nanopowders are usually surface modified with metal oxides such as SiO_2 and Al_2O_3. The sol can also be used to coat the inorganic nanoparticles to improve the properties of the nanoparticles. If $ZnFeO_3$ nanoparticles are added to TiO_2 solution, TiO_2 sol is deposited on the surface of $ZnFeO_3$ nanoparticles to form a coating layer, and its photocatalytic efficiency is greatly improved. Doping and modifying the surface of nano-TiO_2 particles with Cu and Ag significantly improved its bactericidal effect.

7.2.2 Chemical Modification of Nanomaterials

The surface chemical modification method is a method of chemical reaction between the surface atoms of nanoparticles and modifier molecules to change the surface structure and chemical state, and is an important means of dispersion and compounding of nanoparticles. The selection principle of the surface modifier is that it must be able to reduce the surface energy of the particles, eliminate the surface charge and surface attraction of the particles. For surface modification aimed at increasing the adhesion between nanoparticles and other media, the modifier must also have a strong affinity with the particles and media. Commonly used chemical modification methods include coupling agent modification, esterification reaction and polymer surface grafting.

7.2.2.1 Coupling Agent Modification

Coupling agent modification is a chemical coupling reaction between the coupling agent and the surface of nanoparticles. In addition to van der Waals force, hydrogen bond or coordination bond interaction, there are ionic bonds and covalent bonds between the two components. The coupling agent molecule must have two groups: one group chemically reacts with the inorganic nanoparticle surface, and the other group (organic functional group) is reactive or compatible with the organic matrix [4].

The issue of interfacial adhesion becomes quite important when inorganic nanoparticles are composited with organic matrices. In general, the surface energy of inorganic nanoparticles is relatively high, and the affinity with the organic matrix with low surface energy is poor. When the two are mixed, they are not compatible, resulting in voids on the inorganic–organic interface. If such particles are filled with polymers and placed in the air for a long time, the oxygen and moisture in the air will invade the gaps in the interface, causing degradation and embrittlement of the resin at the interface between the organic matrix and the interface, and reducing its ability to resist external stress.

Through chemical or physical action, the coupling agent is covered on the surface of the inorganic nanoparticles. Inorganic nanoparticles treated with it have lipophilic properties, and when used in composite materials, they have better compatibility with organic matrix, which is beneficial to eliminate the interface between filler and matrix. The role of the coupling agent is equivalent to providing a "molecular bridge" between inorganic nanoparticles and organic polymers, forming a combination of organic matrix-coupling agent-inorganic substances. It acts as a link between interfaces that have no affinity or are difficult to be compatible, so that inorganic and organic substances with different properties can show good affinity and improve the performance of composite materials. Taking silicon dioxide as an example, SiO_2 nanoparticles have high surface energy and poor affinity with organic substances with low surface energy. When the two are combined, they are incompatible, and voids are prone to appear on the interface, resulting in easy degradation and embrittlement of the polymer at the interface. If the surface of SiO_2 nanoparticles is treated with a coupling agent, it can have good compatibility with organic matter. For example, when methacryloxypropyl trimethoxysilane coupling agent is used to prepare polymethyl methacrylate/Sis nanocomposite materials, the carbon–carbon double bond in the coupling agent is copolymerized with polymethyl methacrylate, and the propyl trimethoxysilane group is hydrolyzed with tetraethyl orthosilicate to form silica bonds, so that the composite system is uniformly dispersed and stable.

Since the modification operation of the coupling agent is relatively simple and the modification effect is ideal, it is widely used in the surface modification of nanomaterials. Common coupling agents include silane coupling agents, titanium coupling agents and chromium complexes, etc., especially for the modification of inorganic nanoparticles with abundant hydroxyl groups on the surface. Silane coupling agent is a type of coupling agent that was developed the earliest, used in the largest amount and has a good effect.

7.2.2.2 Esterification Method

The so-called esterification reaction is to use the reaction of acid and alcohol to modify the surface of nanoparticles. After the esterification reagent reacts with the surface atoms of nanoparticles, hydrophobic groups (such as long-chain alkyl groups, chain alkyl groups, and cycloalkyl groups) replace the hydroxyl groups on the surface of nanoparticles, so that alkyl groups and the like are firmly combined on the surface of nanoparticles, showing strong hydrophobicity. The method is suitable for nanomaterials with hydroxyl groups on the surface. After the esterification treatment, the dispersibility of the nanoparticle is enhanced, it is not easy to agglomerate, and the hydrophobic effect is good at the same time [5].

The original purpose of surface modification is to modify the surface without changing the bulk properties of the material, endowing it with new functions and improving its properties. But if the structure of the substance itself is changed by the treatment and new functions are given to the surface, then this treatment should not be some kind of obstacle, but is desired. That is to say, it is also a very valuable method to treat the substance in the previous stage of the target object, and to change the structure of the substance itself into the target substance through the treatment, and at the same time to impart new functions on the surface.

When preparing nano metal oxides, add a small amount of polyvinyl alcohol (PVA). Since PVA contains a large number of hydroxyl groups, these hydroxyl groups form chelate bonds with metal ions in aqueous solution, and are tightly wrapped around the metal ions to form a limited structure with a restricted shape of PVA chains, which can effectively control the size of the synthesized nanoparticles and achieve the purpose of surface modification at the same time.

7.2.2.3 Polymer Surface Grafting

Polymer surface grafting is also a commonly used surface modification method. The surface of some inorganic nanoparticles has active points that can undergo free radical reactions. Under appropriate conditions, polymer active monomers can react and graft on the surface of nanoparticles on these active points, and then initiate polymerization. The polymer long chain is grafted on the surface of the nanoparticle, and the long chain containing the hydrophilic group in the polymer plays a three-dimensional shielding role in the aqueous medium through hydration stretching. In this way, the dispersion stability of nanoparticles in the medium depends not only on electrostatic repulsion but also on steric hindrance, and the effect is very obvious. The degree of agglomeration of nanoparticles before grafting is significantly reduced after grafting, and it is not easy to re-agglomerate, and the dispersion stability is increased. After the nanoparticle surface is grafted, its dispersibility in organic solvents and polymers is greatly improved, and composite materials with high nanopowder content and uniform distribution can be prepared. This treatment method gives full play to the respective advantages of inorganic nanoparticles and polymer materials, and can also realize the optimal design of functional materials.

There are active functional groups such as phenolic groups, hydroxyl groups, and alkyl groups on the surface of carbon black and carbon fibers, and active hydroxyl groups also exist on the surface of inorganic nanoparticles such as silicon dioxide, ferrite, and titanium dioxide, which can be used as a place for graft polymerization.

There are generally three types of grafting reactions on the surface of nanoparticles:

(1) When the monomer is completely polymerized under the action of the initiator, it is immediately captured by strong free radicals on the surface of ultrafine particles, so that the polymer chain is chemically connected with the surface of inorganic nanoparticles, and the grafting of the particle surface is realized.

This method is the easiest to do, but it is only suitable for carbon black with strong free radical capture ability, and it is not effective for graft polymerization of other inorganic particles. Wei Li et al. found in the study of grafting acrylic phenol on the surface of carbon black particles that under certain conditions, acrylic acid monomers can be directly grafted on the surface of carbon black particles. From the observation of transmission electron microscopy (TEM), it was found that because the grafted polypropylene long chain contains hydrophilic groups, it can better extend the steric barrier effect in aqueous media, prevent the re-aggregation of carbon black particles, make the carbon black particles uniformly dispersed, and increase the dispersion stability.

(2) Grafting from particles (grafting from surface) In this method, under the action of an initiator, monomers directly start to polymerize from the surface of ultrafine particles, induce growth, and complete particle surface coating with polymer compounds. This method is characterized by high grafting efficiency.
(3) Grafting reaction with the surface of particles. This method uses the functional groups on the surface of particles to directly react with polymers to achieve grafting. It has the characteristics of controllable relative molecular weight and quantity of grafting, and high grafting rate. Yiqing Qian et al. used toluene diisooxyester (TDI) to treat the surface of nano-silica. Because the –NCO group in TDI reacted with –OH on the surface of nano-silica, the connection between silica and polymer bonds was improved, and the polymer separated the inorganic nanoparticles to prevent agglomeration.

7.3 Methods for Surface Modification of Nanomaterials

The surface modification of nanomaterials is to introduce functional groups with certain functions or a coating layer on the surface according to the needs, so that the modified nanopowder can be regarded as a composite powder composed of a "core layer" and a "coating layer". The shell layer can be organic or inorganic. By coating a layer of coating with different chemical composition on the surface of nano powder, it can have good dispersion and compatibility, improve its heat resistance, mechanical and chemical stability, change its optical, electrical, magnetic, catalytic, hydrophilic, hydrophobic and sintering properties, and improve its corrosion resistance, durability and service life. There are many methods for surface modification of nanomaterials, mainly including: sol–gel method, precipitation method, microemulsion method, electroless plating, heterogeneous flocculation method and surface coating modification method.

7.3.1 Sol–Gel Method

The sol–gel method uses compounds containing highly chemically active components as precursors, uniformly mixes these raw materials in the liquid phase, and undergoes hydrolysis and condensation chemical reactions to form a stable transparent sol system in the solution. After aging or drying for a certain period of time, the colloidal particles slowly polymerize to form a continuous three-dimensional space network structure. The network is filled with solvents that have lost fluidity to form a gel. The gel is composed of a solid skeleton and a continuous phase. After the liquid phase is removed, the gel shrinks into a dry gel, and the dry gel becomes a uniform ultrafine powder after simmering [6].

The sol–gel method is a wet chemical synthesis method, which can not only be used to prepare nanopowders, thin films, but also successfully applied to particle surface modification. The nanoparticles are evenly dispersed in a medium with good compatibility, and the surface of the nanoparticles is covered by various reactions such as hydrolysis and condensation of the shell material. Using this method can obtain a uniform surface coating layer, which is widely used. According to the different raw materials used, it can generally be divided into two categories, namely, inorganic salt sol–gel method and alcohol salt hydrolysis method.

(1) The inorganic salt sol–gel method first prepares the sol, and through the control of the inorganic salt precipitation process, the generated particles are not agglomerated into large particles and precipitated to obtain the sol directly; or part or all of the components are first precipitated with an appropriate precipitant, and after decoagulation, the original agglomerated precipitated particles are dispersed into colloidal particles. After the sol is prepared, the coated particles are added, and the nanoparticles coated with a certain thickness of the coating are obtained after gelling.

(2) Alcohol hydrolysis method, in this method, metal alkoxides are first dissolved in organic solvents, hydrolysis reaction occurs when water is added, and then further condensation and gelation occur, and then the gel is dried and roasted to remove organic components, and finally metal oxide nanoparticles are obtained. This is the most common sol–gel reaction.

7.3.2 Precipitation Method

The precipitation method is to coat the particles by adding a precipitating agent (such as sodium hydroxide solution, ammonia water) to the solution or triggering the formation of a precipitating agent in the system (such as the pyrolysis reaction of urea), so that the modified ions undergo a precipitation reaction and precipitate on the particle surface.

The hydrolysis reaction can occur in solution or directly on the surface of nanoparticles. By adjusting the temperature of the reaction system, evaporating solvents and other physical methods to increase the supersaturation of the precipitation product, and changing the pH value of the system to control the hydrolysis reaction of metal ions, the nanoparticles are inorganically coated. The key to coating by precipitation method is to control the ion concentration in the solution and the release rate and dosage of the precipitant, so that the modifier (or its precursor) generated by the reaction has a certain degree of supersaturation in the system and does not exceed the critical saturation concentration, so that the coated particles are precipitated as the nucleus. Otherwise, a large amount of precipitates will be generated instead of being evenly coated on the surface of the particles.

According to different precipitation methods, it can be divided into: co-precipitation method, homogeneous precipitation method, hydrolysis method and hydrothermal method. Co-precipitation method controls the precipitation reaction, so that a variety of metal ions in the solution precipitate on the surface of the coated particles in a certain composition and order to form a coating.

7.3.3 Heterogeneous Flocculation Method

The heterogeneous flocculation method refers to that in the solution, when the surface of heterogeneous nanoparticles has opposite charges, due to electrostatic interaction, they attract each other and coagulate to prepare coated composite nanomaterials. If the particle size of one kind of particle is much smaller than the particle size of another kind of particle with different charges, then during the agglomeration process of these two kinds of particles, the small particle will be adsorbed on the outer surface of the large particle to form a coating layer. The method is based on the theory of a diffuse electric double layer. By adjusting the pH value or adding a surfactant, the charges on the surface of the coating particles and the coated particles are reversed, and the coating particles are adsorbed around the coated particles through the action of electrostatic force to form a coating layer. The key to this method is to find a suitable pH value so that the two powders are charged differently. Generally, finer nanoparticles are selected as the coating material, and the thickness of the coating layer is the particle size of the coating particles. Usually, the coating of small particles on the outer surface of large particles is not very tight and firm by relying on charge attraction.

7.3.4 Heterogeneous Nucleation Method

The non-uniform nucleation method is a method in which the particles of the shell material are non-uniformly nucleated on the coated particle matrix and grown to form a coating layer. Since the kinetics required for inhomogeneous nucleation are lower than for homogeneous nucleation, shell particles preferentially nucleate on the coated substrate. By controlling the pH value of the solution, the concentration of coating ions, the concentration of the shell precursor, temperature and time and other influencing factors, a layer of precursor can be uniformly coated on the particle surface. After cooking, the nanoparticles coated with oxides on the surface are obtained. The key to this coating technique is to control the concentration of modifier species in solution between the critical concentration required for heterogeneous nucleation and the critical concentration required for homogeneous nucleation. In this concentration range, the modifier particles meet the non-uniform nucleation conditions, so that the coated particles are used as the nucleation matrix, and nucleation and growth are preferentially formed on the outer surface of the matrix to coat the particles. An amorphous coating layer is formed between the non-uniform nucleation critical concentration and the homogeneous nucleation critical concentration, and a polycrystalline coating layer is formed between the homogeneous nucleation critical concentration and the critical saturation concentration. Above the critical saturation concentration, a large amount of precipitate is formed, and the particles will not be coated. Compared with polycrystalline phase coating, amorphous coating is easier to achieve a uniform and dense coating layer, so the coating modification based on the precipitation method based on the heterogeneous nucleation mechanism is more meaningful [7].

7.3.5 Microemulsion Method

Microemulsion is a transparent or translucent colloidal dispersion system with low viscosity, isotropic thermodynamic properties and stable thermodynamic properties spontaneously formed by oil, water, emulsifier and co-surfactant in an appropriate ratio, and the size of the dispersed phase is nanoscale. The size of microemulsion droplets is generally 10–100 nm. From a microscopic point of view, the stable water core formed by the surfactant interfacial film can be regarded as a "microreactor" or nanoreactor. Therefore, it is widely used in the preparation of nanomaterials. Because the surface of nanoparticles is coated with a layer of surfactant molecules, the particles are not easy to coalesce; by selecting different surfactant molecules, the surface of the particles can be modified and the size of the particles can be controlled. In the process of preparing nanoparticles by using the microemulsion method, certain reaction conditions are controlled to promote the formation of one kind of nanoparticle on the surface of another kind of nanoparticle, and composite particles with coating structure can be obtained.

7.3.6 Electroless Plating

Electroless plating refers to the process of chemically precipitating metals and coating them on the surface of nanoparticles without applying an external current. There are three specific methods: displacement method, contact plating and reduction method. The reduction method is generally used for particle surface coating, that is, the reducing agent added in the solution provides electrons after being oxidized, and the metal is reduced and precipitated to coat the surface of nanoparticles.

The surface of nanomaterials needs to be pretreated before electroless plating. Its purpose is to remove oil stains and oxide layers on the surface of the material, and to increase the adhesion between the coating and the particles. Sensitization is to immerse the coarsened particles in an aqueous solution of stannous chloride, so that a layer of Sn^{2+} is adsorbed on the surface of the material; activation is to quickly put the sensitized nanomaterials into a $PbCl_2$ solution, and Sn^{2+} reacts with $PbCl_2$ to form a catalytic nucleus for a chemical coating reaction, which is the key to achieving good coating on nanomaterials; reduction is to ensure the stability of the plating solution, and treat nanomaterials with sodium phosphite solution before electroless plating. When performing electroless plating, the plating solution is heated in a constant temperature heater, stirred rapidly, and the powder to be plated is put in for chemical coating. The plating time is determined by the thickness of the desired plating layer. During the overwork process, it is necessary to prevent the decomposition of the plating solution due to local heating.

(1) Electroless plating of carbon nanotubes Carbon nanotubes are one-dimensional nanomaterials with a hollow structure. Due to their highly graphitized structure, their surface reactivity is very low, they do not have catalytic properties, and their diameters are nanoscale and their surface curvature is large. It is difficult to obtain a continuous and dense coating layer by general coating methods such as electroplating and vapor deposition. If the electroless plating method is used, a better coating can be obtained by strengthening the pretreatment of carbon nanotubes such as purification, sensitization, and activation, rationally adjusting the coating treatment process, and performing appropriate post-plating

treatment on the coating layer. At present, Ni, Cu, Ag, Au, etc. have been successfully plated on the surface of carbon nanotubes.

(2) In addition to the characteristics of ordinary carbon fibers, such as low density, high specific strength, and high electrical conductivity, nanocarbon fibers are plated with silver on the surface of nanocarbon fibers. Nanocarbon fibers also have the advantages of less defects, small diameter, and large specific surface area. They can be used as catalyst carriers, lithium-ion secondary batteries, electrode materials, gas storage materials, and reinforcements for advanced composite materials. In microwave absorbing materials, because of its high conductivity and large specific surface area, it increases the ability to scatter and absorb electromagnetic waves. When used in composite materials, it can have both wave absorbing and load-carrying functions. Therefore, it is a potential structural wave absorbing material and has a good application prospect in the field of microwave absorption. In order to improve the microwave absorption capacity of the material, generally speaking, ordinary carbon fibers in composite materials used for microwave absorption need to be specially treated or processed by special processes, such as silver plating on the surface of carbon fibers [8].

(3) There are three main types of conductive fillers for electroless plating of nano-copper powder or conductive fillers for electromagnetic wave shielding coatings: one is copper powder, the other is silver powder, and the third is copper/silver bimetallic powder. Copper powder has the advantages of wide source, low price, and good conductivity, but its oxidation resistance is weak. Silver powder has good electrical conductivity and oxidation resistance, but its resources are increasingly scarce. A large number of studies have tried to improve the performance of copper powder by surface modification. There are two methods: one is coating treatment, and SiO_2 sol is used to treat copper powder. The oxidation resistance of the copper powder treated in this way is improved, but its surface conductivity is greatly reduced; another method is to cover the surface of the copper powder with a layer of silver or gold with good conductivity and oxidation resistance to make bimetallic powder. Weihui Liao et al. used the copper rot plate waste liquid produced by printed circuit board manufacturers as the copper source. In the liquid phase, hydration swelling was used as the reducing agent, and polyvinylpyrrolidone (PVP) and OP-10 were used as the dispersant to prepare nano-copper powder with high purity and excellent performance. After the nano-copper powder is sensitized and activated, a mixed solution of 5% PVP and OP-10 is added to the plating solution as a dispersant for electroless plating to obtain nano-Cu/Ag bimetallic powder. Guojun Mou coated tin on the surface of nano-copper powder in the presence of special copper complexing agent-thiourea and protective agent PVP, and prepared copper/tin bimetallic powder with nano-core–shell structure. In the preparation process, the particle size of the product is controlled by adjusting the particle size of the raw copper powder, the concentration of the reactants, the reaction temperature, and the amount of the protective agent PVP. The oxidation resistance of nano-core–shell copper/tin bimetallic powder is better than that of nano-copper powder of

the corresponding size; when used as an additive in lubricating oil, it can be stably dispersed in lubricating oil, maintain the transparency of lubricating oil, and has anti-wear and anti-friction effects.

(4) Electroless plating of nano-Al_2O_3 powder Al_2O_3 particle-reinforced copper-based composites have good mechanical and physical properties, and these properties are easy to change, such as by changing the volume fraction of the reinforcement, the composition of the matrix, and the heat treatment process. This makes Al_2O_3 particle-reinforced copper-based composites have a wide range of application prospects. For example, Al_2O_3/Cu can improve the thermal conductivity of Al_2O_3 ceramics to meet the requirements of electronic packaging materials; adding Al_2O_3 particles to the copper matrix can increase the copper recrystallization temperature and high-temperature strength without significantly reducing the conductivity of the copper matrix. In the case of the same Al_2O_3 volume fraction, the copper-based composite material reinforced with nano-Al_2O_3 can strengthen the matrix metal more effectively because the Al_2O_3 particles are more dispersed. It has little effect on other properties of the composite material, thus obtaining a material with high strength and high electrical conductivity. Preparation of ceramic particle/copper composite powder by electroless plating method, and mixing with copper powder to prepare particle reinforced copper matrix composite material can change the combination of ceramic particles and matrix. At the same time, direct preparation of nano-Al_2O_3-reinforced copper-based composites by electroless copper plating can avoid the phenomenon of uneven mixing between nano-composite powder and copper powder, and is conducive to improving the uniformity of ceramic particles in the metal matrix. Yuanting Liu and others studied the sintering and densification characteristics of Al_2O_3 electroless copper plating powder with a particle size of 10–20 nm, and analyzed the effects of cleaning of electroless plating powder, hydrogen reduction and other pretreatments, molding pressure, sintering temperature, holding time, repressing and refiring process, etc. on densification. In the case of optimizing each influencing factor, a sample with a relative density of 0.94 was obtained by using a conventional powder metallurgy process on an electroless copper plating powder with a mass fraction of Al_2O_3 of 10%.

7.3.7 Vapor Deposition Method

The vapor phase chemical deposition method is to modify the surface coating of nanoparticles through chemical reactions in the gas phase. The shell material can be converted into gas directly by using gas or by various means, so that it undergoes physical or chemical changes in the gaseous state, and deposits on the particle surface or chemically bonds with the particle surface molecules to form a uniform and dense film coating layer to achieve nanoparticle surface coating. Vapor phase coating methods include physical vapor deposition coating methods and chemical vapor deposition coating methods. The former achieves particle coating with the help of van der Waals force, but the disadvantage is that the core–shell binding force is not strong; while the chemical vapor deposition method uses gaseous substances to react on the surface of nanoparticles to form solid deposits to achieve nanoparticle coating. For example, the nitrogen gas loaded with trimethylaluminum is passed into the fluidized bed reactor in which the core layer particles are suspended, so that the surface of the particles is saturated with trimethylaluminum, and the particles coated with alumina can be obtained after further reaction.

7.3.8 Polymer Surface Coating Modification

Surface coating is mainly aimed at preventing particle growth, solving agglomeration problems and increasing the compatibility of inorganic nanoparticles in organic matrices in nanosynthesis, and has a clear application background. The coated nanoparticles not only eliminate the charging effect on the surface of the particles and prevent agglomeration, but also form a potential barrier so that the particles are not easy to grow up during the synthesis and sintering process (referring to the inorganic coating). Organic coating enables inorganic nanoparticles to reach an infiltration state with organic matter and organic reagents, laying a good foundation for the incorporation of inorganic particles into polymer plastics. For example, adding the particles coated with organic matter on the surface of nano oxides to plastics can increase the strength and melting point of the material. At the same time, the waterproof ability is enhanced, and the light transmittance is improved. If high-dielectric nanoparticles are added, the insulation of the system can also be enhanced, and it has a good application prospect in packaging materials.

7.3.9 Carbon Coating on the Surface of Nanomaterials

Since Ruoff et al. reported carbon-coated materials, because of this unique "nanocapsule" structure coated in multilayer graphite, carbon-coated materials have singular electrical, optical, and magnetic properties, which has promoted the application of nanomaterials. Carbon-coated nano-metal particles are a kind of nano-carbon/metal composite material, in which several layers of graphite sheets are closely arranged around the nano-metal particles to form a "nano-capsule" structure, and the nano-metal particles are in the center. Because the carbon shell can imprison metal substances in a small space, it can avoid the influence of the environment on nano-metal materials, and solve the problem that nano-metal particles cannot exist stably in the air; in addition, due to the existence of carbon coating, it is expected to improve the compatibility between some metals and organisms, so it has broad application prospects in medicine. In addition, depending on the metal particles and carbon matrix, the material is expected to be used as a magnetic recording material, lithium-ion secondary battery anode material, radio wave shielding material, redox catalyst, nuclear waste treatment material, fine ceramic material and antibacterial material, etc.

The main preparation methods of carbon-coated metal nanoparticles include arc discharge method, chemical vapor deposition method, pyrolysis method, liquid phase impregnation method and so on [9].

The arc discharge method is to use DC arc discharge to evaporate the graphite electrode in an inert atmosphere to obtain carbon-coated nano-metal particles in the product deposited on the cathode or the wall of the reaction chamber. The anode is composed of graphite powder and the metal that needs to be coated.

The formation mechanism of carbon-coated metal nanoparticles synthesized by arc discharge method is generally considered to follow the vapor–liquid-solid (Vapourliquid-solid, VLS) growth mechanism. The premise of vapor–liquid-solid growth is the coexistence of vapor, liquid and solid phases, because there are LS interface and VS interface, correspondingly there are two growth mechanisms. The arc center temperature is as high as 4000 K, at such a high temperature, the anode graphite together with the catalyst melts or even vaporizes. The gaseous carbon atom clusters are first deposited on the surface of the liquid catalyst with a large adsorption coefficient, and diffuse along the surface and inside of the catalyst droplet; after a certain period of time, the carbon atoms in the catalyst reach saturation [10]. Then, clusters of supersaturated carbon atoms are precipitated from the surface of the liquid catalyst to become carbon crystals. In the VS system, since the catalyst particles are in contact with the gaseous carbon atoms and the surface is always in an activated state, the gaseous carbon atoms are continuously deposited on the surface of the catalyst particles and solidified directly. In this way, after one shell layer is formed, it gradually transitions to the other shell layer, and finally forms carbon-coated nano-metal particles, that is, the epitaxial growth mechanism. In the LS system, when the liquid carbon atoms dissolved in the catalyst are cooled, the carbon atoms on the surface crystallize first; as the crystallization process proceeds, the carbon atoms advance from the outer shell layer to the inner shell layer in a continuous and orderly manner, forming regular carbon-coated nano-metal particles, which is the endogenous formation mechanism.

7.3.10 Plasma Treatment

The surface treatment of nanoparticles by plasma treatment is a newly developed modification technology. Plasma is by means of a glow discharge plasma system and uses one or more gases as the plasma gas. That is to say, in a vacuum reaction chamber, two metal electrodes, positive and negative, are installed, and a DC voltage of hundreds of volts is applied. At this time, the phenomenon of glow discharge will appear, and some areas with different degrees of light and darkness and different temperatures will be generated, and plasma will be generated. If an appropriate amount of reactive gas is introduced, the gas can undergo a series of chemical reactions and plasma reactions. When molecules dissociate, free radicals are often generated, electrons and neutral atoms are generated, and molecules form negative ions [11]. Therefore, the entire plasma is a mixed state of electrons, positive and negative ion excited state atoms, and free radicals. Because the total charge of the negative and positive ions in the whole system is equal, the collective behavior is shown as an electrically neutral gas in the macroscopic view, so it is called plasma. These include six typical particles: electrons, positive ions, negative ions, excited state atoms or molecules, ground state atoms or molecules, and photons. From the perspective of the usual energy distribution: gas > liquid > solid, plasma has higher energy than gas, and can exhibit characteristics that ordinary gases do not have, so it is also called the fourth state of matter. Because various chemical reactions are carried out in highly excited states, which are completely different from classical chemical reactions. In this way, the nature of plasma atoms or molecules usually changes, and even relatively stable noble gases become highly chemically active. Plasma gas has very high energy, which is equivalent to the bond energy between atoms of organic matter. Therefore, introducing plasma into chemical reactions can not only greatly reduce the reaction temperature, but also make possible reactions that are difficult to occur or speed up reactions that have a very slow reaction rate [12]. Plasma chemical reactions are mainly excited, dissociated, and ionized by high-speed electrons colliding with molecules, and react in a non-equilibrium state. This method has the advantage of allowing for a completely dry process.

There are four ways to use plasma for powder surface treatment: use polymeric gas plasma to treat powder, and form an organic coating film on the particle surface through reaction, thereby changing the properties of the nanoparticle surface and achieving the purpose of surface modification; use non-polymerizable gas (such as Ar, He, H_2O, etc.) If vinyl monomers are introduced, the active radicals on the surface of the powder will initiate graft polymerization [13].

Surface modification of powders using plasma has been applied to the oxidation treatment of carbon black. Due to the low polarity of carbon black, it is not easy to be wetted by water, and its wettability in water is poor. After plasma treatment, the number of acidic functional groups on the surface increases, and the pH value decreases; at the same time, the wetting heat to water increases, and it is easily wetted by water, showing obvious dispersion in water. In order to improve the surface treatment effect, increase the oxygen flow and pressure for treatment. The plasma oxidation of carbon black is mainly due to the reaction between the free radical active points generated on the surface of carbon black and oxygen molecules, and is carried out through the free oxidation reaction of free radicals.

If a reactive gas is added to the system, the plasma gas generated by the discharge contains various active ions, which have different energies, and various physical and chemical effects occur on the solid surface. After plasma treatment, the polymerizable gas can coat or even graft a layer of polymer on the surface of nanoparticles. The principle is to use plasma polymerization technology to activate and activate organic compound monomers to form gas phase free radicals. When the gas phase free radicals are adsorbed on the solid surface, surface free radicals are formed. The surface free radicals react with the original monomers in the gas phase or derived monomers generated in the plasma to form a polymer film with a large molecular weight. The polymerization process of unsaturated organic compounds in plasma is roughly divided into the following steps: Excitation process, under the action of plasma, monomer free radicals and their derivatives necessary for sufficient polymerization reactions are generated. In the process of surface adsorption, monomer free radicals and their derivatives are adsorbed on the solid surface. In the non-uniform growth process, the free radicals on the solid surface react with the original monomers and monomer derivatives in the gas phase and the atomic monomers and monomer derivatives in the solid phase. In the final reaction of polymerization, the polymerized low molecular weight polymer reacts to form a high molecular weight polymer, including various polymerization reactions on the solid surface. Re-excitation process, including the decomposition of surface stable products caused by electrons, positive ions and photons in the plasma [14].

After plasma treatment, the surface contact angle of calcium carbonate powder increased significantly, and the surface properties changed from hydrophilic to lipophilic. Akovali et al. filled the plasma-treated calcium carbonate powder into polypropylene to prepare a composite material, and found that its mechanical properties were greatly improved [15]. They used ethylene-dried plasma to modify the surface of calcium carbonate powder, and added the modified calcium carbonate powder to polypropylene in proportion, and compared it with the materials added to unmodified calcium carbonate powder. It was found that the elongation at break of the composite material filled with calcium carbonate powder after plasma modification was significantly higher than that of the unmodified calcium carbonate powder filled material, an increase of more than 55% [16].

Surface modification has been widely valued in the research of nanomaterials. There are more and more types of surface modifiers used, and the synthesis technology of modifiers has also been greatly developed. But each method has its own limitations and shortcomings, and nanoparticles modified with specific surface modifiers often have specific functions. Modification is for the purpose of application, and most of a method only has a good effect in a certain aspect of application. With the continuous expansion of the application range of nano-powder, the surface modification mechanism, modification method and equipment, modification effect, etc. need to be further improved. Nano-modification has become an extremely important content in the research and development of nano-materials, and is regarded as an important means of preparing new materials in the future. Therefore, in-depth and detailed research on it will have broad application prospects.

References

1. F. Dionigi, T. Reier, Z. Pawolek, M. Gliech, P. Strasser, *ChemSusChem* **2016**, *9*, 962–972.
2. M. Mandal, G. Huang and P. A. Kohl, *J. Membr. Sci.* **2019**, 570–571, 394–402.

3. R. Akiyama, N. Yokota, E. Nishino, K. Asazawa and K. Miyatake, *Macromolecules* **2016**, 49, 4480–4489.
4. J. H. Clark, *Green Chem.* **1999**, *1*, 1–8.
5. S. Chempath, B. R. Einsla, L. R. Pratt, C. S. Macomber, J. M. Boncella, J. A. Rau and B. S. Pivovar, *J. Phys. Chem. C* **2008**, 112, 3179–3182.
6. K. Juodkazis, J. Juodkazytė, R. Vilkauskaitė, V. Jasulaitienė, *J. Solid State Electrochem.* **2008**, *12*, 1469–1479.
7. S. Samad, K. S. Loh, W. Y. Wong, T. K. Lee, J. Sunarso, S. T. Chong, W. R. Wan Daud, *Int. J. Hydrogen Energy* **2018**, *43*, 7823–7854.
8. L. Wang, H. Zhang, G. Cao, W. Zhang, H. Zhao, Y. Yang, *Electrochim. Acta* **2015**, *186*, 654–663.
9. C. Fujimoto, D.-S. Kim, M. Hibbs, D. Wrobleski and Y. S. Kim, *J. Membr. Sci.* **2012**, *12*, 423–424.
10. S. Devaraj and N. Munichandraiah, *J. Phys. Chem. C* **2008**, 112, 4406–4417.
11. K. Izumiya, E. Akiyama, H. Habazaki, N. Kumagai, A. Kawashima, K. Hashimoto, *Mater. Trans. JIM.* **1997**, *38*, 899–905.
12. J. Zheng, *Appl. Surf. Sci.* **2017**, *413*, 360–365.
13. M. R. Hibbs, *J. Polym. Sci., Part B: Polym. Phys.* **2013**, 51, 1736–1742.
14. A. Lim, H.-j. Kim, D. Henkensmeier, S. Jong Yoo, J. Young Kim, S. Young Lee, Y.-E. Sung, J. H. Jang and H. S. Park, *J. Ind. Eng. Chem.* **2019**, 76, 410–418.
15. K. Kordek, H. Yin, P. Rutkowski, H. Zhao, *Int. J. Hydrogen Energy* **2019**, *44*, 23–33.
16. J. E. Bennett, *Int. J. Hydrogen Energy* **1980**, *5*, 401–408.

Chapter 8
Applications of Nanomaterials and Nanotechnology in Military and Aerospace Fields

Nanomaterials refer to ultrafine materials whose grain size is on the order of nanometers (1–100 nm). When the particle size of the material is in the nanometer state, its molecular arrangement, electron distribution and crystal structure change, which is very different from the usual bulk materials in terms of structure and performance. The main four physical effects of nanomaterials include: surface effect, small size effect, quantum size effect and macroscopic quantum tunneling effect. These effects make nanomaterials have a series of excellent physical, chemical, surface and interface properties compared with conventional bulk (granular) materials, and can obtain extraordinary effects when used. Utilizing these unique properties of nanomaterials, traditional materials can be modified to develop materials with higher performance and open up new material production methods to meet the requirements that traditional materials cannot meet, especially to meet the special requirements for material performance in the military and aerospace fields. The application of nanomaterials can reduce the volume and quality of spacecraft electronic components and improve their reliability [1].

8.1 Application of Nanomaterials in the Military Field

When the size of the material is as small as 0.1–100 nm, the properties of the material will change greatly. For example, copper block is processed into nanoscale powder, and then pressed into block shape, its heat conduction speed is several times that of natural copper block; many substances are processed to nanoscale, their conductivity and light absorption ability are increased several times, etc. The technology to study these phenomena is called nanotechnology. Advanced technologies are always the first to be applied in the military field, and so is nanotechnology. When this technology was just emerging, the world's major military powers have successively formulated a large number of military nanotechnology projects. They believe that in future

S. Peng and P. Li, *Nanomaterials and Nanotechnology*,
https://doi.org/10.1007/978-981-97-8433-2_8

wars, nanotechnology will greatly improve battlefield reconnaissance and battlefield command methods, and accelerate the process of miniaturization, informatization and integration of weapons and equipment, and even change the mode of future wars [1].

The application of nanotechnology in the military field mainly includes the following aspects: obtaining continuous information superiority through advanced nanoelectronic equipment; developing a virtual training system based on nanostructured electronic equipment that is more advanced in technology and better in economical affordability; developing nanoautomated weapons and equipment with better performance. The breakthrough and development of nanotechnology has laid a material foundation for the development of weapons and equipment with lower power consumption, lighter weight, stronger performance, and higher combat effectiveness, which will have a comprehensive and far-reaching impact on national defense and the form and combat mode of future wars.

8.1.1 Nanorobots

As shown in Fig. 8.1, nano-robots are designed and manufactured based on biological principles at the molecular level as "functional molecular devices" that can operate in nano-space. Their development belongs to the category of molecular bionics, so nano-robots are also called molecular robots. Nano-robots are the most attractive and important content of nanotechnology. Nano-robots will include nano-computers. Once this device that can conduct man–machine dialogue is successfully developed, it can complete billions of operations within 1 s. This kind of robot can be used to make up for the shortage of manpower in the army, reduce the risk of the army in the use of biological and chemical weapons and nuclear war, increase the mobility, and improve the automation of the army. This will greatly change people's views on the balance of forces in war, and bring about fundamental changes in the future battlefield model and pattern. The research and development of nanorobots has become the frontier and hot spot of today's technology [2].

Entering the twenty-first century, the development of science and technology is in full swing, and military revolution is surging. Looking at it from a new historical starting point, what technology can have a revolutionary impact on the future military? At present, major military powers are actively researching and developing nano-robots, and have successfully developed dozens of components for nano-robots. According to experts' calculations, it is expected that military nano-robots will be successfully developed by 2025, and by then the international political and military situation will also undergo major changes. How do nanobots eliminate or disable enemies? There are mainly the following ways. As shown in Fig. 8.2, first, nano-robots are applied to traditional weapon technology and equipment to enhance the lethality of traditional means of combat. Secondly, develop new means and methods of human combat, such as the development of nano-miniature components that can block the face, nose, mouth, and eyes [2]. Third, develop new types of chemicals

Fig. 8.1 Schematic diagram of nanorobot equipment [2]

or organisms, inject them into artificial or hybrid insects, and spread these lethal chemicals or organisms to the bodies of the enemy's soldiers and civilians through insects; fourth, after nano-robots enter the enemy's body, they can quickly spread in the enemy's camp through self-replication or self-reproduction. In addition, nano-robots can be used not only for offense, but also for defense, and the effectiveness of defense is far stronger than that of offense. Nano-robots are taking on more and more tasks on the battlefield, especially participating in strategic defense, creating conditions for strategic offense.

Fig. 8.2 Nanorobots produced in the United States [2]

Nano-robots are the most attractive content in nano-biology, and the content related to nano-robots can be summarized into the following three aspects:

(1) Understanding the fine structure–function linkages of biomacromolecules at the nanoscale;
(2) Obtain life information at the nanometer scale, such as using scanning tunneling microscope to obtain structural information on cell membrane surface, and using nanosensors to obtain chemical information and electrochemical information of various biochemical reactions;
(3) Development of nanorobots.

New technology is a double-edged sword. Beneficial technological behavior can benefit mankind; otherwise, it will be a disaster. Nanobots are no exception. Therefore, all countries are starting to study the risk and safety of nano-robots from the perspectives of ethics, law and society. How to make nano-robots, the double-edged sword that is beneficial to human beings, sharper, and make its negative side less harmful to human beings, is a problem that needs to be thought deeply [3].

8.1.2 Nano Weapons and Equipment

The application of nanomaterials in the military field can produce more advanced weapons and equipment. Nano-weapon is the integration of miniaturization of weapons, which realizes the ultra- miniaturization of weapon system, condenses the electronic warfare system on-board and airborne to be carried by a single soldier, and has better concealment and higher security. Secondly, nano-weapons have realized the high intelligence of the weapon system, greatly accelerated the information acquisition speed of the weapon equipment control system, and greatly improved the accuracy of reconnaissance and surveillance. As shown in Fig. 8.3, again, nano-weapons realize the integrated production of weapon systems, which reduces the cost of weapon equipment, improves reliability, and shortens the development and production cycle of weapon equipment. Compared with traditional weapons, military weapons made of nanomaterials have the characteristics of ultra-miniaturization, and show a sharp reduction in the weight and volume of weapons [3]. Therefore, the application of ultra-miniature weapons and equipment in training and combat not only reduces the physical burden of soldiers, but also improves the overall combat capability of the team. At the same time, military weapons made of nanomaterials are also highly intelligent, especially in terms of working speed, which is more than a thousand times faster than traditional weapons, while the volume is reduced by nearly one-thousandth. It also supports the acquisition of more information at the same time, creating infinite value for the development of national defense modernization.

The working speed of quantum devices is 1000 times faster than that of semiconductor devices, so replacing semiconductor devices with quantum devices can greatly improve the information transmission, storage and processing capabilities in weapon equipment control systems. The use of nanometer technology can reduce the volume

Fig. 8.3 Schematic diagram of nano-weapon equipment [3]

of existing radars by one thousandth, and can place ultra-high-resolution synthetic aperture radars on satellites for high-precision ground reconnaissance. Nanotechnology can improve the performance of individual soldiers' equipment according to the requirements for the performance and functions of materials. By arranging the atoms of matter on the atomic scale, the soldiers' equipment can be made light and small, stronger and simpler, which can reduce the soldier's load and improve the soldier's mobility and survivability. The birth and extensive use of nano-weapons will have a profound impact on all aspects of the military field.

Nano-weapons have realized the ultra- miniaturization of weapon systems, and the destructive effect of nano-weapons in science fiction movies has condensed the electronic warfare system on board the vehicle to be carried by a single soldier, with better concealment and higher security [3]. The emergence and use of nano-weapons will greatly change people's views on the comparison of war forces, make people re-understand the relationship between quantity and quality in the military field, generate a new concept of war, make the development and production of weapons and equipment more free from the limitation of quantity and scale, and further develop in the direction of quality and intelligence, thus completely changing the face of future warfare.

The production of nano-weapon equipment requires minimal loss and high reliability, and the development and production cycles are greatly shortened. Moreover, nano-weapons are also very convenient to use. With an unmanned aircraft, tens of thousands of micro-electro- mechanical system detectors can be air-dropped to areas where the enemy may be deployed or scattered in the sky, so that it is very easy to grasp the enemy's movements; In addition, nanotechnology can also be used to develop super-invisible coatings, smart military uniforms, etc. The above-mentioned various nano-weapons are combined to form a unique "nano-army".

In the future battlefield, giant weapon systems and micro-weapon systems will exist at the same time. Coordinated operations will have great effects, while small ones will have their own advantages. The emergence and use of nano-weapons will greatly change people's views on the comparison of war forces, make people re-understand the relationship between quantity and quality in the military field, generate a new concept of war, make the development and production of weapons and equipment freer from the limitation of quantity and scale, and further develop in the direction of quality and intelligence, thus completely changing the face of future warfare.

8.1.3 Nanosatellites

A nanosatellite is a spacecraft with a minimum size and a weight of less than 0.1 kg. It is based on the deployment of several nano-satellites in different orbits in space in the form of local star clusters or distributed constellations. They are connected to each other by means of telemetry and remote control to form a constellation with internal tight alarm connections to complete the complete functions of a rich satellite. The core components of nanosatellites are electrostatic micromotors, laser gyroscopes, sensors and transmitters, etc. These components are integrated and installed on semiconductor wafers. After the silicon nanosatellites are assembled, the height is about 10 cm and the diameter is about 15 cm. Thousands of such satellites can all be launched into the sky with a small carrier rocket and form a multi-orbit constellation system, which is sufficient to ensure continuous coverage of any point on the earth. As shown in Fig. 8.4, if the sun-synchronous earth orbit is used, continuous and uninterrupted monitoring of earth targets and information forwarding can be realized [4]. Using about 18 equally spaced orbits and deploying 36 satellites on each orbit, all points on the earth can be covered, and at least one satellite above the ground level of 300 km can be guaranteed at each point.

In the history of aerospace development, due to the limitation of carrying capacity and technical level, the satellites developed in the early stage all adopt the small

Fig. 8.4 Nanosatellite picture [4]

satellite scheme, and its weight is only tens of kilograms. At the end of the 1970s, due to the successful development of high-thrust launch vehicles and the improvement of design and manufacturing capabilities, large-scale multi-functional satellites began to appear, and the volume of satellites continued to increase, and their functions became more and more complex. What followed was rising costs and increasing risks. For example, the value of a Hercules/Centaur carrier rocket together with the reconnaissance satellites launched could reach more than 1.05 billion US dollars. Once the launch failed, it would cause serious losses. For this reason, the aerospace industry has turned its attention to small satellites again. Nano-satellite refers to a small satellite with a weight of less than 10 kg. It is the crystallization of the comprehensive application of contemporary high-tech, and nano-technology is a brand-new science and technology faced by modern human society. Its development is an effective way to solve a series of problems encountered at present, especially major problems such as system integration. The advancement of nanotechnology will directly lead to the improvement of the technical level of nanosatellites [5]. Therefore, vigorously developing nanotechnology is the fundamental driving force to promote various technologies including nanosatellite technology.

Replacing the relevant systems currently used on satellites with application-specific micro-integrated circuits can make micro-satellites and nano-satellites smaller in size and lighter in weight. These satellites have strong survivability, and will not lose all their functions even if they are attacked; the development cost is low, and large-scale experimental facilities are not needed; they are easy to launch, and a small carrier rocket can launch thousands of satellites, and then form a satellite network according to different orbits, which can cover the surface of the earth. As shown in Fig. 8.5, the nanosatellite subsystem can now be integrated including mechanical structure, propulsion power system, attitude control, thermal control, etc., and the power supply, as an independent system, must be integrated at present and in the future, unless there are major innovations. Command and data processing, telemetry tracking and communication are the core parts of satellite technology, and this part is the most likely system part to be integrated into a subsystem at present. And with the rapid development of MEMS, some functions in the mechanical system such as power will gradually be integrated into the core part, so that the multi-purpose of nano-satellites will enter the substantive stage. It can be predicted that with the increase of integrated function density, the payload of nano-satellites will make a leap.

From a military point of view, the cuteness of microsatellites as weapons is that the enemy may not even know that they are lurking next to their own satellites. Once necessary, it can launch an ambush on the target satellite according to ground instructions. Because of their low weight, tiny satellites can hitch a ride quietly into space on civilian space launch missions. In the future, satellites with smaller weight and volume such as "nano-satellites" weighing only 10 kg can also undertake similar approaching and sticking tasks. These satellites will operate in a synchronous orbit more than 32,000 km from the earth. In the same orbit, there are many commercial and civilian satellites operating. It will be very difficult for people on the ground to

Fig. 8.5 Schematic diagram of the Israeli BGUSAT nanosatellite [5]

find the threat of nano-satellites. This situation makes nano-satellites seem to have stealth technology and are more suitable for ambush warfare.

8.1.4 Space Elevator

Space warfare will become a hot spot in future wars. Due to the low carrying capacity of the current space shuttle and space shuttle, the number of launches is limited, and the safety is poor. In the previous stage, American scientists even called for the suspension of the use of the existing space shuttle. On the other hand, some developed countries are stepping up the development of new space vehicles that can meet the needs of space operations. A buckyball is a football-shaped structure formed by the aggregation of 60 carbon atoms. Based on this, carbon nanotubes are developed, which are 100 times stronger than steel and only 1/6 the weight of steel. 50,000 carbon nanotubes lined up together are only equivalent to the diameter of a hair. As shown in Fig. 8.6, this carbon nanotube or similar structure material is envisioned to be used in the manufacture of "space elevators", which will play an important role in future space militarization applications.

Carbon nanotubes are characterized by high strength and light weight. The diameter of a single carbon nanotube is only 1.4 nm, and 50,000 carbon nanotubes arranged side by side may become rational superfibers in the future. One potentially groundbreaking application of carbon nanotubes is in space elevators. As shown in Fig. 8.7, the key to the difference between a space cable made of carbon nanotubes and steel or other materials is that it can support its own mass [6]. This provides a possible

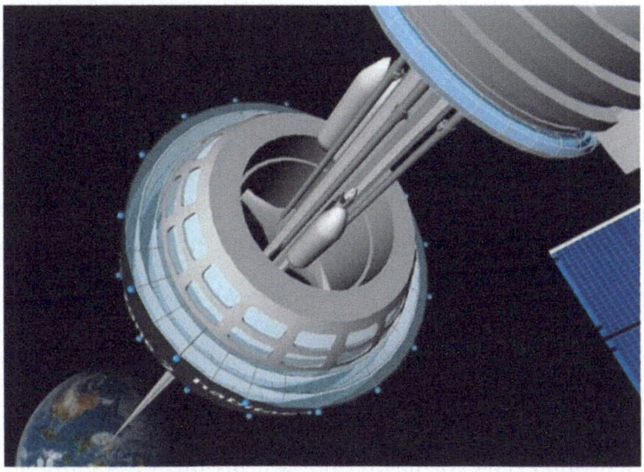

Fig. 8.6 Conceptual diagram of space elevator [6]

Fig. 8.7 Schematic diagram of carbon nanotube structure [6]

method of lifting people or objects into outer space, and may become a rational method for human colonization of alien planets.

A geosynchronous satellite 36,000 km above the ground to the ground base station, and an elevator-type spacecraft will be built along this cable to travel between the earth and space to transport materials. The space elevator can carry various items including large solar generators, nuclear waste, etc., and can also carry people. More importantly, the space elevator does not need to carry a large amount of fuel to the sky, and the estimated energy consumption is only 1% of the spaceship launch.

8.1.5 Prospects for the Development of Nanomaterials in the Military Field

The emergence of nanotechnology marks that humans have gone from the micron level to the atomic and molecular level, enabling humans to finally manipulate individual atoms and molecules according to their own wishes, so as to achieve effective control of the mesoscopic world. As a key technology in the twenty-first century, nanotechnology is still immature in its application research, but because of its obvious military potential, it greatly stimulates people to seek the application of nanotechnology in military affairs. In the military field, with the development of nanotechnology and its continuous application in weapons and equipment, weapons and equipment can be improved and enhanced in many ways [7].

(1) Improve material properties and increase the quality of weaponry

Nanomaterials can significantly improve and improve the performance indicators of weapons and equipment. Nano-ceramics can overcome the fatal weaknesses of traditional ceramics such as brittleness and impact resistance, and are expected to be ideal materials for naval and aircraft turbine engine components to improve engine efficiency, working life and reliability. Nano-ceramic is also the key material for the high-power, low-heat dissipation engine of the main battle tank. The high fracture toughness and impact resistance of nano-ceramics can be pasted or installed on surface ships and other parts that are prone to collisions and blows, and can be used to improve the bullet resistance of main battle tanks and composite armor. The use of nano-ceramic liners in high-firing rate weapons, such as muskets and torpedoes, can improve the ablation impact resistance of weapons and prolong their service life. Nanofibers "woven" with nanomaterial tubes will have very good fiber elasticity, are not afraid of bending, puncturing, and extrusion, and are expected to be made into thin and light bulletproof vests.

(2) Improved stealth performance of weaponry

Stealth performance is one of the notable features of the new generation of weapons and equipment. The pros and cons of stealth not only depend on the structural design of the weapon and equipment, but more importantly, whether the stealth materials it uses have good absorption properties for radar waves and infrared rays. Due to the excellent structure of nanomaterials, the surface, interface effects and quantum effects of materials will have an important impact on the microwave-absorbing performance of weapons and equipment. Utilizing the characteristics that the particle size of nanoparticle materials is much smaller than the wavelength of infrared and radar waves, it is expected to make stealth materials with very high electromagnetic wave absorption rate, thereby greatly improving the stealth performance of aircraft, tanks, missiles, ships, etc.

(3) Enhance information storage and acquisition capabilities

Carbon nanotubes made with nanotechnology can act as a tunnel for electrons to pass through quickly. The diameter of the nanomaterial wire equivalent to the atomic scale is only 1/100 of the diameter of the thinnest circuit on a computer chip. It is used in the fine system of weaponry and equipment, and can make electronic information quickly and accurately transmitted to every corner of the battlefield. Under the action of nano-magnetic functional materials, weapons and equipment can greatly improve the acquisition, transmission, processing, storage and display capabilities of various information such as electricity, magnetism, sound, light, and heat in complex battlefield environments, and provide stronger information support capabilities for the electronic information system of weapon platforms.

(4) Speeding up of weapons and equipment

Due to the surface effects of nanomaterials, some special nanoparticles can be made into catalysts with higher combustion efficiency, thereby improving the combustion efficiency of propellants and explosives. Adding nickel nanoparticles as a catalyst to solid rockets and other combustion agents can increase the combustion efficiency of weapon ammunition charges by 100 times, not only increasing the flight speed of rockets, aircraft, missiles, shells, bullets, etc., but also improving the penetration ability of shells, bullets, etc.

(5) Miniaturization of weapons and equipment

The size of weapons and equipment is an important factor affecting its flexibility, and replacing large-scale integrated circuits with quantum devices will reduce the quality and energy consumption of weapon control systems by thousands of times, and the volume and quality of weapons and equipment will also be reduced. The sophisticated and complex high-performance nano-electronic information warfare system will be carried by individual soldiers instead of tanks.

(6) Intelligent weapons and equipment

With the help of nanotechnology, quantum devices can work 1000 times faster than semiconductor devices. Replacing semiconductor devices with quantum devices can improve the capability of information work in weapon control system. Using nanotechnology, existing radars can be thousands of times smaller and their ability to acquire information hundreds of times better. The ultra-high resolution synthetic aperture radar can be placed on the satellite to carry out high precision ground reconnaissance. The submarine skin made of nanomaterials can sensitively "feel" the extremely small changes in water flow, water temperature and pressure, and feed the obtained information to the central computer in a timely manner.

Nanotechnology is an emerging technology, its emergence marks the human from the micron level to the atomic, molecular level of the nanometer level, so that human beings can eventually manipulate individual atoms and molecules according to their own wishes, in order to achieve effective control of the microcosmic world. After years of development, it has been proved that nanotechnology is a new technology with great military potential, and it has been paid more and more attention by the military departments of developed countries. With the development and application of nanotechnology in the military field, the form of future war will be changed to a large extent. Only by mastering advanced nanotechnology can the army become stronger in detecting and striking capabilities, multi-dimensional reconnaissance and stealth attack capabilities, thus further widening the disparity in national defense forces caused by military science and technology. It can be said that the development of nanotechnology makes the future war mainly win by science and technology. For example, through advanced nanotechnology, it can directly attack the enemy command system, and make the enemy into chaos by destroying the information command system, so as to quickly solve the battle and reduce soldier casualties. Thus, China must increase the research and development and transformation application of nanotechnology in the military field, in order to improve the national military safety protection capability. In addition, the development of nanotechnology will provide new means of deterrence for future wars. The higher the efficiency of a country's research and development and transformation of advanced nanotechnology in the military field, the stronger the military deterrent capability for other countries.

The deterrent power possessed by nanotechnology is not that ordinary ammunition can have, therefore, the development of nanotechnology will change the rules of the game in the future war, in order to achieve the purpose of victory or to form a certain deterrent against the enemy, we must seize the development frontier of nanotechnology, and improve the research and development and transformation of nanotechnology in the military field. In a word, as one of the core technologies of human beings in the twenty-first century, the development trend of nanotechnology has become the focus of attention in the world. Due to the huge impact of nanotechnology on the military field, in the future with the development and application of nanotechnology, the combat performance of weapons and equipment will become stronger and stronger, and the war force comparison will be mainly reflected in science and technology. Therefore, China must pay great attention to the development trend of nanotechnology and its impact in the military field, increase the research and development and application of nanotechnology in the military field, and be ready to meet the challenges of the new military revolution in the world.

8.2 Application of Nano-functional Materials in Aerospace Field

In the twenty-first century, aerospace has shown a broader prospect for development, and high-level or ultra-high level aerospace activities are more frequent, and their role will go far beyond the field of science and technology itself, and have a broader and more far-reaching impact on politics, economy, military and even human social life. It should be pointed out that the great achievements made by the aerospace industry are inseparable from the development and breakthrough of aerospace material technology. Materials are the foundation and forerunner of modern high-tech and industry, and to a large extent are the prerequisite for high-tech breakthroughs. For example, the breakthrough of high-purity silicon semiconductor materials in the 1960s enabled mankind to enter the information age. The development of aerospace materials plays a strong role in supporting and guaranteeing aerospace technology; In turn, the development needs of aerospace technology have greatly led and promoted the development of aerospace materials. Since the twenty-first century, the development of aerospace industry has entered a new stage, which will promote the development of aerospace materials in the direction of higher quality, updated categories, stronger functions and more economic effectiveness.

According to the scope of use, aerospace materials can be divided into structural materials and functional materials. Structural materials are mainly used to manufacture various structural components of aircraft, such as aircraft body, spacecraft bearing cylinder, engine shell, etc., whose role is mainly to withstand various loads, including static loads caused by self-weight and dynamic loads generated during flight [8]. Functional materials mainly refer to materials with special functions in light, sound, electricity, magnetism, heat and other aspects, such as electronic information materials involved in aircraft measurement and control systems (including functional materials for microelectronics, optoelectronics and sensor parts), and such as modern aircraft stealth technology for wave transmission and absorption materials, thermal protection materials on the surface of the space shuttle. The general development trend of structural materials is light weight, high strength, high modulus, high temperature resistance, low cost. Functional materials are developing in the direction of high performance, multi-function, multi-varieties and multi-specifications.

Nanomaterials have a very wide range of application prospects in aerospace and other high-tech fields, such as (a) nanopowder, aerospace nanomaterials use a large number of metal nanoparticles (such as Fe, Al, Mg, Ni, etc.) and non-metallic nanopowder (such as Al_2O_3, TiO_2, etc.); (b). Nanometer coating - aerospace coating materials (such as protective coating, stealth coating, etc.) are increasingly using advanced nanocomposite coating technology; (c). Nano-structured materials, especially nano-composite materials, are widely used in aerospace vehicles, such as fuselage and its auxiliary devices, wings, engines and their components, propellers, rocket nozzle igniter, engines, gas turbines, etc., not only to use superalloys, but also to use fiber and nano-toughened composite ceramic materials; (d). Nano-functional materials have been increasingly applied in aerospace and high-tech fields (including sensors, aviation instruments, communication devices, radar devices, etc.) due to their large specific surface area, fast response speed and high sensitivity [8].

8.2.1 Nano-conductive and Heat-Conducting Materials

Nano-conductive materials have excellent electrical conductivity and can conduct large amounts of current in a relatively small volume. Nano-conductive materials exhibit very high thermal conductivity, which can transfer heat quickly and maintain the stability of the material. In the aerospace field, nano-conductive materials have a wide range of applications, a typical example is the use of nano-silver conductive ink in electronic components. Traditionally, the manufacturing of aerospace electronic components requires a large number of metal wires and connectors, which adds weight and bulk. However, by using nano silver conductive inks, electronic circuits can be printed directly on a thin film substrate, thus achieving lightweight and compactness. This technology not only improves the performance of electronic components, but also simplifies the manufacturing process and reduces costs. Another important application is the use of nano-conductive materials in aerospace structures. The structure of aerospace spacecraft usually requires high temperature resistance and high strength materials, and traditional conductive materials are often difficult to meet these requirements at the same time, and the excellent performance of nano-conductive materials makes it a solution to meet the needs of aerospace spacecraft. For example, carbon nanotubes are a material with excellent electrical conductivity and strength that can be used to make lightweight and high-strength aerospace structures. In addition, nano-conductive materials can also be applied to conductive coatings to provide electrostatic protection and anti-corrosion functions of aerospace spacecraft.

In addition to conductive materials, nano-conductive materials also play an important role in the aerospace field. Thermal management is a key problem in spacecraft design. Traditional heat-conducting materials often have low thermal conductivity, limiting the efficiency of heat transfer. The high thermal conductivity of nano-conductive materials can effectively improve the effect of thermal management. For example, alumina nanoparticles can be used to make thermal conductive adhesives for connecting thermal sensors and radiators in spacecraft, enabling efficient heat transfer and heat dissipation.

The application of nano-conductive and heat-conducting materials has a broad prospect in the field of aerospace. However, it should be noted that there are still some challenges and obstacles. For example, the preparation and processing technologies of nanomaterials still need to be further developed to meet the demand for high-performance and highly reliable materials in the aerospace sector. In addition, the safety and environmental impact of nanomaterials also need to be paid attention to and studied.

In summary, nano-conductive and heat-conducting materials have a wide range of application potential in the aerospace field. Through applications such as improving the performance of electronic components, lightweight aerospace structures, and improving thermal management, nanomaterials have brought many innovations and breakthroughs to the development of aerospace technology. However, further research and development is needed to overcome the relevant technical and safety challenges to achieve widespread application of nanomaterials in the aerospace sector.

8.2.2 Nano-radar and Infrared Stealth Materials

The application of nano-radar and infrared stealth materials in the aerospace field has been paid more and more attention. With the continuous development of radar and infrared technology, nanomaterials have brought new breakthroughs and innovations to the aerospace field. Nano-radar materials have important application value in aerospace field. The special structure and properties of nanoparticles make them have excellent ability of scattering and absorbing electromagnetic waves, which can be used to improve the performance of radar system. For example, nanoparticles can be used as radar absorbing materials, which can effectively absorb incoming radar signals, reduce reflection and scattering, and thus reduce the radar system's ability to detect targets. In addition, nanomaterials can also be used to prepare antennas, conductors and substrates with excellent microwave transmission performance, further improving the sensitivity and resolution of radar systems.

Infrared stealth materials are another key aerospace application area. Infrared radiation is an important electromagnetic radiation generated by aircraft during flight, which has high detection and tracking performance [9]. Nanomaterials can effectively reduce the scattering and absorption of infrared radiation by means of their unique optical properties and structural design, so as to improve the stealth performance of aircraft. For example, materials such as nano-metals, nano-carbon, and nano-oxides can be used to prepare coatings and materials with excellent infrared radiation suppression capabilities, so that the aircraft presents a lower signal signature in the infrared spectral range, reducing the possibility of detection by infrared detectors.

France has successfully developed CoNi nano-wave absorbing material, which is composed of binder and nano-sized microchip filling material, and has good wave absorbing properties. The United States has developed a nano-absorbing material called "super black powder", which can absorb 99% of radar waves. This "ultra-black powder" nano-absorbing material is essentially a graphite-thermoplastic composite material and a graphite epoxy resin composite material made of nano-graphite as an absorber. This material not only has a high absorption rate, but also maintains good toughness at low temperatures [10]. The results show that the reflectance of 3 nm FRP plates prepared by adding trace amount of nano iron powder to phenolic resin can reach 5–8 dB at 2–18 GHz. Ali Ghasemi et al. studied the electromagnetic properties of multi-walled carbon nanotubes (MWCNTs)/doped barium ferrite nanocomposites. Compared with pure MWCNTs and doped barium ferrite, the absorption properties of the composites were significantly improved, and the maximum reflection loss increased significantly with the increase of the volume fraction of MWCNTs. Gu Liu et al. prepared carbon nanotube (CNTs)-SiC/Al_2O_3-TiO_2 composite coating by micro-arc plasma spraying technology. With the increase of coating thickness within a certain range, the wave absorption capacity of the coating was significantly improved, and its resonance frequency kept moving to the low frequency (Table 8.1).

In addition, nanomaterials can be combined with other technological means to further improve the performance of aerospace systems. For example, nanoparticles can be combined with porous materials to prepare infrared wave absorbing materials with multi-scale structures, improving the absorption capacity of the infrared band. Nanomaterials can also be combined with sensor technology to achieve multi-functional integrated monitoring and control systems. These integrated applications make the application of nano-radar and infrared stealth materials in the aerospace field more extensive and flexible.

Table 8.1 Comparison of infrared emissivity of some metals

Metal	Test temperature/°C	Emissivity
Au	38–260	0.02
Ag	38	0.01
Al	25	0.02
Cu	38	0.03
Zn	38	0.02–0.03
Pt	38	0.05
Co	38	0.08
Ni	38	0.05

Therefore, nano-radar and infrared stealth materials have great application potential in the aerospace field. Its characteristics and advantages in radar performance improvement and infrared stealth provide a new solution for the performance improvement and stealth capability of aerospace systems. With the continuous progress of nanotechnology, we can expect more innovative applications of nano-radar and infrared stealth materials in the aerospace field.

8.2.3 Nano-sensing Materials

Because of their large specific surface area and high surface activity, nanomaterials can be widely used as various sensitive materials. The gas sensor made of nanomaterials not only maintains the advantages of coarse crystalline materials, but also improves the response speed, enhances the sensitivity, and can effectively reduce the operating temperature of the element. Sensors are widely used in the military, especially in the field of detection equipment. Although the application research of nano-sensor is still in its infancy, it has shown the advantages that other sensors cannot achieve: high sensitivity, small body, low energy consumption, and multiple functions. As shown in Fig. 8.8, an important class of sensors is chemical sensors, and gas sensors are an important part of chemical sensors. Gas sensors usually use metal oxides that change with the composition of the surrounding gas, resulting in changes in resistance, etc., to detect and quantitatively measure the gas. The smaller the particle size of the gas sensor material, the higher the specific surface area, the greater the interaction between the sensor and the surrounding gas contact, and the higher the sensitivity.

With the rapid development of nanotechnology, nano-sensing materials provide new solutions for the monitoring and control of aerospace systems through their unique characteristics and advantages, and their applications in the aerospace field are becoming more and more important. A key application of nano-sensing materials in aerospace is structural health monitoring, where the structural health of aircraft is critical to aviation safety. Nano-sensing materials have the characteristics of high

Fig. 8.8 Schematic diagram of nanosensor [11]

sensitivity and small size, and can be embedded into aircraft structures for real-time monitoring of structural stress, deformation and damage [11]. For example, nanofiber sensors can be used to monitor the stress distribution of aircraft wings, detect and repair potential structural problems in a timely manner, and improve the safety and reliability of aircraft.

Nano-sensing materials can also be used in the aerospace field for environmental monitoring and control. Aircraft need to face various complex environmental conditions during flight, including changes in temperature, humidity, air pressure and other factors. Nano-sensing materials can be used as sensors to monitor and feed back environmental parameters in real time to help aircraft systems make appropriate adjustments and controls. For example, nano-temperature sensors can be embedded in the engine to monitor temperature changes and adjust the cooling system in time to ensure the normal operation of the engine and thermal management.

In addition, nano-sensing materials can also be applied to gas detection and chemical analysis. In aerospace, the gas environment around the aircraft needs to be monitored in real time to ensure flight safety and personnel health. The high sensitivity and selectivity of nano-sensing materials make them ideal gas sensors. As shown in Fig. 8.9, for example, nanoparticle sensors can be used to detect toxic gas or fuel leaks inside an aircraft, send alerts and take action to ensure the safe operation of the aircraft.

Nano-sensing materials also have potential for space exploration and navigation applications in the aerospace field. Aerospace systems require accurate space perception and navigation capabilities to ensure precision and accuracy in navigation. Nano-sensing materials can be used to prepare high-precision inertial sensors, optical sensors and wireless communication sensors to provide accurate spatial information and navigation data. For example, nano inertial measurement units can be used for attitude control and accurate navigation of spacecraft, ensuring the positioning and navigation of spacecraft in space.

The advantages of nano-sensing materials in structural health monitoring, environmental monitoring and control, gas detection and chemical analysis, as well as space

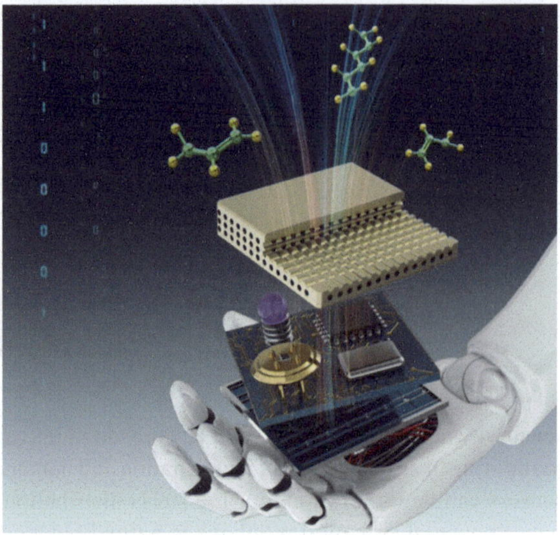

Fig. 8.9 Remote Gas monitoring system based on independent nanosensor [11]

exploration and navigation provide new possibilities for the performance improvement and safety assurance of aerospace systems [11]. With the continuous development and innovation of nanotechnology, the application of nano-sensing materials in the aerospace field will be more and more in-depth and extensive.

8.2.4 New Ceramics and New Glass

(1) New functional ceramics

Infrared ceramics is a kind of polycrystalline ceramic material that can transmit infrared radiation, and it is also a new functional structural material developed for the application of missile technology [12]. In order to accurately hit the target, the missile head is equipped with an infrared automatic tracking device, and can automatically adjust the direction of flight, so that the shell material of the missile head (head cover) should not only be resistant to high temperature and pressure, but also be able to pass through infrared. For the "infrared radar" or "infrared–laser" composite missile hood, it is also required to pass through microwave, laser and other rays. The use of infrared ceramics can also be made into a forward-looking infrared system, monitoring the enemy using scientific and technological means can be made into a variety of transparent ceramics, it is as hard as sapphire, as transparent as glass, resistant to 2000 °C high temperature, corrosion resistance, and high mechanical strength. It can be made into various bulletproof ceramics, missile rectifier, etc. (Table 8.2).

Table 8.2 Comparison of properties of typical bulletproof ceramics

Ceramic material	Density (g/cm^3)	Hardness (Hv)	Fracture toughness (MPa/m$^{1/2}$)	Modulus of elasticity (Pa)
Al_2O_3	3.50–3.90	1400–1800	2.8–4.5	350
B_4C	2.52	2800–3200	2.5–3.0	400
SiC	3.20	2500–2850	2.8–4.3	350–450
Si_3N_4	3.20	1700–1800	5.0–6.0	320

(2) New nano glass

a. Sol–gel glass

Using cheap silicon-based compounds as raw materials, abnormally small suspended colloidal particles (sol) are connected together (gel), thus changing from liquid to solid and forming sol–gel glass. This glass is only half the mass of ordinary glass and can be used as a space telescope to reduce the mass.

b. Nanocrystalline glass

Glass–ceramics are composed of tiny crystals, and the structure of this glass is similar to ceramics, so it is also called "glass ceramics". The strength of glass–ceramics is large, and its bending strength is 7–12 times that of ordinary glass. Good high temperature resistance, softening temperature up to 1000 °C. As shown in Fig. 8.10, nanocrystalline glass is composed of smaller nanocrystalline and has higher performance than microcrystalline glass. Nanocrystalline glass can transmit microwaves to become the eye "mask" of the missile head, and nanocrystalline glass can also be used as structural materials for rockets, satellites and space shuttles.

Fig. 8.10 Picture of glass–ceramics [12]

8.2.5 Solid Rocket Propellant

Solid rocket propellant is an important energy source for spacecraft, and the introduction of nano-functional materials can significantly improve the performance and characteristics of the propellant. Due to the small size and large specific surface area, the proportion of atoms on the surface of the nanoparticles is high, and the suspended chemical is more and more, which increases the activity of the nanomaterials. The existence of a large number of suspension bonds makes the interface polarized, and the high specific surface area causes multiple scattering, which is the main reason for the wave-absorbing property of nanomaterials. Under the radiation of microwave field, the motion of atoms and electrons is intensified, which causes the magnetization to convert electron energy into heat energy, thus increasing the absorption wave of electromagnetic wave. In addition, nano ions have high coercivity and can cause large hysteresis loss. The combined action of these three aspects makes the nanomaterials have good absorbing properties. At the same time, the nanopowder has a plasmon resonance frequency shift of the absorption peak, which is related to the quantum size and changes with the change of the quantum size. Therefore, the displacement of the absorption edge can be controlled by changing the quantum size, so as to control the bandwidth of the absorption band. Therefore, the nanomaterials can be used as absorbers to prepare a wide-band absorbing coating. The structural characteristics of nano-powders make the nano-materials have the characteristics of wide wave absorbing frequency band, good compatibility, light weight and thin thickness, etc., and it is easy to meet the requirements of "thin, light, surprisingly strong" radar absorbing materials. The wave-absorbing materials represented by nanoparticles and composite particles have a good development prospect, and the research in this field is being actively carried out at home and abroad.

Nano functional materials can be used to enhance combustion efficiency in solid rocket propellants. Traditional solid rocket propellants often use metal powder as a combustion agent, but its surface area is relatively small and the reaction rate is limited, while nanoparticles have huge specific surface area, which can provide more reactive sites and accelerate the combustion reaction [13]. For example, nano-aluminum powder can react with oxidizing agents to produce high-temperature and high-pressure gases, thereby enhancing the thrust and energy output of rocket propellants. As shown in Fig. 8.11, nano-functional materials in solid rocket propellants can also be used to adjust the combustion rate and stabilize the combustion process, enhance the mechanical and thermal properties of the propellant, and reduce the toxicity and environmental pollution of solid rocket propellants. Traditional solid rocket propellants often contain harmful metal powders and gaseous emissions that pose potential risks to the environment and human health. Through the introduction of nano-adsorption materials and catalysts, harmful substances can be captured and transformed, reducing emission concentrations and toxicity. For example, nano-zinc oxide and nano-copper can act as catalysts to effectively degrade harmful gases and reduce their impact on the environment.

Fig. 8.11 Actual cross section of AP/HTPB propellant subcharge column containing 7.8% aluminum powder [13]

The application of nano functional materials in solid rocket propellants is of great significance. By improving combustion efficiency, regulating combustion rate, enhancing mechanical and thermal properties, and reducing toxicity and environmental pollution, nano-functional materials provide new breakthroughs and improvement directions for propellant technology in aerospace systems. At present, the application of nanomaterials in solid propellants has become a hot topic at home and abroad. The research focuses on the following two aspects, one is the application of nano-metal powder as high-energy additive in solid propellant, and the other is the application of nano-metal oxide as combustion catalyst in solid propellant. In addition, the preparation of nano-oxidants is also a development direction to improve the combustion performance of propellants.

8.3 Application of Nanostructured Materials in Aerospace Field

High-tech materials are the basis of aerospace technology, and the development of aerospace technology constantly puts forward new requirements for material science. The development of aerospace technology is inseparable from nanotechnology and materials. When nanotechnology is applied in the aerospace field, its consumption can be greatly reduced and the use effect of aerospace equipment can be improved. For example, in many aerospace equipment, a lot of electronic equipment with nanotechnology as the main material is used, and protective coatings with nanomaterials are added. For example, in the field of space satellites, a lot of heat will be generated during their liftoff due to friction, so on the one hand, it is necessary to add thermal insulation coating, and on the other hand, it is necessary to improve the heat dissipation efficiency to achieve the effect of reducing the surface temperature. Using nanomaterials to make a metal skeleton, using copper, gold and other materials with

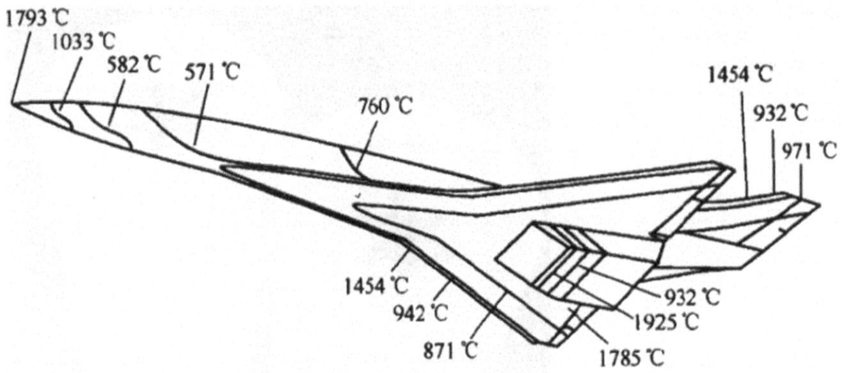

Fig. 8.12 Temperature distribution of different parts of the aerospace shuttle shell [14]

low melting point to fill the gap in the skeleton, can be made into a good protective coating, improve the quality of the aircraft.

Aerospace structural materials require high strength and low density, and some materials also require high temperature resistance, corrosion resistance, friction resistance, high pressure resistance and other properties. As shown in Fig. 8.12, different parts of the aerospace vehicle are subjected to different effects during flight, and the temperature is also different. Some parts of the aerospace aircraft shell have a temperature as high as nearly 3000 °C, while others have a very low temperature, such as the temperature of the liquid hydrogen container is – 253 °C. Therefore, the aerospace shuttle shell should not only be able to withstand high temperature, friction resistance, corrosion resistance, but also have good toughness and stress damage resistance.

The main structural materials for the manufacture of aircraft and spacecraft include metal materials such as aluminum alloy, magnesium alloy, titanium alloy, lithium alloy, alloy steel, and non-metallic materials such as special ceramics, engineering plastics, and various fiber composite materials. Nanomaterials have excellent mechanical properties, such as high strength, high hardness and good plasticity [14]. For example, the yield strength and hardness of metal materials increase with the decrease of grain size, while maintaining good plasticity and toughness.

Because the synthesis and production of nanomaterials is still in its infancy, the current research and application of nano-structured materials in the aerospace field mainly include nano-toughening and strengthening materials, nano-special structural materials and nano-welding materials and technologies. The main purpose is to improve the performance of the original material on the one hand, on the other hand to achieve the special function that the general material can not complete. Nano-toughened and reinforced materials have good application prospects in aircraft and spacecraft shell, engine and other parts because of their high hardness, high toughness, low temperature superplasticity, easy processing and small density.

8.3.1 Metals and Metal Matrix Composites

Grain refinement is one of the most effective ways to improve the strength of metal materials. Method for strengthening metal alloy based materials by adding nano-ceramics. The nano ceramic powder is evenly dispersed in the alloy to improve the nucleation rate of the alloy and inhibit the grain growth, so as to play the role of grain refinement. Inhibit the expansion of microcracks during the use of materials and improve the strength of products. For example, Nano-silicon carbide, nano-silicon nitride, nano-titanium nitride, nano-silicon powder added to the metal matrix (aluminum, copper, silver, steel, iron and other alloys), can produce light weight, high strength, good heat resistance of new alloy materials.

The research of metal and compound nanocomposites has attracted great attention. Nano-oxide dispersion strengthened superalloy has good high temperature strength, excellent oxidation resistance, friction resistance and high temperature corrosion resistance, and has been partially used in guiding blades and turbine blades of aviation generators. As shown in Fig. 8.13, for example, the hardness of nano-Co-WC is more than 1 times higher than that of ordinary Co-WC, and the toughness and wear resistance are significantly improved.

It has been found that nano-iron materials made from 6nm iron crystals are 12 times stronger and 23 orders of magnitude harder than ordinary steel. Nano-iron materials can be used to make special steels with high strength and toughness. The hardness of nano-copper or nano-bulk materials is 50 times higher than that of conventional materials, and the yield strength is also 12 times higher.

(1) Nano titanium nitride is used in alloy steel and iron

Nano titanium nitride has the characteristics of high hardness, high thermal stability, small particle size and good dispersion. In the cooling crystallization process of molten steel, nano titanium nitride becomes the nucleation phase, which can greatly increase the number of nucleation. The grain size is reduced to achieve the effect of refining the alloy grain, and the comprehensive properties of the alloy are greatly improved.

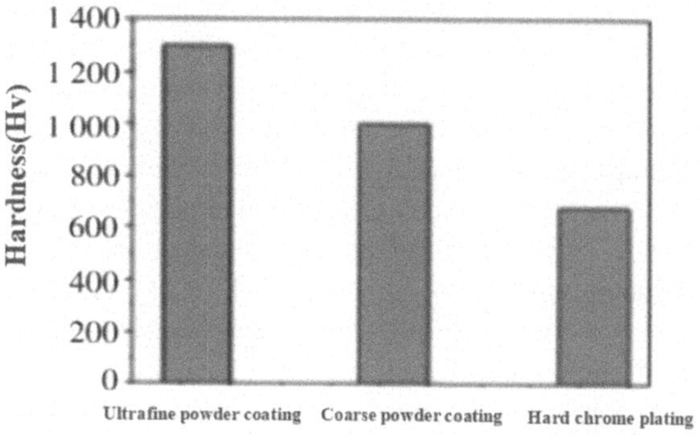

Fig. 8.13 Comparison of hardness of ultra-fine structure Co-WC coating, commercial coarse powder coating and electroplated hard chromium layer [14]

(2) Nano silicon carbide used in silver matrix composites

It is one of the important ways to prepare high strength and high conductivity alloy materials by adding even, fine and stable particles into the matrix to achieve the purpose of dispersing and strengthening the alloy. Nano silicon carbide is an effective reinforcement phase for silver alloys. When the mass percentage content of nanometer silicon carbide is 1%, the strengthening effect is the best. The tensile strength of the material can reach 391 MPa, the relative conductivity is 60.2%, and the strength and wear resistance are improved.

(3) Nanometer silicon carbide dispersion reinforced copper matrix composites

High strength and high conductivity copper matrix composite in integrated circuit lead frame. All kinds of spot welding, hob welding machine electrode, contact material, armature, electric tool commutator and other electronic equipment has a wide range of uses. However, the high strength and high conductivity of copper alloys have always been a pair of contradictory characteristics. Generally, the mechanical properties of copper can only be improved at the expense of electrical and thermal conductivity to obtain high strength. Using nano silicon carbide stable dispersion to strengthen copper based materials is a good way to solve this contradiction. By adding evenly to the matrix. Nano silicon carbide particles with good stability for dispersion strengthening of copper alloys have become a research focus in the preparation of high strength and high conductivity copper matrix composites.

(4) Nanometer zirconium carbide is used in cemented carbide

Nano-zirconium carbide is an important high-temperature structural material with high melting point, high strength and corrosion resistance. Nano-zirconium carbide is used in cemented carbide materials, which can improve the strength and corrosion resistance of materials.

8.3.2 Polymer Based Composites

Plastic is a widely used material, compared with other materials, plastic has light weight, corrosion resistance, high specific strength, good electrical properties, bright colors, easy processing and forming characteristics, so plastic has become one of the important materials in the aerospace field. By filling the plastic modification, not only can improve the mechanical properties of plastics, but also can develop a variety of functional plastics, such as conductive plastics, magnetic plastics, anti-degradation plastics, anti-UV anti-aging plastics and so on.

The polymer nanocomposite is a kind of composite material with the polymer as the matrix continuous phase and the nanoscale filler as the dispersed phase. Polymer-based nanocomposites can generally be divided into three categories: organic organic nanocomposites, organic/inorganic hybrid nanocomposites, and organic/inorganic particle nanocomposites. After the nanoparticle is added to the polymer matrix, its wear resistance, hardness, strength and heat resistance can be improved, and it has the characteristics of easy processing, low friction and wear rate, high surface hardness and low cost, among which, the nano-sized dispersed phase mainly includes metal, semiconductor, inorganic salt nanoparticles, nanofibers, carbon nanotubes, graphene and clay, etc., and this kind of nano-filler is uniform Dispersed in the polymer substrate, polymer nanocomposites are formed. This kind of material gathers the advantages of different material compositions and nanomaterials, producing a synergistic effect, thus giving the material high mechanical properties, electrical conductivity, magnetism, heat insulation and biological compatibility [15]. At present, the synthesis and application of polymer nanocomposites have attracted the attention of many researchers and are in a stage of rapid innovation.

(1) Nanometer aluminum nitride applied to epoxy resin

In the nano-aluminum nitride epoxy resin system, when the amount of nano-aluminum nitride is 1–5%, the glass transition temperature increases and the elastic modulus reaches a maximum value. The composite material prepared by adding nano aluminum nitride to epoxy resin is completely different from the aluminum nitride epoxy resin matrix composite with coarse crystal added: coarse aluminum nitride is generally added as a reinforcing agent, and it is mainly distributed in the chain of the polymer material. Nano-aluminum nitride exhibits strong activity due to its severe lack of coordination and large specific surface area. At the same time, some nano-aluminum nitride particles are distributed in the interspaces of polymer chains. Compared with coarse crystalline aluminum nitride, nano-aluminum nitride has high fluidity, which can greatly improve the strength, toughness and ductility of epoxy resin.

(2) Application of nano silicon carbide in rubber tire

A certain amount of nanometer silicon carbide was added to the rubber tire and modified without changing the formula of the original rubber. As shown in Fig. 8.14, the wear resistance can be increased by 15–30% without reducing the original performance and quality. In addition, nano silicon carbide can also be used in rubber rollers, printer fixing film and other rubber products that need to wear, heat dissipation and temperature resistance.

8.3.3 Engineering Plastics and Other Composites

The composite of nanomaterials and engineering plastics can not only improve the inherent properties of engineering plastics, but also give them high electrical conductivity, high barrier property and excellent optical properties. Therefore, the application of nanomaterials to the modification of engineering plastics can further broaden the application range of engineering plastics.

Fig. 8.14 Schematic diagram of rubber tire with nano-sic added [15]

(1) Engineering plastics/nanoparticle composites

Using nanoparticles to toughen engineering plastics with certain brittleness is an effective method to improve the mechanical properties of engineering plastics such as toughness and strength. As long as the nano-particles are well combined with the matrix resin, the nano-particles can also withstand the tensile stress, and the toughening and strengthening effects are obvious. When a small amount of nano-titanium nitride powder is used to modify thermoplastic engineering plastics, it can play the role of crystallization nucleating agent. Nano-titanium nitride is dispersed in ethylene glycol, and nano-titanium nitride is better dispersed in PET (polyethylene terephthalate) engineering plastics by polymerization, which can accelerate the crystallization rate of PET engineering plastics, make its molding simple, and expand its application range. A large number of nano-titanium nitride particles dispersed in PET can greatly improve the wear resistance and impact resistance of PET engineering plastics [16].

(2) Engineering plastics/nano-magnetic metals and their nitride composites

This composite material has special photoelectric function (special absorption of electromagnetic wave) and excellent magnetic properties and electrical conductivity, and can be widely used in military, aerospace, electronic communications and other high-tech fields. Nano silicon carbide after surface treatment with coupling agent. When the addition amount is about 10%, it can greatly improve and improve the performance of polyimide, polyether-ether-ketone, polytetrafluoroethylene and other special plastics, and comprehensively improve the wear resistance, thermal conductivity, insulation, tensile resistance, impact resistance, high temperature resistance and other properties of the material.

8.3.4 Ceramic Matrix Composite Materials

Nano-ceramic matrix composites are a kind of high performance materials with wide application potential, especially in aerospace field. Ceramic matrix composite material is a ceramic matrix, and a variety of nanomaterials composite materials, including silicon nitride, silicon carbide, etc. These advanced ceramics have high temperature resistance, high strength and hardness, relatively light weight, corrosion resistance and other excellent properties. Its fatal weakness is that it has strong brittleness, which will produce cracks under the action of stress, and even fracture leading to material failure. The composite of nanomaterials and ceramic matrix is an effective method to improve the toughness and reliability of ceramics, and nano-reinforced ceramic matrix composites with excellent toughness can be obtained. Due to its excellent mechanical properties, including high strength, high hardness and excellent wear resistance, this material can be used to manufacture structural parts such as fuselages, wings and tails of aircraft and rockets. For example, as shown in Fig. 8.15, recent studies have shown that aircraft wings made using nano-ceramic matrix composites can reduce weight and increase flight efficiency, resulting in lower fuel consumption and emissions.

The influence of nano-powders on ceramic materials is multifaceted, not only in the toughening and strengthening of ceramics, that is, on the microstructure, microstructure and properties of ceramics, but also on the sintering process of ceramics. Sintering is the process by which a powder system or porous material is transformed into a dense and firmly bonded ceramic body by high temperature heat treatment. Powder sintering process is very complex, it is affected by many factors such as particle size, surface state, impurity and diffusion coefficient. In essence, the main driving force of the sintering process is the surface stress caused by the powder surface energy (of course, the reaction chemical energy should also be considered when reacting to sintering):

$$P = \frac{2v}{r} \tag{8.1}$$

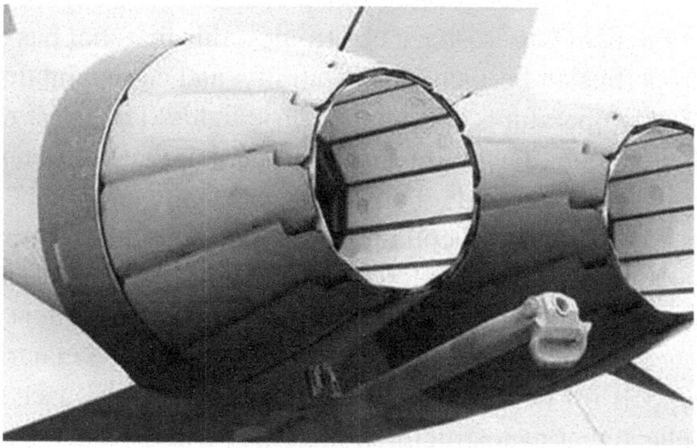

Fig. 8.15 Nano-ceramic based engine [17]

where r is the radius of surface curvature and v is the surface energy. Therefore. The smaller the particle size, the larger P, the easier sintering.

In the sintering process, the key to achieve densification is the migration of materials, that is, diffusion and flow. The flow can only be carried out when there is a liquid phase, so it is very important to introduce additives to promote the formation of liquid phase for difficult-to-sinter ceramics. Diffusion is very important for densification, and it is generally believed that ceramic sintering mainly relies on the diffusion mechanism such as surface and interface diffusion and intracrystalline diffusion. There are many paths of diffusion, which can be carried out through the internal surface boundary of the particle, or through the gas phase or the liquid phase. In the powder system, the rate of diffusion mass transfer is obviously closely related to the contact area between the particles.

Nanoparticles have small particles, large specific surface area and surface energy, and high diffusion rate, so sintering with nano-powders has fast densification speed and can also reduce the sintering temperature. From the perspective of application, the most important thing to develop high-performance nano-ceramics is to reduce the cost of nano-powders. In addition to ensuring the quality of nano-powders in the preparation process, the size and distribution are controllable, no agglomeration, the shape of the particles can be controlled, and the output is large, which will lay a good foundation for the development of new nano-ceramics.

Ceramic matrix composite materials have been practical or will be practical in the field of cutting tools, sliding components, engine parts, energy components and so on. For example, precision ceramic structure devices made of nano silicon nitride doping can be used in the ball and roller of rolling bearings, plain bearings, sleeve valves and structural devices with wear resistance, high temperature resistance and corrosion resistance requirements used in metallurgy, chemical industry, machinery, aviation, aerospace and energy industries.

In the field of aerospace engines, engine components need to withstand high temperature and high pressure environments, and nano-ceramic matrix composites have excellent heat resistance and oxidation resistance, which can effectively resist high temperature airflow and gas corrosion. Recent studies have shown that the application of nano-ceramic matrix composites to components such as engine turbine blades and combustion chambers can improve engine efficiency and reliability [17]. In the field of aerospace electronics, this material has excellent electrical insulation properties and dimensional stability, and can maintain good electrical properties at high temperatures and high electric fields. The latest research shows that electronic devices manufactured using nano-ceramic matrix composites exhibit excellent radiation resistance and reliability in spacecraft, and can be used in key components such as control systems, communication devices and sensors.

Nanostructured materials have the potential for revolutionary applications in the aerospace field. Light weight, strength enhancement and high temperature resistance are the main advantages of nanostructured materials, which provide important support for the design and performance of aerospace spacecraft. Experts generally believe that nanostructured materials will drive technological progress and innovation in the aerospace field, bringing new possibilities for the design and manufacture of future aircraft. Nano-ceramic matrix composites have great application potential in aerospace field. By combining the advantages of nano-ceramics and reinforcement materials, this material shows excellent performance in the fields of structural parts, engine components and electronic devices. With further research and development, nano-ceramic matrix composites will play an increasingly important role in the aerospace field [14, 17].

8.4 Application of Nanocomposite Coatings in Aerospace Field

The development of nanotechnology has led to many innovative applications in the aerospace field. Nanometer coating materials, as an important technical means, have a wide application prospect in the field of aerospace. With the continuous development of aerospace technology, the performance requirements of spacecraft materials are getting higher and higher. Nano-coating materials, as a new technology, have many excellent properties, which make the application of nano-coating materials in the field of aerospace increasingly important.

Coating material is an important part of aerospace material. With the continuous development of aerospace technology and material science and technology, aerospace coating materials have developed from a single function in the past into a multi-functional composite coating material that integrates protection, decoration and special functions, and has developed from a single organic or inorganic coating to a composite functional material coating that combines surface treatment nano modification and organic–inorganic coating technology.

An important application of nano-coating materials in aerospace is friction reduction and wear protection. Friction and wear between many parts in an aircraft is a serious problem that can lead to reduced performance and shorter life. Nano-coated materials have the characteristics of low friction coefficient and high hardness, which can effectively reduce the friction loss between parts and extend the service life of aircraft. For example, nano-diamond coating can be applied to friction parts such as engine blades and bearings, significantly reducing energy loss and wear, and improving engine efficiency and reliability [18].

Another key application of nano-coating materials in aerospace is corrosion resistance and oxidation resistance. Aircraft are often subjected to harsh environmental conditions, exposed to high temperatures, humidity and corrosive media. Nano-coating materials can form a protective layer to effectively prevent oxidation and corrosion. Aircraft at high speed or re-entry into the atmosphere, will be subjected to huge thermal loads and radiant heat. Nano-coating materials have low thermal impedance, which can effectively improve the heat dissipation performance of aircraft materials and improve the thermal management efficiency. At the same time, nano-coating materials can also be used for electromagnetic shielding and electrostatic protection in aerospace. Aircraft need to withstand external electromagnetic radiation and static interference to ensure the normal operation of electronic equipment and the safety of personnel. Nanometer coating materials have good electrical conductivity and can provide electromagnetic shielding and electrostatic protection for aircraft.

8.4.1 Preparation, Classification and Characteristics of Nano Coatings

Material surface modification and coating technology has become an important branch of materials science. By the combination of nano powder and surface coating technology, the surface composite coating containing nano powder can improve the mechanical, physical and chemical properties of the matrix surface, and give the matrix surface new mechanical, thermal, optical, electromagnetic and catalytic sensitivity functions, so as to achieve the purpose of surface and functional materials.

There are many methods to manufacture nanocoatings, including various deposition (including vapor deposition and liquid deposition), spraying (including normal temperature spraying, flame spraying, plasma spraying, etc.) and plating [19]. As shown in Fig. 8.16, the coating of nanocomposite system can be obtained by using existing coating technology and adding nanomaterials according to the properties of the coating. Nanocoatings can be applied to either conventional material substrates, powder particles or fibers for surface modification, coating, modification or adding new properties.

Any traditional surface coating technology can be used or slightly modified to achieve nanomaterial composite coating. Adding nano phase to the wear-resistant coating with high hardness can further improve the hardness and wear-resistant properties of the coating, and maintain high toughness. By adding nanoparticles to the surface coating, the friction coefficient can be reduced, self-lubricating materials can be formed, and even super-lubricating functions can be obtained. In some coatings, the composite C_{60}, bucky tube, etc., can prepare a super lubricating new material. The introduction of nanomaterials into the coating can significantly improve the high temperature and oxidation resistance of the material. Nanometer Ni-La_2O_3

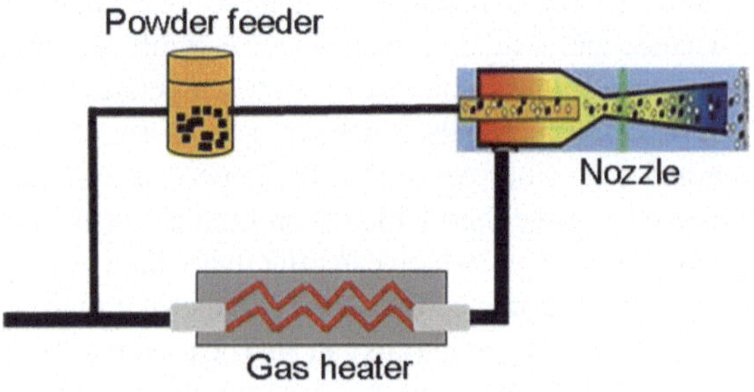

Fig. 8.16 Schematic diagram of working principle of cold spraying [19]

layer deposited on Ni surface can prevent the short-circuit diffusion of Ni ion and improve the growth mechanism and mechanical properties of the oxide layer.

Nano surface coating and surface modification technology formed by combining nanomaterials with surface technology is one of the main application functions of nanomaterials. According to the composition of nanocoatings, they can be divided into three categories: a. A single nanomaterial system; b. A composite system composed of two (or more) nanomaterials; c. Nano-coating of the composite system with nano-materials refers to the use of surface treatment technology to coat part or all of the material containing nano-powder on the matrix, and give the matrix new properties or achieve the purpose of modification due to the role of nano-powder. The matrix materials suitable for nanocoatings are diverse, from steel and non-ferrous metals to plastics, glass, ceramics and so on. There are also many types of nano-coating materials, including metal and alloy nano-coating materials, inorganic non-metallic materials and ceramic nano-material coatings, plastics and polymer composite coatings.

Because of the particularity of its surface and structure, nanomaterials have excellent functions that are difficult to obtain from general materials, which provides favorable conditions for improving the composition and performance of coating materials, and makes the functionalization of materials have great possibilities. Coating materials can be divided into structural coatings and functional coatings according to their use. Structural coatings include: a. High-strength, high-hardness and wear-resistant coatings; b. Self-lubricating coating; c. Heat, high temperature and oxidation resistant coating; Corrosion, protective and decorative coatings. Functional coatings include: a. Thermal coatings; b. Optical coating; c. Electrical coating; d. Magnetic coating; e. Catalytic sensitive coating, etc. Aerospace nano coating materials not only contain a large number of high strength, friction resistance, high temperature resistance, corrosion resistance, oxidation resistance of nano coating, such as various oxides, carbides, nitride coatings; Moreover, it also contains a large number of functional coatings or functional structural coatings, such as optical and stealth coating materials.

8.4.2 Protective Coating

Aircraft are made from a variety of metal and non-metal materials. The deterioration and destruction caused by the chemical or electrochemical action between the metal material and the environmental medium is called metal corrosion. Non-metallic materials such as engineering plastics, rubber and other environmental media will also have some effects and cause deterioration and destruction, generally called aging. Corrosion is not limited to metals, all kinds of polymers on an aircraft are damaged by chemicals and various liquid media on board. Stress cracking is when a polymer cracks on its surface under stress. The thinning of the ozone layer accelerates the degradation of polymers and coatings by ultraviolet rays.

Corrosion forms of aerospace materials are various, generally including the following: a. Concentration battery corrosion or gap corrosion; b. Fretting corrosion;

c. Corrosion of galvanic battery; d. Intergranular corrosion; e. Surface corrósion or uniform corrosion; f. Linear corrosion; g. Atmospheric corrosion; h. Stress corrosion; i. Denudation; j. Microbial corrosion. Therefore, all kinds of new, high-performance protective coatings are the basic measures to prevent the corrosion of metal materials and delay the aging of composite materials, so as to ensure the safe flight of aircraft and extend the life of aircraft.

The composition design of nano-coating materials is related to the composition, structure and form of the materials. To obtain rational surface coating properties and functions, it depends on the form of the nanoparticles present in the coating, the number, size (or size), shape and distribution of the particles, as well as the performance of the particles, the role of the particles, and the effects of the particles and other materials.

China's aircraft skin coatings [20] began in the 1950s, and the alkyd coatings mainly based on C01-7 long oil alkyd coatings have been widely used. In the 1970s, China's aviation coatings began to develop. Zhonghao North paint Institute (former paint industry Research Institute of the Ministry of Chemical Industry), Beijing 621 Institute, Tianjin paint factory have invested a lot of manpower and material resources to research. Acrylic varnish, acrylic acid modified polyurethane enamel, polyurethane enamel and silicone modified polyurethane coating products have been successfully researched and widely used in the surface coating of new military aircraft. At present, the domestic is still to use epoxy primer and polyurethane (polyester, acrylic and various modified polyester) topcoat system of aircraft skin coatings as the main varieties, its technical performance to meet the GJB and MIL standards.

As shown in Fig. 8.17, there is still a certain gap between the overall technical level of domestic aircraft skin coating primer and top paint and similar foreign products, such as insufficient serialization, weak pertinency, narrow application range of products and poor process stability. Although there are also some applications of fluorosilicon and fluorocarbon skin coatings, their technical maturity is insufficient, and their performance such as stain resistance cannot meet the daily maintenance requirements of aircraft, and they cannot be mass-produced and applied, and are currently limited to small batch trials. With the continuous increase of aircraft speed, the airframe aerodynamic heat will increase, therefore, the common aircraft skin coating that can withstand the high temperature above 220 °C for a long time is an important development direction.

In general, organic coatings are mainly used for aircraft skin and other parts, while inorganic coatings are mainly used for rocket and aircraft engine parts and combustion chambers [21].

(1) Nano-coating materials for metals and alloys

Nano-coating materials of metals and alloys are usually used by electrolysis, reduction, spray and other methods to generate nano-powder of metal or alloy, and then as a separate metal (or alloy) coating, metal composite coating or metal matrix composite coating as required.

Fig. 8.17 Schematic diagram of aircraft skin coating layer [21]

Metal nanocoatings are mainly nickel, copper, iron, cobalt, etc., and based on these metals, add other elements such as aluminum, chromium, carbon, boron, silicon, phosphorus, tin, tungsten, etc., to form iron, cobalt, nickel, copper-based alloys. In addition, there is the use of metal or alloy as a matrix material, composite inorganic non-metallic material particles, through sintering, spraying or deposition methods to form a metal matrix composite coating, and the addition of metal powder to polymer materials to form a composite coating.

Ni, Ni-P, Ni-Zn, Ni-Cr-Cu and other metal or alloy nanocoatings can obviously improve the corrosion resistance and wear resistance of the materials. The nano-Fe-Ni-Cr composite coating also has good anti-stress corrosion cracking performance. Nano Fe, Co, Ni and WC, SiC_4 and Al_2O_3 can be combined to form super-hard, high wear resistance Fe-WC, CO-WC, Ni-WC, Ni-SiC_4, Ni-Al_2O_3 composite coating materials. Nano-metal aluminum powder is added to zinc sulfate galvanizing solution as dispersion particles to form zinc-aluminum composite coating, whose corrosion resistance is better than that of electrogalvanizing and hot-dip galvanizing [22].

(2) Nano-coating of ceramic materials

Nano-coating of inorganic nonmetallic materials and ceramic materials is the main part of nano-coating materials. It can take into account the advantages of metal, non-metal and composite materials, and play a very favorable role in the design and functional properties of composite materials, so the development speed is very fast. In aerospace materials, this kind of coating is widely used in high temperature resistance, corrosion resistance, oxidation resistance, wear resistance, high strength, electrical insulation and other key parts. Inorganic nonmetallic nanocoatings mainly include oxide coatings, non-oxide coatings and cermet composite coatings.

a. Nano oxide coating material

Nano-oxide coating materials have the advantages of high melting point, high temperature resistance and oxidation resistance, low thermal conductivity, wear resistance, high chemical stability, corrosion resistance, electrical insulation, etc., and have become the most widely used nano-coating materials. Commonly used oxide nano-coating materials are Al_2O_3, TiO_2, ZrO_2, Cr_2O_3, Ir_2O_3, SiO_2 and so on. A binary or multi-component composite coating can also be formed between the oxides. In aerospace materials, the main research and application of high temperature resistance, wear resistance, oxidation resistance, corrosion resistance of nano-oxide coatings are Al_2O_3, ZrO_2, Al_2O_3-TiO_2, ZrO_2-Ir_2O_3, ZrO_2-Al_2O_3 and other oxides and their composite coatings.

b. Non-oxide nano coating materials

The non-oxide coating mainly includes carbide coating, nitride coating, boride coating and so on. Non-oxide coatings have better mechanical properties, high temperature resistance and chemical erosion resistance than oxide coatings, but some non-oxide coatings have poor high temperature oxidation resistance. Commonly used non-carbide nanocoatings include SiC, WC, TiC_4, B_4C, Si_3N_4, TiB_2, etc. [23].

c. Cermet nanocomposite coating materials

In order to better play the function and role of nano-coating, nano-composite coating materials are often formed. Cermet nanocomposite coating is a good example. For example, adding WC to Fe, Co and Ni can form cemented carbide materials. Compared with traditional cemented carbide materials, nano cemented carbide coating materials have high hardness, wear resistance, and lower brittleness.

Nanometer cermet composite coating materials can be obtained by sintering, spraying and plating. They can be divided into two categories: cermet coating materials and coating materials that combine metal matrix and dispersed phase nanoparticles. Commonly used cermet include alumina series, silicon nitride series, boron nitride series, silicon carbide series, zirconia series, tungsten carbide series, carbon (nitride) series and so on. The coating materials combined with metal matrix and dispersed phase nanoparticles include Fe, Co, Ni, Cu, Zn, Cr, Pb, Au, Ag and other metals and their alloys as the matrix materials, and various composite coating materials with Al_2O_3, WC, Si_3N_4, BN, SiC and other nano-dispersed particles as the dispersion particles [24].

(3) Plastic and polymer nano composite coating materials

In aerospace coating materials, in addition to a lot of research on various types of inorganic coatings, nano coatings are also used in plastics and polymer coatings, by adding composite nano powder to the plastic or polymer material matrix to form plastic or polymer based nano coatings. For example, TiO_2 and SiO_2 nano-filling materials are added to the resin. With the solidification of the coating, the nanoparticles can strengthen and toughen [25].

The combination of organic polymer nano or ultra-micro rice particles with good affinity to coatings can not only enhance the bonding strength of coatings, but also improve the corrosion resistance of coatings. In general, tanks, armored vehicles, ships, aircraft and other surfaces should be coated with paint, especially the primer of these equipment is mainly made of neoprene, dioxy resin or epoxy resin as the main raw material. These resins and rubber-based polymers easily age and become brittle under ultraviolet radiation in sunlight, causing the coating to fall off. If the surface coating of these weapons and equipment can be added to absorb ultraviolet nanoparticles can play a good role in protecting the primer, thus greatly delaying the process of aging and brittleness of resin or rubber.

Plastic products are more and more widely used in the military, and the plastic surface is coated with a transparent coating of nanoparticles, which has a strong absorption effect on 300–400 nm wavelength ultraviolet light, so that it can obviously prevent the aging of plastics from becoming brittle.

The nanomaterial coating can improve the corrosion protection ability of the substrate and achieve the purpose of surface decoration. Adding nanoparticles to the coating can further improve its protective ability, which can resist atmospheric and ultraviolet damage, so as to achieve anti-degradation, anti-discoloration and other effects; In addition, nanomaterial coatings can also be used in sanitary ware, indoor space, utensils, etc., to produce sterilization and cleaning effects.

8.4.3 Nanometer Microwave Coating

(1) Microwave transparent material coating

Microwave wave transmittance materials refer to materials with a transmittance greater than 70% for electromagnetic waves with a wavelength of 11,000 mm and a frequency ranging from 0.3–500 Hz. Such materials can be used to make radomes, window materials for high-energy gyroscopes, microwave diagnosis and treatment, and microwave communication equipment [26].

Microwave wave-transmitting materials belong to the category of optical, electrical, magnetic and structure–function composite materials. They are functional materials integrating structure, heat protection and wave-transmitting. Among various radomes, missile radomes have the highest performance requirements. As shown in Fig. 8.18, In order to protect the antenna system in the missile to work normally under high temperature flight conditions and not affect the antenna frequency modulation radiation, the radome should not only have the wave transmission performance of coupling with the frequency used by the radar antenna of the aircraft, but also have the performance of withstand the aerodynamic load of the aircraft, the erosion of the environmental hot air flow and rain flow and the vibration impact of the load, and the supersonic millimeter wave transmission antenna The requirements are higher, and the integrity of the structure and the transmittance of high temperature and high frequency electromagnetic waves must be maintained under harsh conditions.

With the rapid development of electronic countermeasures, stealth materials and stealth technology, as well as the continuous improvement of aircraft and aircraft performance, modern war requires that guided weapons can adapt to various complex war conditions and harsh climatic conditions, and need to have strong anti-jamming and anti-stealth capabilities, good identification capabilities and high guidance accuracy, resulting in infrared/microwave, infrared/millimeter wave and other dual modes Development of guided weapons [27].

Fig. 8.18 Schematic diagram of foam absorbing materials [27]

(2) Microwave absorbing material coating

In contrast to the effect of a wave-transmitting material, a wave-absorbing material is a material that can reduce the radar cross section of a target through its own absorption. The basic principle is to convert radar waves into other forms of energy (such as mechanical energy, electrical energy and heat) and consume them. Through reasonable structure design, impedance matching design and proper forming process, the wave absorbing material can almost completely attenuate and absorb the incident electromagnetic wave energy. In view of the important position of absorbing materials in radar countermeasures, stealth, anti-stealth and other military technologies, the research related to it has increasingly become the focus of attention [28].

Absorbing materials are generally composed of matrix materials (or adhesives) and loss media composite, the current focus of research includes matrix materials, loss media and molding process design, in which the loss of media performance, quantity and matching selection is an important environment in the design of absorbing materials at present, has been developed and successfully applied in absorbing materials in the loss of media up to dozens of, and is still continuing Develop new varieties. According to the different absorption mechanism, the loss medium in the absorbing material can be divided into two categories: electrical loss type and magnetic loss type. The former includes a variety of conductive graphite powder, smoke toner, silicon carbide powder or silicon carbide fiber, special carbon fiber, carbon particle metal short fiber, barium titanate ceramics and various conductive polymer, etc. Its main feature is that it has a high electrical loss tangent value, relying on the medium The electron polarization, particle polarization, molecular polarization or interface polarization attenuation, absorption of electromagnetic waves. The latter includes a variety of iron carbon powder, carboxyl iron powder, ultrafine metal powder and nanomaterials, etc., with high magnetic tangent value, attenuation and absorption of electromagnetic waves rely on magnetic polarization mechanisms such as hysteresis loss, domain wall resonance and natural resonance, and after-effect loss.

Structural wave absorbing material is the main kind of wave absorbing material. Through a variety of special fibers, while improving the mechanical properties of materials, but also make it have a certain wave absorption performance, stealth and bearing dual functions, which is the main direction of the development of wave absorbing materials. In order to make submarines and warships not found by the enemy sonar system, it must be coated with a layer of materials that can absorb ultrasonic waves, so that naval facilities such as submarines and warships are covered with a "stealth cloak".

8.4.4 Nanometer Optical Coating

Nanocoatings have a wide range of optical properties, and their transmission spectrum can be extended from ultraviolet to far infrared. Nanometer multilayer composite coatings show fluorescence in the visible range after treatment, and are widely used in various optical applications, such as sensors and other devices [29]. The nanomaterial coating is applied on the surface of various signs to make them luminous and reflective signs. By changing the composition and characteristics of nano-coatings, photochromic, thermochromic and electrochromic effects can be obtained, and special anti-counterfeiting and identification means can be produced. 80 nm yttrium oxide can be used as an infrared shielding coating with high thermal reflection efficiency [30]. Coating the surface of glass and other products with nanomaterial coating can reduce the effect of light transmission and heat transfer, and produce heat insulation effect; Adding nanomaterials to coatings can play the role of flame retardant, heat insulation and fire prevention.

Full nanomaterial coatings are still a long way from commercialization and have only military applications. However, with the help of traditional coating technology, the addition of nanomaterials can make the function of the traditional coating be improved by a leap, and the technical cost does not need to increase too much. This nano-added composite system coating can soon go to the market to show a strong application momentum. Nanometer optical coatings play an extremely important role in military and aerospace fields, especially the application of nanometer optical and microwave composite functional coatings in stealth technology and other fields is the focus of competition among countries.

(1) Nanometer optical coating

The nanomaterial coating has a wide range of optical properties and its optical transmission spectrum can extend from ultraviolet band to far infrared band. Nanometer multilayer composite coatings are treated to fluoresce in the visible range and are used in a variety of optical applications, such as sensors. Nanometer material coating is applied on the surface of various signs to become luminous and reflective signs; The effect of photochromism, thermochromism and electrochromism can be obtained by changing the composition and characteristics of the nano-coating, which can produce special anti-counterfeiting and identification means. The 80 nm vanadium oxide can be used as an infrared shielding coating, which has a high efficiency of reflecting heat. Coating the surface of products such as glass with nanomaterial can achieve the effect of reducing light transmission and heat transfer, resulting in thermal insulation; The addition of nanomaterials to coatings can play a role in flame retardancy, heat insulation and fire prevention [29].

Because of the small size effect, nanoparticles have optical characteristics that conventional bulk materials do not have, such as optical nonlinearity, light absorption, light reflection, and energy loss in the process of light transmission, which are strongly dependent on the size of nanoparticles. The research shows that various optical materials prepared by the special optical properties of nanoparticles will be widely used in daily life and high-tech fields. At present, the research on this aspect is still in the laboratory stage, and some have been popularized and applied.

The membrane material of nano particles has a good application prospect in the light bulb industry. High-pressure sodium lamps and various iodine arc lamps used for photography and photography require strong lighting, but 69% of the electric energy is converted into infrared, which indicates that a considerable amount of electric energy is converted into heat energy is consumed, and only a small part is converted into light energy to illuminate. At the same time, lamp heating will also affect the life of the lamp. How to improve the luminous efficiency and increase the illumination has always been a key problem to be solved. The birth of nanoparticles provides a new way to solve this problem. Since the 1980s, nano-SiO_2 and nano-TiO_2 particles have been used to make multi-layer interference films [31], with a total thickness of micrometers, lined on the inner wall of the lamp cover with a filament, resulting in not only good light transmittance, but also strong infrared reflection ability. It is estimated that this bulb can save about 15% of the electricity when the brightness is the same as that of traditional halogen lamps.

(2) Nano infrared coating

Nanoparticles are mainly used to make thin films and multilayer films on infrared reflective materials. Structurally, the conductive film is the simplest, single-layer film, low cost metal-dielectric composite film and dielectric multilayer film are multi-layer films, the cost is slightly higher. In terms of performance, metal-dielectric composite film has the best infrared reflection performance, and the heat resistance is below 200 °C. Dielectric multilayers have good infrared reflectivity and can be used at very high temperatures (<900 °C). Although conductive film has good heat resistance, its infrared reflection performance is slightly poor [32].

Infrared absorbing materials have important application prospects in daily life and national defense. Some economically developed countries have begun to use fibers with infrared absorption function to make military uniforms for the use of troops, this fiber has a good shielding effect on the infrared released by the human body. If the infrared radiation in this frequency band is not shielded, it is easy to be found by very sensitive mid-infrared detectors, especially at night, personal safety will be threatened. In this sense, it is necessary to develop clothes with infrared shielding for the human body [33]. Nanoparticles are small, easily filled into the fiber, do not block the nozzle when pulling the fiber, and some nanoparticles have a strong absorption of mid-infrared frequency bands. The composite powder of nano-Al_2O_3, nano-TiO_2, nano-SiO_2 and nano-Fe_2O_3 has this function. Nano-added fiber also has a characteristic, that is, it has a strong absorption effect on the human body infrared ray, which can increase the thermal effect, reduce the quality of clothing, some estimates, with the addition of infrared absorption nano powder fiber made of clothing, its quality can be reduced by 30%.

(3) Nanometer UV protective coating

The quantum size effect of nanoparticles makes them blue shift in the absorption of certain wavelengths of light. The absorption of nanoparticle powder to various wavelengths of light has a broadening phenomenon. Ultraviolet-absorbing materials for nanoparticles take advantage of these two properties. Generally, the nano-particle ultraviolet absorption material is the dispersion of nano-particles into the resin to make a film, this film has the absorption capacity of ultraviolet light, this absorption capacity depends on the size of the nanoparticles and the amount and composition of the nanoparticles in the resin. At present, several materials with good UV absorption capacity are: 30–40 nm TiO_2 nanoparticle resin film; Polysterol resin film with Fe_2O_3 nanoparticles [34]. The former has a strong absorption capacity for ultraviolet light below 400 nm wavelength, and the latter has a good absorption capacity for light below 600 nm, and can be used as an ultraviolet filter for semiconductor devices.

It has recently been found that nano-Al_2O_3 has a strong absorption capacity for ultraviolet light below 250 nm, which can be used to improve the service life of daylight lamps. We know that daylight lamps use the ultraviolet spectrum line of mercury to excite the phosphor on the wall of the lamp tube, resulting in high brightness lighting. In general,185 nm short-wave ultraviolet light has an impact on the life of the lamp, and the ultraviolet leakage of the lamp is harmful to the human body, which has been a key problem plaguing the solar lamp industry. If the particle size of a few nanometers of Al_2O_3 powder mixed into rare earth phosphor, the use of nano ultraviolet absorption of blue shift phenomenon is possible to absorb this harmful ultraviolet light, and does not reduce the luminous efficiency of the phosphor, in this area of the experimental work is being carried out. At present, there are many examples of using nanoparticles in combination with resins for ultraviolet absorption. For example, sunscreen and cosmetics are commonly used with nanoparticles. We know that the ultraviolet light in the atmosphere is mainly in the 300–400 nm band, and the ultraviolet light that the sun has harm to the human body is also in this band. Sunscreen and cosmetics are used to select nanoparticles that have a strong absorption of this band. Recent studies have shown that nano-TiO_2, nano-ZnO, nano-SiO_2, nano-Al_2O_3, nano-mica, etc. have the characteristics of absorbing ultraviolet light in this band. It also needs to be emphasized here that the particle size should not be too small when the nano is added, otherwise it will block the sweat pores, which is not conducive to health. If the particle size is too large, the ultraviolet absorption will deviate from this band. In order to solve this problem, nanoparticles with strong ultraviolet absorption should be coated with a layer of harmless polymers [35], and this complex should be added to sunscreen and cosmetics, which both plays the role of nanoparticles and improves the performance of sunscreen. Plastic products are easy to age and become brittle under ultraviolet irradiation, if the plastic surface is coated with a transparent coating containing nanoparticles, this coating has a strong ultraviolet absorption performance in the range of 300–400 nm, so that the aging of plastics can be prevented. The surface of cars and ships need to be coated with paint, especially the primer is mainly made of neoprene rubber, double grease or epoxy resin as the main raw material, these resins and rubber polymers are easy to age and become brittle under the ultraviolet radiation of the sun, resulting in the coating falling off, if the top paint can be added to strongly absorb ultraviolet nanoparticles can play a role in protecting the primer. Therefore, it is also very important to study coatings with ultraviolet absorption by adding nanoparticles.

(4) Nano stealth materials

The word "stealth", by definition, means concealment. In recent years, with the development of science and technology, various detection methods are becoming more and more advanced. For example, aircraft can be detected by electromagnetic wave emitted by radar; An infrared detector can detect objects emitting infrared light. At present, in order to meet the needs of modern war to improve the strength of military confrontation, countries around the world also take stealth technology as an important research object, among which stealth materials play an important role in stealth technology [36]. Nanometer stealth coating is a kind of composite structure coating which includes nanometer microwave coating, nanometer optical coating and infrared coating.

Stealth technology is one of the high and new technologies in the contemporary military field, which is called the three revolutions of modern military technology together with laser and cruise missile, so it is paid attention by all countries in the world. The current stealth technology mainly includes anti-sonar detection technology, anti-radar detection technology, anti-optical detection technology and anti-infrared detection technology. As shown in Fig. 8.19, In the Gulf War, the first day of the United States fighter jets eluded Iraq's strict radar surveillance network, quickly reached the capital of Baghdad, directly destroyed the telegraph building and other military targets, in the 42 days of fighting, the mission of the aircraft reached 1270, so that 95% of the Iraqi army's important military targets were destroyed, while the United States fighter jets were not damaged. This high-tech war once shocked the world. Why is Iraq's radar defense system helpless against American fighter jets? Why did American missiles hit Iraqi military targets so accurately? Why do air-to-surface missiles have such a high hit rate on Iraqi tanks? An important reason is that the surface of the F117A fighter body is covered with infrared and microwave stealth materials, which has excellent broadband microwave absorption capacity and can evade radar surveillance. Iraqi military targets and weapons such as tanks did not have stealth materials to protect against infrared detection, and were easily detected by sensitive infrared detectors on American fighters and accurately hit by advanced laser-guided weapons [37].

The stealth material on the skin of the F117A aircraft of the United States contains a variety of ultrafine particles, which have a strong absorption ability of electromagnetic waves of different bands. Why do ultrafine particles, especially nanoparticles,

Fig. 8.19 F117A Stealth fighter of the United States [37]

have stealth effects on infrared and electromagnetic waves? There are two main reasons: on the one hand, because the size of the nanoparticle is much smaller than the wavelength of infrared and radar, the transmittance of the nanoparticle material to this wave is much stronger than that of conventional materials, which greatly reduces the reflectivity of the wave, making the reflected signal received by the infrared detector and radar become very weak, so as to achieve the role of stealth; On the other hand, the specific surface area of nano-micro materials is 3–4 orders of magnitude larger than that of conventional coarse powder, and the absorption rate of infrared light and electromagnetic waves is much larger than that of conventional materials, which makes the reflected signal strength of infrared detectors and radar greatly reduced, so it is very accurate to find the detected target and play a stealth role.

At present, although stealth materials have broad application prospects in many aspects, the stealth materials that really play a role at present are mostly used in aerospace and other fields that are closely related to the military. There is a requirement for aerospace materials to be light in weight, and nanomaterials have obvious advantages in this regard, especially nanomaterials composed of light elements are widely used in aviation stealth materials. There are several nanoparticles that are likely to play a role in stealth materials, such as nano-alumina, iron oxide, silicon oxide and titanium oxide composite powder combined with polymer fiber has a strong absorption performance in the middle infrared band, and this complex has a good shielding effect on the infrared detector in this band. Nano-magnetic materials, especially ferrite-like nano-magnetic materials, are put into coatings, which have not only excellent wave-absorbing properties, but also good performance of absorbing and dissipating infrared rays, coupled with relatively low density, and have obvious advantages in the application of stealth [22]. In addition, this material can also cooperate with the signal control device in the cockpit, emit interference through the switch, change the reflected signal of the radar wave, make the waveform distortion, or make the waveform change, can effectively interfere with and confuse the radar operator, to achieve the purpose of stealth. Nanoscale borides and carbides, including nanofibers and carbon nanotubes, will also be useful in the application of stealth materials.

8.4.5 *Other Nano Functional Coatings*

Nano-composite wear-resistant coating is a commonly used coating material in aerospace and weapons equipment. Carbon can obtain more advanced heat-resistant materials and superhard materials by transforming metals into nano-metal carbides. Nano iron carbide particles are commonly used as coatings on cemented carbide or steel weapons to improve the surface hardness, wear resistance and corrosion resistance of weapons and equipment, and are ideal coating materials for aircraft, missiles and rockets [38].

The nano-composite coating has excellent electromagnetic properties. Nano titanium oxide, chromium oxide, iron oxide, zinc oxide and other particles with semiconductor properties are added to the resin to form a coating, which has good electrostatic shielding performance; $BaTiO_3$ of 80 nm can be used as high dielectric insulation coating, and Fe_3O_4 of 40 nm can be used as magnetic coating. Nanostructured multilayer film systems produce giant magnetoresistive effects and are expected to be used as read heads in memory systems [39]. The chemical activity of nano powder lays the foundation for its widespread use as a catalyst, using the surface catalytic sensitivity of nano powder as an electrode material to accelerate chemical reactions and improve reaction efficiency, such as the formation of nanocrystalline porous titanium dioxide electrode coating. Nano powder coating can also be used for oil and gas cracking and various chemical reactions, can be used as temperature, gas, humidity, environment and other sensitive devices, can also be used for sewage treatment, automobile exhaust purification, etc.

Nanomaterial coatings and their technology develop with the development of nanomaterials, and the characteristics and potential functions of surface coatings mainly rely on nanomaterials to create and improve on the basis of continuous progress in the preparation and synthesis technology of nanomaterials and the improvement of basic theories, nanomaterial functional coatings will have faster development, and applications will cover many fields.

Nano-coating materials have great potential in aerospace applications, and their advantages in friction reduction and wear protection, corrosion and oxidation resistance, thermal management and thermal radiation protection, electromagnetic shielding and electrostatic protection provide new solutions for aircraft performance improvement and reliability [40]. With the continuous development of nanotechnology, we can expect a wider application of nano-coating materials in the aerospace field.

8.5 Development Prospects of Nanomaterials in the Aerospace Field

Nanoscience and technology are on the eve of a major breakthrough, it has achieved a series of results caused concern about the future world development of scientists to think, people are watching the field of nanoscience and technology continue to emerge strange phenomena and new progress. By using the unique physical and chemical properties of nanomaterials, new materials with high performance can be developed to meet the special requirements of material properties in the aerospace field. At present, nanomaterial technology has been applied to radar and infrared stealth materials, solid rocket propellants, structural materials, special sealing materials, coating materials and sensors. However, whether in theory or in practice, there are still many problems in nanotechnology, which need further exploration and improvement.

We believe that with the development of nanotechnology, nanomaterials will be more and more applied in the field of aerospace, and nanotechnology will make great contributions to the development of China's aerospace industry. Nanotechnology whether in theory or in the actual use of the process, there are still many problems, need our further exploration and improvement, we believe that with the development of nanotechnology, nanomaterials in the field of space will be more and more application, nanotechnology will promote the development of China's space industry to make a major contribution.

References

1. X. Y. Wang, X. C. Lu, Y. Ren, *Nucl. Phys. B.* **2012**, *192*, 470–474.
2. S. M. Douglas, I. Bachelet, *Science.* **2012**, *335*, 831–834.
3. Sultana, Nazima, Raul, Prasanta, K, Goswami, Diganta, Das, Bodhaditya, *Environ. Chem. Lett.* **2018**, *16*, 123–125.
4. X. U. Wei, *J. Power Sources,* **2009**, *33*, 1017–1018.
5. O. R. Polo, S. Esteban, L. Cercos, P. Parra, *Acta Astronaut*, **2014**, *93*, 94–105.
6. M. Kozako, H. Nakatsu, H. Toda, S. Yamada, M. Takei, *IEEJ Trans. Fundam. Mater.* ***2012**, 132,* 773–779.
7. L. Long, F. Jie, L. Shuang, W. Xiaoliang, M. Cha, L. Yanqin, *New J. Chem.* **2013**, *41*, 187–190.
8. M. Jie, S. Shu-Qing, L. I. Ting, M. A. Xiang-Hui, F. Jun-Gang, *Chem. Eng. J.* **2007**, *51, 4682–4691.*
9. R. Li, L. Qiu, Z. Meng, S. Wang, *Optik.* **2019**, *180*, 894–899.
10. Y. Guo, C. Bao, L. Song, B. Yuan, Y. Hu, *Ind. Eng. Chem. Res.* **2011**, *50*, 7772–7783.
11. J. Li-Ying, Y. Fei-Fei, R. Jing-Ying, Z. Fa-Quan, H. E. Zhen-Dong, Guang-Zhao, *Microsyst. Technol.* **2009**, *69,* 125–128.
12. F. Maglia, I. G. Tredici, U. Anselmi-Tamburini, *J. Eur. Ceram. Soc.* **2013**, *33*, 1045–1066.
13. H. Y. Alfarra, M. Omar, *Int. J. Biol. Sci.* **2013**, *3*, 49–67.
14. E. Tylianakis, K. Georgios, D. Santiago, M. A. Jose, D. E. George, *ChemComm.* **2011,** *47,* 2303–2305.
15. M. Brahmakumar, C. Pavithran, R. M. Pillai, *Compos Sci Technol.* **2005**, *65*, 563–569.
16. T. Tsuda, H. Hayashi, T. Yamamoto, A. Abe, *J. Mech. Eng.* **2012**, *6*, 711–720.
17. O. Yilmaz, N. Gindy, J. Gao, *Robot. Comput. Integr. Manuf.* **2010**, *26*, 190–201.

18. X. T. Huang, G. Cai, *J. Hazard. Mater.* **2010**, *26–28*, 843–846.
19. T. Liu, F. Zhang, C. Xue, L. Li, Y. Yin, *Surf. Coat. Technol.* **2010**, *205*, 2335–2339.
20. G. Zhu, S. Wang, W. Cheng, Y. Ren, D. Wen, *Ocean. Sci.* **2020**, *132*, 106475.
21. S. Li, Y. Huang, L. Ding, M. Zong, N. Zhang, X. Qin, P. Liu, *J. Magn. Magn. Mater.* **2019**, *486*, 165–259.
22. X. Hong, W. Guo, H. Yuan, J. Li, Y. Liu, L. Ma, Y. Bai, T. Li, *J. Magn. Magn. Mater.* **2004**, *269*, 95–100.
23. Chen, Xun, Pease, R., F., & W. *J Vac Sci Technol B.* **1996**, *14*, 3980–3984.
24. Ramezanzadeh B, Attar M M. *Prog. Org. Coat.* **2011**, *71*, 242–249.
25. Poyraz. S, Zhang. L, Schroder. A, Zhang. X. ACS Appl. *Mater. Interfaces.* **2015**, *7*, 22469.
26. C. Zhou, Q. Fang, F. Yan, W. Wang, K. Wu, Y. Liu, Q. Lv, H. Zhang, Q. Zhang, J. Li, *J. Magn. Magn. Mater.* **2012**, *324*, 1720–1725.
27. M. Yunus, M. S. Alsoufi, *High Temp. Mater. Process.* **2017**, *21*, 37–52.
28. U. R. Lima, M. C. Nasar, R. S. Nasar, M. C. Rezende, J. H. Araújo, *J. Magn. Magn. Mater.* **2008**, *320*, 1666–1670.
29. L. I. Jian, L. Jiayi, Z. Yangde, *Chin J Optom. Ophthalmol Vis. Sci.* **2010**, *12*, 214–217.
30. E. Y. Sun, P. F. Becher, K. P. Plucknett, C. H. Hsueh, M. E. Brito, *J. Am. Ceram. Soc.* **2005**, *81*, 2831–2840.
31. S. Bruynooghe, D. Tonova, M. Sundermann, T. Koch, U. Schulz, *Surf. Coat. Technol.* **2015**, *267*, 40–44.
32. Q. Qin, H. Lu, S. N. Zhu, C. S. Yuan, Y. Y. Zhu, N. B. Ming, *Appl. Phys. Lett.* **2003**, *82*, 4654–4656.
33. K. R. Rodriguez, H. Tian, J. M. Heer, J. V. Coe, *J. Phys. Chem. C.* **2007**, *111*, 12106–12111.
34. T. Sun, H. Fan, Z. Wang, X. Liu, Z. Wu, *Mater. Des.* **2015**, *87*, 10–16.
35. S. R. Allahkaram, S. Golroh, M. Mohammadalipour, *Mater. Des.* **2011**, *32*, 4478–4484.
36. Q. Tong, X. G. Shu, X. M. Lu, W. Y. Li, G. B. Wang, ***Acta Anat. Sin.*** **2009**, *40*, 7–12.
37. X. Wei, H. C. Yin, P. K. Huang, Z. H. Xiao, G. T. Yin, *J. Syst. Eng. Electron.***2006**, *28*, 492–494.
38. M. Lekka, N. Kouloumbi, M. Gajo, P. L. Bonora, *Electrochim. Acta.* **2005**, *50*, 4551–4556.
39. L. H. Chen, S. Jin, T. H. Tiefel, R. Ramesh, *Appl. Phys. Lett.* **1994**, *64*, 1039–1041.
40. S. R. Taylor, K. Sieradzki, *Prog. Org. Coat.* **2003**, *47*, 169–173.

Chapter 9
Applications of Nanomaterials and Nanotechnology in Electronic Information

9.1 Nanomagnetic Materials

Magnetic materials have been the national economy, national defense industry is an important pillar and the basis of a wide range of applications, especially in the information storage, processing and transmission has become an indispensable part of a wide range of telecommunications, automation, communications, household appliances and other fields, in the microcomputer, mainframe computers in the application of an important position. The general trend of the development of information technology is to small, light, thin and multi-functional direction. Magnetic materials have gone through the crystalline, amorphous, nano-microcrystalline, nanoparticles and nanostructured materials development stage. Nowadays, the importance of nanomagnetic materials in the field of information technology has been increasingly shown. The properties of nanomagnetic materials are related to the size of the particles, grains or thickness of the film, and when their scales are comparable to the size of magnetic single domains, the critical size of superparamagnetism, the length of exchange interactions, and the physical length of electrons' mean free travel, they tend to show new properties different from those of traditional magnetic materials. Therefore, it is very important to control the particle size or film thickness when preparing nanomagnetic materials.

9.1.1 Magnetic Properties of Nano-materials

The small size effect of the magnetic performance of nano particles, quantum size effects, surface effects, etc. make it have the magnetic specialty that conventional coarse grain materials do not have. The main magnetic characteristics of nano particles can be summarized as follows.

S. Peng and P. Li, *Nanomaterials and Nanotechnology*,
https://doi.org/10.1007/978-981-97-8433-2_9

1. Superparamagnetism

When the size of nanoparticles is small enough to reach a certain critical value, the nanoparticles enter into superparamagnetic state, for example, α-Fe, Fe_3O_4 and α-Fe_2O_3 become paramagnetic when the particle sizes are 5 nm, 16 nm and 20 nm, respectively.

The origin of the superparamagnetic state can be attributed to the following reasons: at small sizes, when the anisotropic energy is reduced to a level comparable to that of the thermal motion energy, the magnetization direction is no longer fixed in an easy magnetization direction, and the easy magnetization direction changes irregularly, which leads to the appearance of superparamagnetism. The critical size of superparamagnetism is different for different kinds of magnetic nanoparticles.

2. Coercivity

Nanoparticle sizes above the critical size for superparamagnetism usually exhibit high coercivity H_c. For example, Fe nanoparticles prepared by inert gas evaporation and condensation show a decrease in saturation magnetization strength M_s as the particles become smaller, but a significant increase in coercivity.

There are two explanations for the origin of the high coercivity of nanoparticles: the consistent rotation mode and the spherical chain reversal magnetization mode. The basic content of the consistent rotational magnetization mode is that when the particle size is small enough to a certain size, each particle is a single magnetic domain, for example, for Fe and Fe_3O_4 single magnetic domains, the critical sizes are 12 nm and 40 nm, respectively. Each single domain nanometer particle actually becomes a permanent magnet. To remove the magnetism of this magnet, the magnetic moment of each particle must be reversed as a whole, which requires a large reverse magnetic field, that is, a high coercivity. Another explanation is that spherical nanoparticles form chains due to magnetostatic effect, which results in high coercivity of nanoparticles.

3. Curie temperature

The curie temperature T_c is an important parameter of the magnetic properties of a substance, usually proportional to the exchange integral J_e and related to the atomic configuration and spacing. In the case of thin films, theoretical and experimental studies have shown that the curie temperature decreases as the thickness of the ferromagnetic film decreases. For nanoparticles, the intrinsic magnetic properties of nanoparticles change due to small size and surface effects, resulting in lower curie temperatures. For example, for Ni particles with a particle size of 85 nm, the curie temperature is about 623 K, which is slightly lower than that of the conventional bulk Ni (631 K) due to the obvious peak of the magnetization at the curie temperature, as measured by the magnetization-temperature relationship in a low magnetic field.

4. Magnetization rate

The magnetic properties of a nanoparticle are closely related to the parity of the total number of electrons it contains. The electrons of each particle can be viewed as a system, and the number of electrons can be odd or even. One-valent metal micropowder, half of the particles for odd, the other half for even, two-valent metal particles for even, the number of electrons for the odd or even number of particles with different temperature characteristics of magnetism. The magnetization rate of the particle aggregate with an odd number of electrons obeys the Curie–Weiss law, and the quantum size effect makes the magnetization rate obey the d^{-3} law, while the system with an even number of electrons, $\chi \propto k_B T$, and obeys the d^2 law. They are paramagnetic at high fields. The value of χ of nanomagnetic metals is 20 times higher than that of conventional metals.

9.1.2 Giant Magnetoresistive Materials

Magnetic metals and alloys are generally characterized by the phenomena of magnetoresistance and giant magnetoresistance. The so-called magnetoresistance refers to the phenomenon that the resistance of a material changes under a certain magnetic field. The so-called giant magnetoresistance refers to a certain magnetic field under the material resistance decreases dramatically, the general magnitude of the decrease than the usual magnetic metals and alloys magnetoresistance value is about 10 times higher. The giant magnetoresistance effect is a new phenomenon discovered in the last 10 years. The discovery of the giant magnetoresistance effect is attributed to the development of nanotechnology in the 1970s and 1980s. Ultra-thin metal crystals or polycrystalline films with a thickness of a few atomic layers can be prepared by molecular beam epitaxy (MBE) or sputtering in a vacuum environment. Compared with bulk materials, these ultrathin films show many novel physical phenomena. 1986, scientists at the Jülich Research Center in Germany, Greenberg et al. used MBE technology to discover that in a "non-magnetic/ferromagnetic (Fe/Cr/Fe) sandwich thin film structure", when the thickness of non-magnetic thin film Cr is about 0.9 nm, the thickness of the thin film can be reduced to 0.9 nm, and then the thickness of the non-magnetic thin film Cr can be increased to 0.9 nm. 0.9 nm, through the medium of the Cr layer, two layers of ferromagnetic film layer will appear between the "anti-ferromagnetic exchange coupling" effect, that is, the two layers of ferromagnetic film between the performance of anti-parallel magnetization state. Later, this research result was recognized as the key first step in the discovery of the GMR effect.

In 1988, the research team led by Professor Phil of the University of Paris in France and Professor Greenberg of the Ulrich Research Center in Germany independently found that in the Fe/Cr metal Superlattice multilayers prepared by the MBE method, the application of an external magnetic field can lead to a sharp change in the size of the electrical resistance of the structure. The role of the external magnetic field is to change the relative orientation of the magnetic moment direction between adjacent magnetic layers, this causes a change in the resistance value, which is called the "giant magnetoresistance effect". The Greenberg group studied the Fe/Cr/Fe three-layer film structure and obtained a giant magnetoresistance signal of about 6% [1].

The above research results are considered two groundbreaking works in discovering the GMR effect. The discovery of the GMR effect quickly attracted the attention of the scientific community, but due to the traditional view that magnetism has not contributed much to conductivity for a long time, many scholars internationally still have doubts about the GMR effect one year after the research results were published. With the deepening of research work, it has been found that except for the Fe/Cr system, the GMR effect can be observed in many systems where Fe, Co, Ni and their alloys are used as ferromagnetic layers, and Cu, Ag, Au or Cr is used as non magnetic layers. Only then did people reach a consensus on the generation of the GMR effect.

In the subsequent research, IBM's Parkin and others made outstanding contributions. First of all, they did not follow the MBE technology to grow thin films, but used a simple, fast magnetron sputtering method to prepare GMR multilayer thin film structure and achieved success, so that the film preparation process from the original several days to a few hours to complete, which is crucial in the subsequent practical application of the study. Meanwhile, they also investigated the coupled GMR effects of Fe/Cr, Co/Cu and Co/Ru in detail, and found that the magnetoresistance showed periodic oscillations with the thickness of the nonmagnetic layer. The results show that, for the $(Co/Cu)_n$ multilayer film system, in the absence of an external magnetic field, the magnetic moment orientations of the two neighboring magnetic layers alternate between parallel and antiparallel states with the thickness of the nonmagnetic layer (Cu), and when the thickness of the Cu layer increases to the extent that the interlayer coupling strength is not strong enough to modulate the orientation of magnetic moments between the neighboring layers, the disorder of the magnetic structure within the neighboring magnetic layers increases, which can lead to the disappearance of the GMR signals. This can lead to the disappearance of the GMR signal. Because of this, this type of giant magnetoresistance effect is a coupled GMR effect. Accordingly, Greenberg, Phil and Parkin won the Hewlett Packard European Physics Prize in 1997.

9.1.2.1 Applications of the Giant Magnetoresistance Effect

1. Application of GMR effect to computer hard disk drives (HDD)

Due to its large capacity, small size, fast read/write speed, and high data transfer rate, HDD remains the preferred device for computer external storage and the main support for various operating systems and application software. In order to further expand capacity and meet the requirements of computer network and multimedia computer development, IBM first used GMR effect spin valve structure readout heads in HDDs in 1994, achieving 1 billion bits per square inch (1 Gb/mch^2) The world record for HDD surface density. In 1995, IBM announced the production of 3 billion bits per square inch (3 Gb/mch^2) The GMR SV readout head has once again set a world record. Due to the enormous potential of GMR magnetic heads in improving HDD surface density, world records have been repeatedly broken, and by 1996, the world record for HDD surface density had reached 5 billion bits per square inch (50 Gb/mch^2). The National Storage Industry Association (NSFC) of the United States and the Information Storage Association of Japan (SRC) have both organized a consortium of government, industry, academia, and research to achieve HDD area density targets of 10–20 billion bits per square inch around 2000. By then, the capacity of HDDs will reach over 100–200 billion bytes. At present, the breakthrough in HDD capacity caused by the GMR effect is still moving towards a deeper level, and basic research has begun to achieve 400 billion or even 1 trillion bits per square inch.

With the rapid growth of surface density, the area occupied by each information bit on the recording medium will rapidly decrease. Taking the surface density of 1 billion bits per square inch (1 Gb/mch^2) as an example, each area has entered around 0.5 um^2, and by the beginning of the century around 2000, it had reached 10 Gb/mch^2, with each area decreasing to around 0.05 um^2. The decrease in area means that the remaining magnetic channels per unit of information are weakened, and the signal of using thin film induction (TFI) magnetic migration is very small, making it impossible to guarantee the reliability of HDD [2]. Therefore, improving the readout sensitivity of the magnetic head has become very important. When the surface density exceeds 0.6 Gb/inch^2 or more, TFI magnetic heads will be replaced by MR magnetic heads; When the surface density exceeds 3 Gb/inch^2, the MR head will be replaced by the GMR head.

2. Giant magnetoresistive readout head

GMR magnetic materials were first commercialized as readout heads for computer hard disks. GMR magnetic materials were first commercialized as readout heads for computer hard disks. Figure 9.1a shows a simple model of the GMR readout head. When the residual magnetic field on the recording medium is applied to the head, the direction of the magnetization intensity of the free layer of the spin valve changes, and the change in resistance is read out through the current of the head.

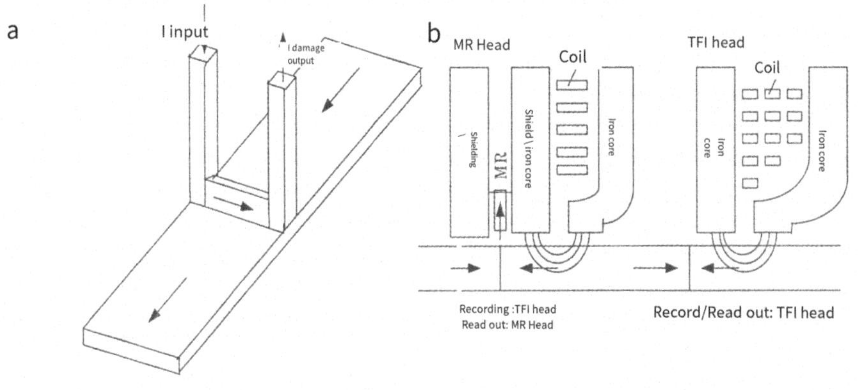

Fig. 9.1 **a** Simple model of GMR readout head, **b** schematic structure of TFI head and MR head

Traditional thin-film inductive (TFI) heads use the principle of electromagnetic induction to record and read out information. When recording, the current through the coil generates a recording magnetic field at the front gap of the core, magnetizing the recording media to store the bit information. Readout is on the recording media, representing the bit information of the residual magnetic flux, through the core of the magnetic head through the coil induction readout voltage, the magnetic head using thin film materials similar to semiconductor integrated circuits manufactured using the production process. Since the readout voltage of TFI is proportional to the number of coils and the rate of change of magnetic flux, in order to detect the weak residual magnetic flux corresponding to high-density bit information, it is necessary to increase the number of coils and to increase the linear speed of the hard disk. The increase in the number of coils not only increases the difficulty of the manufacturing process, but also causes the growth of the head impedance, which is detrimental to the improvement of the magnetic recording density, and with the miniaturization of the hard disk, hard disk line speed decreases accordingly, and the amplitude of the readout signal will become smaller. In addition, in the design, TFI head read and write process using the same size of the front gap, it is impossible to achieve the best read and write efficiency. Therefore, this type of head is limited in improving readout sensitivity.

MR head (shown in Fig. 9.1b) is a composite of TFI head and MR readout head. TFI head is only used as a recording head, no need to increase the number of coils, not only to simplify the process, but also conducive to the improvement of the efficiency of the recording. MR head is used as a readout head, the use of the MR element in the read shielding between the readout of the residual flux on the recording media in order to obtain the optimal sensitivity of the readout. Although MR heads are the best choice for HDD heads after TFI heads, MR heads are gradually exposing more and more shortcomings as HDD capacity continues to increase, and GMR heads are the key technology for HDD capacity to jump to 100 billion bytes (100 GB).

American scientists Parkin using conventional vacuum sputtering film production methods, found by a variety of ferromagnetic layer and non-magnetic layer (including 3d, 4d and 5d non-magnetic metal) alternate growth and the composition of the magnetic multilayer film, most of them have the GMR effect, aroused the application of the scientific community's great interest, and the first to obtain the application of the magnetic head in the HDD.

GMR head and MR head in the structure of the difference is not very important, the most important change is to use GMR components instead of AMR components, in order to achieve a greater rate of change of magnetoresistance, and to overcome the inherent Barkhausen noise of the MR head, increasing the sensitivity and reliability of the head. Currently, the low-field GMR SV material magnetoresistance rate of change of more than 7%, much higher than the MR head of 2%, so the GMR SV head readout voltage than the MR head is 3–5 times larger, greatly improving the readout sensitivity of the head. GMR SV film consists of antiferromagnetic layer (Fe, Mn, etc.), magnetic pinned layer, non-magnetic spacer and magnetic free layer. When the magnetization direction between the two magnetic layers is opposite, it shows high impedance, and when it is the same, it is low impedance. When the residual magnetic field of the information bits on the recording media acts on the free layer, it produces a change of 0–180 degrees, causing a change in the magnetoresistance of the GMR element, thus reading out the information stored on the recording media. Since the free layer is rotating when it changes the direction of magnetization due to the external field, it overcomes the Barkhausen noise and helps to overcome the signal-to-noise ratio of the magnetic head.

IBM has been the dominant international production of hard disks and heads, Japan's hard disk and head technology is also at the forefront of the world, while China's research in this field started late.

3. GMR Random Memory (RAM)

Currently, the widely used RAM are semiconductor dynamic memory (DRAM) and static memory (SRAM). The capacity of DRAM has been reported to be 4 billion bits per chip (4 Gbit/chip), with a chip area of approximately 1000 mm^2, each occupying an area of 0.23 um^2, and a fetch speed of 15–20 ns. The capacity of SRAM has reached 4 million bits per chip (4 Mbit/chip), with a chip area of 132 mm^2 and an area of 12.77 um^2 per chip. The speed is one order of magnitude faster than DRAM, reaching 2 ns. However, both DRAM and SRAM are volatile, with poor radiation resistance, which brings great inconvenience to use. Semiconductor nonvolatile memory, such as Erasable Programmable Read Only Memory (EPROM), currently has a small capacity of 512 K^{-1} megabytes, a chip area of 29–86 mm^2, and each occupies a size of 2.9 μm^2, with poor radiation resistance and high production costs. Therefore, developing non volatile RAM with small size, fast speed, large capacity, and low production cost is crucial for promoting the development of computers.

The use of magnetoresistance effect to form MRAM was reported in the 1970s, but at that time, AMR material was used, with low magnetoresistance change rate and small output signal, making it difficult to make MRAM. After the discovery of the GMR effect, it undoubtedly gave new impetus to the development of MRAM. Compared with existing semiconductor RAM, MRAM has the greatest advantages of non volatility, radiation resistance, long lifespan, and low cost. Due to the use of GMR materials, the reduction of each size does not affect the sensitivity of the readout signal, achieving maximum storage density and a simple structure, reducing the production process steps. At the beginning of this century, it was possible to achieve a storage density of 10 $GBIT/cm^2$ and an access speed of 0.5 ns. Therefore, the adoption of MRAM will be a revolution in computer memory storage chips. With new progress in TMR research, making MRAM using TMR will achieve better performance.

MRAM in the defense and civil industry have extremely important value and application prospects, such as computer BIOS chips, cellular phones, fax machines, solid-state video recorders, PDAs and large-capacity electronic memory have good application prospects. Especially radiation resistance, in the military and aerospace applications are of great significance.

Magnetic sensors are mainly used to detect the presence, strength, direction and change of magnetic fields. There are many types of magnetic sensors, such as induction coils, flux gates, Hall effect devices, superconducting quantum interference, and magnetoresistance. In addition to the direct measurement of magnetic field, magnetic sensors also provide a wide range of solutions for the measurement of other physical quantities, such as electric current, linear displacement, linear velocity, angular displacement, angular velocity, and so on, and thus magnetic sensors have a wide range of applications.

Application of the previously mentioned AMR material made of sensors, with a small size, high sensitivity, low impedance, resistance to harsh environments, low production costs, etc. However, due to the low change of AMR magnetoresistivity, the detection of weak magnetic field is still subject to great limitations, the use of GMR and TMR effect sensors, the magnetoresistive sensors to expand the scope of measurement and application.

Today, in household appliances, automobiles and automatic control involving angular velocity, rotational speed, acceleration, displacement and other physical quantities, can be made using GMR magnetic sensors to make high sensitivity, high resolution of the magnetic sensor device to control. Because the magnetic sensor sensitivity is not affected by the speed of change of the object, coupled with the advantages of resistance to harsh environments and long life, so that it has a great competitiveness in various types of motion sensors, such as video recorders, precision pitch MR sensors (FPM), in the automotive industry to monitor the rotational speed of magnetic sensors to change the brakes of the car's braking system (ABS), and so on [3]. GMR sensors can also be used for automotive GMR sensors can also be used in automobiles, driverless systems, toll collection systems, and satellite positioning systems. GMR sensors can also be made into security checking systems, full electronic compasses, and so on.

9.1.3 Magnetic Recording Materials

Magnetic recording is an information technology with wide and important applications, including magneto-optical recording technology and magnetic technology. As one of the most important recording media for sound, image and data recording, magnetic recording materials have been widely used in radio, television, satellite broadcasting, engineering control, scientific research, office automation, and various aspects of people's daily lives. As long as the application of tape recorders, video recorders, computers, there must be a corresponding audio tape, video tape, computer tapes, disks supporting the use. Modern life is popular in all kinds of new cards, such as credit cards, transportation cards, telephone cards, parking cards, medical cards, key cards, etc., are also widely used in magnetic recording materials. Magnetic recording equipment and magnetic recording materials are mutually supporting and promoting each other, constituting the overall development of magnetic recording technology. The main sign of the progress of modern magnetic recording technology is high-density recording and miniaturization. In recent years, magnetic recording materials in the product of high-grade, miniaturization, recording high-density, application of simple, etc., have made significant progress. Despite the serious challenge of new recording methods such as optical recording, magnetic recording technology and its materials still have broad prospects for development.

9.1.3.1 Applications of Magnetic Recording Materials

1. Tunneling Magneto Resistive Materials

Generate Tunneling Magneto Resistive (TMR) effect of the magnetic tunnel junction for the ferromagnetic film (a few nanometers thick)/insulating film (about 1 nm thick); when the upper and lower two ferromagnetic film's magnetic moments are in a parallel or antiparallel state, respectively, there is a current through the magnetic multilayer film or the magnetic tunnel junction, they will be respectively When the current passes through the magnetic multilayer film or the magnetic tunnel junction, they show low and high resistance states respectively. Such low and high resistance states can correspond to the two data recording states of 0 and 1.

TMR materials and their magnetically sensitive devices can be used in magnetic readheads and magnetic random memories (MRAM) and other fields, can increase the magnetic recording density to 100 Gbit/in^2 level, single-pole magnetic head of the magnetic channel width (MWW) and the magnetic head of the magnetic channel width (MRW) are lower than 0.28 μm and 0.18 μm, respectively, the gap is 0.07 μm, the resistance of the TMR element is 2350. The resistance of the TMR element is 2350. The writing element is a Stitched Pole type made of high saturation magnetic susceptibility material. The characteristics of the components have reached the target of 100 Gbit/in^2. TMR materials will bring unprecedented upgrading and huge market as well as far-reaching impact to the computer and information technology (IT) industries which have the needs of high speed, high density, high stability and low cost, and it is an example of many nanomaterials and technologies that can be rapidly transformed into industries.

At present, the United States, Japan, the European Union and other countries of the research unit and its first-class many large companies, have carried out the GMR and TMP materials and its magnetic read head and magnetic random memory and other spintronics device development, intended to maintain high-tech and economic income of the leading position. China's Chinese Academy of Sciences Institute of Physics, Institute of Solid State Physics, Semiconductor Institute, Microelectronics Center and Nanjing University, Shandong University, University of Science and Technology of China are also carrying out research on spintronics functional materials. It is expected that in the future, with the participation of some electronic enterprises with strength and foresight, high-tech products with intellectual property rights can be developed.

2. Co-C nanoparticle film

A series of Co-C nanoparticle films were prepared on monocrystalline silicon wafers by magnetic filtration pulsed vacuum arc co-deposition, and the prepared films were subjected to vacuum annealing at different temperatures, and the chemical components, morphology, particle size, microstructure, and magnetic domain structure of the films were measured and analyzed by X-ray photoelectron spectrometry, transmission electron microscopy, and magnetomicroscopy; the magnetic properties of the thin films were measured by a superconducting quantum interferometer and magnetometer. The magnetic properties of the films were measured by superconducting quantum interference magnetometer. The results show that the as-prepared films have a granular amorphous structure with a particle size of 3–4 nm, and no obvious magnetic domain structure is observed, and the samples exhibit soft magnetic properties; after vacuum annealing at 250–350 °C, the films exhibit a substable structure with the coexistence of densely-packed hexagonal Co and crystalline CoXC, and the films are gradually transformed from soft magnetism to hard magnetism. When the annealing temperature reaches 350–400 °C (related to the components), the CoXC decomposes into hcp-Co and graphitic C, and the size of Co particles increases to 5–7 nm, at which time, the obvious long-range magnetic domain structure and typical ferromagnetic features can be observed. The soft magnetic properties of the films in the prepared state are determined by their amorphous structure, while the magnetic hardening of the annealed state samples is due to the Co granularity in the films. This nanoparticle thin film system has a potential application as a high-density magnetic recording medium.

3. Magnetic Powder

With the development of information technology, the requirement of large information storage capacity, fast processing speed, prompting the magnetic particle size for magnetic recording tends to be ultra-fine. Magnetic powder is the main raw material for all kinds of magnetic devices, in the particle-type magnetic storage media, the size of the magnetic particles from the micron, submicron to the nanometer scale direction of the transition, the nanotechnology of this micro-miniaturization has created conditions for high-density magnetic storage.

At present with metal magnetic powder (20 nm or so of ultra-fine magnetic particles) made of metal magnetic tape, foreign countries have been commercialized, per centimeter can be recorded 4 million to 40 million information units, compared with the ordinary magnetic tape, it has a high-density, low noise and high signal-to-noise ratio and other advantages.

Magnetic recording materials made of nano-powder, can make the tape and hard and soft disk recording density increased by tens of times, and can greatly improve the fidelity performance. Cobalt nanopowder can improve the performance of magnetic tapes and large-capacity hard and soft disks by utilizing the advantages of high recording density, high coercivity, high signal-to-noise ratio and good oxidation resistance. Using the advantages of high coercivity, high saturation magnetization strength (up to 1477 km^2/kg), high signal-to-noise ratio and good oxidation resistance of iron nanopowder, the performance of magnetic tapes and large-capacity hard and soft disks can be further improved.

Fe_3O_4 is a widely used magnetic material at present, and its nanosized product efficacy such as wave-absorbing effect, catalytic effect and so on are being studied and a large number of experiments have been carried out, and the application prospect is very promising.

4. Nanoparticle magnetic recording materials

In recent years, the amount of information has been increasing rapidly, and the amount of information that needs to be recorded is also increasing, which requires high performance of recording materials, especially high density recording. Recording materials with high recording density are closely related to ultrafine particles. For example, each square millimeter can record more than 100,000 pieces of information. Then, a piece of information required to be recorded in 1–10 mm^2, at least 300 stages of hierarchical recording, in 1–10 mm^2 must have at least 300 recording units, so that the recording density is greatly improved.

Magnetic nanoparticles have the characteristics of single domain structure and high coercivity due to their small size. It can be used to make magnetic recording materials to increase the signal-to-noise ratio and improve the image quality. As a magnetic recording unit of the size of the magnetic particles must meet the following requirements; the length of the particles should be much smaller than the recording wavelength, the width of the particles should be much smaller than the depth of the record; a unit of the recording volume, there should be as much as possible, the magnetic particles.

In short, as magnetic recording particles require single-domain needle particles, its volume requirements as small as possible, but not less than the critical size of the superparamagnetic into (about 10 nm). At present, the size of the magnets used in video tapes is 100–300 nm (long), 10–20 nm (short diameter) of the ultra-micro particles. The magnetic ultrafine particles generally used in magnetic tapes are needle-like particles of iron or iron oxide.

5. Magnetic recording films

Magnetic storage technology plays a pivotal role in modern technology. Since the recording density of magnetic signals depends largely on the width of the magnetic head gap, the flight height of the magnetic head and the thickness of the recording medium, in order to further improve the density and capacity of magnetic storage, it is necessary to continuously reduce the size of the magnetic head, while also reducing the thickness of the magnetic recording medium. Therefore, thin-film magnetic head materials and thin-film magnetic storage media is one of the main directions of the current development of magnetic materials.

First of all, compared with γ-Fe_2O_3 magnetic powder coated magnetic recording media, thin-film recording materials are generally composed of alloys formed by the transition elements, and thus have a higher saturation magnetization strength. Moreover, since they do not contain any adhesive, they allow for smaller thicknesses and more uniform properties of the magnetic media. These are favorable to increase the density of data recording and reduce costs. At the same time, the exploration of new thin-film magnetic head materials with high magnetic induction strength, good soft magnetic properties and excellent abrasion resistance is also of great significance to further increase the recording density and capacity.

There can be two ways of reading and writing magnetic records, namely, parallel recording and perpendicular recording. In the case of parallel recording, the magnetic moment vectors in the magnetic recording medium are arranged along the direction parallel to the plane of the recording medium, and the change of the magnetization direction corresponds to the change of the recorded data. In the read/write head and the magnetic recording medium at the same time relative motion, or the head constantly changes the magnetization state to change the magnetization direction of the magnetic recording medium, that is, write data. Or due to the relative motion in the head coil to generate an induced potential, that is, read out the data. Obviously, for the magnetic recording medium and the magnetic head of the two material performance requirements are not the same. For the magnetic head material, the need for its typical soft magnetic properties, that is, high saturation magnetization strength, low coercivity, high permeability, low coefficient of magnetostriction, allowing high frequency. For magnetic recording media, typical hard magnetic properties are required, i.e., high saturation magnetization, high residual magnetic induction, and an appropriate level of coercivity.

6. Composite and thin film heads

For the electromagnetic induction principle of reading and writing magnetic head, data reading and writing can be done through the same head. This head uses a high permeability sintered ferrite, the advantage is that it has good soft magnetic properties and wear resistance, and high resistivity, and therefore good high-frequency characteristics. However, its magnetization strength is much lower than that of alloy soft magnetic materials, such as Bomol alloy (79% Ni-Fe) or 85Fe-9Si-54Al alloy (Sendust).

In order to further improve the performance of the magnetic head material, on the one hand, you can use electroplating or sputtering, evaporation and other methods, in the magnetic head gap deposition of a layer of several microns thick, soft magnetic properties of the alloy film, made of ferrite-alloy thin-film composite magnetic head. On the other hand, it is also possible to use thin-film technology completely, the magnetic material and magnetic field coils are deposited on a specific substrate, constituting the so-called thin-film magnetic head. Compared with ferrite heads, the heads prepared by thin-film technology will have higher sensitivity, and at the same time, the head size can be effectively reduced and the magnetic recording density can be increased.

In order to further improve the sensitivity of the magnetic head, continue to improve the saturation magnetization strength of magnetic thin film materials is very important. Pure iron saturation magnetization strength of up to 2.1T, is more than two times the above thin film materials, but its magnetostriction coefficient is very large, and in the atmospheric environment is easy to be corroded, not suitable for cooperation with the magnetic head material. For a long time, people have been looking for a new type of magnetic head material with high saturation magnetization strength. In recent years, the method of molecular beam epitaxy in GaAs substrate epitaxial production of $Fe_{16}N_2$ this iron substable nitride, it is said that its saturation magnetization strength reached 2.6 T, is a very promising magnetic head material.

Another important form of magnetic head is based on the magnetoresistance effect, that is, the external magnetic field changes at the same time, the use of the resistivity of the material to produce a response to the phenomenon of the magnetic head. This type of head does not have a write function and is therefore a read-only head. The use of magnetoresistance effect to produce magnetic heads with high sensitivity to signal readout, and signal strength is not affected by the speed of the head movement, without the need to prepare the advantages of induction coils, and therefore the use of it and the combination of thin-film head technology, can effectively reduce the size of the head, improve the density of magnetic recording. BMU not only has good soft magnetic properties, but also has a certain level (about 2.5%) of magnetoresistance effect, so it has been used to make this kind of magnetic head.

7. Magnetic recording medium thin film and its manufacturing technology

Compared with magnetic powder coated magnetic recording media, magnetic thin film recording media has a higher saturation magnetization strength, which is conducive to reducing the thickness of the medium, increasing the recording density and reducing costs.

In the case of parallel recording mode, the magnetization vectors of the material are required to be arranged along the film plane, that is, the film is required to have in-plane magnetic anisotropy. In order to meet the above performance requirements, in the preparation of Fe-Co-Cr parallel recording thin film media, can be used to tilt the evaporation deposition at a certain angle to obtain the appropriate direction of growth of the film columnar crystals, or in the plane of the film to apply a certain strength of the magnetic field to produce a certain degree of induced anisotropy.

One of the problems of the above parallel recording method is that there is a strong interaction between the magnetization vectors of adjacent magnetized regions, which limits the size of each magnetized region can not be too small, and thus the storage density is not easy to increase, while in the case of perpendicular magnetic recording method, the magnetization vectors in the recording medium are either upward, or downward, is perpendicular to the media plane orientation. In this case, the magnetization vectors in the film are required to be perpendicular to the film plane, i.e., the film itself is required to have perpendicular magnetic anisotropy. However, not every material can have perpendicular magnetic anisotropy. This is because, when the direction of the magnetization vector is perpendicular to the plane of the film, the magnetization vector itself will be subjected to the strong effect of its own magnetic field, which tries to turn the magnetization vector into the plane. Has been found with perpendicular magnetic anisotropy of the thin film material CoCu, CoCrX (X = Rh, Pd, Ta) alloy film, the preparation method includes a variety of PVD technology.

The manufacture of magnetic recording media requires the solution of a variety of material problems, such as substrate materials, substrate coatings and surface protection coatings. Substrate materials can be used according to its needs of hard or soft materials, that is, the daily referred to as hard disk and soft disk.

8. Magneto-optical recording

Magneto-optical storage technology on the magneto-optical recording medium puts forward new requirements, magneto-optical storage technology relies on two properties of magnetic materials, namely: when the temperature changes, the material magnetization state produces a corresponding change in the thermo-magnetic effect; the material magnetization state so that the polarization of polarized light reflected back from the surface of the direction of polarization changes in the Kerr magneto-optical effect. In a magnetic disc, the direction of the magnetization vector is perpendicular to the film plane, either upward along the normal direction, or downward in the opposite direction. During the writing process, the laser beam locally heats the magnetic medium to a temperature at which the ferromagnetism disappears. In the temperature drop, ferromagnetism re-emerges at the same time, write the magnetic head to apply a certain magnetic field, it makes this region of the magnetization vector in accordance with the information to be recorded in order to arrange, so as to complete the writing process.

Directly related to magneto-optical storage technology is the magneto-optical Kerr effect. Magneto-optical information recorded on the media, mainly using the magneto-optical Kerr effect to read out the information. Magneto-optical Kerr effect refers to a beam of linearly polarized light in the magnetized media surface reflection, the reflected light will be elliptically polarized light, and the long axis of the ellipse as the symbol of the "polarization surface" relative to the incident polarization of the polarized light polarization surface rotated by a certain angle. This angle is often referred to as the magneto-optical Kerr angle of rotation, denoted θ_k, as shown in Fig. 9.2.

The Magneto-optic Kerr effect includes three cases: (1) polar Kerr effect, that is, the Kerr effect when the magnetization M is perpendicular to the surface of the medium; (2) Transverse Kerr effect, that is, the Kerr effect when M is parallel to the surface of the medium but perpendicular to the incident plane of light; (3)

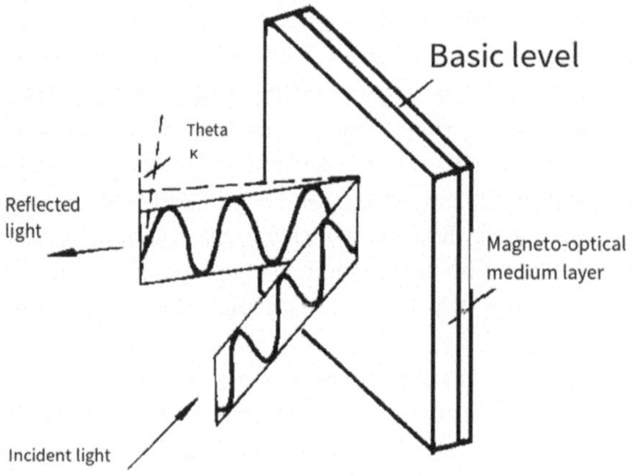

Fig. 9.2 The polarization plane turns by an angle when the incident polarized light is reflected by the surface of the magneto-optical medium

Longitudinal Kerr effect, that is, the Kerr effect when M is parallel to both the surface of the medium and the incident plane of light. The polar Kerr effect is mainly used in magneto-optical storage technology.

Therefore, in addition to the requirements for general magnetic storage media mentioned above, the requirements for magneto-optical storage films also require the following two points: an appropriate magnetic transition temperature, which not only ensures that the laser power required for information rewriting is not too high, but also ensures that the magnetization state of the film has sufficient stability; The strong Magneto-optic Kerr effect means that the polarization direction of polarized light changes greatly when the magnetization direction of materials is different.

The currently used magneto-optical storage film materials are mainly amorphous alloys of rare earth elements and transition group elements. The temperature of the compensation point is a characteristic temperature where the magnetization of such alloys approaches zero, which is related to the heating temperature required for laser writing, while the magnetic anisotropy constant is in direct proportion to the Coercivity level that can be obtained by the materials. Due to the presence of different elements with significantly different properties in such films, various sputtering methods are often used for material preparation.

9.1.4 New Magnetic Fluids

Nano-magnetic liquid, also known as magnetic fluid, ferromagnetic fluid or magnetic colloid, nano-magnetic liquid is made of 10 nm below the nanometer scale magnetic solid particles such as Fe_3O_4, $MeFe_2O_4$, Fe, Co, Ni, FeCo or NiFe alloy, etc., through the surfactant highly uniformly dispersed, suspended in the base liquid, and its mixing and the formation of a solid–liquid stable colloidal liquid mixed with the material, with liquid carrier fluidity, lubrication and sealing. Material, with liquid carrier fluidity, lubrication and sealing. The magnetic particles in the magnetic liquid are so small that they exhibit chaotic Brownian motion in the base liquid. This thermal motion is sufficient to counteract the settling effect of gravity and weaken the electromagnetic coalescence between particles, and can be stabilized under the action of gravity and electromagnetic fields without precipitation and coalescence.

The structure of magnetic liquids is shown in Fig. 9.3, which mainly consists of three components, namely the base liquid or carrier liquid, nanoscale magnetic solid particles [4], and interfacial active agents (also known as surfactants or dispersants) coated on the surface of nanoscale magnetic solid particles. In its usual state, it is a highly stable viscous colloidal solution. In order to improve the performance of magnetic fluids, oiliness agents, antioxidants, preservatives, and tackifiers are often added.

The main characteristic of magnetic liquids is that they can be magnetized under the action of a magnetic field and can move under the action of a magnetic field, but at the same time they are liquids with the fluidity of liquids. Under the action of a static magnetic field, magnetic particles will form orderly arranged cluster chains

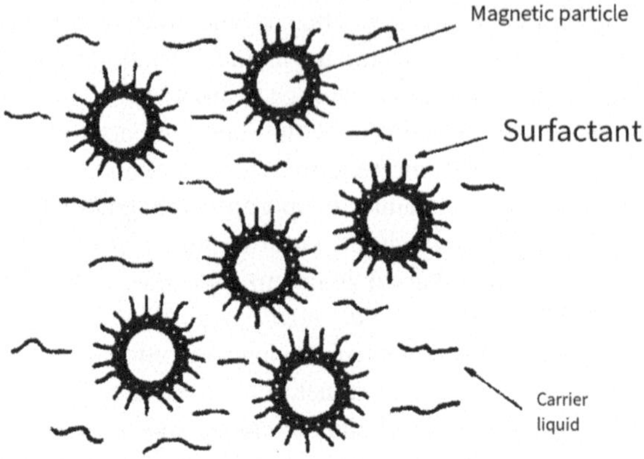

Fig. 9.3 Basic structure of a magnetic nanofluid

along the direction of the external magnetic field, resulting in the liquid becoming an anisotropic medium. When light waves and sound waves propagate within it, they produce Faraday rotation of light, birefringence effect, dichroism, and anisotropy of ultrasonic propagation speed and attenuation. In addition, under the action of a static magnetic field, the dielectric properties of magnetic liquids also exhibit anisotropy. These unique properties that are different from ordinary liquids have laid the foundation for the development of several novel magnetic devices.

Due to the combined magnetism of magnets and the fluidity of liquids, nano-magnetic liquids have unique and widespread applications in industries such as electronics, instrumentation, machinery, chemical engineering, environment, and healthcare.

1. Increase speaker power

Dropping magnetic liquid into the gap between the voice coil and the magnet, due to its high thermal conductivity, people successfully solved the problem of heat dissipation and cooling of the voice coil as early as the mid to late nineteenth century, preventing failure due to instantaneous overheating, and increasing input power by two times. At the same time, the addition of magnetic fluids has a significant impact on the low-frequency part of the frequency response curve. Usually, according to the structure of the speaker, selecting a suitable viscous liquid can make the speaker have a better frequency response curve.

2. Industrial sealing and lubrication

Magnetic fluid dynamic sealing technology is a new sealing technology that is different from traditional methods, and is also the most mature commercial application of magnetic fluid. It is the first developed and commercialized magnetic fluid product. This technology has been applied to semiconductor processing, electron microscopy, mass spectrometry, ion and plasma etching, lasers, heat treatment furnaces, computer hard disk dust prevention, avionics equipment, single crystal furnaces and other crystal growth equipment. Compared with various other rotating dynamic sealing technologies, magnetic fluid sealing technology has excellent performance such as zero leakage, no solid friction, low energy consumption, and long service life. The use of magnetic fluids for sealing oil has broader application prospects, but due to the difficulty, there have been few public reports so far. The main difficulty of sealing oil with magnetic fluid lies in the mutual dissolution or emulsification phenomenon between the magnetic fluid and the sealed oil under high-speed rotation disturbance, which reduces the performance of the magnetic fluid and the pressure resistance, leading to sealing failure. The magnetic fluid used for rotating shaft seal oil must have a base fluid that is not miscible with the sealed oil, does not undergo destructive chemical reactions, and has a lower saturated vapor pressure. In the 1990s, the Solid Microstructure Laboratory of Nanjing University developed a new type of oil repellent magnetic fluid and discussed the application of this magnetic fluid in the dynamic seal of rotating shaft, and achieved good results.

Magnetic fluid is a new type of lubricant. The size of magnetic particles in magnetic fluid is only 10 units, so its unique advantage is no wear, and it can resist acceleration, Centripetal force, gravity, etc. in the lubrication process. It can be fixed by magnetic force. As a lubricant, it basically has no leakage. As long as a suitable magnetic field is used, magnetic lubricating oil can be constrained in the required area. At present, it has been widely used in developed countries in dynamic lubrication journal bearings, thrust bearings, gear transmissions, hydraulic systems, and any complex motion mechanism where two surfaces come into contact with each other.

3. Environmental protection

Over the past 20 years, research on high gradient magnetic separation technology has been achieved both domestically and internationally in areas such as water and coal desulfurization. Compared with traditional separation technologies, it has enormous advantages and can solve previously unimaginable problems. For example, the separation of fine particles, weakly magnetic or even non magnetic particles. The application of different specific gravity for mineral separation, the magnetized magnetic fluid is equivalent to increasing the magnetic pressure [5], and the objects in the magnetic fluid will float, just as the apparent density of the magnetic fluid increases with the increase of the magnetic field. This principle can be used to design a magnetic fluid Hydrometer. Magnetic liquids separate objects of different densities by controlling the appropriate magnetic field strength, which can cause objects below a certain density value to float and objects above this density value to sink. In principle, they can be used for mineral separation. For example, separating high-density metals from low-density sand and gravel can also be used to separate metals from non-metals in urban fertilizers. In recent years, due to substantial progress in the research of high gradient magnetic separation (HGMS) technology both domestically and internationally, magnetic liquids have been widely used in the treatment of industrial waste, coal desulfurization, water supply and drainage, as well as the separation of strong and weak magnetic and even non magnetic particles. According to reports, Shenhong Zhu and others from Qingdao Institute of Architecture and Engineering in Shandong Province have achieved satisfactory results in using this technology and chemical beneficiation to jointly treat pyrite cinder from chemical plants.

4. Medical treatment

In recent years, the applications of nanomagnetic liquids in biomedical fields have received extensive attention, and they include magnetic liquid intracellular thermotherapy, magnetic targeted drug delivery system, biomagnetic separation, etc. As a new type of opaque material, magnetic liquid [6] has been used in NMR and X-ray diagnosis. Since its position can be controlled by an applied magnetic field, it is also used for drug delivery during cerebral thrombosis treatment and for blocking blood in blood vessels during cardiovascular surgery.

5. Biotechnology

Because magnetic fluid can move directionally under the action of magnetic force, it has been applied in the fields of biotechnology, such as immunoassay, Immobilized enzyme, cell separation, DNA separation, biological missile, etc. The typical ones are HGMS/F, aqueous two-phase extraction, mixed/clarified magnetic swimming, etc. The UK has captured red blood cells from whole blood using HGMS technology.

In addition, magnetic fluids can also be applied in dampers in instruments, silent and fast magnetic printing, magnetic fluid generators, and directional quenching.

9.1.5 Soft Magnet Materials

Nanomagnetic materials, as a new type of material, exhibit unique physical and chemical properties that differ from conventional magnetic materials due to their unique physical and chemical properties, such as quantum size effects, small size effects, surface and interface effects, and macroscopic quantum tunneling effects. With the development of electronic products towards miniaturization and the continuous expansion of application fields, the preparation technology of nano soft magnetic ferrite materials has become one of the hotspots of international research. Soft ferrite refers to a type of ferrite that is easy to magnetize and demagnetize. Its structure is divalent metal oxide (Fe_2O_3), and its microstructure belongs to the spinel type of Cubic crystal system. The development of soft magnetic materials has gone through the process of crystalline, amorphous, and nano microcrystalline states. Due to the high symmetry of crystals, low crystal magnetic anisotropy, significantly higher resistivity than metal based soft magnetic materials, and low eddy current losses, magnetic cores, mainly used as inductance components, are widely used in industries such as electronics, communication, and information. They are a widely used, high-yield, and low-cost basic material in the electronic and electromechanical industries, and have been applied in switching power supplies Transformers, sensors, and other devices can achieve miniaturization, lightness, high-frequency, and versatility. In recent years, they have developed rapidly and play a crucial role in various fields such as new materials, energy, information, and biomedicine, directly related to the development of electronic information industry, home electronics industry, computer and communication, environmental protection, and energy-saving technology.

Nanomaterials are known as new materials facing the twenty-first century and are also an inevitable direction for the development of magnetic materials. The development of nanotechnology has also created good conditions for the traditional magnetic industry to improve the quality of products, increase the added value of products, improve its competitiveness and export earning capacity, thus making it possible for China to carry out large-scale industrial production directly from a higher starting point by leaps and bounds.

9.1.6 Nano-magnetic Cooling Media

In short, magnetic refrigeration is the use of magnetic thermal effect, also known as magnetic card effect of refrigeration. Magneto-caloric effect refers to the magnetic refrigeration workpiece in the isothermal magnetization to the outside world to release heat, and adiabatic demagnetization when the temperature decreases, from the outside world to absorb heat phenomenon. Magnetic refrigeration technology in the refrigeration medium is a solid magnetic material. As we know, matter is composed of atoms, atoms are composed of electrons and nuclei, electrons have a spin magnetic moment and orbital magnetic moment, which makes some material atoms or ions with magnetic moment [7]. Paramagnetic material ions or atomic magnetic moments in the absence of an external magnetic field is disordered, plus an external magnetic field, the magnetic moments of the atoms along the external magnetic field orientation arrangement, so that the magnetic moments of the order, thereby reducing the magnetic entropy of the material, and therefore will be released to the outside of the heat; and once the removal of the external magnetic field, the magnetic order of the material system decreases, magnetic entropy increases, and therefore will be absorbed from the outside of the heat. Magnetic entropy is a function of temperature and magnetic field, if the two adiabatic demagnetization caused by the heat-absorbing process and adiabatic magnetization caused by the exothermic process with a cycle connected through the external magnetic field, consciously control the magnetic entropy, can make the magnetic material continuously from one end of the heat absorption and exothermic at the other end, so as to achieve the purpose of refrigeration. Compared with the usual compressed gas refrigeration, it has the advantages of high efficiency, low power consumption, low noise, small size, no pollution. Magnetic refrigeration development trend is from low temperature to high temperature development, in the 1930s the use of paramagnetic salt as a magnetic refrigeration workpiece, the use of adiabatic demagnetization method to successfully obtain the low temperature of the mK scale; in the 1980s, the use of $Gd_3Ga_5O_{12}$ (GGG) type of parabasic garnet compounds were successfully applied to the magnetic refrigeration of the 1.5–15 K. In the 1990s, the use of magnetic Fe ions to replace some of the non-magnetic Gd ions, due to the existence of super-exchange interaction between Fe ions and Gd ions, the local magnetic moments are ordered, constituting magnetic nanoclusters, whose magnetic entropy change is higher than that of GGG when the temperature is greater than 15 K, thus becoming the optimal magnetic refrigeration workpiece in the temperature region of 15–30 K [8].

In 1976, Brown firstly used metal Gd as the magnetic refrigeration work material, and realized the room temperature magnetic refrigeration test under 7 T magnetic field, which could not be commercialized due to the superconducting magnetic field [9]. Until 1996, Youwei Du et al. obtained the breakthrough of magnetic entropy greater than that of Gd in $RmnO_3$ chalcogenide compounds. It is reported that in 1997, the United States has developed a magnetic refrigerator with Gd as the magnetic refrigerant. If the magnetic refrigeration agent is nanosized, it may be used to widen the temperature range of refrigeration.

With the rapid development of nanotechnology and the continuous improvement of synthesis technology, the control of nanoparticle shape, size and composition is becoming more and more mature. However, how to precisely control the synthesis conditions to realize large-scale production of high-quality magnetic nanoparticles is still a challenging task for researchers. Therefore, we must develop new and more advanced industrial technologies, otherwise we will not be able to utilize the attractive properties of nanomaterials.

9.2 Nano-optical Materials

9.2.1 Optical Properties of Nanomaterials

An important feature of nanomaterials is their quantum size effect. When the particle size of nanoparticles is comparable to the superconducting coherence wavelength, the Bohr radius, and the de Broglie wavelength of electrons, the quantum size effect of small particles is very significant. At the same time, the large specific surface area causes significant differences in the behavior of atoms and electrons on the surface compared to those inside small particles. This surface effect and quantum size effect have a significant impact on the optical properties of nanoparticles, giving them unique optical properties that are not possessed by bulk materials of the same material.

1. Broadband strong absorption

Large metals have different color luster, which indicates that they reflect and absorb various colors (wavelengths) in the visible light range differently. When the size is reduced to the nanometer scale, almost all metal nanoparticles are black in color. Their reflectance of visible light is extremely low, for example, 1% for platinum nanoparticles and less than 10% for gold nanoparticles. This low reflectivity and strong absorption of visible light leads to the black coloration of the particles.

Nano SiN, SiC, and Al_2O_3 powders exhibit a broadband strong absorption spectrum for infrared radiation. This is because the large specific surface area of nanoparticles leads to the decrease of average coordination number and the increase of unsaturated bonds. Different from conventional bulk materials, there is no single and preferred bond vibration mode, but there is a wider distribution of bond vibration modes. Under the action of infrared light field, there is also a wider distribution of their infrared absorption frequency, which leads to the broadening of the infrared Absorption band of nanoparticles.

Many nanoparticles, such as ZnO, Fe_2O_3 and TiO_2, have strong absorption of ultraviolet light, while the submicron TiO_2 has almost no absorption of ultraviolet light. The absorption of UV light by these nano-oxides mainly comes from their semiconductor properties, i.e., under UV light irradiation, the electrons are excited by the valence band to the conduction band in the process of the UV light absorption.

2. Blue-shift and red-shift phenomena

Compared with bulk materials, the absorption bands of nanoparticles are generally "blueshifted", i.e., the absorption bands are shifted to the short wavelength direction. For example, the peak infrared absorption frequencies of SiC nanoparticles and bulk SiC are 814 and 794 cm^{-1}, respectively, and the infrared absorption frequency of SiC nanoparticles is blueshifted by 20 cm^{-1} compared with that of bulk materials.

There are several explanations for the "blue shift" of the absorption band of nanoparticles, which are summarized in two aspects. One is the quantum size effect, due to the particle size decreases the energy gap becomes wider, which leads to the light absorption band shifted to the short-wave direction. Ball et al. [10] gave a universal explanation for this blue-shift phenomenon: the width (energy gap) between the molecular orbital energy levels that have been occupied by the electrons and the molecular orbital energy levels that have not been occupied increases with the decrease in the diameter of the particle, and this is the fundamental reason for the blue-shift. This explanation applies to both semiconductors and insulators. The other is the surface effect, where the large surface tension distorts the lattice and reduces the lattice constant due to the small diameter of the nanoparticles. Studies of oxide and nitride nanoparticles have shown that the distance between the first and second nearest neighbors becomes shorter, and the shortening of the bond length leads to an increase in the bond eigenvalue vibration frequency of the nanoparticles, resulting in a shift of the infrared absorption band to higher wavelengths.

3. Quantum limiting effect

Semiconductor nanoparticle particle size $r < a_B$ (a_B is the exciton Bohr radius), the average free range of the electron is limited by the small particle size, confined to a very small range, the hole is easy to form an exciton with it, resulting in the overlap of the electron and hole wave functions, which is easy to produce exciton absorption bands.

4. Luminescence of nanoparticles

When the size of nanoparticles small to a certain value can be in a certain wavelength of light excitation luminescence. 1990, Japan's Canon Research Center Tabagi found that the particle size of less than 6 nm of silicon by the ultraviolet excitation can be triggered by the emission of visible light. When the particle size is larger than 6 nm, this light emission phenomenon disappears, Tabagi believes that the luminescence of silicon nanoparticles is caused by the quantum confinement effect of carriers.

5. Optical properties of nanoparticle dispersion system

Nanoparticles are dispersed in a dispersing medium to form a dispersion system (sol), where nanoparticles are also called colloidal particles or dispersed phase. Due to the high dispersion and inhomogeneity of colloids in the sol, the dispersed matter system has special optical characteristics. For example, if a beam of aggregated light passes through this dispersion, a luminous cone can be seen in the vertical direction of the incident light, which is called the Tyndall effect. This cone is called the Tyndall cone. The Tyndall effect is related to the size of the dispersed particles and the wavelength of the projected light. When the diameter of the dispersed particles is larger than the wavelength of the projected light wave, the light projected onto the particles will be reflected. If the diameter of the particles is smaller than the wavelength of the incident light wave, the light wave can propagate in all directions around the particles, scattering occurs, the scattered light, the so-called milky light. Since the diameter of nanoparticles is much smaller than the wavelength of visible light, the nanoparticle dispersion system should be based on the role of scattering.

9.2.2 Nanocrystalline Solar Cells

Solar energy is a natural energy given to mankind by nature, inexhaustible, is one of the important energy sources that people have been researching and developing. However, the current utilization cost of solar energy is still very high, therefore, the development of new solar energy conversion materials will play a pivotal role in the utilization of solar energy. To make solar cells become a real alternative energy, we must improve the conversion efficiency of the battery, reduce manufacturing costs, reduce the unit cost of electricity. To this end, people have been exploring solar cells from the materials, structures and processes.

9.2.2.1 Nanocrystalline Photoelectrochemical Cells

1. Nanoparticle membrane electrode

The photocurrent originates from the effective separation of electron–hole pairs generated at the semiconductor nanoparticle-electrolyte interface after the absorption of light by ultrafine particles. Unlike bulk materials, the photogenerated charge separation in semiconductor nanocrystals does not depend on the built-in spatial electric field, but is a kinetically controlled process determined by the rates of oxidation and reduction reactions at the nanoparticle-electrolyte interface. Nanoparticle membranes exhibit n-type or p-type properties. The problem of energy dissipation during conversion can be solved by depositing nanosized ultra-particles, such as cadmium sulfide and lead sulfide, on a porous nanocrystalline electrode. This porous electrode has a large specific surface, the deposition of less than a layer of cadmium sulfide ultra-particle film is sufficient to absorb a large amount of incident light, light absorption carriers generated can also be quickly transferred to the titanium dioxide electrode. Semiconductor nanoparticle film as the working electrode of the photoelectrochemical cell research is still in the study, from the practical use of a greater distance.

2. Dye-sensitized nanocrystalline particulate film cells

Unlike conventional solar cells, this nanocrystalline solar cell uses a composite organic and inorganic system, and its working electrode is a nanocrystalline semiconductor porous film. In the film's microporous adsorption of a single layer of organic dye molecules as a photosensitizer. After illumination, light-induced charge separation occurs at the interface between the particles and the electrolyte, and the photosensitizer absorbs the incident light and injects electrons into the conduction band of the semiconductor particles, which are themselves positively charged. This is also a kinetically controlled process (electrons are injected into the particles and semiconductor particles at different rates from those injected into the photosensitizer or electrolyte) with a quantum yield close to one.

The electrolyte usually consists of an organic solvent containing an oxidation–reduction pair. After the loss of electrons, the photosensitizer is reduced by the reducing agent in the electrolyte, and the reducing agent is regenerated by the corresponding oxidizing agent that gets electrons at the counter electrode, and the electrons form a photocurrent through the external circuit without a net chemical change. Dye sensitization method of photoelectrochemical solar cells, not only can overcome the shortcomings of the broad-band semiconductor itself only absorbs ultraviolet light, but also can make the cell on the visible spectrum of the absorption is greatly increased. The performance of the photovoltaic cell can be further improved by changing the type of dyes, so as to obtain the ideal photoelectrochemical solar cells. The number of dye molecules absorbed on the surface of TiO_2 nanoporous electrode is more than 50 times more than the number of dye molecules adsorbed on the surface of ordinary electrode, and almost every dye molecule is in direct contact with the titanium dioxide molecule, and the photogenerated carriers are transferred quickly at the interface, so the cell has excellent light absorption and photoelectricity conversion characteristics.

9.2.2.2 Dye-Sensitized Solid-State Heterojunction Nanocrystalline Solar Cells

The liquid electrolyte used in photoelectrochemical cells and its redox pairs are corrosive and make the packaging of the cells difficult. It is hoped to replace the redox electrolyte with a p-type semiconductor running through the n-type TiO_2 nanoparticles, which is the dye-sensitized solid-state heterojunction nanocrystalline solar cells. p-type semiconductors play a role in the dye molecules injected electrons into the TiO_2 particles, the oxidation of the dye molecules in the state of the positive charge (holes) transferred to the counter electrode, so that the dye molecules regeneration. The current solid charge transfer materials are: amorphous organic hole conductors, p-type semiconductors CuI and CuSCN.

9.2.2.3 Nanocrystalline Organic Photovoltaic Cells

One of the major obstacles to the development of photovoltaic cells based on nanostructures is the preparation of large arrays of nanostructures at the desired cycle time and size at low cost. One approach is the electrochemical synthesis of semiconductor nanostructures on porous aluminum templates obtained by anodic electrochemical etching. This technique can be widely used in electrochemical manufacturing because it is inexpensive, reliable, suitable for large-scale preparation, and can be applied to a wide range of substrates and semiconductor materials. If the substrate is silicon, nanostructures can be integrated on silicon wafers. The method is to first deposit a thin layer of lead on the substrate and then anodize it to obtain a porous aluminum film with a planar hexagonal arrangement. This results in a film with controllable pore size and length. Subsequently, the semiconductor materials required for electrochemical deposition, such as CdS, ZnS, CdS, and CdSe, can be selected in the pores, and CdS_xSe_{1-x}, Cd_xZn_{1-x}, and GaAs can be prepared at present. Since the thickness of the deposited semiconductor material can be controlled, it is possible to deposit a layer of n-type material followed by a layer of p-type material to obtain a nano-p-n junction structure. Repeating the above process can lead to stacked multilayer p-n junction arrays, conjugated polymer-nanocrystalline structures of photovoltaic devices, such as polystyrene (PPV)-TiO_2, which are particularly attractive because of their all-solid-state properties, ease of preparation, and chemically tunable nanocrystalline sizes, shapes, and surfaces.

9.2.3 Nanoluminescent Functional Materials

9.2.3.1 Silicon Based Semiconductor Nano Luminescent Materials

The twentieth century was an era of microelectronics technology leading the way and shining brightly. Silicon materials and devices are the semiconductor technology with the most in-depth research, mature technology, and widespread application among them. It can be said that without silicon materials, modern electronic industry cannot start. With the development of the times, microelectronics technology is becoming increasingly inadequate due to the limitations of the "electronic bottleneck" effect in information transmission, storage capacity, and processing speed. In contrast, optoelectronic technology stands out with its unique advantages and has become the most eye-catching key technology in today's information systems and networks. It will trigger a revolution beyond electronic technology and bring a greater impact to the entire industry and society than electronic technology. In this context, people can imagine that if mature silicon semiconductor processing technology is combined with luminescent materials and technologies to achieve optoelectronic integration, it will inject new vitality into the development of silicon materials and bring new growth points to the development of the silicon industry. The rise of porous silicon in the 1980s and the subsequent research on various nano luminescent materials based on silicon materials have flourished in pursuit of this goal.

Due to the considerable maturity of silicon integrated circuit technology, the key to achieving optoelectronic integration on silicon based materials is to obtain luminescence on silicon materials. There are two basic ways to solve this problem: one is to modify monocrystalline silicon, which usually involves doping luminescent centers such as rare earth ions and silicon homologous atoms into monocrystalline silicon; The second is to prepare luminescent materials on silicon materials to obtain composite systems with luminescent properties.

1. Porous silicon

The luminescence of porous silicon reported by the Cabham research [11] group in 1990 was a milestone in silicon-based nanomaterials. The research group used silicon as an anode for electrochemical corrosion in hydrofluoric acid, resulting in a porous structure with nanoscale, which produces visible light emission from silicon materials that typically emit light in the infrared region. This phenomenon has attracted great attention, and in subsequent research, people have utilized porous silicon to obtain various frequencies of luminescence in the visible and even ultraviolet regions. There are two main physical models to explain the luminescence mechanism of porous silicon: one is the quantum confinement model. As the corrosion time prolongs, the crystal silicon line or dot size of porous silicon decreases, the quantization energy increases, and the luminescence peak shifts blue; The second is the surface state model proposed by Koch et al. [12], which suggests that the surface state plays a decisive role in the luminescence of porous silicon. By modifying and passivating its surface, the luminescence characteristics of porous silicon can be improved. Porous silicon luminescence has shown people the promising prospects of optoelectronic integration, but it is disappointing that the problem of poor luminescence stability of this material has not made breakthrough progress, and research interest in it has gradually decreased. Nevertheless, the theoretical research on nano luminescent materials triggered by porous silicon still has important academic value and has driven the development of other related nano luminescent materials. In order to achieve the goal of optoelectronic integration, some scientists are still persistently working on porous silicon and have achieved some exciting results.

In addition to studying the luminescence of porous silicon itself, some researchers have also used its nanostructures to prepare luminescent materials for composite systems, resulting in some new physical phenomena. For example, by embedding originally non luminescent C_{60} molecules into porous silicon, bright photoluminescence can be observed; Embedding laser dyes or other organic molecules into porous silicon can reveal narrowing and blue shift of the emission spectrum of organic molecules.

2. Silicon based quantum structured luminescent materials

Silicon based quantum structure light-emitting materials refer to quantum dots, Quantum wire, quantum wells, Superlattice and other light-emitting materials and related devices with nanometer size. Silicon quantum dots commonly refer to crystalline silicon particles with a size less than 5 nm. When excited by ultraviolet light, luminescence in the visible region can be observed, and the luminescence wavelength is related to the preparation method and passivation treatment of the nanosilicon.

There are several commonly used methods for preparing nano silicon quantum dots.

1. Injecting Si into SiO_2 matrix material using ion implantation method, and forming nano silicon grains in it after heat treatment;
2. Microwave plasma dissociation was performed on the mixture of SiH_4 and H_2, and the silicon particles generated after dissociation were deposited on the substrate material;
3. SiH_4 and H_2 are thermally dissociated at high temperatures to produce silicon particles, which are oxidized to form a 1–2 nm thick oxide layer on the surface. Finally, ethylene glycol liquid is introduced to form a colloid containing silicon particles;
4. Directly evaporate silicon in an argon atmosphere, deposit nano silicon on the substrate material, and perform appropriate passivation treatment;

A multi-layer structure was obtained by plasma enhanced chemical vapor deposition (PECVD) method, and nanosilicon was formed by laser crystallization in an amorphous environment. Although the luminescence intensity of this type of nanosilicon quantum dot is weak, its stability has been greatly improved compared to porous silicon, and its process is fully compatible with existing silicon processing technology. With the continuous deepening of research and the continuous improvement of luminescence intensity, this type of luminescent material has great prospects for making progress in optoelectronic integration.

Silicon based Quantum wire light-emitting material refers to the light-emitting material with linear structure with radial size in the nanometer range formed on the basis of silicon material. The nano silicon pillar in porous silicon can be considered as a Quantum wire. In addition, other people have prepared GeSi, silicon polymer and other quantum line structures by various methods, and obtained the corresponding luminescence. Silicon based quantum wells and Superlattice are nano multilayer periodic structures. In this structure, one material is used as a luminescent material, while the other material acts as a limiting barrier [13]. Superlattice light-emitting structures such as Si/SiO_2 and SiGe/Si have been prepared by Molecular-beam epitaxy, and strong red and orange light has been obtained.

9.2.3.2 Nano Powder Luminescent Materials

Nano powder light-emitting materials are important components of light-emitting materials, which refer to light-emitting materials with particle size less than 100 nm. It covers a wide range of nano semiconductor luminescent materials, as well as rare earth ion and transition metal ion doped nano oxides, sulfides, composite oxides, and various inorganic salt luminescent materials.

Using these materials, we can obtain some superior luminous properties than traditional light-emitting materials, and improve the resolution of light-emitting devices. Similar to other nanomaterials, further development and application of nano luminescent materials still require more efforts in improving dispersion, stability, and other aspects.

9.2.4 Photonic Crystals and Photon Storage

9.2.4.1 Photonic Crystals

The discovery of this material can extend to animals in the deep sea and butterflies flying in the sky. These colors and luster are not produced by pigments, but by reflection or refraction of sunlight, and are called Structural coloration. Based on the inspiration of these natural photonic crystals, people have begun to pay attention to the artificial preparation of photonic crystals to achieve various needs [14].

Photonic crystals are ordered structures composed of materials with different dielectric constants arranged periodically, and the periodic structure in space can trigger resonant Bragg diffraction, forming a "photonic band gap (PBG)" within a certain frequency range. Photonic band gaps can prevent the propagation of light waves at specific frequencies, thus producing obvious optical phenomena. According to Bragg's law of diffraction, the diffraction wavelength (γ), lattice spacing (d), refraction coefficient (n), Bragg diffraction level (m) and diffraction angle (θ) in any one of the parameters of the change will cause Bragg diffraction wavelength changes, such as when stimulated by the external environment (humidity, temperature, force, electromagnetic field and light irradiation, etc.), the diffraction peaks of the photonic crystalline material will be changed, which will lead to structural color change, which can be recognized by the naked eye [15]. In addition, due to the periodic photonic banding effect, photonic crystal materials can produce signal enhancement effect, which is conducive to the establishment of high-sensitivity detection methods; detection can be self-expressed response signal, without additional reagents or labeling, which becomes an ideal sensing material, and has a good potential for rapid qualitative and quantitative detection [16], as shown in Table 9.1, which summarizes the application of different photonic crystal materials in food safety detection. applications in food safety detection.

Traditional photonic crystal preparation methods mainly use mechanical processing and etching and other physical methods, which require fine processing technology, and its process is cumbersome and unable to realize large-area preparation. In recent years, in order to solve these problems, researchers have developed self-assembly, inkjet printing, microphase separation, and other simple and efficient methods for the preparation of photonic crystals.

Self-assembly is a process in which orderly assembly can be realized without any direct external action. This method is simple, without expensive equipment and precise processing technology, and is suitable for the production of photonic crystals on a large area. According to the different inducing factors in the self-assembly process of colloidal particles, it can be categorized into field-induced self-assembly, evaporation-induced self-assembly and template-directed self-assembly.

Field induced self-assembly method refers to the method of arranging and assembling colloidal particles under the action of gravity, centrifugal force, magnetic field, electric field, etc. The most commonly used external fields are magnetic and electric fields. Yin et al. [17] orderly assembled monodisperse magnetic iron trioxide (Fe_3O_4) microspheres under the action of a magnetic field, and prepared photonic crystals with reversible band gap regulation throughout the entire visible spectrum range by adjusting the magnetic field intensity to control the distance between colloidal particles. Gong et al. [18] based on the combination mechanism of induced dipole interaction and hydrodynamic flow, the colloidal particles are rapidly arranged into three-dimensional photonic crystals under the action of electric field.

Evaporation induced self-assembly is a method of regulating the evaporation of colloidal particles at the gas–liquid interface and gas–liquid solid interface to achieve ordered assembly of colloidal particles. This preparation technology has the characteristics of simplicity, high efficiency, strong controllability, and good repeatability. According to the evaporation method, it can be divided into vertical deposition method, spin coating method, spray coating method, and scraper method. Vertical deposition method is the process of vertically or obliquely inserting a substrate into the dispersion of colloidal particles, utilizing solvent evaporation and capillary forces between colloidal particles to assemble colloidal particles. Zhao [19] et al. proposed an improved vertical deposition method: by adding tetraethyl orthosilicate precursor into the silica lotion, a large-area crack free photonic crystal was produced, and then further expanded the method, using polystyrene colloidal particles instead of silica colloidal particles to prepare crack free polystyrene/silica mixed photonic crystal films, as shown in Fig. 9.4.

Small-size colloidal particles that cannot be deposited but can self-assemble under centrifugal force can be self-assembled by spin-coating. Jiang et al. [20] prepared large-size colloidal photonic crystals by spin-coating and improved the strength of the

Table 9.1 Applications of photonic crystal materials in food testing

Application area		Detect target object	Analytical methods	Analyze performance
Pretreatment materials	Enrichment and purification	OTA, AFB_1, ZEN	High-performance liquid chromatography	Rate of recovery: 70.01–100.12%
		DON	Liquid chromatography	Rate of recovery: 60.63–69.20%
	Reinforced substrate material	OTA, FB_1, DON	SERS immunoassay	OTA: 2.46 pg/mL, 0.001–10 ng/mL
				FB_1: 0.20 pg/mL, 0.001–10 ng/mL
				DON: 68.98 pg/mL, 0.1–1000 ng/mL
		AFB_1, OTA, ZEN	SERS immunoassay	AFB_1: 0.82 pg/mL, 0.001–0.1 ng/mL
				OTA: 1.43 pg/mL, 0.01–10 ng/mL
				ZEN: 1.00 pg/mL, 0.001–0.1 ng/mL
		AFB_1, OTA	SERS Immunoassay	AFB_1: 0.36 pg/mL, 0.01–100 ng/mL OTA: 0.034 pg/mL, 0.001–10 ng/mL
Testing materials	Coding materials	Hg^{2+}, Ag^+	Fluorescence detection method	Hg^{2+} rate of recovery: 93.0–110%
				Ag^+ rate of recovery: 97.0–105%
		AFB_1, FB_1, OTA	Fluorescence detection method	Detection limit: 0.70 pg/mL
				Detection limit: 80.4–120.1%
	Unmarked testing materials	SEB	Reflectance spectroscopy	0.103×10^{-6} ng/mL
				1×10^{-6}–1 ng/mL
		SEB	Reflectance spectroscopy	2.820 fg/mL
				10^{-2}–10^3 pg/mL
		KANA	Reflectance spectroscopy	1.10 pg/mL
				5 pg/mL–5 μg/mL
	Visual inspection materials	Hg^{2+}	Diffraction spectroscopy	5×10^{-16} g/L
		α-amanitine	Diffraction spectroscopy	5.0×10^{-10} mg/L
				10^{-9}–10^{-3} mg/L

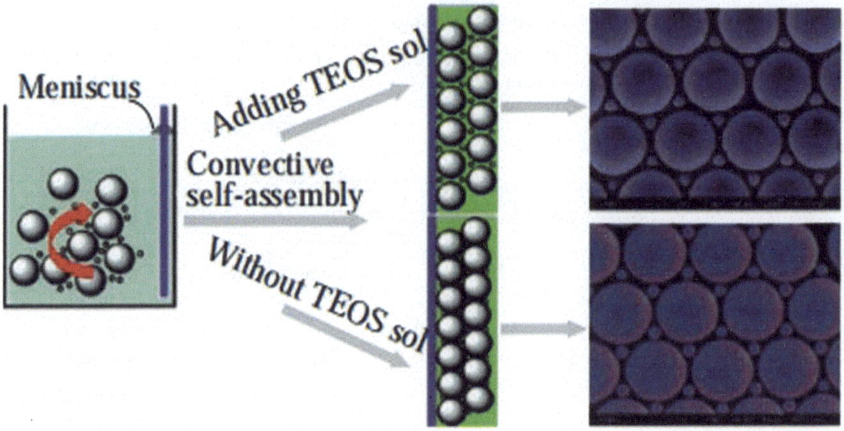

Fig. 9.4 Preparation of crackless polystyrene/silicon dioxide thin films by improved vertical deposition method

films by adding a polymer binder (triacrylate monomer). Spraying is a method of self-assembly of colloidal particles on substrates by spraying the dispersion containing colloidal particles in a spray gun. Gui et al. [21] used the spraying method to prepare large-area tightly arranged ordered photonic crystal structures on different substrates with the help of hydrogen bonding between the carboxyl groups on the surface of the colloidal particles. The colloidal particles in the solution are the core–shell structure composed of hydrophobic polystyrene microsphere nuclei and hydrophilic polyacrylic acid shells, and after drying, the microspheres with different particle sizes were assembled in an orderly manner and showed different colors on the surfaces of different substrates.

Template-directed self-assembly is to control the crystallization direction and distribution of crystals by templates to obtain photonic crystal patterns with fewer defects and larger sizes. Xia et al. [22] confined colloidal particles in the gap between two glass substrates, thereby limiting the growth of colloidal photonic crystals. They also prepared a uniform colloidal photonic crystal microsphere array with controllable size, shape, and structure using silicon etching templates. Huang et al. [23] proposed a new method for self-assembly of centimeter scale narrow stop band colloidal photonic crystals on a low adhesion superhydrophobic substrate, as shown in Fig. 9.5 When the emulsion suspension evaporates, there is a backward three-phase contact line on the substrate, which releases tensile stress caused by latex shrinkage and completely eliminates cracks in the photonic crystals.

In addition to the use of solid templates for domain-limiting, liquid membranes or gas–liquid interfaces can also be used for assembly. Vogel et al. [24] prepared binary ordered monolayers of photonic crystal membranes formed by close-packed large (L) and small (S) spheres arranged at intervals at the gas–liquid interface of a Langmuir tank with monodispersed hydrophobic polystyrene microspheres and an ethanol solution. A new type of injectable photonic hydrogel

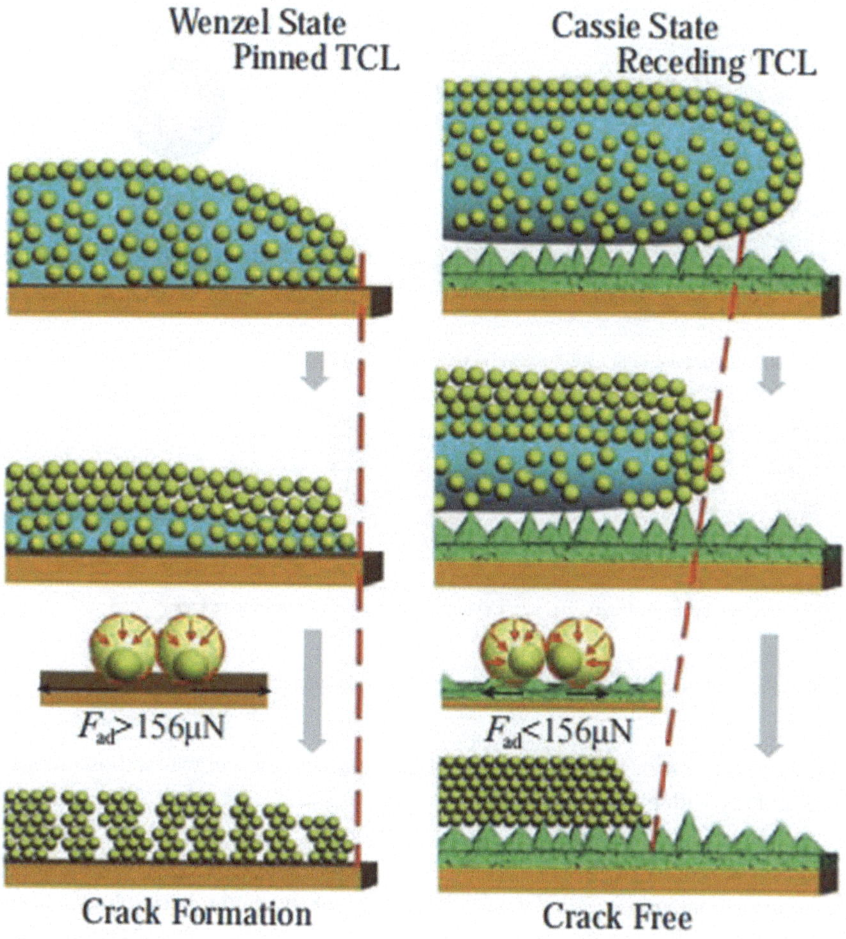

Fig. 9.5 Preparation of large-area crack-free photonic crystals on low-adhesion superhydrophobic substrates

was self-assembled by combining microfluidic-derived photonic hyperspheres with supramolecular hydrogels using microfluidic emulsion polymerization, and self-repairing amorphous photonic crystals were prepared in a large scale and rapidly. As shown in Fig. 9.6, firstly, core–shell colloidal microspheres were composed of a polystyrene core and a poly(2-hydroxypropylacrylic acid-N-vinylimidazole) copolymer shell, and then photonic hyperspheres were constructed by microfluidic technology using the core–shell colloidal microspheres as the assembling units, and finally, supramolecular hydrogels, which were made by the combination of β-cyclodextrin/poly(2-hydroxypropylacrylic acid-N-vinylimidazole) with gelatin methacrylate, were combined as the scaffolds, with photonic crystals of self-repairing

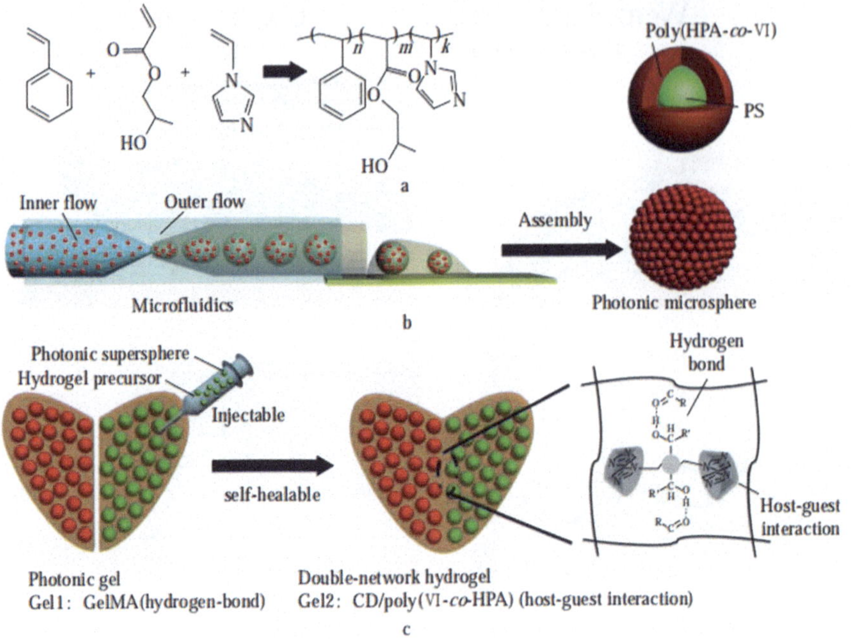

Fig. 9.6 Preparation process of self-healing amorphous photonic crystal

amorphous type. As a scaffold, an injectable photonic hydrogel was self-assembled by combining with photonic hyperspheres [25].

1. Inkjet printing method

The solution containing colloidal particles as ink, inkjet printing technology can be convenient, fast, non-contact preparation of colloidal photonic crystal patterns, but also to prepare optical devices with special properties. Therefore, inkjet printing is the most promising method for building micro/nano periodic structures of optical devices in the next generation.

2. Microphase separation method

The microphase separation method does not require monodisperse polymer latex particles as the basic assembly unit, and photonic crystals can be prepared by microphase separation between different components of the block copolymer. Fink et al. [26] separated two components of styrene-b-isoprene group copolymer based on the microphase separation method, as shown in Fig. 9.7a, and obtained three-dimensional ordered body-centered cubic structures, hexagonal arrangements, micropillar arrays, double network structures, double diamond structures, and one-dimensional lamellar periodic structures. microcolumn array, double network structure, double diamond structure and one-dimensional lamellar periodic structure. Subsequently, the group obtained a more complex double gyro periodic structure by microphase separation of polystyrene-polybutadiene-poly (methyl methacrylate) ternary block copolymer as shown in Fig. 9.7b.

Laser is the heart of optical communication system. Traditional lasers have some problems, such as laser emission wavelength changes on the wavelength division multiplexing system has a negative impact; with the increase in laser power, laser linewidth saturation, and began to re-expansion, which is very unfavorable for coherent fiber optic communications; laser radiation angle has been very small, but

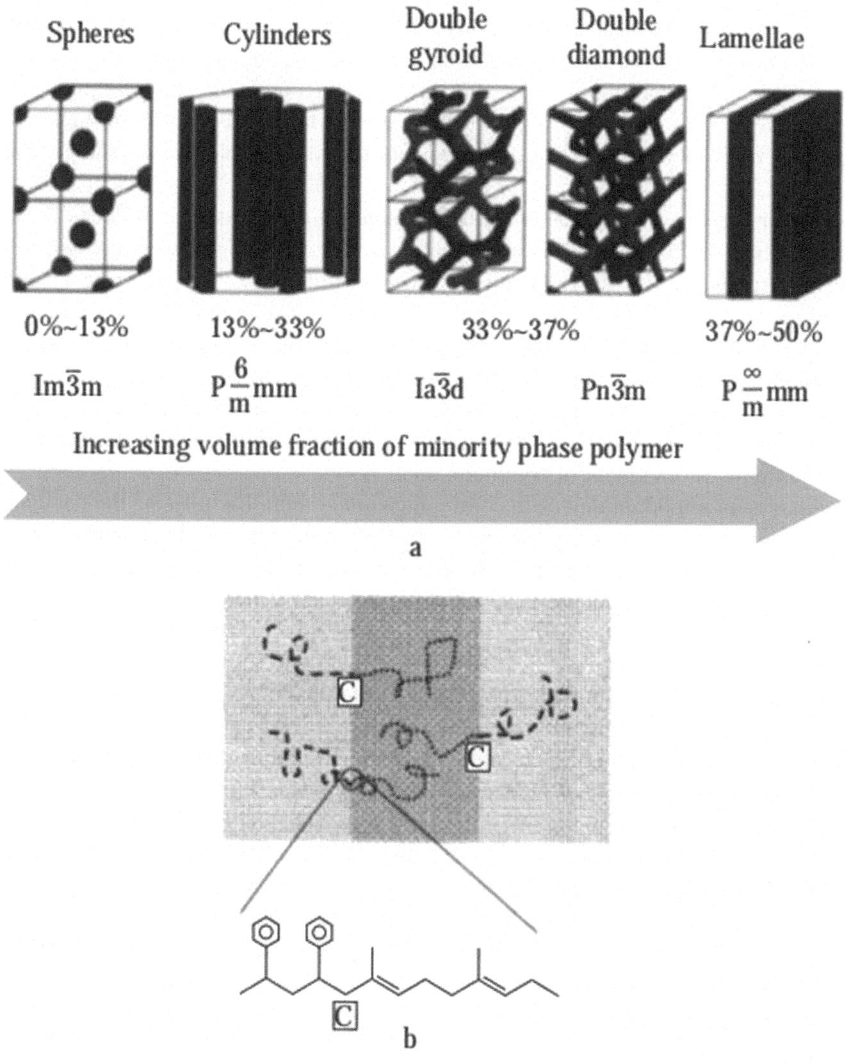

Fig. 9.7 Ordered structures of photonic crystals prepared by microphase separation of block copolymers

is still not an ideal beam, which makes the coupling efficiency of the laser can only be 30–50%; photonic crystal lasers can have excellent frequency stability, linewidth and radiation angle characteristics, completely able to avoid the need to use the laser for optical communication systems. 50%; photonic crystal lasers can have excellent frequency stability, linewidth and radiation angle characteristics, completely able to avoid the problems of traditional lasers.

Existing lasers due to the existence of spontaneous radiation, only when the drive current reaches a certain threshold to produce laser; in the laser with defects added to the photonic crystal, so that the defective state formed by the same direction of the optical waveguide and the emission direction, and the defective state of the energy and the energy of spontaneous radiation, so that the spontaneous radiation of the energy can be almost fully utilized, thus greatly reducing the threshold of the laser; At present, people in the threshold of the reduced Photonic crystal laser threshold has done a lot of work, and the development of zero threshold laser is the goal of scientists.

Photonic crystal waveguide has excellent bending effect. In general fiber waveguides, when the turning radius is too small, some of the light wave energy will be lost, resulting in a decrease in transmission efficiency. The photonic crystal bending waveguide utilizes the principle of resonance matching of defective modes in different directions. In principle, as long as the mode matching is achieved, even if the bending angle is 90°, high transmission efficiency can be achieved. The excellent bending transmission characteristics of photonic crystals are crucial for optical path integration and optoelectronic integration technology, and are of great significance for the future development of communication systems.

Filter is an important device in communication system. By utilizing the bandgap characteristics of photonic crystals, broadband bandstop filters and very narrowband frequency selective filters can be fabricated to obtain excellent filtering performance [27]. The bandwidth of photonic crystal filters can be very large, realizing broadband filtering from very low frequencies (almost 0 Hz) to the infrared. This wide range of filtering effect is impossible to realize with conventional filters.

Communication systems are inseparable from antennas. Conventional electromagnetic wave transmitting antennas are large in size and cannot achieve high efficiency. And with the photonic crystal made of small dipole plane microwave transmitter antenna, can be almost 100% of the electromagnetic waves emitted into the space, for the antenna into the integrated circuit to create the conditions.

Photonic crystals can also be used to make many key devices in optical communication systems, such as optical amplifiers, optical memory, optical limiters, optical frequency converters, polarizers, optical switches and optical switches. The performance of communication devices made of photonic crystals will be much better than that of the corresponding electronic devices.

Photonic crystals also create conditions for the ultra-miniaturization of semiconductor lasers and circuits. Combined with two-dimensional photonic crystal thin films and other technologies, including visible light, infrared and microwave bands can be realized in the integrated optical circuit, the formation of photonic crystal-based device system, the establishment of optical frequency domain electronics [28].

This is of great value and significance for the development of optical communication technology and the whole information industry.

Compared with the dye color, photonic crystal structure color has the characteristics of high saturation, good stability, not easy to fade, green and so on, and the fluorescence enhancement characteristics of photonic crystal make photonic crystal display devices completely do not need a backlight, so photonic crystal is a new generation of green materials, can be widely used in the field of imaging display.

Fan [29] and others prepared a smart-responsive structured color film for the first time by partially embedding a monolayer of polymer microspheres in the adhesive polyacrylate layer of a transparent tape, as shown in Fig. 9.8a. The smart-responsive structural color film undergoes interference and total internal reflection on the surface of the embedded microspheres and the surface of the unembedded microspheres, respectively, resulting in a structural color of the reflected light, and at the same time, under coaxial or non-coaxial illumination and observation conditions, iridescent or non-iridescent colors with a wide viewing angle can be observed, respectively, as shown in Fig. 9.8b, c. This intelligent, energy-saving, and adjustable structured color film can be used in the fields of intelligent display and traffic safety signs (Fig. 9.8d), which effectively improves the safety of traffic at night. Yang et al. [30] significantly improved the brightness and saturation of the pattern by adding refractive index-matching layers such as polymethylmethacrylate and dimethylsulfoxide to the super-surface of monocrystalline silicon on an alumina substrate, which significantly extends the color gamut area. color gamut area.

Periodic nanostructures and tightly arranged responsive materials make photonic crystals have optical properties that are difficult to replicate, and the color of photonic crystals comes from their periodic structure rather than pigments or dyes. Therefore, if suitable materials are selected to construct photonic crystal structures, their color and optical properties will be more stable, can be used for a long time, and the cost will be lower. Therefore, photonic crystal materials have become an indispensable part of the anti-counterfeiting field.

Choi et al. used nano imprinting technology to form a grating structure composed of epoxy resin, and then coated polyacrylamide hydrogel on the grating structure to prepare a water responsive photonic crystal grating, as shown in Fig. 9.9a. When the hydrogel coating expands when encountering water, the refractive index of the hydrogel changes, while the refractive index of the epoxy resin remains unchanged, and the photonic crystal structure shows a strong Iridescence color, as shown in Fig. 9.9b. After drying, the microcolumn bends, the refractive index of the hydrogel layer is close to that of the grating layer, and the surface is transparent without producing Structural coloration, as shown in Fig. 9.9c. The creative combination of hydrogel materials and two-dimensional grating structure provides a new and simple strategy for realizing reversible, durable and fast response cryptography, which has far-reaching significance for the improvement of anti-counterfeiting technology. Li et al. used a simple inkjet printing method and Computer-aided design to design patterns with physical non replicable functions, and produced structural color anti-counterfeiting labels with reliable random photon structures and convenient AI authentication.

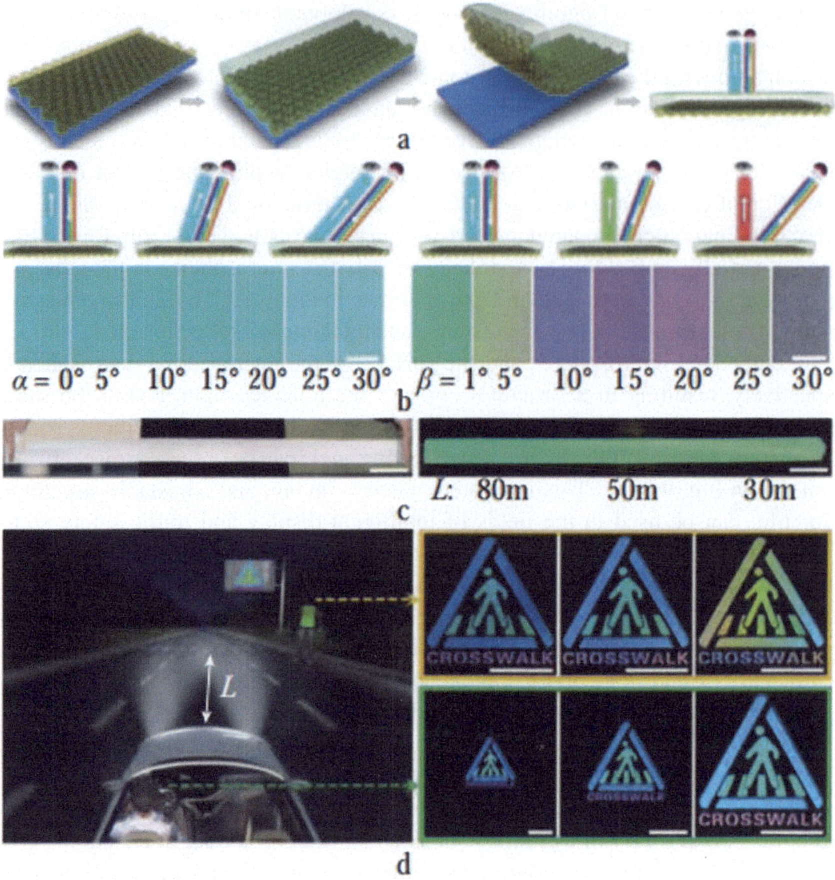

Fig. 9.8 Responsive photonic crystal structure color film for smart display

Photonic crystals not only have the optical properties of photonic band gap and photon localization, which can selectively enhance the fluorescence effect, but also can be combined with Photochromism materials to achieve Intersectionality response. Therefore, efficient ion detection can be realized based on the change of photonic crystal color.

Qin et al. prepared a photonic crystal sensor chip containing a Photochromism molecule (spirooxazine). Spirooxazine metal complexes display different fluorescence intensities under dark, ultraviolet or visible light stimulation, which provides a variety of fluorescence signals and is helpful for the selectivity and Intersectionality reaction detection of Photochromism sensor chips. Using similar working principles, Huang et al. assembled the poly (styrene Methyl methacrylate acrylic acid) nanoparticles to form a multimodal photonic crystal sensor chip, as shown in Fig. 9.10a. Based on the principle that the Structural coloration produced by photonic crystals

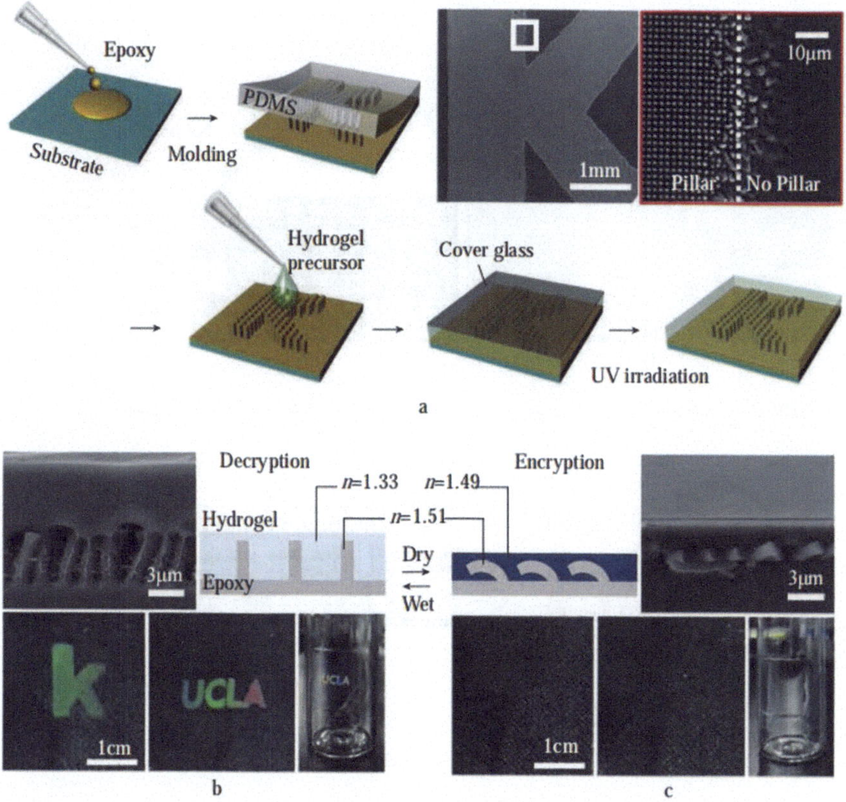

Fig. 9.9 Anti-counterfeit labels prepared by water-responsive photonic crystal gratings

is angle dependent, different forms of photonic crystals show selective fluorescence enhancement at different locations, providing rich differential fluorescence information, which makes it easy to identify 14 kinds of metal cations, and successfully identify 12 groundwater samples with similar components, as shown in Fig. 9.10b–c.

In addition to the detection of ions by color changes, the property of photonic crystals to amplify optical signals can be applied to the detection of amines or drugs in cells, thus enabling the introduction of cross-responsive photonic crystals into the field of biomedical detection. Ren et al. fabricated a sensor chip with a graphene oxide and a poly (styrene-methylmethacrylate-acrylic acid) bilayered photonic crystal structure to distinguish between 10 biogenic amines and 7 pharmaceutical amines. As shown in Fig. 9.11, the graphene oxide aerogel acted similar to olfactory cilia, effectively capturing analytes and generating rich sensory signals. Photonic crystals act as olfactory spheres, and their periodic structure selectively enhances the fluorescent signals, allowing for further signal processing. The integrated sensors with responsive signal manipulation units are useful for multi-analyte

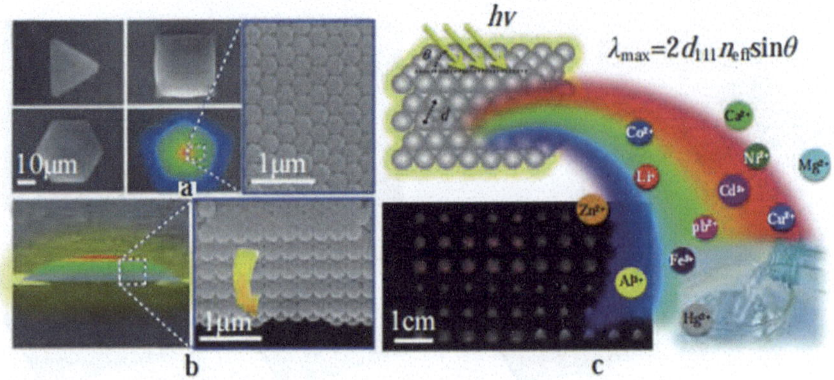

Fig. 9.10 Polymorphic photonic crystal sensor chip that can detect groundwater samples

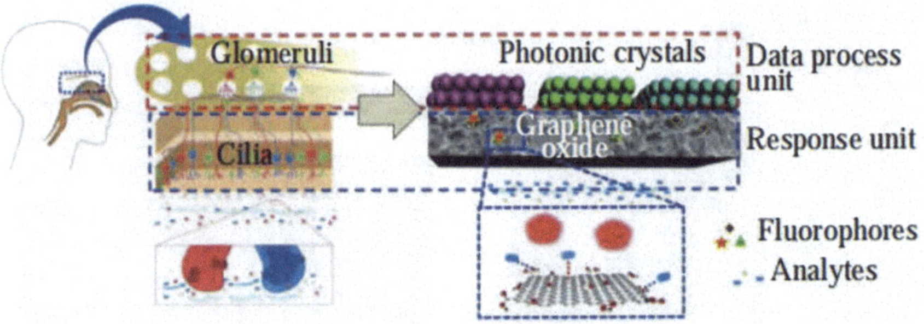

Fig. 9.11 Dual-layer sensor inspired by the olfactory system

detection and complex system analysis, and will solve problems such as clinical diagnosis or difficult pathology analysis.

In addition, photonic crystal sensors have extensive research and application in the fields of temperature, electricity, force, magnetism, pH value, humidity, etc. Moreover, some multi-responsive sensors can realize the response to multiple external stimuli. Moreover, some multi-responsive sensors can also realize the response to a variety of external stimuli. Liao et al. prepared a multi-responsive film constructed by responsive elastic copolymer nanoparticles, which can not only control the hiding or displaying of encrypted color information, but also respond to external stimuli such as solvents, strains, and pH value in a reversible manner.

Photonic crystal fiber is a kind of two-dimensional photonic crystal, and it is also the most researched photonic crystal. The photonic crystal fiber can be formed by introducing a wire defect as the core and restricting the light in the core of the fiber.

Photonic crystal fiber has many extraordinary characteristics: endless single-mode transmission characteristics, controllable nonlinear characteristics, excellent dispersion characteristics and birefringence characteristics; at the same time, through the physical structure of the optical fiber or fiber material changes can be achieved by changing a certain characteristic of the fiber or achieve a specific combination of certain characteristics [31]. Therefore, photonic crystal fiber will be widely used in communication systems.

The core of the photonic crystal fiber for the air holes, will make the input light wave in the surrounding material of the forbidden effect of the propagation along the center hole, which will greatly reduce the transmission loss, so that the output information increased by hundreds of times. The solid-core photonic crystal fiber is rich in nonlinear effects, in the optical soliton generation and transmission, optical frequency conversion and other important applications.

The traditional single-mode fiber only supports single-mode transmission within a certain frequency range, and multimode transmission will occur when the frequency is higher. As long as the adjustment of the air hole diameter and hole spacing ratio, photonic crystal fiber can be designed to support single-mode transmission in the entire optical frequency range, which is the so-called "Endlessly Single-mode Transmission" characteristics (Endlessly Single-mode). A well-structured photonic crystal fiber has the characteristic of supporting single-mode transmission in the wavelength range from 337 nm to more than 1550 nm.

The Endlessly Single-mode property of photonic crystal fibers does not depend on the absolute size of the fiber, which suggests that the mode field area of the fiber can be designed according to specific needs. By changing the effective area by changing the structural parameters, photonic crystal fibers can easily obtain an effective area of the mode field close to 150 nm^2, which enables low-loss transmission of optical signals.

The zero dispersion point of conventional single-mode fiber is usually at 1310 nm. By reasonably adjusting the aperture size and spacing of the photonic crystal fiber, the zero-dispersion point can be moved to below 800 nm, and the zero-dispersion point is moved to short wavelengths, making the photonic crystal fiber able to obtain anomalous dispersion (positive dispersion) at wavelengths lower than 1.3 μm, which cannot be achieved by conventional step fibers. This anomalous dispersion property provides the possibility of short-wave optical soliton transmission for the first time. In addition, photonic crystal fibers with finer air-hole diameters in the cladding can achieve positive, near-zero, and negative flat dispersion in a wide wavelength range around 1550 nm by optimizing the air-hole diameters and spacings, which makes photonic crystal fibers promising for applications in WDM transmission systems.

9.2.4.2 Photon Storage

The information age requires highly advanced information storage technology, and people need to deal not only with data, text, sound and images, but also with moving images and high-definition images. The capacity of information will be measured in terabytes, and the flow of information will be measured in terabytes per second (TB/S). People use the means of information is also from the simplest text, images and data storage of paper records, micrographic storage to magnetic storage. The earliest form of optical storage for micrographics, from the late 1960s the United States IBM and ECD jointly developed the first CD-ROM since the optical storage technology will be an unexpected speed of rapid development. Compared with magnetic storage, optical storage technology has the following characteristics: (1) high storage density, large capacity, easy to carry. (2) long storage life, multi-functional. (3) non-contact read, write and erase. (4) High carrier-to-noise ratio of information. (5) low price per unit of information. In spite of this, the optical storage technology to have new improvements and development, but also in the storage density and data transfer rate must have sufficient growth, so improve the optical storage density to become the development direction and development trend of optical storage.

Photonic storage is based on the direct interaction of photons with the recording medium, i.e., the recording medium first absorbs photons. This light is absorbed by the material process is very fast, up to the femtosecond scale, the result is first of all, leading to the formation of an ultra-short (nanoseconds to femtoseconds) excited state, and then state or structural changes, that is, from the "1" state to "0" state, and vice versa. This direct interaction between photons and matter or recording medium based on the storage method, determines its greatest advantage lies in the ultra-high recording speed (easily reach nanoseconds and picoseconds scale), ultra-high density and ultra-high resolution. Theoretically, the recording resolution can reach the molecular level, but the actual resolution will be determined by the recording system, including the laser wavelength, drive device, signal processing and conversion processes.

This photon with the material or recording medium on the basis of the direct action between the storage method, determines its greatest advantage lies in the ultra-high recording speed (can easily reach nanoseconds and picoseconds scale), ultra-high density and ultra-high resolution. Theoretically, the resolution can reach the molecular level, but the actual resolution will be achieved by the recording system, including the laser wavelength, drive device, signal processing and conversion processes. In the current technical conditions, two-dimensional (planar) recording density limit is still in the 108–109 bits/cm^2.

Photon storage and recording method has the following four kinds.

This is currently the most commonly used way, that is, the laser through a circular lens will be focused into a spot, the role of the recording medium, by the interaction between the photon and the medium to complete the information recording. A spot that is a bit of information, after the completion of a spot and then go to record another spot, by a certain number of spots to achieve a complete record of information. The size of the spot and the laser wavelength, the shorter the wavelength, the smaller the spot, thus increasing the information density per unit area.

This format is a vector-parallel system with a high bandwidth that extends from a spot to a line. This vector parallel system is characterized by high recording density and high processing speed. It is 3 orders of magnitude higher than the first method in terms of data parallel processing and information bandwidth. Obviously, this vectorial approach is realized by laser one-dimensional array.

This approach is to expand from recording a line to record a flat image at the same time. In the computer is becoming more and more popular, the information superhighway is gradually realized today, the requirements of more and more large-area image storage and reproduction. This laser two-dimensional array through the realization of the recording format, just to meet the high-capacity, high-speed information superhighway in the information storage, transmission and processing requirements.

This method realizes holographic optical storage through multifunctional functions such as angle (phase), wavelength and space. This is also a multi-dimensional storage method, is a high-density storage method, its storage density can reach 10^{12}–10^{13} bits/cm^3. According to the literature, more than 160,000 pieces of information can be stored in a 1 cm^3 crystal.

The basic principle of photonic storage is based on the direct interaction between photons and matter, resulting in changes in the physical and chemical properties of the recording medium that can be recognized, thus achieving the purpose of information recording. According to the characteristics of bistable state before and after recording, both timely storage and permanent memory can be realized. The four different mechanisms currently being studied are described.

Photonic burn holes can be roughly divided into two categories, namely, photochemical burn holes and photophysical burn holes, now focusing on photochemical burn holes.

Photochemical burn-in refers to the light reactive molecules to a single molecule dispersed in a solid medium, low temperature, in the laser-induced photochemical reaction with position selectivity, resulting in a non-uniform broadband absorption spectral band selectively produce a uniform spectral hole. The reason for this phenomenon should be attributed to the micro-uniformity of the light-recording medium, that is, the light-active molecules in the inhomogeneous micro-environment, so that the absorption frequency of the existence of weak differences, which leads to a non-uniform absorption band, can be used to tune the laser burned out a number of uniform narrow bands. A is non-uniform Absorption band, b is uniform Absorption band. When the laser irradiates on the non-uniform Absorption band, a photochemical reaction occurs, that is, a hole is burned out to realize information recording. The absorption of its photoproducts is c. This is a high-density storage method that can add a frequency dimension to the original two-dimensional storage, thereby improving its optical storage density.

Electron capture optical storage is based on photoinduced electron transfer reactions in storage materials. The material consists of at least two structural units, atoms, ions, molecules or clusters with different redox potentials. One of them is an electron donor (D) and the other is an electron acceptor or hole (A or H^+). Theoretically the reaction is simple and straightforward, however, in terms of practical requirements, the biggest difficulty in preparing such storage materials is how to achieve bistability. That is to say, the forward Electron transfer, that is, the energy of photons required in the information writing process should be greater than the energy of photons required in the reverse Electron transfer or erasure process. This change can be illustrated in Fig. 9.12.

E_D is the ground state of the electron donor. When excited by the write light ($h\nu_1$), an electron is excited by a photon from the valence band (E_D) of D to the Excited state (E_D^*) of D. In this state, it is associated with the electron acceptor, and its Electron transfer rate constant is k_{ET}. The Excited state electron emits heat

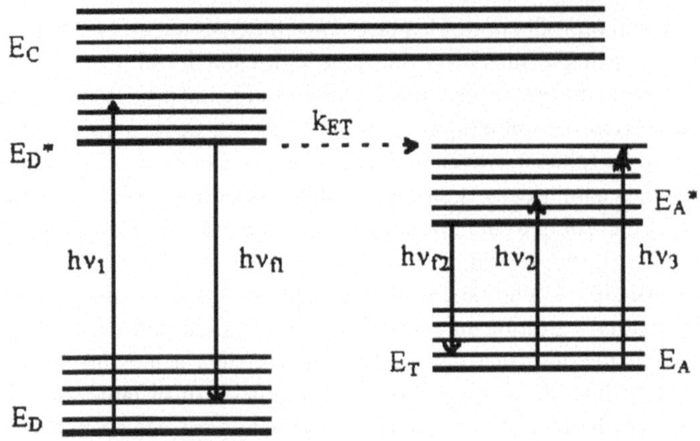

Fig. 9.12 Schematic of energy levels of electron capture storage materials

or fluorescence ($h\nu_{f2}$) in the Excited state (E_A^*) of the electron acceptor to reach its ground state (E_A), and the electron is trapped in this state. Therefore, this state can also be called the Electron capture state (E_T, traveling level) to complete the writing process. Its readout process uses another beam of light ($h\nu_2$), which can only reactivate the trapped electrons to the Excited state (E_A^*) of the receptor. The appropriate system must be selected so that the Excited state energy of the receptor is lower than the Excited state energy of the donor, $E_A^* < E_D^*$. When the readout light is cancelled, the excited electrons return to the ground state E_A in the form of heat or emitted fluorescence ($h\nu_{f2}$), thereby achieving non-destructive readout. When the energy of the used readout photon ($h\nu_3$) is too large, the electron of A can be excited to the higher vibration level of its Excited state, so that it is in the same energy state as the Excited state of the electron donor, that is, the degenerate state. As a result, the electron can either return to E_A or reach E_D^* through inter system crossover, and then release heat or fluorescence ($h\nu_{f1}$) to return to E_D. If this situation occurs, it is the erasure of written information, which is destructive reading. Therefore, the most ideal operation process of Electron capture optical storage is to use three beams of light: one is to write light ($h\nu_1$); The second is readout light ($h\nu_2$); The third is erasing light ($h\nu_3$).

The conduction band energy level (E_c) of the Electron capture material is also shown in Fig. 9.12. Whether in the process of writing, reading or erasing, if the photon energy selected is too large, it is possible to excite the electrons in the valence band to the conduction band, and no result is obtained from the perspective of information storage. From this, it can be seen that the optical storage process must select photons with appropriate energy based on the characteristics of the material.

The photorefractive effect in inorganic materials has been discovered for more than 20 years, but research on organic photorefractive materials began only in 1990 and has developed rapidly since the 1990s. The potential applications of such materials are wide-ranging, including high-density optical information data storage, a variety of image-processing techniques, and the simulation of neural networks and associative memories. The photorefractive effect is based on photophysical-photochemical processes in matter. When a charge carrier is subjected to spatially modulated light intensity, a system of unequal spatial charge distribution is generated, and the light-generated charges are captured separately, thus achieving the purpose of long-term charge separation. A photorefractive material is subjected to light, resulting in the generation of an internal space-charge electric field, a change in the reflectance index of the material, resulting in a phase grating effect to achieve the purpose of optical information storage.

This storage technology uses the phenomenon of photochromism for storage. Photochromism refers to the phenomenon that certain inorganic and organic compounds undergo reversible changes in their absorption spectra under the action of light. Since photochromic materials, especially organic compounds, are changed by photons in molecular units, they have a very high resolution and have the possibility of being used as reversible storage media for high-density information storage. In the ideal model of photochromic storage, the storage medium has two absorption bands, and under the irradiation of the wavelength λ_1, the state is changed from 1 to 2. Similarly, under the irradiation of the wavelength λ_2, the medium is changed from state 2 to 1. Firstly, by irradiating with the light of the wavelength λ_1 (the erase light), the storage medium is changed from state 1 to 2. When recording, the part of the storage medium irradiated with the light of the wavelength λ_2 (the write light) is written with the binary encoding, and the portion of the storage medium irradiated with λ_2 is changed from state 1 to 2, and the portion of the storage medium irradiated with λ_1 is changed to state 2. When recording, the binary code is written by the light of wavelength λ_2 (writing light), so that the part irradiated by λ_2 is transferred from state 2 to state 1 and the binary code "1" is recorded; the other part which is not irradiated by the light of λ_2 is still in state 2, which corresponds to the binary code "0".

Information can be read out either by reading the change in transmittance or by reading the change in refractive index. The method of reading out the change in transmittance is to measure the change in transmittance by irradiating light with a wavelength of λ_2 and reading out the information. This method has the disadvantage of destroying the original data. However, in 1994, Yokoyama et al. succeeded in developing an optical disc prototype using indole-type captured arginic anhydride, and solved the key problem of non-destructive readout of organic photochromic materials by utilizing the relationship between the quantum yield of achromatic reaction and wavelength. Read out the refractive index change is the use of wavelength is not in the two absorption spectrum of light irradiation, measurement of its refractive index change and read out the information, which requires thickening of the thickness of the recording medium, write the energy density and power of the light should be increased by several times.

In short, photochromic materials in the photonic storage will play an increasingly important role, can develop a write multiple readout devices, but also can develop write -read-erase recycling devices.

Theoretically, the size of an optical storage information recording domain and the recording wavelength used, that is, limited by the diffraction of the light beam. In the field of optical disk is still commonly used recording wavelength is the near-infrared laser, with the gradual development of laser technology, short-wave lasers will be gradually introduced, the recording density will also gradually increase.

Rentzepis firstly introduced the idea of two-photon three-dimensional bulk phase storage, and utilized the principle of photochromism to study spiropyrans as model compounds. However, the fatal drawback of this kind of photochromic material is the poor fatigue resistance, and the second is the poor preservation of information and the short lifetime at room temperature.

Since 1995, Meigong Fan and Rentzepis's research group to work closely together, the development of new photochromic materials, is expected to be developed in the next five years at room temperature bistable performance, fatigue resistance is good, can be of practical value of the three-dimensional "massive" optical storage materials and devices, the storage density of up to 10^{12} bits/cm^3.

Optical holographic data storage is a kind of storage mode with large capacity and high storage density that appeared with the development of laser holography in the 1960s. This large capacity is achieved by using Fourier transform holograms to create small holograms with a diameter of 1 mm, arranged in an array, or arranged on a rotating disk like a record. In addition, holographic information storage has the advantages of high reliability and recording and reproduction. In the optical path of the storage system, lens L_1 plays a collimating role, and the input information is placed on the front focal plane of lens L_2, where the hologram is recorded. The reconstruction is done through lens L_3. If the original optical path is used for reconstruction, the original image corresponds to the + 1 diffraction light passing through the hologram and passing through L_3 to form a real image.

To reproduce and amplify the real image optical path, the input surface can also be located between the front double focal length of the lens and the focal point, and the recording medium can be placed on the conjugate image surface of the illuminating light source. Its lighting source can be either a directional light or a spherical wave.

In short, optical storage is a key and important research field in Information science. Photonic storage represents the forefront and development direction of information storage. In the past 10 years, a large number of human and material resources have been actively invested in research and development at home and abroad. Photochromism optical storage materials have been on the eve of application and development in two-dimensional optical disks, and significant progress has been made in the research of three-dimensional storage principles and devices.

9.3 Application of Nanotechnology in Electronic Devices

9.3.1 Nanoelectronics

Nanoelectronics is a science that studies the motion laws of individual quantum or quantum waves within nanostructures (quantum dots) at the scale of 0.1–100 nm, or detects, recognizes, controls them, as well as the artificial assembly and self-assembly techniques of individual atoms and molecules; It is a science to study the devices, circuits and systems that are used for the generation, transmission and exchange of information and their applications in Information science, nano biology, nano Surveying, nano microscopy, nano mechanics, etc. Nanoelectronics, also known as quantum functional electronics.

Individual quanta include individual electrons, individual electron pairs, individual flux quanta, individual atoms, individual molecules, solitons, photons, excitons, exciton pairs, polarizers, and so on. Quantum waves include electron waves, light waves, magnetoacoustic waves, charge density waves, spiral waves, spin waves, etc. Nanoelectronics can integrate semiconductor electronics, superconducting electronics, atomic electronics, molecular electronics and so on, which brings infinite vitality to the development of new devices and application technologies.

Based on different development stages and research perspectives, nano-electronic devices with the characteristic size of 0.1–1.0 nm are called sub-nanometer devices, which are essentially atomic and molecular scale electronic devices, and the corresponding electronics is sub-nanometer electronics. 1.0–10 nm nano-electronic devices are called nano-devices, and mainly belong to the quantum structure of quantum dots (also known as man-made atoms), quantum wires, etc. Electronic devices, and the corresponding electronics are sub-nanometer devices, and the corresponding electronics is sub-nanometer electronics. They are mainly quantum dots (also called artificial atoms), quantum wires and other artificial quantum structures, and the corresponding electronics is called nanoelectronics. The nanoelectronic devices of 10–100 nm are called quasi-nano devices, mainly the current nano-CMOS devices and resonance tunneling effect devices of III–V compound semiconductors, etc., and the corresponding electronics is called quasi-nano electronics [32]. Single-electron electronics and single-electron pair electronics can be realized in sub-nanoelectronics, nanoelectronics, and quasi-nanoelectronics, but only the permissible operating temperature is different.

Artificial atomic devices, artificial molecular devices, single-atom devices, and single-molecule devices have the potential to lead to new concepts in the design of computer systems and communication systems due to the new concepts of quantum network design that will be opened up in circuit and system design.

Nanoelectronics research focuses on nanodevice physics, new concepts of device and circuit and system design and process technology, simulation, fabrication and testing of nanodevices and circuits and systems, and applied research. The key research areas of device physics include: the structure of quantum surfaces, quantum wires, quantum dots, quantum rings, quantum dot contacts, and atomic and molecular self-assembly; various quantum effects such as quantum tunneling effect, quantum coherence, quantum volatility, ballistic quantum transport, quantum size effect, Coulomb blocking, single-electron oscillations, Bloch oscillations, and singular electrical conductivity; and the heterogeneous interfaces of superconductors-conductors-semiconductors-insulators physical effects of quantum waves and heterojunction quantum dots; singular properties of high-temperature copper-oxide superconductor normals, electron waves, electron-pair waves, and flux-line motion effects due to two-dimensional electrical conductivity; single-electron versus single-electron-pair electronics and single-flux quantum electronics; and mechanisms of atomic and molecular self-assembly [33]. The main research content of new concept devices, circuits, and systems includes: (1) quantum surface: resonant tunneling effect → resonant transistor, circuit, and system; (2) Quantum wire: high mobility microelectronic gas → ultra-high speed logic switches, circuits and systems; (3) The principle of electron harvesting in quantum dots → maximum capacity memory; (4) Coulomb blocking effect, single Electronic oscillation and single electron tunneling effect of quantum dot contact → Single-electron transistor, circuits and systems (including single electron switch and single electron memory); (5) Single electron pair tunneling effect and Bloch oscillation in quantum dot contact → single electron pair transistor, circuit, and system (including single electron pair switch and single electron pair memory); (6) Quantization of magnetic flux of two-dimensional superconductor quantum plane → single flux quantum transistor, circuit and system (including single flux quantum switch and single flux Quantum memory); (7) Quantum dot electronic systems (such as wire less integrated circuits) → 4 quantum dots form a unit → multiple units are connected together → electronic motion between units is carried out in a coupled manner → signal transmission is achieved; (8) Single atom switches and single atom memories; (9) Molecular wires, molecular switches, and molecular memories. The main research content of process technology includes: thin layer growth techniques such as atomic layer epitaxy, MBE, MOCVD, and lithography techniques such as X-ray, electron beam, and soft X-ray; Ultra micro machining technology using Scanning probe microscopy as tool.

Generally speaking, the main content of nanoelectronics includes two aspects. On the one hand, it is to design and fabricate new generation of quantum devices according to new effects, and on the other hand, it is to develop new integrated circuit fabrication methods. New nanoelectronic devices are: (1) the use of quantum effect of electron well resonance tunneling devices and the use of the device negative resistance effect of microwave oscillators and superlattice devices; (2) the use of quantum dot contact Coulomb blocking, single electron tunneling effect of quantum dot field effect tube, single electron transistor (SET); (3) the use of quantum lines of one-dimensional electron gas of the design of high electron mobility effect of the ultra-high speed Logic switching devices; (4) the use of quantum dots can accumulate the principle of electronic design of ultra-high density information memory. New integrated circuit production methods such as the use of scanning probe microelectronics technology (SPM) production of "wire-free integrated circuits" (Japan ToshibaCo), that is, four quantum effect box composed of a basic unit, coupled to each other with the movement of electrons between the units to achieve the transfer of information, due to the absence of wires and the stability of circuit performance, and very small power consumption. And power consumption is very small, it will be combined with molecular devices, superconducting devices, neuron photonic devices and vacuum microelectronic devices, constituting the basis of the sixth or seventh generation of calculators.

So in short, nanoelectronics refers to the study of physical, chemical, acoustic and electronic phenomena and their laws of motion within the scope of nanometer scale, the construction of nano and quantum devices, the realization of nano-integrated circuits, and thus the establishment of quantum-based computers and quantum communication systems, information computing, transmission, processing of a science and technology.

Now, nano-electronics technology is in a period of vigorous development, and its ultimate goal is to be based on the latest physical theory and the most advanced technology, break through the traditional physical dimensions and technical limits, develop the potential information and structure potential of matter, design and manufacture nano devices and construct electronic systems according to new concepts, so as to make the electronic system's ability of storing and processing information to realize a revolutionary leap.

At present, people use nanoelectronic materials and nanolithography, has developed many nanoelectronic devices, such as electronic resonance tunneling devices (resonance diode, triple resonance tunneling, single-electron transistors, metal-based, semiconductors, nanoparticles, single-electron static meter, single-electron, semiconductor Memory, silicon nanocrystal memory, nano-floating-gate memory, nano-silicon microcrystalline thin-film devices and aggregates). electronic devices, etc.

Nanoelectronic devices are the next generation of microelectronic devices, but they have different materials, processing and assembly techniques and working mechanisms from microelectronic devices [34]. So far, the following three types of nano-electronic devices have been developed, i.e., quantum dots (QD), where the island of the device is made to confine electrons with zero-dimensional degrees of freedom; resonant tunneling devices (RTD), where the island is made to confine electrons with one- or two-dimensional degrees of freedom; and single-electron transistors, where the island is made to confine electrons with three-dimensional degrees of freedom. The minimum size of islands in nanoelectronic devices is about 5–100 nm.

Microelectronic limit leads to a new type of quantum devices, namely tunneling transistor, its size is only the size of the traditional device of the company has developed a single-electron transistor, only control an electron to complete a specific function. From the control of a large number of electronic transistor development to control a single electron transistor, device response speed can be increased by 1000–10,000 times, while power consumption can be reduced to 1/1000–1/10,000, its superiority is obvious.

Based on the current development and future predictions, if the main nano-electronic devices are further classified, nano-CMOS devices mainly include silicon-on-insulation MOSFETs (SOI MOSFETs), silicon–germanium heterojunction MOSFETs, low-temperature MOSFETs, dual-gate MOSFETs, intrinsic silicon-channel tunneling MOSFETs and so on; quantum effect devices include quantum interference devices, quantum Point devices and resonant tunneling devices, and resonant tunneling devices will include transverse resonant tunneling devices, longitudinal resonant tunneling devices and tunnel barrier modulation transistors, etc.; single-electron devices include single-electron box, capacitively-coupled and resistively-coupled single-electron transistors, single-electron neural network transistors, single-electron junction arrays, single-electron pumping, single-electron traps and single-electron rotary gates, etc.; single-atom devices and single-molecule devices Including single-atom switches, single-atom point contact devices, single molecular switches, molecular wires, quantum effect molecular devices, electrochemical molecular devices.

As mentioned earlier, in the nanoelectronic circuit and system design, due to the advantages and disadvantages of various types of nanoelectronic devices, it is possible to use a single type of nanocircuit, but also may constitute a hybrid nanocircuit of their respective advantages [35], such as nano-CMOS circuits and III–V compound semiconductors resonance tunneling effect circuits, nano-CMOS circuits and single-electronic circuits, nano- CMOS circuits and superconducting single flux quantum circuits, nano-CMOS circuits and carbon nanotube circuits, nano-CMOS circuits and artificial atom circuits and artificial molecule circuits, nano-CMOS circuits and DNA circuits, nano-CMOS circuits and nano-metallic-based spin circuits, and other mainstream nanocircuits, which have opened up a completely new development prospect for nanoelectronics.

9.3.2 Processing Technology for Nanostructures

Nanotechnology is an emerging discipline and field. The rapid development of nanotechnology in the twenty-first century will play an extremely important role in the progress of civilization and the development of human society, and may bring about the fifth technological revolution. All developed countries in the world are trying to seize the strategic high point of science and technology for this basic technology in the twenty-first century.

The strong vitality of nanotechnology lies in the nanoeffects (such as quantum effect, huge surface and interface effect, etc.), which can make many properties of matter change qualitatively, and the key to realize the nanoeffects is to have nanostructures in the first place, and any nanotechnology has to rely on the nano-processing technology to process the objects to the nanometer scale [36]. Therefore, nanostructure processing technology is the core foundation of the whole nanotechnology, and it is one of the urgent problems to be solved in the current world scientific research.

Nanoelectronics is one of the most important subfields of nanotechnology. The future development of electronics will aim at "smaller, faster and cooler". "Smaller" is to further improve the integration degree of the chip, "faster" is to realize higher information computing and processing speed, and "colder" is to further reduce the power consumption of the chip. Only in these three aspects have been synchronized development, nanoelectronics technology can make new major breakthroughs. To realize the above goals, the size of electronic devices will inevitably enter the nanometer scale range, that is, less than 100 nm, which indicates that, in the near future, microelectronic devices will certainly be transitioned to nanoelectronic devices, so that it becomes the core of the twenty-first century information age. Since the current lithography technology can process the minimum line width of 130 nm, can not meet the processing requirements of nanoelectronic devices, it will be replaced by more advanced processing technology, such as scanning tunneling microscope (STM) and atomic force microscope (AFM) nanofabrication technology.

STM can be used for nanoscale machining of surfaces in addition to inspection of the surface morphology and physical properties of the samples. The reason it can be used for machining is that when the STM is operated, a highly spatially confined electron beam is generated between the probe and the sample. This electron beam can be used for microfabrication of the sample surface in the same way as a normal focused electron beam. Since the distance between the tip of the needle and the sample is only on the order of nanometers, the area over which it acts on the sample is also on the order of nanometers. It is possible to etch the sample surface directly, photolithograph surfaces coated with resist, induce precipitation in liquid and gas atmospheres with the electron beam generated by the STM, manipulate and transport particles and atoms adsorbed on the surface of the sample, etc. The electron beam generated by the STM can also be used for microfabrication of sample surfaces.

The electron beam generated by the STM can also be used for electron beam lithography on surfaces coated with sufficiently thin (tens of nanometers) homogeneous resist films, provided that the substrate is electrically conductive.

There are van der Waals and electrostatic forces between the tip of the STM and the sample. By adjusting the bias pressure applied between them or by adjusting the position of the tip of the needle, the magnitude and direction of the forces can be varied and individual atoms can be moved.

In short, the emergence of STM provides an extremely important new tool for humans to understand and transform the micro world. STM's nano processing technology is a newly emerging and currently not yet fully developed process, but it has extremely broad potential functions: it is expected to reduce the LSI line width from micrometers to nanometers (which can process extremely uniform nano oxidized fine lines). Becoming an important means for future device processing (nanoelectronics) and molecular cutting (nanobiology). It is possible to detect surface morphology while repairing surface structural defects using induced deposition and etching methods; Capable of manipulating atoms and atomic clusters, producing single molecule, single atom, and single electron devices to significantly increase information storage capacity; It can also be used for species regeneration in life sciences and the creation of new atomic structural materials in materials science, achieving the goal of material design.

With the increasing improvement of the theoretical and experimental techniques of STM, it is bound to be widely applied in various research fields such as single atom manipulation and nanotechnology [37]. At that time, perhaps today's supercomputers will be as small as Pin. Even the newly developed Minisupercomputer in the United States with a peak speed of 12 trillion times per second will be small enough to be put in your pocket.

AFM is a nano probe device similar to STM. Unlike STM in detecting tunnel currents, AFM detects the micro cantilever deflection of the nanotip during scanning on the sample surface. This is important because it means that for AFM, the sample does not need to be conductive. Although AFM cannot manipulate atoms by changing voltage, it provides a way to push atoms. At present, an AFM based nanoetching technology has been developed and metal oxide field-effect transistors (MOSFET) with a characteristic size of 60 nm have been prepared.

Chemical synthesis is another way to prepare nanoscale electronic devices, which is to use Chemical process to "bottom up" the material units of the micro system to assemble nanoscale devices. Due to the difficulty of assembling a large number of nanostructures and devices simultaneously using nanoprobes for mechanical synthesis, it is very important to study chemical synthesis methods.

Chemical self-assembly refers to the spontaneous orientation of a large number of molecules to form the lowest energy Supermolecule structure without human intervention. The most common is self-assembly through non Covalent bond between molecules [38]. The information of Supermolecule structure is implied in the shape and functional groups of molecular units.

Compared with other processing methods for preparing nanostructures, an important advantage of chemical self-assembly is its inherent self correction process in the assembled chemical nanostructures, which allows the self-assembly process to create very similar copies of nanostructures and make the system quite stable. The presence of a large number of identical sites in the solution may lead to parallel processing of a large number of nanostructures, greatly improving processing efficiency.

Focused Ion Beam (FIB) technology focuses ion beams on the submicron or even nanometer scale under the action of electric and magnetic fields. By controlling the ion beam through deflection and acceleration systems, it achieves the detection and analysis of fine patterns and maskless processing of nanostructures.

In recent years, FIB technology has made significant progress, with its main feature being that ion beams can range from a few square micrometers to nearly 1 mm^2 Digital raster scanning can be carried out in the area of, which can collect secondary charged particles through Microchannel plate detector or channel electron multiplier to collect patterns; Remove unwanted materials through high-energy or chemically enhanced sputtering; Submicron patterns of deposited metal, carbon, or dielectric thin films.

Specific applications of FIB technology.

It can not only repair ordinary optical element masks, but also repair X-ray masks and advanced optical phase shifting masks. The main advantage of FIB repair is that its process has a high spatial resolution (which can be lower than 25 nm) and can repair various defects, including phase defects.

The etching and deposition capabilities of FIB technology can be utilized to modify failed or improved integrated circuits, resulting in significant reductions in design and production cycles. For example, undesired connections between circuits can be broken, or nano-connections can be formed between circuits with line widths of tens of nanometers and line lengths of tens of microns.

Recently developed focused ion beam/electron beam dual-beam nanofabrication system, you can use high-intensity focused ion beam nanofabrication of materials and scanning electron microscope observation in real time, opened up from the bulk of the material manufacturing nano-devices, nanofabrication of a new way. This nano-processing center has been used in the semiconductor integrated circuit production line, direct repair, processing integrated circuits, a substantial increase in productivity.

With the rapid development of VLSI technology, the structure of devices is becoming more and more complex, and the failure analysis of devices is becoming more and more important. FIB etching can be used to open up the profile of the failure area and to observe and analyze it.

Nanomechanical structures (cantilever beams with rectangular and U-shaped cross-section and nanocups) can be prepared by localized surface doping with FIB ions and ion milling in combination with KOH wet etching, and FIB ion milling can also be used in MEMS and sensors.

In addition, FIB technology has been used in the preparation of many optoelectronic devices, super-quantum interference devices and nanobiological devices. In conclusion, the main advantage of FIB technology is to realize complex microstructures with high precision, but the disadvantage is the slow processing speed. Therefore, FIB technology is mainly used for microstructures with relatively small size and less time-consuming, and is most suitable for post-processing and prototyping of small-scale structures.

Excimer laser with its high resolution, photon energy, cold processing, "direct writing" features, no environmental pollution and a wide range of adaptability to the processing material, making it an important MEMS and nanofabrication technology.

In the excimer laser nanofabrication system, most of the mask projection processing, but also can not use the mask, the direct use of focused spot etching workpiece materials, integrated laser beam scanning motion and the relative motion of the x-y working platform and the z-direction of the micro-feeding, in order to realize the three-dimensional microstructure processing. The excimer laser emits a continuous laser beam, an acousto-optic modulator controls the laser on/off, an attenuator regulates the laser beam energy, a beam homogenizer homogenizes the laser intensity distribution, and a microscope is used to focus the beam. The excimer laser can be used to produce MEMS (such as microfluidic devices) and MOEMS.

In order to meet different processing batch and structure shape requirements, according to the relative relationship between the mask and the workpiece (worktable) in the processing process can be divided into three categories of excimer laser nanofabrication system: static mask and workpiece, dynamic mask and workpiece, dynamic mask and workpiece (synchronized).

The nanoimprint lithography (NIL) technology, which was born in the mid-1990s, has recently been referred to as one of the "top ten emerging technologies that will change the world" abroad. The concept of NIL technology can be said to originate from the act of stamping seals in our daily lives, which can imprint the graphics originally on the seal onto the surface of another object. Compared with other lithography technologies, nanoimprinting technology has obvious advantages, such as high resolution (up to several nanometers), high efficiency (which can easily replicate various nanostructures on a large area substrate, prepare multiple parts through parallel processing), and low cost (which does not require extremely complex and expensive optical lenses, optical systems, and electronic focusing systems). NIL is expected to quickly overcome the world's technological challenges in the production of tens of nanometer-width IC, and has strong competitiveness, showcasing broad prospects for the production of nanodevices.

SII is most similar to the traditional Chinese concept of stamping, which combines top-down graphic definition and bottom-up self-assembly molding, and is the best combination of two nanometer-sized technologies. The engraved patterns have been successfully realized up to 50 nm level. Soft engraving technology is generally divided into five categories, namely, micro-contact printing, micromolding in capillaries (MIMIC), replica molding (REM), microtransfer casting (μIM) and auxiliary solvent-assisted micromolding (SAMIM). Among them, the development of a more mature process for the μCP method and MIMIC method, SIL technology is the key to the elastic stamp or elastic compression mold how to transfer the pattern to the substrate.

In the soft engraving technology in the preparation process of elastic stamps, first of all, using electron beam photolithography, in the silicon wafer with relief patterns on the production of masters, the relief part of the material can be SiO_2, Si_3N_4, metal, resist or wax, and then exposed to $CF_3(CF_2)_6(CH_2)_2SiCl$, steam for about half an hour to make it silanized. The elastic liquid prepolymer was poured onto a master plate and separated from the master plate after curing to obtain an elastic stamp. Typical values of h, d and l of the elastic seal are (0.2–20) μm, (0.5–200) μm, (0.5–200) μm. Because of the many advantages of PDMS (polydimethylsiloxane), the general elastic seal material is mostly used PDMS.

HEL process is proposed to make up for the SL process PDMS template material is easy to deformation of the shortcomings of the HEL process template material using a silicon wafer, the template of the nanopatterns through the electron beam or optical photolithography method to obtain the substrate selection of SiO_2/Si. HEL process first in the substrate coated with a layer of photoresist, such as PMMA (polymethylmethacrylate), after the adjustment of the relative position of the template and the substrate (positioning). The relative positions of the template and the substrate are adjusted (positioning). Then the template is pressed on the photoresist film (applied pressure of 4–10 MPa) heated to the photoresist glass phase transition point temperature T above (generally 200 °C). Keep a period of time after cooling, when the temperature drops to the T point near the pressure release, so that the template pattern is transferred to the photoresist film. Next, the PMMA is subjected to oxygen reactive ion etching (RIE). Finally, etching and other processes to transfer the pattern to the substrate. HEL's biggest advantage is the high efficiency of the processing pattern; shortcomings are the template processing cycle time, high cost. HEL is mainly used in the preparation of photovoltaic devices and optical devices, and the most suitable for grating processing.

9.3.3 Single Electron Transistor

Single-electron transistor (SET) is a nano electronic device based on Coulomb blocking effect and single electron tunneling effect. Compared with traditional transistors, SET has advantages such as high frequency, high speed, high sensitivity, low power consumption, and high integration. Therefore, it has a wide range of potential applications in fields such as high-speed high-density integrated circuits, ultra large capacity memory, ultra high speed micro power ultra large scale quantum functional devices, quantum functional computers, and ultra high sensitivity infrared radiation detectors. Due to its high sensitivity to small charge changes in its neighborhood, SET is also suitable for fixed-point, highly sensitive charge detection at the micro/nano scale. As an ultra sensitive charge meter, SET can measure electricity as small as one thousandth of an electronic charge.

SET can be used to make ultra sensitive charge meters, but quantitative analysis of SET charge detection has not yet been achieved. This article simulates the process of using SET for charge detection. A charge detection model was established using the electrostatic field module of COMSOL software; By changing the charge and detection distance [39], the potential distribution of Coulomb Island in SET was obtained; By analyzing the changes in detection distance and distance between charges, the electrostatic coupling coefficient was obtained, and based on the obtained coupling coefficient, the changes and distribution of SET conductivity during scanning detection of two-dimensional charges were studied.

SET consists of two extremely small Josephson junctions (–50 nm), Coulomb Island, and gate. Set the gate capacitance as C_g, the junction capacitance of the two Josephson junctions as C_1 and C_2, the gate voltage as V_g, and the bias voltage as V_b.

The total capacitance of the system can be expressed as:

$$C = C_1 + C_2 + C_g \tag{9.1}$$

The electrostatic energy E of the coulomb island charge Q is:

$$E = -QV_g + Q^2/2C \tag{9.2}$$

The charge $Q_0 = CV_g$ attracted by the gate voltage is introduced into Eq. (9.2) to obtain electrostatic energy:

$$E = (Q - Q_0)^2/2C \tag{9.3}$$

Among them, the value of charge Q can only be taken as an integer multiple of the electronic quantity e, changing V_g so that $Q_0 = Ne$, at which point the electrostatic energy is taken as the minimum value. Change V_g, so that $Q_0 = (N + 0.5)\cdot e$, when $Q = Ne$ and $Q = (N + 1)\cdot e$ of the minimum energy is simple, that is, from the source pole into the Coulomb island of electrons do not need to cross $e^2/2C$ of the potential barrier. Josephson junction, when the junction area is large, the charging energy $E_c = e^2/2C$ is small, when the junction area is small, the charging energy E_c is large, SET requires a large charging energy, so the need for a small area of the junction, generally produced by electron beam exposure process, the junction size is generally less than 50 nm. SET in normal operation, must meet the $kT < e^2/2C$, i.e., the charging energy is greater than the electron heat. In addition, the total resistance R should be greater than the quantum resistance $R_k \approx 26$ kΩ to avoid the impact of quantum fluctuation on the SET work.

The component of SET is the Josephson junction, which is composed of two superconductors sandwiched by a thin insulating layer. The method of dual angle evaporation is an important method to obtain the Josephson junction. In order to ensure that the actual size of the Josephson junction is the same as the design size, the angle and parameters of dual angle evaporation should be modeled and calculated. The calculation model for dual angle evaporation is shown in Fig. 9.13, where h and H represent the thickness of the upper and lower layers of the electron beam adhesive, W represents the width of the suspended part, X represents the transverse design size of the Josephson junction, and M represents the opening width of the upper layer of the electron beam adhesive. Equations can be listed based on geometric relationships:

$$M - X = h \tan \theta \tag{9.4}$$

$$W + 2X + 2h \tan \theta = X + 2(h + H) \tan \theta \tag{9.5}$$

According to the above two equations, given the adhesive thickness h and H, as well as the evaporation angle θ. In this case, any size of Josephson junction can be designed. There are generally two typical layout design methods for dual angle

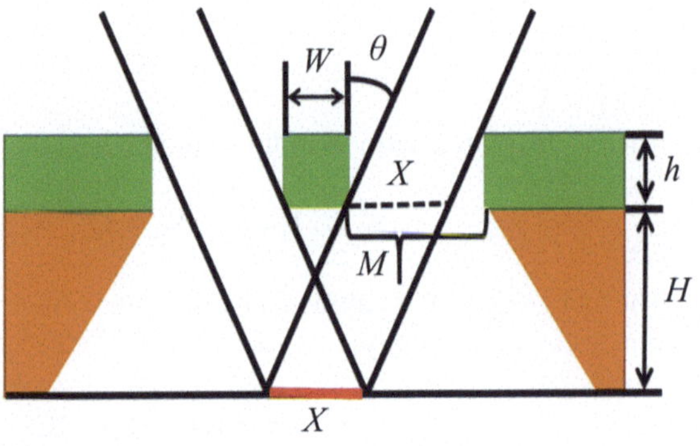

Fig. 9.13 Schematic diagram of dual angle evaporation

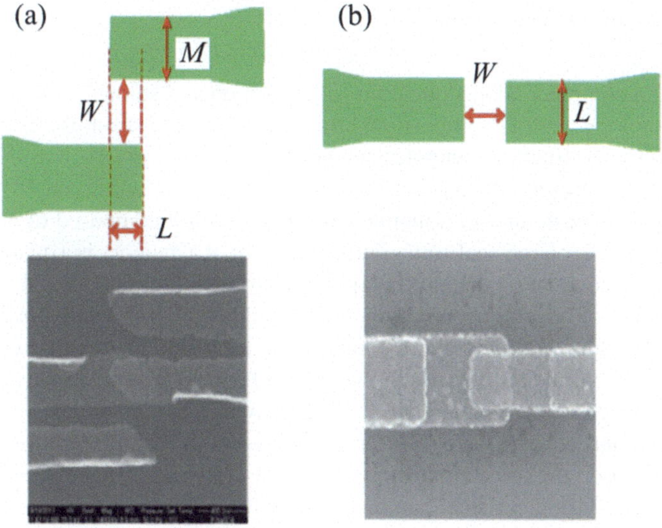

Fig. 9.14 **a** Three parameter evaporation layout and SEM diagram, **b** two parameter evaporation layout and SEM diagram

evaporation, with slight differences in design parameters. One method requires three design parameters, while the other method only requires two design parameters. Their layout and SEM images of the evaporated Josephson junction are shown in Fig. 9.14a, b

Due to the small Coulomb Island of SET, the three parameter evaporation method was chosen. Aluminum is a relatively common type of single material superconductor, and the preparation process is relatively mature. Therefore, selecting aluminum as the preparation material for SET is simple and feasible. The size of Josephson junction in SET is about 50 nm, and Electron-beam lithography process and double angle evaporation process are required for preparation. In order to form an undercout structure, two layers of adhesive need to be sequentially thrown onto the silicon wafer. The upper layer of adhesive is ZEP520-A7 with a thickness of 300 nm, and the lower layer of adhesive is Copolymer (MMA/MAA) EL11 with a thickness of 600 nm. These two types of adhesives can be developed using different developer solutions, so the size of the undercout can be flexibly controlled. Firstly, evaporate the first layer of aluminum film at an angle, with a thickness of 30 nm. To ensure the uniformity and quality of the aluminum film, the evaporation rate is maintained at 0.2 nm/s. Next, oxidation is carried out, and the thickness of the oxide layer is a key parameter determining the characteristics of the Josephson junction [40]. In order to accurately control the thickness of the oxide layer, a volume ratio of argon and oxygen of 9:1 is selected for oxidation. The oxidation time is 20 min, and the pressure of the mixed gas is 60 mTorr. Finally, perform a second angle of evaporation with a thickness of 50 nm.

9.3.4 Quantum Computer

The quantum computer adopts quantum superposition and storage-calculation integrated technology, and its composition structure is different from the five parts of the von Neumann structure computer, but consists of four parts, such as the storage calculator, the controller, the input device, the output device and so on.

Memory: The basic unit of a memory is a quantum bit composed of a quantum binary system. After large-scale integration, the form of a memory is a quantum chip, such as a superconducting quantum chip, a quantum chip, etc. This is the physical implementation of quantum computing, following the laws of quantum mechanics.

Controller: Quantum bits require precise regulation, and the regulation methods include microwave and laser, forming two types of controllers: microwave measurement and control system (microwave measurement and control chip) and laser measurement and control system (laser measurement and control chip). This part of the composition is classic and follows the theory of electromagnetic fields.

Input device: Using classical computers, quantum algorithm software is used to generate data and instructions for computational problems and input them to the controller, and the output device is notified.

Output equipment: classical computer is used to analyze the precision measurement results of Quantum state with quantum software, and quantum calculation results are obtained by combining calculation problems and quantum algorithms.

Clarifying the four components of Quantum computer and their functions has important guiding significance for Quantum computer engineering development, organization and division of labor.

Compared to classic computers, it has two significant core advantages.

In terms of storage methods, classic computers use a series of 0 and 1 to store information. Each unit in the 0 and 1 series is called a bit, and 1 bit can be set to 0 or 1. The Quantum computer uses qubits to store information, and each qubit can be set to 1 or 0, as well as 1 and 0. The above content can be simply understood as that each unit of Quantum computer stores more information [41]. Quantum computer not only has powerful storage capacity, but also has powerful parallel computing capacity. Just like turning on a light in a room, light can pass through all the cracks on the wall in a moment.

It is understood that the speed of Quantum computer is much faster than that of traditional computers. A 30 bits Quantum computer is already very powerful, while a 50 bits Quantum computer will surpass the current strongest supercomputer when dealing with some specific problems. Compared with simple computing power, it takes about 150,000 years for the current trillions of classic computers to process 3 million Big data, while it only takes 1 s to use Quantum computer.

Relevant experts pointed out that when quantum computation is compiled on the electron atom, it will behave very naturally in simulation, such as the discovery of new materials, biomedical drug synthesis and so on. However, when the use of classical supercomputers to study the microscopic world of quantum mechanics, the original strong computing power immediately becomes overstretched, thus showing the super strong simulation ability of quantum computers. Once it is put into application, in the research and development of new materials, biomedical synthesis, financial analysis, weather forecasting and even the development of artificial intelligence, quantum computers will play an important role.

Scientists have pointed out that quantum computing has the potential to dramatically change the world's status quo by providing solutions to key problems that modern supercomputers are unable to solve. Taking weather forecasting as an example, it is difficult for meteorologists to predict the weather with 100% accuracy, especially when the weather is volatile and the information available is limited. Current climate models can only predict regional weather, such as snowstorms and hurricanes, but not more localized weather, such as thunderstorms. Therefore, we need computational superpowers to monitor the entire planet and predict when a simple storm will become dangerous somewhere. The advent of quantum computers makes all this possible. Hartmut Nevin, Google's director of engineering, points out that quantum computers can help humans build better climate models, which will give them a deeper understanding of how humans affect the environment.

Looking around the world, quantum computing is currently in a state of white-hot competition. As early as 2019, the quantum computer "Suzuki" developed by Google successfully completed the problem that traditional computers need 10,000 years to deal with in 3 min and 20 s, which is a landmark event. In the field of quantum computing, China is certainly not lagging behind. Public information shows that China's self-developed prototype of superconducting quantum computer and optical quantum computing prototype continues to break the world's new records in the field of quantum computing. 2020 December, the journal of Science announced that a Chinese research team to build a quantum computer "Nine chapters", the realization of boson-sampling problem of the rapid solution of the calculation speed than the then-fastest supercomputer. The computational speed was one trillion times faster than the fastest supercomputer at the time. According to reports, "Nine Chapters" is a 76-photon quantum computing prototype constructed by the University of Science and Technology of China (USTC) in cooperation with the Shanghai Institute of Microsystems and Information Technology (SIIMIT) of the Chinese Academy of Sciences (CAS) and the National Center for Parallel Computer Engineering and Technology (NCPET).

In 2021, Pan Jianwei team of CSU successfully developed a 62-bit programmable superconducting quantum computing prototype, "Zuchongzhi", and on this basis to realize the programmable two-dimensional quantum walk, only 200 s to complete the supercomputer 2.5 billion years to complete the huge amount of data, its computational speed far beyond the United States. The speed of computation far exceeds that of the U.S. "Suzuki" quantum computer.

January 31, 2023, China's first quantum chip production line in Hefei, Anhui Province, debuted to the public, China's newest quantum computer "Wukong" was launched here. "Wukong core" is specialized for quantum computer "Wukong" supporting quantum chip, performance is more than a thousand times the traditional chip. The latest product "Wukong core" launched by this source of quantum, higher than the 64-bit number, and its equipment is the latest quantum computer in China's quantum computer.

References

1. G. Binasch, P. Grünberg, F. Saurenbach, W. Zinn, *Phys. Rev. B* **1989**, *39*, 77–81.
2. R. D. Mcmichael, R. D. Shull, L. J. Swartzendruber, L. H. Bennett, R. E. Watson, *J. Magn. Magn. Mater.* **1992**, *111*, 29–33.
3. C. Scherer, A. F. M. Figueiredoneto, *Brazilian J. Phys.* **2005**, *35*, 718–727.
4. B. Ando, A. Ascia, S. Baglio, N. Pitrone, *IEEE. Instrum. Meas. Mag.* **2006**, *9*, 44–47.
5. D. K. Kim, Y. Zhang, W. Voit, K. V. Rao, M. Muhammed, *J. Magn. Magn. Mater.* **2001**, *71*, 133–137.
6. J. B. Shukla, D. Kumar, *J. Magn. Magn. Mater.* **1987**, *65*, 375–378.
7. Y. Z. Shao, J. K. L. Lai, C. H. Shek, *J. Magn. Magn. Mater.* **1996**, *163*, 103–108.
8. R. D. Shull, R. D. Mcmichael, J. J. Ritter, *J. Nanostructures* **1992**, *286*, 205–211.
9. J. Zhang, Y. Shao, *T. Nonferr. Metal. Soc.* **1996**, *14*, 86–91.
10. K. Yoshinaga, I. Toyofuku, K. Yamashita, H. Kanehara, K. Ohkubo, *Colloid. Polym. Sci.* **2000**, *278*, 481–484.
11. Y. Chen, C. Li, H. Lai, S. Chen, *Nanotechnology* **2010**, *21*, 115207.
12. A. Irrera, P. Artoni, F. Iacona, E. F. Pecora, G. Franzò, M. Galli, B. Fazio, S. Boninelli, F. Priolo, *Nanotechnology* **2012**, *23*, 075204.
13. Q. Li, T. Y. Luo, M. Zhou, H. Abroshan, J. Huang, H. J. Kim, N. L. Rosi, Z. Shao, R. Jin, *Acs Nano* **2016**, *54*, 8385–8393.
14. A. A. Zhokhov, V. M. Masalov, N. S. Sukhinina, D. V. Matveev, P. V. Dolganov, V. K. Dolganov, G. A. Emelchenko, *Opt. Mater.* **2015**, *49*, 208–212.
15. R. V. Nair, R. Vijaya, *Prog. Quantum. Electron.* **2010**, *34*, 89–134.
16. P. Zhao, B. Li, Z. Tang, Y. Gao, H. Chen, *Smart. Mater. Struct.* **2019**, *28*, 713–725.
17. J. Ge, Y. Yin, *J. Mater. Chem. A* **2008**, *18*, 5041–5045.
18. T. Gong, D. T. Wu, D. W. M. Marr, *Langmuir* **2003**, *19*, 5967–5970.
19. L. Wang, X. S. Zhao, *J. Phys. Chem. C* **2007**, *111*, 64–77.
20. Y. Fang, B. M. Phillips, K. Askar, B. Choi, P. Jiang, B. Jiang, *J. Mater. Chem. C* **2013**, *1*, 2113–2136.
21. L. Cui, Y. Zhang, J. Wang, Y. Ren, Y. Song, L. Jiang, *Macromol Rapid Commun* **2010**, *30*, 598–603.
22. B. Gates, Y. Xia, *Adv. Mater.* **2000**, *12*, 1329–1332.
23. Y. Huang, J. Zhou, B. Su, L. Shi, L. Jiang, *J. Am. Chem. Soc.* **2012**, *134*, 17053.
24. N. Vogel, L. D. Viguerie, U. Jonas, C. K. Weiss, K. Landfester, *Adv. Funct. Mater.* **2011**, *21*, 3064–3073.
25. Z. Meng, Y. Wu, J. Ren, X. Li, S. Zhang, S. Wu, *ACS Appl. Mater. Interfaces* **2022**, *14*, 12562–12570.
26. Y. Fink, A. M. Urbas, M. G. Bawendi, J. D. Joannopoulos, E. L. Thomas, *J. Light. Technol.* **1999**, *17*, 1963–1969.
27. S. Wu, J. Nan, Y. Wu, *ACS Appl. Mater. Interfaces* **2022**, *14*, 141–164.
28. B. Hatton, L. Mishchenko, S. Davis, K. H. Sandhage, J. Aizenberg, *Proc. Natl. Acad. Sci. U.S.A.* **2010**, *107*, 10354–10359.
29. W. Fan, J. Zeng, Q. Gan, D. Ji, L. Wu, *Sci. Adv.* **2019**, *5*, 8755.
30. G. Fu, X. Zhang, X. Chu, Y. Zhou, Y. Song, *Dyes Pigm.* **2021**, *195*, 109747.
31. J. He, X. Shen, H. Li, *ACS Appl. Mater. Interfaces* **2022**, *14*, 7121–7159.
32. S. P. S. Badwal, J. Drennan, *J. Mater. Sci.* **1987**, *22*, 3231–3239.
33. J. Goldberger, A. I. Hochbaum, R. Fan, P. D. Yang, *Nano Lett.* **2006**, *6*, 697–717.

34. Y. Wang, D. Maspoch, S. Zou, G. C. Schatz, R. E. Smalley, C. A. Mirkin, *Proc. Natl. Acad. Sci. U.S.A.* **2006**, *103*, 2026–2031.
35. J. H. Seol, I. Jo, A. L. Moore, L. Lindsay, Z. H. Aitken, M. T. Pettes, X. Li, Z. Yao, R. Huang, D. Broido, *Science* **2010**, *328*, 213–216.
36. L. M. Jiang, Z. F. Liu, J. Tang, L. Zhang, K. Shi, Z. Q. Tian, P. K. Liu, L. N. Sun, Z. W. Tian, *J. Electroanal. Chem.* **2005**, *581*, 153–158.
37. H. K. Chang, Y. K. Kim, *Sens. Actuator A Phys.* **2000**, *84*, 342–350.
38. Michael, D., Austin, Haixiong, Ge, Wei, Wu, Mingtao, Li, Zhaoning, *Appl. Phys. Lett.* **2004**, *84*, 5299–5299.
39. J. J. L. Morton, D. R. Mccamey, M. A. Eriksson, S. A. Lyon, *Nature* **2011**, *479*, 345–353.
40. Devoret, H. Michel, Schoelkopf, J. Robert, *Nature* **2000**, *96*, 1433–1459.
41. L. Postler, S. Heuβen, I. Pogorelov, M. Rispler, T. Feldker, M. Meth, C. D. Marciniak, R. Stricker, M. Ringbauer, R. Blatt, *Nature* **2022**, *605*, 5422–5438.

Chapter 10
Application of Nanomaterials and Nanotechnology in the Field of Energy and Environmental Protection

Since nanomaterials and technology to now gradually mature, its use involves almost to all disciplines and industries, become an important technology to meet the sustainable development, national policy also supports all walks of life large-scale use of nano materials and technology, but also support research center and key laboratory continue to study and improve the nanotechnology. As one of the emerging scientific technologies with the most market potential, the potential importance of nanotechnology is beyond doubt. Nanotechnology is becoming the focus of the scientific and technological circles of various countries, which is worthy of the three pillars of contemporary science and technology (basic science, biomedical science, information technology). At present, nanomaterials and technology may be applied to information storage, energy development and utilization, environmental protection, manufacturing, national defense science and technology and other fields, to further promote the transformation and upgrading of traditional industries. For example, in the energy field, nanomaterials technology can provide new, efficient and environmentally friendly energy alternatives (such as solar cells, fuel cells) and the core technology in energy utilization [1]; in the field of environmental protection technology, nanotechnology will be used to control water pollution, air pollution, sound pollution and light pollution that are previously difficult to control. The practical application of nanomaterials and technologies in the field of energy conservation and environmental protection gives full play to the role of energy conservation and environmental protection. In China, many scientific research institutes and institutions of higher learning also organize scientific research forces to carry out the research work on nanotechnology, and have achieved certain research results. It was born more than ten years has made great achievements and influence and penetration into various fields, especially in the shortage of mineral energy, environmental pollution, and more and more plagued the human survival and development of the today, nano materials in environmental protection, environmental governance and energy conservation, has presented a thriving scene.

S. Peng and P. Li, *Nanomaterials and Nanotechnology*,
https://doi.org/10.1007/978-981-97-8433-2_10

10.1 Application of Nanomaterials in Energy Storage

Nanomaterials also have extensive applications in the field of energy storage and can be used to make more efficient energy storage devices. For example, nanomaterials can be used to make high-performance lithium-ion batteries that can be charged in a shorter time and have a higher energy density. Nanomaterials can be used to store solar energy, enabling thermal energy storage and performance enhancement. In addition, nanomaterials can also be used to make supercapacitors, which can store a large amount of charge in a short time, and have the characteristics of rapid charging and discharging. These applications can not only improve the efficiency of energy storage, but also reduce the cost of energy equipment.

10.1.1 Nanomaterials for the Storage of Hydrogen Energy

With the development of industry and the improvement of human material life and spiritual civilization, the consumption of energy is also increasing day by day. The energy consumption of the world in the last 25 years is equivalent to the consumption of the past 100 years. The storage of primary energy of natural minerals on the earth, such as oil, coal and natural gas, is rapidly reduced, and the rapid increase of energy consumption leads to the deterioration of human living environment. Therefore, the development of new energy must consider the efficient use of energy and reduce the pollution to the environment as much as possible. In addition to renewable energy such as solar energy, water energy, wind energy and biological energy, hydrogen energy is widely valued for its unique advantages and rich resources. Hydrogen is the most abundant element in the universe, accounting for 75% of the total mass of the universe. Hydrogen is also found as inexhaustible hydrogen on earth. The combustion product of hydrogen is water, without any pollution to the environment. Hydrogen combustion in the air produces 80 percent lower nitrogen oxides than petroleum-based fuels. Another valuable aspect of hydrogen is its chemical simplicity. Hydrogen rapidly releases energy quickly breaks and forms relatively few bonds and has high reaction rate constants and faster electrode process dynamics. Hydrogen plays an important role in alternative energy systems in this century because of its purity and simple production advantages. Hydrogen has advantages as aircraft fuel. According to the energy per unit mass, hydrogen is 3.5 times that of jet fuel, so it can greatly reduce the fuel load of the aircraft and increase the cargo capacity and range of the aircraft. In the mid-1970s, lockheed researchers studied with hydrogen fuel business routes, evaluated hundreds of factors, found that the superiority of liquid hydrogen fuel aircraft: lighter, shorter wing, engine number, push-better than quality, shorter runway, less noise, more effective use of energy.

The first problem with using hydrogen energy is the storage of H_2. Hydrogen extraction as an energy source is a wise choice, but the problem of hydrogen storage method and heat storage of the heat generated must be solved. H_2 storage is the bottleneck of hydrogen energy development and utilization. The density of H_2 is very low, and it is easy to explode in O_2. It is very important to achieve high capacity storage of H_2. In the utilization of hydrogen energy, the safe storage and transportation of H_2 has always been the focus of the research of scientific workers in various countries. Many countries attach great importance to the research of hydrogen storage technology while conducting the research on hydrogen production technology and hydrogen energy application technology. The simplest way is to compress H_2 or store it as liquefied gas [2]. The pressure required to compress H_2 is high, and the storage cylinders can usually withstand 15 MPa, which can only be met with expensive carbon fiber reinforced materials. Hydrogen liquefaction is also a way to store hydrogen, but the liquid hydrogen can only be stored at a temperature of 20.3 K, so the gas storage tank used must be vacuum adiabatic, the conditions are harsh, inconvenient, and not too safe. At first, people used ordinary activated carbon adsorption. Due to the wide pore size distribution of activated carbon and the small micropore volume, even at low temperature, the hydrogen storage amount is very low, less than 1% (quality), and lower at high temperature. Later, people used super activated carbon with larger surface, smaller pore size and more uniform. However, the disadvantage of this method is that the performance is good at low temperature. With the increase of temperature, the hydrogen storage is lower and lower, and the adsorption capacity at room temperature is only 0.4% (mass).

In the past, nanomaterials mainly focus on metal hydride hydrogen storage, but these metal materials themselves have large quality, small hydrogen storage per unit mass, the adsorption capacity of hydrogen is not more than 7%, far less than gasoline and diesel oil (about 17%), so their application is greatly affected. Recent studies have turned to more efficient materials such as nanocrystalline and composites. The cheap and light FeTi fails when exposed to the air and becomes brittle when encountering hydrides. The solution is to use nanocrystalline particles coated on the surface [3]. Pd was coated on FeTi films of amorphous or nanocrystalline forms. Upper, the Pd film can hinder the oxidation of FeTi in the air before 150 °C, filling H/M (hydrogen/metal) up to 0.8 at 101 kPa, heating hydrogen release at 150 °C, H/M may drop to 0.1, and the filling and discharging process can be repeated 15 times.

Since the discovery of carbon nanotubes in the 1990s, as shown in Fig. 10.1, attention has been focused on nanocarbon-based materials, such as carbon nanotubes and nanocarbon fibers. Carbon nanotubes have become the most promising hydrogen storage materials due to their pipe structure and graphite-like layer gaps between multi-walled carbon tubes. Compared with activated carbon, which has a larger specific surface area and has some other specific effects and properties of nanomaterials, using carbon nanotubes to absorb H_2 should be a more effective way. Carbon nanotubes have a strong adsorption effect on H_2, and the hydrogen release speed is fast, which can be analyzed at room temperature. Different from metal hydrogen storage, the early metal or alloy hydrogen storage materials for hydrogen storage, that is, first dissociate the surface of the metal alloy into 2H atoms, and enter the human alloy in the atomic state to form metal hydride. However, carbon nanotubes can perform direct adsorption of H_2 due to their unique nanoscale size and rich microporous structure and huge specific surface.

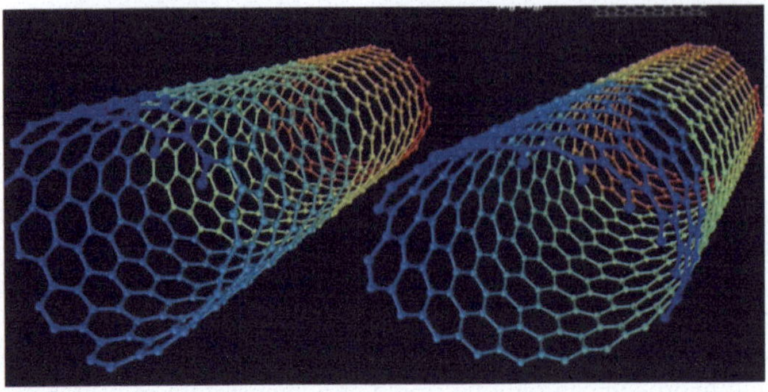

Fig. 10.1 Carbon nano-tube

Many scientists have used computer simulations to simulate the capacity and mechanism of hydrogen storage by carbon nanotubes. The Monte Carlo simulation of the CNTs with a pore size of 1.957 nm, and the arrangement between the CNTs had a great influence on the overall adsorption of the material. The carbon nanotubes were simulated at a low temperature of 77 K, and the hydrogen storage in the gap of the carbon nanotubes reached up to 14% of the total hydrogen storage. The results of density function theory show that the hydrogen storage amount of single-wall carbon nanotubes can be greater than 14%, that is, the average of each carbon atom adsorbs one hydrogen molecule, and the volume hydrogen storage density is much higher than the experimental data, which is more than 2 times the density of liquid argon.

Carbon nanotube hydrogen storage has a profound application prospect in the automotive industry. The rapid development of the automobile industry has facilitated people's life and promoted the production and development of human beings, but it has intensified the consumption of oil and also brought serious environmental pollution [4]. There are both concerns about oil depletion and other concerns about environmental pollution caused by car emissions. The ideal is to build fuel cell vehicles powered by hydrogen, where carbon nanotubes can store more than 10% of the hydrogen, making it possible to use hydrogen as their power. Hydrogen fuel is convenient and safe to store in carbon nanotubes, and this hydrogen storage method is reversible. When H_2 runs out, it can be "inflated" to store the large volume of H_2 at room temperature in small carbon nanotubes, used as hydrogen fuel drive vehicles, which is the main development direction of future vehicles to achieve green fuel drive.

10.1.2 Nanomaterials for the Storage of Solar Energy

Solar energy is a huge, safe and clean energy source that does not pollute the environment. One possible way is to use the photoluminescent properties of nanomaterials of several nanometers to absorb natural light and store it during the day and directly shoot it where needed at night. German scholar Brandt M.S. believes that porous silicon can glow for several hours at low temperatures. Phase change materials (PCM) have been used in different solar systems to achieve thermal energy storage and performance enhancement. Improved heat transfer from the PCM leads to reductions in charging and discharge duration, which makes them more suitable as storage units. Dispersting conductive solid materials with nanosize is a practical way to enhance their thermal properties. Several nanosized materials have been added to the PCM to improve their thermal/cold storage capabilities [5]. Research in this field shows that adopting nanotechnology is an effective method for improving heat transfer. The results of studies on the application of nanotechnology in PCM show that adopting nanotechnology can significantly improve the heat transfer rate, which results in reduced charging and discharging time of storage units equipped with PCM. The improved thermal performance of nanoscale PCM is mainly attributed to the increase in thermal conductivity and the decreased melting latent heat, which accelerates the melting/solidification process.

10.1.3 Nanomaterials Used in Lithium-Ion Batteries

In recent years, lithium-ion batteries have developed very rapidly because of its excellent performance. The performance of lithium ion battery depends to a large extent on its positive and negative electrode active materials, and is closely related to the material selection, performance and preparation method. Nanotechnology provides a novel research method for battery materials. Nanostructured materials make the contact area of battery active materials and electrolyte large, shorten the ion diffusion path, have obvious advantages, and improve the system performance. Therefore, researchers have been working to find new, simple, and low-cost synthetic aways to make new nanostructured materials, and to explore their applications in lithium-ion batteries, lithium-s batteries, water batteries, and supercapacitors.

High-performance silicon anodes suffer from large volume changes when cycling in lithium-ion batteries [6]. Using molecular dynamics simulations and ab initio electronic structure calculations, Hwang report the structural, mechanical and electronic properties of graphite amorphous carbon-like coatings on bulk silicon to improve the durability of the silicon anode. Chen studied the $Li_4Ti_5O_{12}$(LTO)/Si composites with different weight ratios. It can be found that electrodes with moderate silicon content have stable capacity and good cycling performance even at very high current density. The improvement in specific capacity and rate performance is a direct result of the synergy between LTO and Si. Focusing on the cathode, Manzi designed the nanostructured $LiCoPO_4$ showing the effects of different Co^{2+} sources, solution acidity and reaction time on the crystal growth of $LiCoPO_4$ particles by various technical means. The lithium excess and nanoscale $Li_{2+x}Mn_{1-x/2}TiO_4$ ($x = 0, 0.2, 0.4$) cathode materials prepared by sol–gel method showed that the charge–discharge performance of the nanostructured materials was significantly improved with the increase of lithium content. Das studied numerical simulations of lithium-ion batteries with heterostructures of silicon-coated carbon nanofiber core and shells as an anode. Sun used a simple hydrothermal strategy to prepare finger-like Co_3O_4 nanorods using H_2O solvent [7]. Chang shows different sizes of ordered mesoporous carbon nanospheres (diameter of 46–130 nm) as the active anode material of lithium ion battery. Rapid synthesis of various forms may be one of the most promising strategies for high performance anode materials. Bismuth has low redox potential and high specific capacity, is a negative material suitable for water-based batteries. As part of the effort of Zuo and colleagues in this regard, a bismuth electrode membrane was directly grown by using a simple hydrothermal method and tested in LiOH, NaOH and KOH electrolytes. The results show that the Bi electrode films have smaller Rs and faster ion diffusion coefficient in the KOH electrolyte [8]. Song designed a stand-alone Si-Ti-Ni (STN) alloy particle/reduced graphene oxide/single-walled carbon nanotube composite film. The improved performance is due to the excellent mechanical properties of the reduced GO-SWCNT network inhibiting the pulverization of STN active materials and the enhanced dynamics associated with electron and lithium ion transport.

10.1.4 Nanomaterials Used in the Supercapacitors

Due to their high aspect ratio and thin monolayer graphene sheets, carbon nanotubes have recently been widely studied as electrode materials for supercapacitors (SC), as shown in Fig. 10.2, because their ratio of the capacitance to the electrode surface area is $C \propto \frac{A}{d}$, SC is a kind of energy storage device, and its technology makes up for the gap between the two by combining some working mechanisms of conventional energy storage technologies such as parallel plate capacitors and batteries. In SC, the electrodes are usually immersed in the electrolyte like a cell, and the electrolyte and the electrode charge accumulate electrostatically at the electrode/electrolyte interface like a parallel plate capacitor. The SC has no dielectric medium; however, these two layers are separated by a rather thin (0.5–1 nm) solvent molecular monolayer (called the Holmholtz layer) to create a high capacitance [9].

Fig. 10.2 Supercapacitor

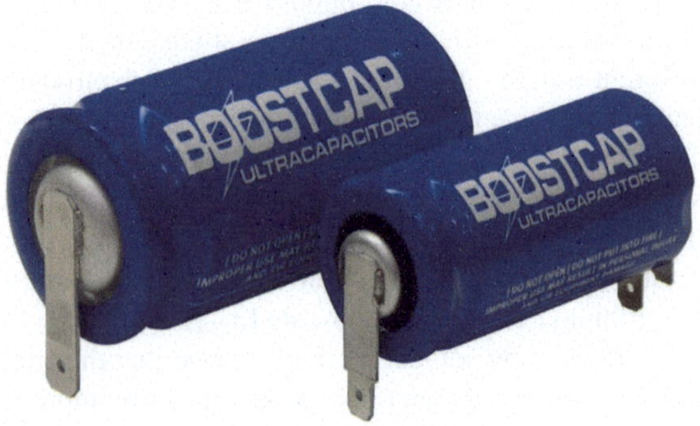

SC has the outstanding characteristics of rapid storage/energy release, about 1 million charge and discharge cycles, and eco-friendliness. The power density of the supercapacitor (100 kW/kg) is much higher than that of conventional batteries; however, the energy density (10 Wh/kg) is an order of magnitude smaller than the battery (100 Wh/kg) [10]. Ragone diagram shows the power density (specific power) and energy density (specific energy) of different energy storage/conversion technologies. See Fig. 10.3 SC is expected to be used in portable electronics, hybrid vehicles, heavy machinery, air vehicles and even satellites. Moreover, SC can help improve the life of existing batteries. In fact, the demand for high-power pulses from conventional batteries leads to battery deterioration and reduced life span. However, when coupled to the SC, the life will be increased by providing the pulse of the SC and the average power of the cell. In addition, the existing SC with moderate energy density can be used to require high power in a short time, such as closing the bus door, the initial push of the hybrid bus, backup power such as uninterruptible power supply (UPS) rapid response; However, their aim is to use supercapacitors as an independent energy source to drive buses by further increasing the energy density.

The selection of suitable electrode materials for double layer capacitors (EDLCs) has a great influence on the capacitance of double layer capacitors. As shown in Fig. 10.4, a hollow "carbon nanocage" is an important electrode material [11]. It is generally believed that the regular spherical shape facilitates sufficient contacts between the electrode and the electrolyte, the cavity can store the electrolyte, and the porous shell can promote rapid charge transfer. The hierarchical pore structure is very suitable for supercapacitors, where the micropore is the main storage site for electrolyte ions, the middle pore is the fast channel for ion transport, and the large pore is the memory for electrolyte transport. On the other hand, modifying the electrode surface with heteroatoms according to the wetting and capacitive properties is another effective way to improve their performance. Therefore, it is important to explore the efficient preparation of doped hollow "carbon nanocages" with improve

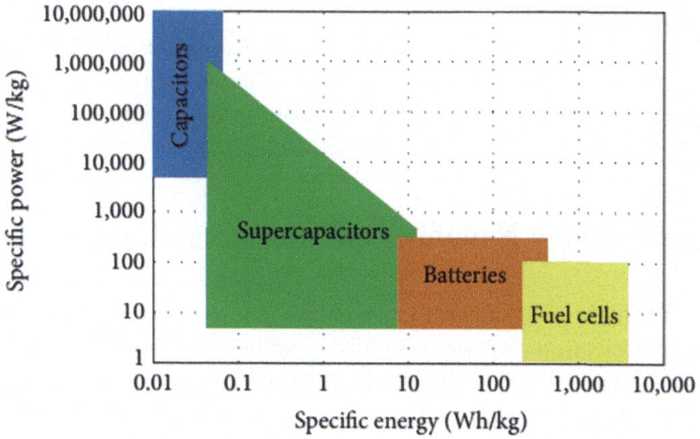

Fig. 10.3 Specific energy and specific power diagram of the energy storage conversion device

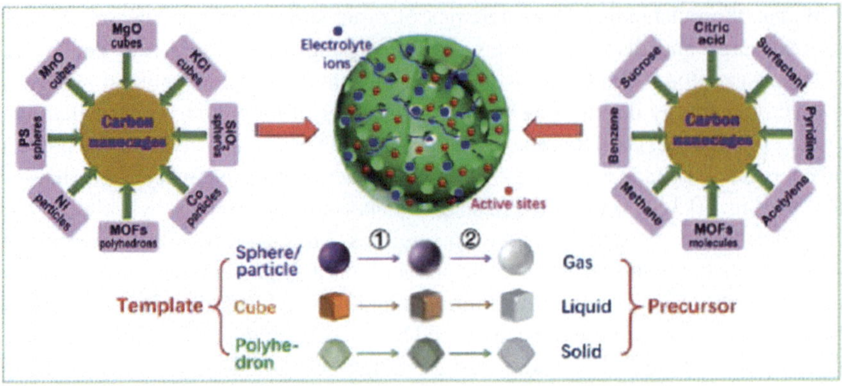

Fig. 10.4 Schematic synthesis of hollow "carbon nanocages" (HCNCs)

electrochemical properties. At the same time, the nanotopography and the crystal structure also have a great influence on the electrochemical properties of the hollow "carbon nanocage" [12] (For example: amorphous carbon nanocages, graphene-like carbon nanocages, and hollow porous (micro/mesoporous) carbon nanospheres).

Faraday False Capacitors (PCs) The electrode materials of Faraday False Capacitors include transition metal oxides, hydroxides, sulfides, etc. These materials have the problem of low conductivity, which leads to low multiplier performance and poor cycle stability during high current charge and discharge. Therefore, researchers usually improve the rate and cycle performance of pseudo-capacitive materials by using carbon support composite design. Hollow carbon nanocages are ideal carbon carrier materials for improving the properties of pseudocapacitor materials and the overall performance of electrodes [13]. For example, ultra-thin MnO_2 nanofibers are grown vertically on the outer surface of the graphite hollow carbon sphere to produce composite electrode materials with excellent electron transport, rapid ion penetration, rapid reversible Faraday reaction, and excellent rate performance. A new hollow carbon microsphere/MnO_2 nanosheet composite was prepared by in situ self-limiting deposition under hydrothermal conditions, showing promising prospects for high-rate electrochemical pseudocapacitive energy storage applications.

10.2 Application of Nanomaterials in Energy Conversion

Compared with traditional materials, nanomaterials have some remarkable novel features, such as electronic energy level discontinuity, quantum size effect, surface effect, quantum wear, dielectric limit domain effect, macroscopic quantum tunneling effect, and these new properties further attract people's research. The small size and high performance of nanomaterials provide a new idea for solving the problem of energy conversion.

10.2.1 Nanomaterials Used in Nanogenerators

The application fields of piezoelectric materials can be simply divided into two categories: the application of converting electric energy into mechanical energy, such as underwater acoustic transducers, ultrasonic transducers, speakers, sonar, piezoelectric actuators, etc.; the application of converting mechanical energy into electric energy, including pressure sensors, ultrasonic sensors, nanogenerators, and microphones. Here want to emphasize particularly is nano generator, its purpose is to put the tiny and scattered mechanical energy (such as ultrasonic, noise, air flow, blood flow, large instrument vibration, body movement, etc.,) into electricity, especially for power consumption devices, for the development of micro and nano devices put forward the new direction.

There is an increasingly urgent need for portable, lightweight personal mobile electronics, and it is well known that one of the biggest problems that plague and block these products is power supply. And nanogenerators were born to solve these problems. Since the concept of nanogenerators was proposed by Professor Zhonglin Wang in 2006, they have been widely studied and underwent a series of optimization and innovations, which not only improved the output power of nanogenerators, expanded their application scope, but also derived some new designs of nanogenerators [14]. Professor Zhonglin Wang further proposed that the self-energy supply system composed of nanogenerator and nano devices will be the development trend of micro and nano technology in the future. The output voltage of nanogenerator has been increased from the initial few millivolts to tens of volts in 2011, and the power has also been increased by 500 mW/cm^3. In 2012, Long Gu et al. obtained a high output of 209 V nanogenerator. Nanogenerator includes piezoelectric nanogenerator, friction nanogenerator, thermoelectric nanogenerator and composite nanogenerator, the most studied and mature is piezoelectric nanogenerator. The principle of the original nanogenerator is shown in Figs. 10.5 and 10.6, which utilizes the piezoelectric properties, flexibility, and the precious metal–semiconductor Schottky effect of the ZnO nanowires. Figure 10.5a is a free ZnO nanowire with one end fixed at the other end of the substrate, which is electrically neutral in equilibrium when it is not subjected to an external force. When moving its free end with a probe, it is equivalent to applying an external lateral force, thus creating a corresponding strain inside the ZnO nanowire, One stretched side is compressed, and the central surface is no stretch nor compression, stress is zero, by calculating the ZnO nanowire internal stress distribution as shown in Fig. 10.5b, due to piezoelectric effect, strain piezoelectric field, Fig. 10.5c, and piezoelectric potential, Fig. 10.5d platinum plating probe dial ZnO nanowire make it bent, the first contact is its tensile surface, the voltage is positive, because Schottky effect, the external circuit charge accumulation in the probe head, but the probe head and ZnO nanowires is not ohmic contact, so there is no current flow between the two. However, when the probe head is moved to the compressed side of the ZnO nanowire, as shown in Fig. 10.5f, the charge originally accumulated in the probe head instantly flows back through the external circuit, generating a large current (Fig. 10.6).

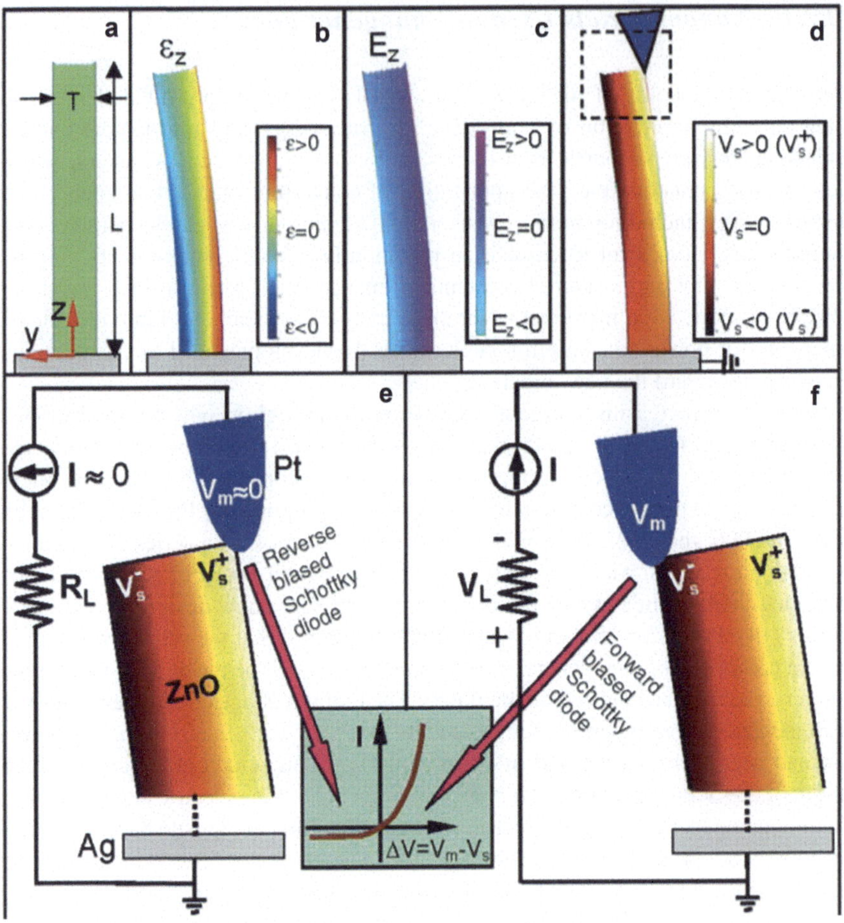

Fig. 10.5 Schematic diagram of the ZnO base piezoelectric nanogenerator

In 2008, Dr. Qin Yong et al. used this principle to propose a clever design that integrate nanogenerators into the fabric [15]. The principle is shown in Fig. 10.7. First, the ZnO nanowire rod grows radially along the Kevlar fiber, and the ZnO nanowire on the fiber is plated with Au as an electrode, and then wound between the two fibers, the ZnO nanowire is driven by the ZnO electrode of Au to generate a current in the external circuit. Provide a good idea for nanogenerator applications.

The remarkable feature of the above nanogenerator is that a fixed end of the piezoelectric nanowire is moved, which determines that their output voltage and current are unidirectional (both positive or both negative), and is often known as a DC nanogenerator. Unlike another design, where the ZnO nanowire is stretched or compressed along the axial direction to produce a potential difference at both ends of the nanowire, Dr. Yang first reported this design [16], The principle is shown

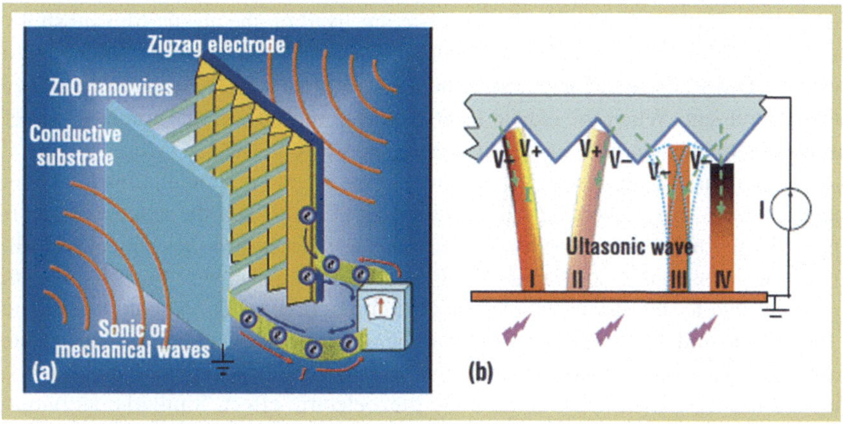

Fig. 10.6 Ultrasonic-driven nanogenerators

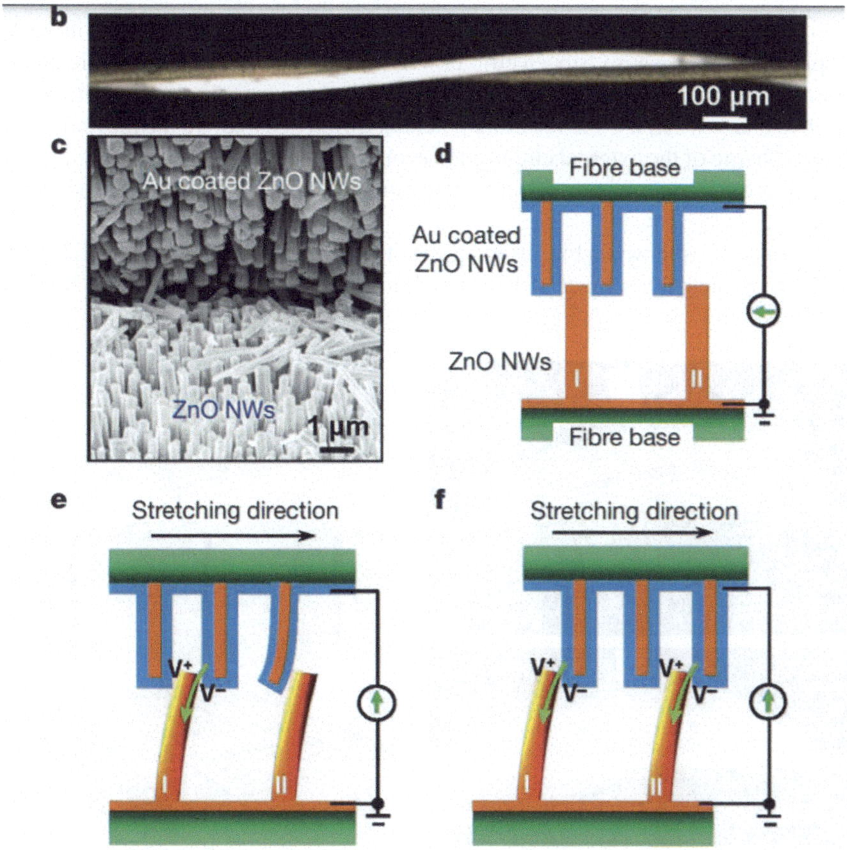

Fig. 10.7 Schematic diagram of a micron-fiber nanowire power generator

in Fig. 10.8. Both ends of a single nanowire are fixed on the elastic film substrate. The fixing material is silver slurry, which also acts as an electrode, on which a layer of PDMS is sealed for reinforcement to lead the wire from the silver slurry for measurement. When the elastic base is bent under an external force, its upper surface is stretched and the lower surface is compressed. The ZnO nanowires fixed to the upper surface are also stretched, and the piezoelectric potential generated at both electrodes drives the external circuit charge flow. When the external force is removed, the ZnO nanowire follows the elastic base to return to the original state. In this process, the charge accumulated at both ends of the nanowire flows back from the circuit due to the loss of the electric field force, producing one output opposite to the stretch, and the power is the same in the two processes [17]. If one side of the fixed ZnO nanowire is compressed, the two ends will produce a potential difference opposite to the stretch, depending on the piezoelectric effect. Similarly, an output signal is generated when the external force removes the elastic substrate is released. The advantages of this design are simple and convenient integration, so it is widely used.

In addition to piezoelectric nanogenerators, triboelectric power also inspires people to use it, because triboelectric power is so common in daily life. Here, we rethink the concept of a generator. As long as the device can separate the positive and negative charges and use the resulting potential difference to drive the movement of the free charge of the external circuit can become a generator. Charge separation can be achieved through electromagnetic induction, thermoelectric effect, piezoelectric effect, etc.

Its principle as shown in Fig. 10.9, a piece of Kapton film and Polyester (PET) film together, on the outer surface of two film plating Au as electrode, because Kapton

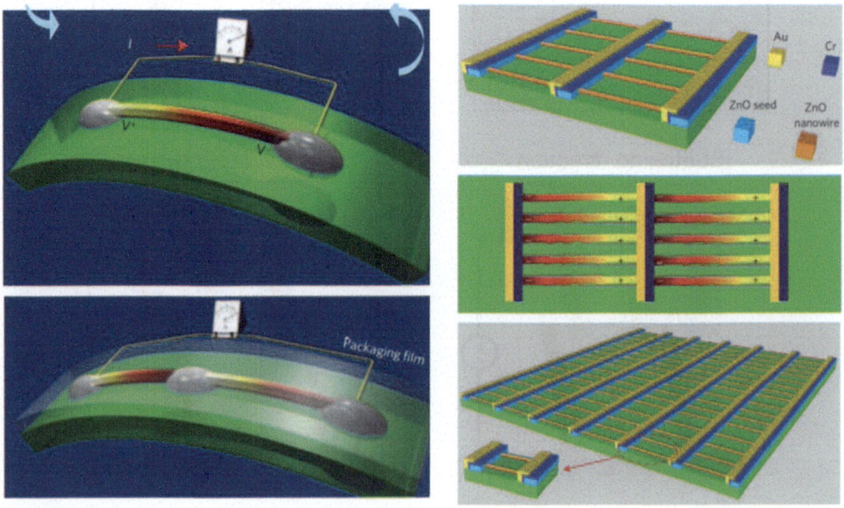

Fig. 10.8 Of the AC nanogenerators

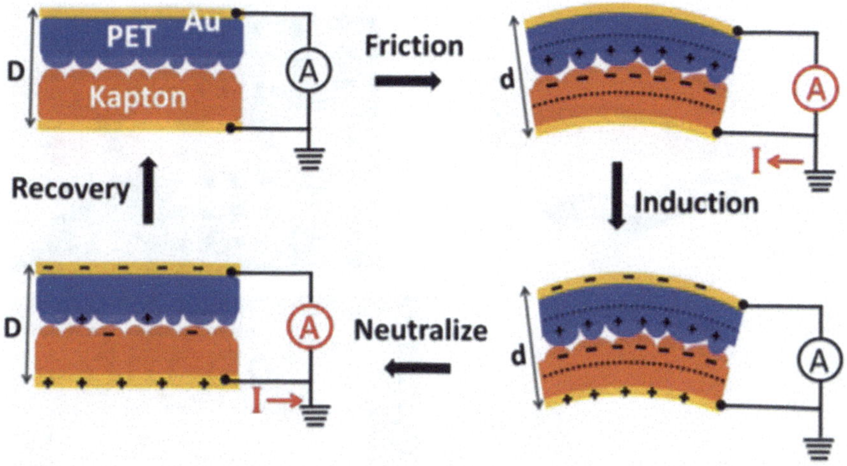

Fig. 10.9 Frictional nanogenerators

membrane and PET membrane on the nanoscale is not smooth, they are under the external force extrusion relative displacement, friction, and makes the PET membrane positive, Kapton membrane with negative electricity, between potential difference, driving external circuit charge accumulation on Au electrode, thus generate current in the external circuit. When the external force is withdrawn and the two membranes are neutralized, the original charge accumulated at both ends of the Au electrode is reflux, producing an opposite output [18]. Repeated, providing continuous output, they got a voltage output of 3.3 V and a power density of 10.4 mW/cm^3. Usually there is a variety of energy in the environment, such as mechanical energy, thermal energy, solar energy, biological energy may exist simultaneously. In order to use energy efficiently, the composite nanogenerator emerged. Its design idea is that a composite device is composed of multiple parts, and different parts are responsible for collecting different kinds of energy. For example, in Fig. 10.10, Yang Ya et al. reported a composite nanogenerator integrating thermal energy, mechanical energy and solar energy. The lower PVDF is used as a thermoelectric piezoelectric nanogenerator unit, and the upper ZnO array P3HT- film is used as a solar cell unit to realize the conversion of heat energy, mechanical energy and solar energy into electric energy at the same time.

10.2.2 Nanomaterials Used in Solar Cells

The principle of solar cell is to convert solar energy into electric energy through photovoltaic effect. As long as certain light conditions are met, it can output voltage and generate current in the case of loop formation. It is an important way for human

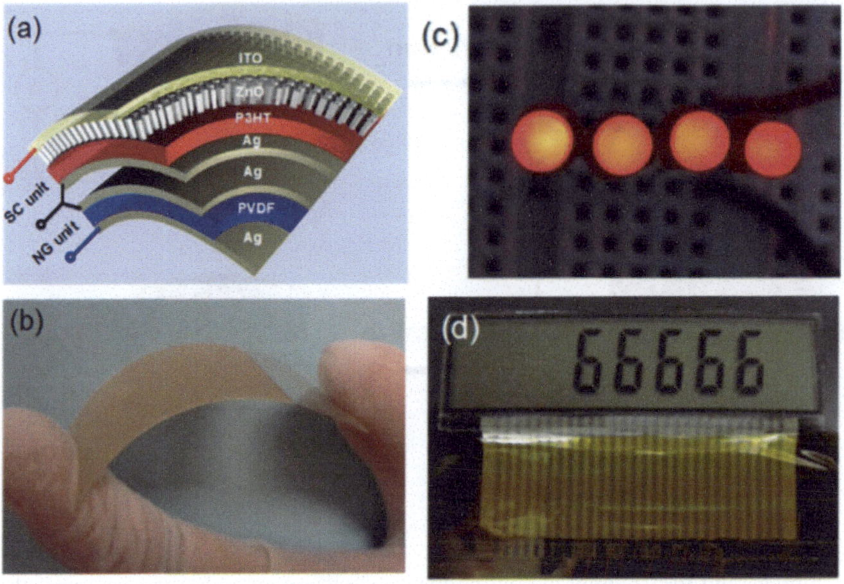

Fig. 10.10 A composite nanogenerator

beings to use solar energy. It has light, simple process, green safety, clean, environmental protection, material sources, strong plasticity advantages, large potential in industrial production, however, the development of organic solar cells is restricted by low energy conversion rate, cannot meet the requirements of industrialization, improve the energy conversion efficiency of organic solar cells has become the focus of the research. In order to improve the energy conversion efficiency of organic solar cells, nanomaterials are applied to organic solar cells, and the energy conversion efficiency of the absorption layer of organic solar cells is studied. The structural unit size of nanomaterials is within the range of 1–100 nm, and the size coherence length is close to that of the electron coherence length. Its properties will change greatly due to the self-organization brought by the strong coherence. Therefore, it is important to study the energy conversion efficiency of the absorption layer of nanomaterial organic solar cells.

The energy conversion efficiency of the absorption layer of nanomaterials organic solar cells was studied experimentally, and the energy conversion efficiency of the organic solar cell with 16 pm concentration was demonstrated. Therefore, the surface plasmon resonance effect of nanomaterials can improve the light absorption of solar cells and effectively improve the conversion efficiency of solar cells [19].

Solar energy is an inexhaustible and renewable energy, it will not produce any environmental pollution, is a clean energy. And the advent of solar cells gives us hope to solve the energy problem. The power generation principle of solar cells is based on photovoltaic effect (PV), which is interacted by the quantum of light and semiconductor materials to produce electromotive force, that is, the interaction between light and materials directly generates electricity, which is a renewable resource without any pollution to the environment. The unique advantages of solar cells, surpassing wind energy, hydropower, geothermal energy, nuclear energy and other resources, are expected to become the main pillar of future power supply. Therefore, solar cells are considered as the best way to solve the two problems of energy resources and clean energy [20].

At present, a very important factor restricting the development of solar cells is the cost of solar cells. The cost of solar cell modules is much higher than the cost of traditional power generation methods. The cost of solar cells depends on the modules of solar cells, the storage and conversion of electricity, the support and erection of battery modules, and the cost of land, site and equipment, etc. Of these costs, the cost of materials is the largest, even the relatively cheap crystalline silicon solar cells, materials account for about 60 to 80 percent of the total cost. Therefore, the selection of semiconductor materials and the improvement of the production process are the development direction of solar cells [21].

Silicon is a good semiconductor material, rich in reserves, is the second largest stored element on the earth, and the performance is stable, non-toxic, so silicon solar cells have become the main material in the research, development, production and application of solar cells. Monocrystalline silicon solar cells currently have a maximum conversion efficiency of 24.7%, but they are too expensive. With the rise and development of nanotechnology is expected to greatly reduce the cost of silicon solar cells, the use of emerging nanotechnology and nanomaterials to design and manufacture high conversion efficiency and low price of silicon solar cells has become another focus of concern and research [22]. Silicon thin film solar cells is a new generation of solar cells after solar cells. Silicon thin film solar cells include monocrystalline silicon thin film solar cells, polycrystalline silicon thin film solar cells and amorphous silicon thin film solar cells.

1. Monocrystalline silicon thin-film solar cells

As shown in Fig. 10.11, in recent years, solar cells are developing more and more ultra-thin. In industry, cells with a thickness of less than 200 μm have been produced, and cells with 40 μm thickness have been prepared in the laboratory, which greatly reduces the demand for raw materials for a single solar cell. Monocrystalline silicon thin film solar cells have many obvious advantages, with a very thin thickness can be prepared to produce efficient solar cells, but also can choose cheap substrate materials. The preparation method of monocrystalline silicon film is mainly transfer technology, that is, excellent FZ monocrystalline silicon wafer is used as the substrate material to epitaxial grow monocrystalline silicon film, and then peeled off from the substrate and transferred to a cheap substrate material such as glass. FZ monocrystalline silicon substrate can be recycled many times for the growth and separation of monocrystalline silicon films. The complexity of film making process is the main factor that restricts the development of monocrystalline silicon thin film solar cells. Therefore, the development of new film making methods using nanotechnology may be the main direction to break through the restriction.

2. Polysilicon silicon thin film solar cells

Polysilicon solar cells generally use low-grade semiconductor polysilicon, or cast polysilicon and other materials specially produced for solar cells. Compared with monocrystalline silicon solar cells, polysilicon solar cells have low cost, and the conversion efficiency is close to monocrystalline silicon solar cells, which is one of the main products of solar cells. With the improvement of long crystal technology, the efficiency of solar cells has been improved. In the cost of photovoltaic modules, about 50–60% of the cost of photovoltaic comes from silicon raw materials. If a thin film solar cell is used, a silicon film is deposited on an cheap substrate as an active layer, a 40 μm thick silicon film can absorb 80% of the sunlight. In recent years, polysilicon thin film solar cells have developed rapidly. Polycrystalline silicon thin film cells are composed of many small particles of different sizes and different crystal surface orientation. The photoelectric conversion rate of polysilicon thin film solar cells is also higher than that of amorphous silicon solar cells, and it has the advantages of monocrystalline silicon and amorphous silicon solar cells. It is considered to be the next generation of solar cells most likely to replace monocrystalline silicon and amorphous silicon cells, so it has become a research hotspot in the international photovoltaic industry.

As shown in Fig. 10.12, polysilicon thin film solar cells have the advantages of high efficiency, stable performance and low cost, which are an important direction for the development of solar cells. It has now been mass-produced. At the same time, the polysilicon film low temperature deposition quality is poor, the film grain size is small, low battery efficiency; high temperature deposition energy consumption,

Fig. 10.11 Monocrystalline silicon solar cells

and the substrate material is also limited, limiting the application of polycrystalline silicon film to some extent [23]. Therefore, how to reduce energy consumption and reduce environmental pollution has become the focus of current research.

3. Amorphous-silicon, thin-film solar cells

Different from the amorphous silicon and crystalline silicon structure, the atomic arrangement is not very regular, which is a semiconductor of amorphous crystal structure. Amorphous silicon is a direct ribbon material with a high absorption coefficient of sunlight, requiring only a 1 μm thick film to absorb 80% of the sunlight. Amorphous silicon thin-film solar cells were introduced in 1976, and the amorphous silicon thin-film solar cells have a low cost and are convenient for large-scale production.

Due to the shortage of silicon raw materials and the expensive price, it promotes the efficient use of silicon technology to a certain extent. The low cost of amorphous silicon thin film cell makes up for its lack of photoelectric conversion efficiency, and will occupy an important position in the photovoltaic power generation industry in the future. The amorphous silicon thin film cell research is the main direction of the combined with microcrystalline silicon, generate amorphous silicon/microcrystalline silicon heterogeneity combined with solar cells, this kind of battery not only inherited the advantages of amorphous silicon cells, and can delay the efficiency of amorphous

Fig. 10.12 Polysilicon silicon solar cells

Fig. 10.13 Amorphous silicon solar cell

silicon cells with the speed of light attenuation, the highest amorphous silicon thin film cell conversion efficiency is 17.4%.

As shown in Fig. 10.13, amorphous silicon thin film solar cells and crystalline silicon thin film solar cells each have their own advantages and weaknesses, and the use of their layered structure can obtain high stability and high efficiency battery modules, which is also one of the research hotspots of researchers [24]. At present, the main problem is how to develop advanced film making process to achieve the large-scale and technological production of thin film, and prepare thin film with high stability and high conversion solar cells, so as to realize the commercialization of silicon thin film solar cells.

10.2.3 Nanomaterials Used in the Fuel Cells

A fuel cell is a device that uses chemical reactions to convert fuel into electricity. Nanomaterials can also be used in fuel cells. Fuel cells fabricated from nanomaterials can improve their efficiency and stability. In addition, nanomaterials can also improve the electrocatalyst activity of fuel cells, thus reducing their cost and extending their service life. These advantages make the application of nanomaterials in fuel cells very promising. Fuel cells are widely considered as one of the important energy options for the future because they can provide consumers with efficient and clean electricity, and fuel cells emit only environmentally friendly substances such as water. Compared with conventional fuel engines, fuel cells have higher efficiency and less carbon dioxide emissions [25]. However, in order to achieve higher efficiency and lower cost, the continuous improvement of the fuel cell components is critical: the application of nanomaterial-based modeling technology is a practical solution. Different types of nanomaterials, such as platinum, silver, gold, diamond, nickel, etc., have been applied in fuel cells, and have achieved varying degrees of success. For example, some researchers have used flat-shaped nanoparticles to prepare oxygen reduction reaction catalysts, and have found that flat-shaped silver nanoparticles have higher catalytic activity than platinum nanoparticles, thus achieving higher battery efficiency.

Nanotechnology can be applied to many aspects of fuel cells, including electrode catalysts, electrolytes, nanoporous materials, to improve the efficiency and performance of fuel cells. (1) Electrode catalyst: Traditional fuel cell electrode catalysts mostly use platinum and other precious metals, resulting in high cost. Nanotechnology can prepare nanoparticles, improve catalytic efficiency, reduce the use and reduce the cost. Moreover, the high specific surface area of the nanocatalyst can increase the catalytic reaction rate and accelerate the reaction of hydrogen and oxygen [26]. (2) Electrolyte: Electrolyte is one of the core components of fuel cells. The traditional electrolyte is mainly a solid polymer film, which has high conductivity and thermal stability, but there are also problems such as large electric polarization voltage. Nanotechnology can produce electrolytes with high porosity and large specific surface area through nanomaterials, and improve the electrical conductivity and the efficiency of fuel cells. (3) Nanoporous materials: Nanoporous materials can be used as fluid carrier for fuel cells, providing a large amount of surface area, increase the active site involved in the reaction, and improve the efficiency of electron transport. At the same time, nanometer porous materials have high specific surface area and high porosity, which can improve the adsorption and diffusion capacity of fuel and accelerate the reaction rate.

Cheng used ethanol electrocatalytic oxidation as a model system to prepare upright gold nanowires (v-AuNWs) with Flammulina mushroom structure, which was used as a powerful electrocatalytic platform for stretchable fuel cells. As shown in Fig. 10.14, unlike the conventional fuel cell electrodes, both sides of the v-Au NWs film have a Janus morphology and are highly malleable. Comparative studies show that the v-Au NWs stretchable electrode can achieve high catalytic activity in ethanol electrooxidation due to the high surface area nanowire direct exposure to fuel. Moreover, the v-Au NWs stretchable electrodes still exhibit high electrocatalytic activity in various mechanical deformation states (including bending, twisting and stretching) [27]. Therefore, this work provides a powerful power for the next generation of flexible wearable and implantable electronic devices. Two different forms of vertical v-Au NWs nanowires were prepared by using ethanol electrocatalytic oxidation as a model system. In the test, the caudal side of the v-Au NWs electrode showed higher electrocatalytic activity than the head side under various mechanical deformation states (including bending, twisting and stretching). The stretchable fuel cells prepared using the traudally exposed interleaved electrodes have excellent comprehensive performance, which results from the direct exposure of the high surface area nanowires to the fuel. By using a sponge as a fuel container, a tattoo-like ultra-thin fuel cell can be designed, which can easily achieve wearable applications and has good performance under a variety of deformations. The planar extensible fuel cell promises to power the next generation of flexible wearable and implantable electronic devices [28].

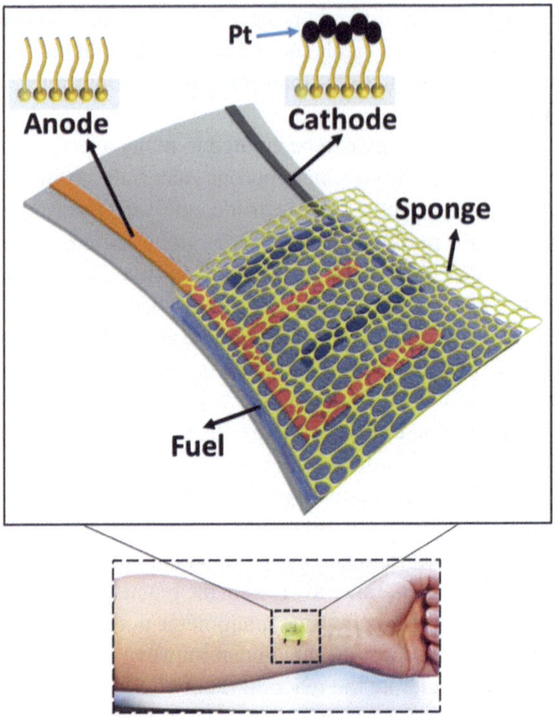

Fig. 10.14 A stretchable wearable fuel cell device based on v-Au NWs

Cathodic oxygen reduction reaction (ORR) dynamics is the main bottleneck of proton exchange membrane fuel cell application, pt based catalyst is the most effective ORR catalyst, with relatively low overpotential and high current density, but still has many disadvantages, including rarity, instability under electrochemical conditions, vulnerable to the influence of cross effect. Through a first-principles comparative simulation between BN co-doping and isolated B or N doping in graphene nanoribbons (GNR) and carbon nanotubes (CNT), Zou et al. not only identified the active sites in these doped carbon nanomaterials, but also clarified the potential mechanism process of ORR. Although CC bond deformation in carbon nanotubes causes firm O_2 binding, the spin density distribution along the edge plays a key role in enhancing the adsorption strength of O_2 on the GNR. As shown in Fig. 10.15, the effective adsorption of O_2 promotes the following basic reduction reactions. Based on thermodynamic analysis, O_2 adsorption and electron and proton transfer to the O atom were identified as critical rate determination steps. For CNT, a good linear ratio is found between the adsorption energy of the key intermediate and that of atom O, thus the latter serves as a good description of ORR activity. Moreover, showing N-doping high quality CNT showed the best performance. For GNR, the linear relationship is broken due to the edge effect, which promotes the catalytic activity in the case of BN co-doping [29].

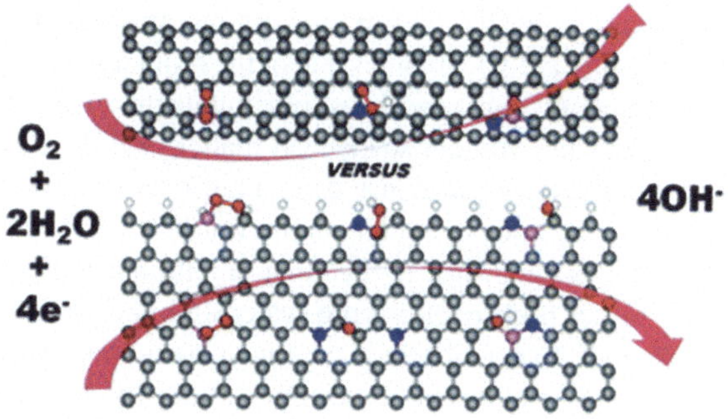

Fig. 10.15 An atomic-level mechanism of oxygen reduction reaction (ORR) catalytic reaction by doped nanomaterials

10.3 Application of Nanomaterials in Wastewater Treatment

On earth, all the activity of life originates from water. However, with the progress of science and technology and the development of society, the problem of environmental pollution has become increasingly prominent. The "three wastes" of industry, various kinds of agricultural pesticides and the discharge of urban wastes, the water environment is polluted by all kinds of heavy metals, organic substances, radioactive substances, viruses and bacteria. Many rivers in China are seriously polluted, and the water environment pollution is becoming more and more serious. The water quality in most parts of the country, especially in the economically developed and densely populated areas, is deteriorating continuously. Industrial wastewater is the most important water pollution source, the composition of such wastewater is more complex, often containing a variety of organic dyes and their intermediates, its toxicity is strong, difficult to degrade, and the components changes greatly, so it has always been the difficulty in industrial wastewater treatment. The pollution of water sources leads to the obvious increase of toxic and harmful substances in drinking water, which affects and even seriously hinders People's Daily life and work. According to the World Health Organization survey, 80% of global diseases and 50% of child mortality are related to poor water quality [30]. Statistics from 78 cities in the United States show that water quality accounts for about 95% of the causes of infectious diseases, and about 12.4 million people die from drinking unclean water every year, which shows that it has caused serious impact and harm to human health.

Sewage treatment is to remove the toxic and harmful substances, suspended substances, sediment, rust, odor pollutants, bacteria, viruses and other substances from the water. Due to the low efficiency, high cost and secondary pollution of the traditional water treatment methods, the sewage treatment has not been well solved. The development and application of nanotechnology is likely to completely solve this problem. The precious metals in the sewage are extremely harmful substances to the human body. It is lost from the sewage, and it is also a waste of resources. New nanotechnology can completely extract precious metals such as gold, nails, palladium and platinum in sewage, turning harm into treasure. A new type of nanoscale water purification agent has a strong adsorption capacity, its adsorption capacity and flocculation ability are 10–20 times that of the ordinary water purifier aluminum trichloride, so it can completely adsorb and precipitate the suspended matter in sewage. First, the water does not contain suspended matter, and then use nano magnetic substances, fibers and activated carbon and other purification devices, effectively remove the rust, sediment and odor pollutants in the water. After the first two purification processes, the water is clear, has no peculiar smell, and the taste is also good. However, in order to get high quality pure water and completely drinkable, it is necessary to go through a special water treatment membrane with nanopaperture and a treatment device composed of ceramic spheres with different nanoaperture to remove 100% of the bacteria and viruses in the water [31]. The purpose of using nanoporous membranes and ceramic spheres is that the diameter of bacteria and viruses is larger than that of nanopheres. When passing through membranes and ceramic spheres of nanopaperture, they are filtered out, and water molecules and minerals and elements below the diameter of water molecules are retained. Liver and kidney failure people drink this kind of water, will greatly reduce the burden of liver and kidney. The technology has also been used in the medical field of hemodialysis, and has the reputation of "in vitro kidney".

Nanomaterials are composed of nanoparticles, which have many new characteristics such as adsorption and catalysis. At present, the most commonly used nanomaterials is TiO_2. As shown in Fig. 10.16, TiO_2 plays a very important role in sewage treatment. Among them, due to the nano TiO_2 has a strong reduction ability, so in the organic sewage treatment, can be high oxidation state silver, platinum and other precious metal ions adsorption on the surface of the material, through optoelectronic strong reduction ability of metal particles to tiny metal crystals, not only remove the toxicity of sewage, also conducive to the recycling of precious metals. However, in the inorganic sewage treatment, nano-TiO_2 acts as a photocatalyst. In sunlight, it can catalyze the oxidation of organic pollutants in sewage, so that it is quickly and completely degraded into water, CO_2 and other harmless substances [32]. According to relevant statistics, nano TiO_2 can handle more than 80 kinds of organic toxic substances.

10.3.1 Nanotechnology in Sewage Treatment

10.3.1.1 Photocatalytic Oxidation Technology

Photocatalytic oxidation technology can effectively treat cyanide, metal particles and various organic acids and other substances, so that the pollutants in sewage are finally oxidized and degraded to form H_2O and CO_2. According to relevant statistics, it has been found that hundreds of organic pollutants can be treated by photocatalytic oxidation technology. The key to this technology is its photooxidation catalyst, and TiO_2 is considered to be the most effective catalyst. Because of nano-TiO_2, photocatalytic oxidation technology has the characteristics of no secondary

Fig. 10.16 Nano TiO_2

pollution, not only high degradation efficiency, no selectivity, but also its oxidation reaction conditions are mild, so it is almost suitable for all sewage treatment.

10.3.1.2 Nanofiltration Technology

Nanofiltration (NF) is a membrane separation technology between ultrafiltration and reverse osmosis. Its interception molecular weight is in the range of 80–1000, and its pore size is a few nanometers, so it is called nanofiltration. Nanofiltration technology belongs to the pressure process driven, the key of this technology lies in the nanofiltration membrane. Nanofiltration membranes can replace electrochemical and adsorption methods to treat pollutants in pulp and paper industry wastewater, and can remove chloride and dark substances produced from the process of wood pulp bleaching. In addition, the nanofiltration membrane can also be used for the treatment of bleach water in the process of fiber processing to control the discharge of pollutants. With its special advantages, nanofiltration membrane water treatment technology has won the general attention of water treatment workers from all over the world, and has made gratifying achievements in the research and development of water treatment technology. Nanofiltration principle: source water → Source pump → mechanical filter → carbon filter → ultrafilter → high-pressure pump → Nanofiltration main filtration system.

10.3.1.3 Nano-flocculation Technology

Nanoflocculation technology replaces the traditional flocculant (such as SiO_2). Due to the strong adsorption capacity of nanoparticles, the pollution substances can be removed by the flocculation effect such as adsorption bridge and rolling flocculation method, and the related precipitation substances have the characteristics of easy dehydration.

10.3.2 Treatment of Wastewater with Nanoscale Photocatalytic Materials

Photocatalytic degradation is a new and promising wastewater treatment technology, which refers to the decomposition of pollutants under light and through catalyst. Nanoparticles have higher catalytic activity due to nanoeffects that are not available by conventional particles. The commonly used catalysts are nano-TiO_2, ZnO_2, CdS, SnO_2, Fe_2O_3 and so on. Semiconductor catalysis has been valued since its discovery, because this effect has important applications in environmental protection, water quality treatment, organic matter degradation, failure pesticide degradation and other aspects. The advantage of this method is that it can completely decompose pollutants, compared with traditional techniques such as activated carbon and gas stripping methods only to change pollutants from one state to another state, through the photocatalysis can be organic and inorganic compounds and even microorganisms degradation or converted into very small harmful substances. Organic pollutants in environmental pollution, such as hydrocarbons, alcohols, ketones, aldehydes, aromatic groups, halogenates, heterocyclic compounds, organic acids, organic complexes, etc., can be decomposed into non-toxic CO_2, H_2O and harmless organic acids under light conditions by using photocatalysts. Photocatalysis can solve the pollution problem of mercury, chromium, lead and other metal ions in water. Using photocatalysis in the presence of citrate ions, Hg^{2+} from the oxygen-containing solution. As photocatalysts, nanoparticles have many advantages, first, small particle size, large specific surface area and high photocatalytic efficiency. In addition, most of the electrons and holes generated by the nanometer particles do not rebind until they reach the surface. Therefore, the number of electrons and holes that can reach the surface is large, and the chemical reactivity is high.

TiO_2 Photocatalytic oxidation technology has so far found that hundreds of organic pollutants can be degraded into harmless CO_2 and H_2O by the photocatalytic oxidation of nano-TiO_2 [33]. Its action principle is, under the ultraviolet light, nano TiO_2 surface will produce strong oxidation capacity of hydroxyl radical (·OH), make the organic pollutants in the water thoroughly oxidation degradation for CO_2 and H_2O. The technology has fast degradation speed, mild oxidation conditions, no selectivity, no secondary pollution, wide application range, specific application research mainly has the following aspects:

(1) The treatment of phosphorus wastewater organophosphorus pesticides account for more than 80% of the pesticide output in China, and it is large in the production process. The amount of toxic wastewater is produced, and the content of organophosphorus in the wastewater treated by the traditional biochemical

method is still as high as 30 mg/L, so there is no ideal solution at present. In China, the nano TiO_2-SiO_2 loaded composite photocatalyst can rapidly enrich the organophosphorus pesticide on its surface, and the light is 80 min, and the test parasites can be completely degraded.

(2) Chlorinated organic wastewater treatment in Japan showed that the 3-chlorophenol in the wastewater treated with nano-TiO_2 photocatalyst and ozone, and the ceramic tube coated with the inner surface treated the trichloroethylene solution. Using artificial lighting and nano-TiO_2, Britain has developed a new room temperature photocatalysis technology to fully decompose PCBs in industrial waste liquid and contaminated groundwater into CO_2, H_2O and HCl.

(3) The oily wastewater contains aliphatic hydrocarbons, polycyclic aromatic hydrocarbons, organic acids and phenolics, which are difficult to degrade themselves. Using the photocatalytic oxidation technology of nanometer TiO_2, these organic matters can be rapidly degraded. Chinese Academy of Sciences using sunlight and nano TiO_2 powder to phenol water and sodium water test, the results show that in cloudy and cloudy conditions, sunlight exposure after 12 h, the concentration of 0.5 mmol/L phenol has been completely degraded, the concentration of 1 mmol/L sodium dodecyl benzene sulfate is basically complete degradation, high clarity and no secondary pollution.

(4) Wool spinning dyeing wastewater treatment surface coated with nano TiO_2 film glass filler filled in glass reactor, wastewater circulation in the reactor for photocatalytic oxidation treatment, organic matter in waste water can quickly decompose into H_2O and CO_2, and catalyst can be continuous use, without separation recovery, has the characteristics of high efficiency, energy saving, no secondary pollution, convenient for industrial applications.

(5) Photocatalytic antibacterial effect of nano TiO_2: TiO_2 as a photocatalyst, more and more attention. Using TiO_2, ZnO and other semiconductors to photocatalytic degradation of organic pollutants, and finally generate non-toxic and tasteless CO_2, H_2O and some simple inorganic substances, is gradually becoming an industrial technology, which opens up a broad prospect for the elimination of environmental pollution. The commonly used fungicides such as Ag and Cu can deactivate the cells, but the endotoxin produced after the bacteria are killed cannot be eliminated. Endotoxin is a deadly substance that can cause typhoid fever, cholera and other diseases. After photocatalysis, nanometer TiO_2 can undergo biochemical reactions with bacterial cells or intracellular components, resulting in cell death and decomposition of the endotoxin produced by bacteria after death. The experiment proved that nano-TiO_2 has strong killing ability against E. aeruginosa Escherichia coli, Staphylococcus aureus, Salmonella, bud rod and Aspergillus. It has been possible to replace the traditional water treatment sterilization by using nano-TiO_2 and sunlight.

Nano TiO_2 is photoactive in the UV region (40 nm), and the valence band electrons are excited to the conduction band to form a hole-electron pair. Under the electric field, the electrons and holes separate and migrate to different positions on the particle surface. The holes on the surface can oxidize the OH and H_2O molecules adsorbed on the TiO_2 surface to ·OH radicals, so the TiO_2 is highly oxidized. The highly active electrons on the surface of TiO_2 have strong reduction ability, which can reduce and remove metal ions in water. The particle size of nano TiO_2 is small. After absorbing light energy, the number of electrons and holes reach the surface is large, the speed is fast, and the photocatalytic efficiency is much higher than that of ordinary TiO_2 powder. With the deepening of basic research on nanomaterials and nanotechnology and the development of practical process, the application of nanomaterials in environmental protection shows a thriving scene. Compared with traditional materials, nanomaterials have many unique properties. In the future, more nanomaterials will be used for environmental protection in the future. Many environmental problems such as air pollution, sewage treatment and municipal waste will be solved. We will fully enjoy the clean environment brought by nanotechnology.

The specific surface area of nanomarticles is large, which is very suitable as a catalyst with high efficiency. Its surface effect and volume effect determine its good catalytic performance and selectivity. As a typical representative of nanomaterials in photocatalytic applications, nano-TiO_2 has become the key research object of photocatalytic wastewater treatment. The removal of soluble organic pollutants in water has always been a big problem in the water treatment process. Photocatalytic oxidation of nanomaterials can completely convert organic matter into simple inorganic substances such as carbon dioxide and water, so as to avoid secondary pollution. Using ultraviolet light excitation catalytic ($UV/TiO_2/H_2O_2$) process, can be disinfection by-product of chlorochloride organic chloride direct oxidation into non-toxic gas ions and carbon dioxide, light excitation catalytic process of hydroxyl radical (·OH) for the removal of organic matter has a wide spectrum, so the nanoscale oxide as a catalyst can remove a variety of organic matter in water, make the water quality improvement. So far, known nano TiO_2 can treat more than 80 kinds of toxic pollutants, it can be hydrocarbons, halogenated hydrocarbon/acid, surfactants, organic dyes, nitrogen containing organic matter, organophosphorus (bromine) pesticides, wood preservatives and fuel oil, heterocyclic aromatic hydrocarbons, replace aniline soon completely oxidized into water and carbon dioxide and other harmless substances. For refractory organic matter such as detergent, polychlorinated biphenyls, polychlorinated dioxin, polyvinyl alcohol, phenolic compounds also showed good degradation performance.

At present, using the normal temperature catalytic technology of artificial lighting and nano TiO_2, the PCB in industrial waste liquid and polluted groundwater can be decomposed into water and carbon dioxide, thus eliminating the pollution to the environment. After 12 h of sunlight irradiation with nano-TiO_2 powder and sunlight, the aqueous sodium dodecyl benzene sulfonate solution with a concentration of 1 mol/L was basically degraded, with high clarity and no pollutants. Heller The fixed TiO_2 photocatalyst attached to the hollow glass ball can float on the water surface; the bul titanate can prepare the floating catalyst with nano TiO_2 powder. Under the direct exposure of ultraviolet light and sunlight, it can not only effectively degrade the oil pollution on the water surface, but also inhibit the formation of harmful copolymers in the process of natural oxidation. Using medium pressure mercury lamp as light source, nano TiO_2 can effectively photocatalytically degrade chloro-DB-p-dioxin

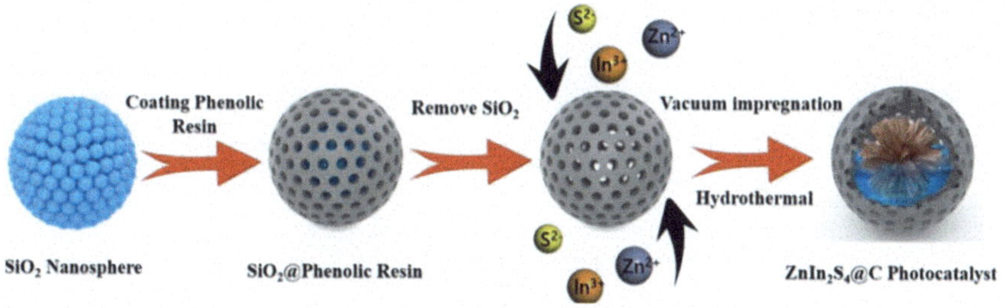

Fig. 10.17 $ZnIn_2S_4$@C Synthesis of the photocatalyst

(CDDS, including CDD, DCDD, PeCDD, OCDD), OCDD, 87.2%, PeCDD and 91.2% respectively within 4 h. At room temperature, the reaction temperature and TiO_2 concentration are the main control factors of the reaction.

Yang reported a nanoconfined $ZnIn_2S_4$@C photocatalyst that maximizes the water inside the MCN and stabilizes its water molecules in the nanocavity, changing the localized electron distribution of water molecules on the $ZnIn_2S_4$ surface. As shown in Fig. 10.17, this change indicates that the chemisorption ability of the photocatalyst to the inner water has a sensitive dependence on the nanostructure, thus further improving the photocatalytic performance. The diffusion and adsorption characteristics of water molecules in the active center play a crucial role in the performance of heterogeneous photocatalytic reactions. Compared with the main water system, the water in microporous carbon nanocages (MCN) can form an ideal system to facilitate the adsorption of these nanoconfined water-restricted molecules on the internal metal active center. However, the nature of the water molecules adsorbed in the nanoconfined cavity remain unknown and how they rely on the microporous and chemical functional groups of the carbon nanocage to facilitate the adsorption of water molecules in the active center [34].

10.3.3 Nanofiltration Membrane Treatment of Wastewater

As shown in Fig. 10.18, nanofiltration, hereinafter referred to as nanofiltration, is a new membrane separation process between reverse osmosis and ultrafiltration developed in the mid-1980s. It is a new membrane separation process driven by pressure. The application of nanofiltration in water softening, wastewater treatment, and the concentration and recovery of valuable ingredients in food, beverage and pharmaceutical production has attracted more and more attention, and has become a new research hotspot in the field of membrane separation.

If you want to get high quality pure water, it is necessary to go through a special water treatment membrane with a nanopore size and a treatment device composed of ceramic spheres with different nanopore sizes to remove 100% of the bacteria and viruses in the water. The main characteristics of nanofiltration membrane are reflected

Fig. 10.18 Nano filter membrane

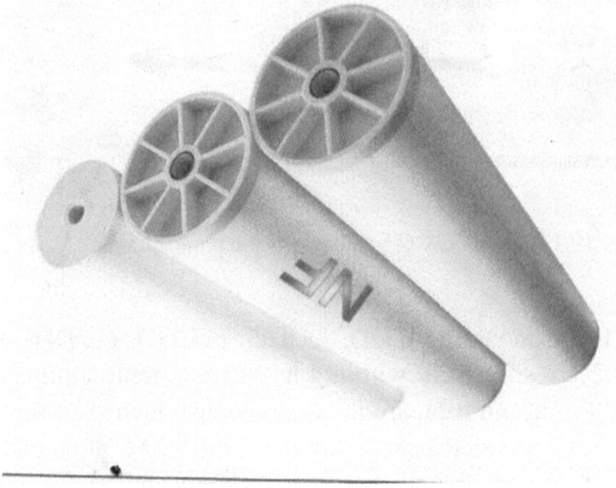

in two aspects: (1) the interception relative molecular mass range of nanofiltration membrane ranges from 200 to 2000, Its surface separation layer is composed of a polyelectrolyte, Therefore, it has a certain interception rate of inorganic electrolyte, Allow some inorganic salts, In particular, the monovalent salt ions and some solvent pass through the membrane, The presence of genuine micropores, The pore size is in the nanoscale range, Its retention effect on different ions of valence states, Low interception rate for the monovalent ions, The interception rate of bivalent and multivalent ions is significantly higher than that of monovalent ions, By allowing most of the monovalent ions to pass freely, Make the nanofiltration membrane use only a lower operating pressure, Generally in 0.5–1.0 MPa; (2) High nanofiltration membrane flux, In contrast to reverse osmosis, Nanofiltration has the advantages of low equipment investment and low energy consumption. This unique function of nanofiltration membrane is unable to be replaced by reverse osmosis membrane and ultrafiltration membrane. Nanofiltration membrane can be applied to the treatment of wastewater in sugar making, paper making, textile, electroplating, mechanical processing and other industries with its special separation performance. In metal processing and alloy production, there are quite high concentrations of heavy metals (Ni, Fe, Cu, Zn, etc.). In order to make these heavy metal-containing waste water meet the discharge requirements. The general measure is to treat these heavy metals into hydroxide precipitation and separate them. If the nanofiltration membrane technology is used, not only more than 90% of the wastewater can be recycled and purified, but also the heavy metal ion content can be concentrated by 10 times, and the concentrated heavy metals have the value of recycling. If the conditions are properly controlled, the nanofiltration membrane can also separate the different metals in the solution.

10.3.4 Treatment of Wastewater with Nanoadsorbent Materials

(1) Strcolumn clay mineral is a nanocomposite that has been widely studied in recent years. It is the use of layered clay can break into the structure of nano area, the crystal spacing generally in a few to ten nanometers, can make some polymer embedded in the size of the sandwich space, form "embedded nanocomposite", or by polymer, form high length ratio of single sheet inorganic, evenly dispersed in the polymer matrix, forming "layer from the nanocomposite". The large specific surface, organic group and aperture size of the layer column clay make it have good adsorption and exchange properties. Using hydroxyl aluminum cross-linked bentonite adsorbent to treat phosphorus-containing wastewater, under certain conditions, the adsorption efficiency is 99.7%, the adsorption capacity is 3.26 mg/g (measured by P), and the adsorption capacity of hydroxyl aluminum cross-linked montmorillonite on the phosphate solution is 5.56 mg/g, the removal rate of phosphorus is close to 100%.The adsorption capacity of Pb^{2+} by natural diatomite and MnO_2 modified diatomite was 24 mg/g and 99 mg/g, respectively. The study shows that after the modification of bentonite with hydroxyl iron, the removal rate of weakly acidic deep blue GR is up to 99% under suitable conditions, and the adsorption mechanism is mainly surface adsorption, ion exchange and distribution.

(2) According to research reports, as shown in Fig. 10.19, nano-SiO_2 powder has a good adsorption effect on alcohols, amides, and ethers. Electrically neutral molecules can be adsorbed on the particle surface by weak electrostatic gravity of hydrogen bonding, van der Waals force and dipoles. It is mainly combined in the form of hydrogen bonding. For example, in the adsorption process of alcohols, amides and ethers, the contact between the silica oxide particles and the organic reagent is the silane alcohol layer, which plays an important role in the adsorption. O or N in the above organic reagent forms an O–H or N–H hydrogen bond with H in the hydroxyl group (OH group) of the silanol, so as to complete the adsorption of SiO_2 particles to the organic reagent. Only one hydrogen bond can be formed between an alcohol molecule and the silanol hydroxyl group on the surface of the silicon oxide, so the binding force is very weak and belongs to physical adsorption; For polymer oxides, such as the adsorption of polyethylene oxide on silicon oxide particles also passes through hydrogen bonds. Due to the formation of a large number of O–H hydrogen bonds, the adsorption force becomes very strong. This adsorption is chemical adsorption, physical adsorption is easy to desorption, chemical adsorption desorption is difficult. Adsorption is affected not only by the surface properties of the particles, but also by the properties of the adsorption phase. Even if the adsorption phase is the same, the adsorption amount is different due to the different solvent types. For example, when straight-chain fatty acids are used as the adsorption phase and benzene and n-hexane solutions as the solvent, the adsorption amount of

straight-hexane acids on the surface is more than when benzene is used as the solvent, because few hydrogen bonds are formed in the case of benzene.

(3) The surface effect of nanomaterials has high surface activity, high surface energy and high specific surface area, so nanomaterials show great potential in the preparation of high performance adsorbents, and the research in this aspect is just beginning. At present, the new nano-scale water purifier has a strong adsorption capacity, its adsorption capacity and flocculation ability are 10–20 times that of the ordinary water purification agent aluminum trichloride. Therefore, it can completely adsorb and precipitate the suspended solids in the sewage, first make the water does not contain the suspended solids, and then the use of nano magnetic substances, fiber and activated carbon purification device, can effectively remove the rust, sediment and odor in the water. After the first two purification processes, the water is clear, no smell, taste is good. After the special water treatment film with the nano aperture and the treatment device with the ceramic sphere assembly with different nano aperture, it can be 100%. Remove the bacteria and viruses in the water, get high quality pure water, can drink. This is because the diameter of bacteria and viruses is larger than that of nanometers. When they pass through the membrane and ceramic balls, they are filtered out, and the minerals and elements below the diameter of water molecules are retained. It is reported that a company uses nanomaterials for sewage treatment purification agent, flocculants and bactericidal disinfectant. The multi-composite new ultra-efficient water treatment agent is formed by these nanomaterials, which not only has good pollution control effect, but also shortens the process and reduces the cost of medicine.

Fig. 10.19 Nano-SiO_2 morphology

10.4 Application of Nanomaterials in Exhaust Gas Purification

The development of modern industry not only creates great wealth for human beings, but also brings serious pollution to the ecological environment. The quantity and type of gas fuels used in industrial production and gases produced in the production process are increasing. These toxic gases and combustible gases pollute the environment, and explosion and fire may occur, endangering the safety of people and property. In China, the ownership of cars has also increased on a large scale, from 150,000 in 1979 to 1.4 million in 1993, with an average annual growth rate of 16%. According to statistics, automobile exhaust emissions in developed countries are three times that of other gas pollution sources. In China, more than 40% of NO_x, more than 80% of CO and more than 70% of hydrocarbons come from automobile exhaust emission pollution. Moreover, gasoline and diesel oil used in industrial production, as well as gasoline and diesel oil as automobile fuel, will produce SO_2 gas when burning, and excessive sulfur dioxide (SO_2), carbon monoxide (CO) and nitrogen oxide (NO_x) in the air are harmful to human health. Therefore, the treatment of air pollution is also a problem to be solved by governments of various countries. It is of great significance to develop alternative fuels or study materials to control and reduce automobile exhaust and air pollution to purify the environment of human existence.

In recent years, the use of nano-photocatalytic technology to remove gaseous pollutants from the air is a new pollution control technology, which has many advantages and has attracted wide attention.

10.4.1 Nanomaterials Used as Catalysts for Vehicle Exhaust Gas Purification

Excessive sulfur dioxide, carbon monoxide, and nitrogen oxides in the air are harmful gases that affect human health. The application of nanomaterials and nanotechnology can ultimately solve the problem of pollution sources that produce these gases. The latest research results show that the nanoscale powder of composite rare earth compounds has extremely strong redox performance, which is incomparable to any other automobile exhaust purification catalyst. Its application can completely solve the pollution problem of carbon monoxide and nitrogen oxide in automobile tail gas. The automobile exhaust purification catalyst with activated carbon as the carrier and nano $Zr_{0.5}Ce_{0.5}O_2$ powder as the catalytic active body. due to the existence of Zr^{4+}/Zr^{3+} and Ce^{4+}/Ce^{3+} on its surface, electrons can be transferred between its trivalent and tetravalent ions, so it has strong electronic gain and loss ability and redox activity. Plus nanomaterials its large surface, space suspension bond adsorption capacity, so it reduces the nitrogen oxides in oxidized carbon monoxide at the same time, make them into harmless to the human body and the environment gas

CO_2 and N_2. And update a generation of nano catalyst, will play a catalytic role in the car engine cylinder, make the gasoline in combustion does not produce CO and NO_x, without tail gas purification treatment.

The experimental study found that the nanoscale rare earth perovskite composite oxide ABO_3 has a good catalytic conversion effect on the CO and NO_x emitted by automobile exhaust gas. The ternary catalytic catalyst made of honeycomb cyanolite carrier is good and cheap, which can replace the expensive precious metal catalyst. In recent years, many rare earth perovskite composite oxides have been put on the market for the treatment of automobile exhaust.

Nanomaterials are used in the excessive alarm and purifier of automobile exhaust gas to reduce the emission of harmful gases. Vehicle exhaust sensors for the preparation and assembly of nanomaterials, through the monitoring of vehicle exhaust emissions, timely alarm of excessive emissions, and adjust the appropriate air–fuel ratio, reduce oil-rich combustion, to achieve the purpose of reducing harmful gas emissions and fuel consumption. This part of the semiconductor gas sensor mainly uses the resistance of the material with the change of the ambient atmosphere concentration, and the environmental atmosphere can be obtained through the change value. According to the matrix can be divided into metal oxide type and organic high molecular type, Metal oxide semiconductor gas-sensitive materials can be divided into simple oxides and composite oxides represented by SnO_2, ZnO and Fe_2O_3. The use of nanomaterials can significantly improve the gas sensitivity. α-Fe_2O_3 belongs to the corundum structure with good stability. The sensitivity to combustible gas is low, if the sol–gel method, chemical vapor deposition method and other advanced materials to synthesize nanoscale α-Fe_2O_3. In the absence of any doping, it showed a high sensitivity and responsiveness to combustible gases such as acetylene, acetone, and ethanol. At the same time, the sensitivity, stability and selectivity can be improved by the following methods. (1) Add additives, that is, add different amounts of precious metals, metal oxides or rare earth oxides and other different kinds of additives in the matrix material. TiO_2 Semiconductor oxygen sensor has the characteristics of small size, simple structure, low cost, no need for reference electrode, easy integration, and has good oxygen sensitivity characteristics. It is found that the TiO_2 electrical sensor of the doped catalyst (such as H) has better sensitivity than the undoped catalyst, and the response time drops from 400–900 ms to 20–600 ms, and the recovery time is also reduced to about a few seconds to tens of seconds. (2) Use filtration equipment or air permeable membrane to obtain selectivity. (3) Control the fine structure of the material, and the literature recently reported that nano CeO_2 coated TiO_2 composite oxygen sensor, its oxygen sensitive factor can reach about 3.7.

In addition, automobile exhaust purifier can also be prepared with nanomaterials, such as ultra-fine Fe, Ni and γ-Fe_2O_3 mixed by sintering instead of precious metal as automobile exhaust purifier, which can significantly reduce the cost and improve the efficiency, as shown in Fig. 10.20.

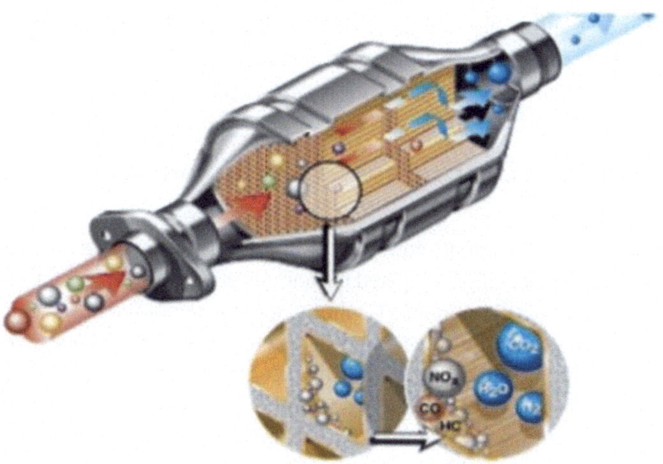

Fig. 10.20 Automobile exhaust catalytic device

10.4.2 Nanomaterials Used as Catalysts for Petroleum Desulfurization

It is far from enough to solve the problem of pollution source from the root cause by adopting the alarm and monitoring measures of automobile exhaust emission. Industrial and automotive fuel oil is the largest source of SO_2 pollution. The sulfur-containing compounds in the fuel oil will produce SO_2 after combustion, so the petroleum refining industry has a desulfurization process to reduce the sulfur content of gasoline and diesel oil. If the desulfurization process is paid attention to in the petroleum refining industry to reduce the sulfur content in the fuel, then the source of harmful gas SO_2 will be truncated, so the environmental governance can get twice the result with half the effort. Nanomaterials will play an important role in improving the desulfurization efficiency in the desulfurization process. Nanotitanate diamond ($CoTiO_3$) is a very good catalyst for petroleum desulfurization. It is reported that the catalytic living body of cobalt titanate synthesis with radius of 55–70 nm, with porous silica gel or Al_2O_3 ceramics as the carrier, has extremely high catalytic efficiency. It can solve the problem of air pollution caused by the combustion of gasoline and diesel oil. The white spherical zinc titanate powder with particle size of about 30–60 nm prepared by precipitation dissolution method has large specific surface area and high chemical activity. It is used as adsorption desulfurizer, which is significantly better than the zinc titanate powder prepared by solid phase sintering method. The sulfur content in the catalyzed oil is less than 0.01%, reaching the international standard.

Similarly, coal combustion will produce SO_2 gas, in the fuel combustion, then add a nanoscale burn catalyst, not only can make coal fully combustion, does not produce sulfur dioxide gas, improve energy utilization, and can make sulfur into solid sulfide, without producing sulfur dioxide gas, so as to eliminate the production of harmful gas.

10.4.3 Nanomaterials Used for Indoor Air Purification

There are three kinds of popular indoor air cleaning technologies in the world: negative ion generation technology, activated carbon filtration and absorption technology and the most advanced nanomaterials photocatalytic decomposition of pollutants technology. Although the first two technologies can increase the density of negative oxygen ions and filter some harmful substances respectively, they are not as good as photocatalytic technology, without trace and natural decomposition without the concern of media saturation. The new photocatalytic technology of nanomaterials will be widely used in the field of indoor air improvement to bring a "urban forest" to the Chinese people who pursue a healthy life. The first batch of nanotechnology applied to the production of air conditioners in China has opened the chapter of healthy air conditioning in the twenty-first century with the international technical background of soil air conditioning in air purification and water treatment. Nanomaterials used for air purification and filtration are synthesized from a variety of rare earth metals, rare metals, and a variety of oxides through high-tech methods, among which special nanomaterials are added. Under the combined action of nanomaterials and a variety of rare earth metals, it is a microcitation activity center with good removal effect of various organic pollutants.

Nano crystal formaldehyde removal is a kind of "nano crystal air purification agent", which is used in indoor air purification, decoration pollution treatment, home odor removal and car air purification. The full name of nano crystal is "nano crystal air purifying agent", which is a pure natural mineral attapulgite and porous mineral material particles, because the pore diameter of the internal pore between 0.27 and 0.98 nm, is in crystal arrangement, so it is referred to as "nano crystal". Nano crystals are widely used in indoor air purification, decoration pollution control, home odor removal and air purification in cars, which is the result of the development of new materials in recent years. Its microstructure has a unique crystal regular structure, a large number of nanoscale pores in the crystal, pore size of 2.7–9.8 Å. Because its nanosize can selectively adsorb benzene, xylene, formaldehyde, carbon dioxide, sulfur dioxide, ammonia and other toxic and harmful substances, it can be made into nanopore air purification agent through a certain process, which not only has a better effect than activated carbon, but also has no negative effect on the environment. The industry is called the creator of the human beings to maintain their own health gifts. The microstructure of nanoreamite is close to the 3, 4 and 5 Å molecular sieve, and the pore diameter is about 3 Å, which is also called the mineral crystal, porcelain and crystal.

10.5 Application of Nanomaterials in Waste Solid Treatment

Solid waste treatment is an important issue in the field of environmental protection, and nanomaterials also have potential applications in solid waste treatment. For example, nanosilicon wafer can be used as a landfill cover to prevent the leakage of harmful substances from waste, and nanocellulose can be used as a substitute for biodegradable plastics to reduce the environmental impact of plastic waste. The degradable starch and non-degradable plastic are crushed to the "nano level" for physical mixing modification. With this new raw material, it can produce 100% degraded agricultural mulch, disposable tableware, a variety of packaging bags and other similar products. After 4 to 5 years of field experiments, agricultural ground film shows that within 70 to 90 days, starch is completely degraded into water and carbon dioxide, plastic becomes fine particles harmless to soil and air, and also completely degraded into water and carbon dioxide within 17 months, which is a substantial breakthrough to completely solve white pollution.

The application of nanotechnology and materials in solid waste treatment, its advantages are mainly reflected in the following two aspects: First, the speed of nanoscale treatment agent degradation of solid waste is fast. For example, nano-TiO_2 degrades solid waste 10 times faster than conventional TiO_2 [35]. Second, the use of nanotechnology can make rubber and plastic products, waste printed circuit board into ultrafine powder, remove the impurities, and recycle them as recycled raw materials. In Japan, waste rubber tires are made into powder for laying sports fields, roads and roadbeds of the Shinkansen. Therefore, nanotechnology can greatly alleviate the huge pressure brought by solid waste to the environment and also reduce the secondary pollution caused by traditional methods such as landfill.

10.5.1 Cellulose-Nanocrystal-Reinforced Bioplastic Composites

In the 1970s, people began to look for sustainable packaging materials that could replace traditional plastics. Biodegradable plastic is a new kind of environmental protection material emerging on this basis. Compared with traditional plastics, it has good biocompatibility and thermal processing performance, and under the same degradation conditions, biodegradable plastics can be completely digested and degraded by microorganisms into water and carbon dioxide within a certain period of time, thus greatly reducing the environmental problems and social problems brought by traditional plastics. However, due to the disadvantages of high cost, poor thermal stability and narrow processing window of biodegradable plastics, which seriously limit the practical application of biodegradable plastics, therefore, choosing appropriate materials to modify them is the main method to solve the problem. Cellulose nanocrystals are natural polymer materials, and cellulose nanocrystal are widely used as reinforcing materials for their low density, high polymerization, large surface area, high Young's modulus, high strength and degradable properties. The use of cellulose nanocrystals for enhanced bioplastics can not only improve the mechanical properties of composite materials, improve the processing properties of bioplastics, but also maintain the biodegradation characteristics of bioplastics.

Due to the large structural differences between cellulose nanocrystal and bioplastics, cellulose nanocrystal has a large number of hydroxyl groups, large polarity and high hydrophilicity. However, bioplastics are weak and not hydrophilic, and it is difficult for cellulose nanocrystal to disperse uniformly in nonpolar solvents, resulting in the binding difference of the interface and the weakening of the interaction between molecular chains, which affects the comprehensive performance of composite materials and limits the enhancement of cellulose nanocrystals. In order to enhance the binding force between the two phases, the modification of cellulose nanocrystal and bioplastics can be modified [36].Common modification methods of cellulose nanocrystals include chemically grafted long chains or special groups, silanization, and TEMPPO-oxidation. Compared with modified bioplastics, it is easier to modify cellulose nanocrystal, so cellulose nanocrystal is also the first choice of modified objects.

(1) Effects of cellulose nanocrystals on the mechanical properties of bioplastics

Because of cellulose nanocrystal has high transparency, high young's modulus and high hardness, make it a kind of can effectively improve the mechanical properties of biological plastic material, in addition, both composite, cellulose nanocrystal as an effective nucleation agent of biological plastic, can speed up the crystallization of biological plastic, reduce the biological plastic spherulites size. The addition of nanofibers has little effect on the transparency of bioplastics. Figure 10.21 shows the comparison of transparency before and after the addition of nanoclose. It is found that improving the mechanical properties of bioplastics has little effect on the transparency of PLA. Ten by solution casting method using cellulose nanocrker (CNW) enhanced PHBV, choose tensile test and expansion test two methods to determine the mechanical properties of PHBV, when CNW dosage is 5%, the two methods of young's modulus increased by 77% and 91% respectively, tensile strength, energy storage modulus and toughness increases with the increase of cellulose nano whisker [37]. However, for different types of bioplastics, cellulose nanocrystalline rice crystals have different strengthening effects, and poor modified composite materials are often supplemented by modifying agents to improve their strengthening effect [38]. Li et al. used citric acid as a crosslinking agent to modify polyvinyl alcohol (PVA), and prepared PVA/CNC nanocomposite membrane and citrate cross-linked PVA/CNC nanocomposite membrane. Compared with pure PVA membranes, the modified PVA films showed the thermal stability of all, but the mechanical properties were sensitive to ambient humidity, while the mechanical properties of citrate cross-linked PVA/CNC nanocomposite films did not change with ambient humidity. The crosslinker acts as a molecular bridge in the composite material, connecting the cellulose nanocrystal with PVA, enhancing the binding between the two phases, and thus improving the mechanical properties. Hu et al. added acetylated cellulose nanocrystal

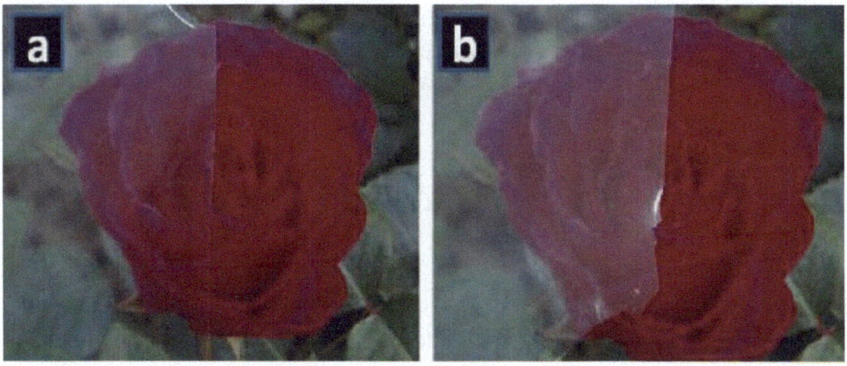

Fig. 10.21 Appearance and morphology of pure PLA and CNFs/PLA composite materials. **a** PLA, **b** wood powder CNFs/PLA

to PBS, the mechanical properties and crystallinity of the foaming polymer, and the morphology and density of the pores were improved, which also fully shows that cellulose nanocrystal is a highly crystalline enhancement material with high hardness, and can be well dispersed in the PBS matrix.

(2) Effects of cellulose nanocrystals on the thermal stability of bioplastics

Adding cellulose nanocrystal into bioplastics, the nanocrystals will limit the movement of the molecular chain of bioplastics in the composite material, which is manifested by the increase of the peak temperature of the material, that is, the increase of the glass transition temperature and the thermal stability of the material. rise. The PLA/CNC nanocomposite felt can be prepared by electrospinning. By adding CNC to the system, the mechanical properties of bioplastics are improved and the thermal stability is also improved. CNC can improve the crystallinity of PLA, and can be evenly dispersed in the bioplastic matrix under certain process conditions, increasing the energy barrier required for PLA melting. In addition, the addition of CNC accelerates the decomposition rate of the composite pad in the phosphate buffer, realizing the ideal decomposition process [39]. However, when cellulose nanocrystals are used to enhance bioplastics, the thermal stability of not all bioplastics is improved. Due to the internal scattering of cellulose nanocrystals is small, it has good thermal conduction properties, which makes the thermal stability of composite materials reduced.

10.6 Application of Nanomaterials in Noise Control

At present, noise is no longer a small problem that can be ignored, it has seriously affected people's work and life, and has become a contemporary worldwide problem. With the continuous development of industry, the continuous increase of the number of vehicles, urban construction, the problem of noise pollution will be more serious, so it is very urgent and necessary to study the sound absorption material with high sound absorption performance, the effective control of noise pollution.

10.6.1 Nanomaterials of Sound Absorption and Sound Insulation Materials

Many nanomaterials have different studies on sound absorption and sound insulation materials, mainly divided into the following types: (1) nanofiber, nanofiber into nanofiber felt or nanofiber film, and then combined with the textile sound absorption material, so as to enhance the sound absorption properties of sound absorption materials; (2) nanoparticles prepare nanoparticles into the fabric, improve the sound absorption performance.

Nanofiber materials gradually attract attention in noise control because of their small fiber diameter and large specific surface area. Studies show that nanofiber film has very good sound absorption effect, especially in its low frequency band sound absorption performance is good, the sound absorption principle is mainly: when the sound wave into the material, the sound micropore friction and nanofiber film, and the interaction between fiber and the resonance of membrane of acoustic energy loss, thus have the effect of sound absorption. However, the nanofiber felt or nanofiber film made of nanofibers is usually very thin and usually does not have a good sound absorption effect when used alone. Therefore, the nanofiber film or nanofiber felt is usually combined with other fibrous sound absorption materials to improve its sound absorption effect [40]. The study shows that the sound absorption of nanometer sound absorbing composite materials combined with nanomaterials and porous sound absorbing materials will be affected by many factors, and its sound absorption performance is related to the preparation parameters, thickness, membrane surface density, whether there is cavity and cavity depth, and whether there is filler in the cavity.

At present, the most commonly used method for the preparation of nanofibers is electrospinning. The different parameters of electrospinning will directly affect the sound absorption performance of nanofiber felt or nanofiber film, thus affecting the sound absorption effect of the nano-sound absorption composite material. At present, the main parameters affecting its sound absorption performance are: (1) different spinning liquid concentration and mass ratio. In the process of nanofiber spinning, when the voltage and receiving distance unchanged, the spinning fluid concentration and solution of mixing mass ratio, will affect the diameter of the spinning nanofiber and uniformity, and affect the performance of nanofiber felt and porosity, causing the difference of sound absorption performance. Therefore, it is necessary to determine the best spinning parameters according to the specific situation during the experiment. (2) Spinning time. The thickness of the nanofiber felt or nanofiber film is directly related to the time of spinning. When the nanofiber felt is too thin, its sound absorption effect is not obvious, only when it reaches a certain thickness, its sound absorption effect begins to appear, and within a certain range, its sound absorption effect increases with the increase of thickness.

After nanofiber membrane sound cavity can effectively improve the sound absorption effect, especially for its low frequency sound absorption performance, within a certain range, the low frequency sound absorption coefficient will increase with the cavity depth, because increasing the cavity is equivalent to increased the thickness and density of the material. Whether in the cavity filling porous material will also affect the sound absorption properties of sound absorption material, after the nano film cavity adding porous material will make the sound absorption properties has been improved significantly, this is because when the sound into the material, the porous medium fiber, air and sound waves between vibration, and produce friction, thus have the effect of energy consumption. Adding porous material in the cavity will make its sound absorption performance has been improved significantly, this is because when the sound wave into the material, porous medium can have the effect of energy consumption, on the other hand, in the range of low frequency, the surface mask layer because very thin, basic did not increase the thickness of the porous material, but can enhance the sound absorption performance of porous material especially the low frequency effect, and the surface of the film of the surface of the surface of the sound absorption material, so the nanomaterials and porous sound absorption material composite way is effective.

Nanoparticles have a large specific surface area, surface atomic number, surface energy and surface tension, and their effect increases sharply as the particle size of the nanoparticles decreases. Due to the small size of nanoparticles, it has many special properties compared with ordinary substances with the same composition, such as small size effect, surface effect, quantum size effect and macroscopic quantum tunneling effect, which also makes nanoparticles widely used in light, electricity, heat, magnetism and other aspects. In recent years, some researchers have proposed to apply nanoparticles in the field of sound absorption to improve the sound absorption effect of sound absorbing materials. Compared with ordinary particles, nanoparticles have a large specific surface area and small size effect, makes them more sensitive to external sound, when the sound into the material, material multiphase, interface structure and the subtle pores between countless capillary particles and nanometer particles will produce intense movement between the friction increases, the energy of the phase interface or nanoparticles interface dissipation, the effect of sound absorption. Polymer/nanoparticle composites have been successfully used in microwave absorbing materials, and the U. S. Navy Laboratory uses a variety of microparticle polymer composites, which are equivalent to a foot of concrete at only 6 cm.

With the increasing severity of noise pollution, In addition to having less noise at the sound source, developing good sound-absorbing materials, reducing the harm during the propagation of noise is imminent, although the traditional porous material has a good sound absorption effect in the high-frequency area of the sound wave. However, the sound absorption effect in the low-frequency area is not ideal to enhance the low-frequency sound absorption performance can only increase its thickness. But this method will make the sound-absorbing material become thick, large floor area.

10.6.2 Nanoparticle Additives Used in Lubricants

Ceramic enterprises generally use air compressor, blower and fuel combustion equipment, its noise can reach hundreds of decibels, easy to cause harm to the human body. However, when the machinery and equipment are miniaturized by nanotechnology, the alternating mechanical force generated by the collision and friction will be greatly reduced, and the noise pollution can be effectively controlled. The lubricant developed by nanotechnology can not only form a permanent solid film on the surface of the object, produce excellent lubrication effect, can greatly reduce the noise during the operation of machinery and equipment, but also prolong its service life. Research believe that the mechanism of nanomaterials is mainly realized through the following three ways: (1) similar function of "microbearing" to reduce friction resistance and friction coefficient; (2) under friction condition, nanomaterials particles form a smooth protective layer on the friction surface under friction condition; (3) fill the micro pits and damage parts on the friction surface to repair.

Some scholars believe that the mechanism of nano particles additive is different from traditional additives, and its itself has the nano effect, in the process of friction, because of friction surface local temperature is high, especially under high load, nano particles (especially particles like n-TiO_2) is likely in the melting, half melting or sintering state, thus forming a layer of nano membrane. In addition, the nanometer particles have extremely high diffusion force and self-diffusion ability (more than ten orders of magnitude higher than the body phase material), and they are easy to form a permeability layer or diffusion layer with excellent friction resistance on the metal surface, showing the principle of in-situ friction chemistry. This mechanism holds that the lubrication of nanoadditives, especially under high load conditions, no longer depends on whether the small additive elements are chemically active for the matrix, but largely depends on whether they form diffusion layers or osmotic layers and solid solution with the matrix components.

In short, with the development of the deep and practical process of basic research on nanomaterials and nanotechnology, especially the further organic combination of nanotechnology and environmental protection and environmental governance, many environmental problems will be solved, and we will fully enjoy the clean environment brought by nanotechnology to human beings.

References

1. J. Bao, X. Zhang, B. Fan, J. Zhang, M. Zhou, W. Yang, X. Hu, H. Wang, B. Pan, Y. Xie, *Angew. Chem. Int. Ed.* **2015**, *54*, 7399–7404.
2. D. Deng, X. Pan, H. Zhang, Q. Fu, D. Tan, X. Bao, *Adv. Mater.* **2010**, *22*, 2168–2171.
3. H. Duan, N. Yan, R. Yu, C. R. Chang, G. Zhou, H. S. Hu, H. Rong, Z. Niu, J. Mao, H. Asakura, T. Tanaka, P. J. Dyson, J. Li, Y. Li, *Nat. Commun.* **2014**, *5*, 3093.
4. Z. Fan, X. Huang, C. Tan, H. Zhang, *Chem. Sci.* **2015**, *6*, 95–111.
5. L. Guardia, M. J. Fernández-Merino, J. I. Paredes, P. Solís-Fernández, S. Villar-Rodil, A. Martínez-Alonso, J. M. D. Tascón, *Carbon* **2011**, *49*, 1653–1662.

6. X. Li, X. Sun, *Nano* **2016**, *6*, 1-3.
7. A. Gupta, T. Sakthivel, S. Seal, *Prog. Mater. Sci.* **2015**, *73*, 44–126.
8. Y. Hernandez, M. Lotya, D. Rickard, S. D. Bergin, J. N. Coleman, *Langmuir* **2010**, *26*, 3208–3213.
9. S. E. Hosseini, M. A. Wahid, *Renew. Sust. Energ. Rev.* **2016**, *57*, 850–866.
10. A. M. Saleem, V. Desmaris, P. Enoksson, *J . Nanomater.* **2016**, *2016*, 1–17.
11. Z. Li, B. Li, C. Yu, H. Wang, Q. Li, *Adv. Sci.* **2023**, *10*, 2206605.
12. X. Huang, H. Li, S. Li, S. Wu, F. Boey, J. Ma, H. Zhang, *Angew. Chem. Int. Ed.* **2011**, *50*, 12245–12248.
13. S. Jayabal, G. Saranya, J. Wu, Y. Liu, D. Geng, X. Meng, *J. Mater. Chem. A.* **2017**, *5*, 24540–24563.
14. M. J. Kory, M. Worle, T. Weber, P. Payamyar, S. W. van de Poll, J. Dshemuchadse, N. Trapp, A. D. Schluter, *Nat. Chem.* **2014**, *6*, 779–784.
15. Y. Qin, X. Wang, Z. L. Wang, *Nature* **2008**, *451*, 809–813.
16. S. Xu, Y. Qin, C. Xu, Y. Wei, R. Yang, Z. L. Wang, *Nat. Nanotechnol.* **2010**, *5*, 366–373.
17. R. Yang, Y. Qin, L. Dai, Z. L. Wang, *Nat. Nanotechnol.* **2009**, *4*, 34–39.
18. F.-R. Fan, Z.-Q. Tian, Z. Lin Wang, *Nano Energy* **2012**, *1*, 328–334.
19. M. Alhuyi Nazari, A. Maleki, M. E. H. Assad, M. A. Rosen, A. Haghighi, H. Sharabaty, L. Chen, *Sol* **2021**, *228*, 725–743.
20. F. Lei, Y. Sun, K. Liu, S. Gao, L. Liang, B. Pan, Y. Xie, *J. Am. Chem. Soc.* **2014**, *136*, 6826–6829.
21. Y. Lin, T. V. Williams, J. W. Connell, *J. Phys. Chem. Lett.* **2009**, *1*, 277–283.
22. P. Mei, J. Kim, N. A. Kumar, M. Pramanik, N. Kobayashi, Y. Sugahara, Y. Yamauchi, *Joule* **2018**, *2*, 2289–2306.
23. M. Naguib, V. N. Mochalin, M. W. Barsoum, Y. Gogotsi, *Adv. Mater.* **2014**, *26*, 992–1005.
24. L. Spanu, S. Sorella, G. Galli, *Phys. Rev. Lett.* **2009**, *103*, 196401.
25. J. N. Tiwari, R. N. Tiwari, K. S. Kim, *Prog. Mater. Sci.* **2012**, *57*, 724–803.
26. X. Wang, G. Sun, P. Routh, D. H. Kim, W. Huang, P. Chen, *Chem. Soc. Rev.* **2014**, *43*, 7067–7098.
27. J. Xie, H. Zhang, S. Li, R. Wang, X. Sun, M. Zhou, J. Zhou, X. W. Lou, Y. Xie, *Adv. Mater.* **2013**, *25*, 5807–5813.
28. Q. Zhai, Y. Liu, R. Wang, Y. Wang, Q. Lyu, S. Gong, J. Wang, G. P. Simon, W. Cheng, *Adv. Energy Mater.* **2019**, *10*, 1903512.
29. X. Zou, L. Wang, B. I. Yakobson, *Nanoscale* **2018**, *10*, 1129–1134.
30. Y. Xue, Q. Zhang, W. Wang, H. Cao, Q. Yang, L. Fu, *Adv. Energy Mater.* **2017**, *7*, 1602684.
31. T. Yu, B. Lim, Y. Xia, *Angew. Chem. Int. Ed.* **2010**, *49*, 4484–4487.
32. X. Y. Yu, X. W. David Lou, *Adv. Energy Mater.* **2018**, *8*.
33. J. Yun, S. H. Hwang, J. Jang, *ACS Appl. Mater. Interfaces* **2015**, *7*, 2055–2063.
34. H. Li, B. Chong, B. Xu, N. Wells, X. Yan, G. Yang, *ACS Catal.* **2021**, *11*, 14076–14086.
35. A. Ananthanarayanan, Y. Wang, P. Routh, M. A. Sk, A. Than, M. Lin, J. Zhang, J. Chen, H. Sun, P. Chen, *Nanoscale* **2015**, *7*, 8159–8165.
36. F. Arcudi, L. Dordevic, M. Prato, *Angew. Chem. Int. Ed.* **2016**, *55*, 2107–2112.
37. E. Ten, J. Turtle, D. Bahr, L. Jiang, M. Wolcott, *Polymer* **2010**, *51*, 2652–2660.
38. X. Dong, Y. Su, H. Geng, Z. Li, C. Yang, X. Li, Y. Zhang, *J. Mater. Chem. C* **2014**, *2*, 7477–7481.
39. Z. Gan, H. Xu, Y. Hao, *Nanoscale* **2016**, *8*, 7794–7807.
40. V. C. Hoang, K. Dave, V. G. Gomes, *Nano Energy* **2019**, *66*, *104093.*

Chapter 11
Applications of Nanomaterials and Nanotechnology in Biomedicine

Biomedical research is a very important aspect of nanotechnology. In recent years, among the patents related to nanotechnology in the United States, biomedical patents account for more than half, indicating that nanotechnology in the field of biomedicine has a very important position and a bright future [1].

Using nanotechnology, biodegradable and biocompatible polymers can be combined with drugs to make nanomedicine, which can be directly introduced into the organs of the focal site as targeted drug preparations. Tissues and even cells, to improve the efficacy of drugs, reduce toxicity. Using nanomaterials as drug carriers can increase the gastrointestinal absorption and bioavailability of certain drugs. Using nanomaterials as carriers can be used for gene delivery and therapy [2]. The application of nanomaterials as biological materials such as tissue repair and artificial organs also has a good prospect. In addition, nanomaterials have a wide range of applications in the diagnosis and monitoring of diseases. In the above fields, the research of nanomaterials has made rapid progress, and some nanomaterials have become products and entered the practical application stage; Some nanomaterials have entered the clinical research stage: more nanomaterials are in different experimental research stages [3]; Nanomaterials will bring surprises to mankind in the near future With the continuous emergence and improvement of nanotechnology and nanomaterials, it will bring new changes and rapid development to the field of biomedicine [4].

The application and progress of nanomaterials in biomedicine will be introduced in detail below.

S. Peng and P. Li, *Nanomaterials and Nanotechnology*,
https://doi.org/10.1007/978-981-97-8433-2_11

11.1 Common Biological Materials and Their Characteristics

Biological materials must have good biocompatibility, absorbability, non-toxicity and non-accumulation. It is a functional non-living material used for medical purposes in contact with living tissue. Biological materials include metal materials, inorganic materials and organic materials three categories. Among them, the main body is biomedical polymer materials used in living tissues to diagnose, treat or replace body tissues, organs or enhance their functions [5]. Medical biomaterials include natural, synthetic and the combination of the two, but also a living cells, natural tissues and non-living materials composite of hybrid polymer or polymer complex. Medical biomaterials can be divided into biodegradable and non-degradable properties. According to whether the degradation products can be metabolized and absorbed by the body, the degradable polymers can be divided into bioabsorbability and bioinabsorbability. According to the influence of the material on blood composition and performance after contact with blood, it can be divided into blood compatibility polymer and blood incompatibility. According to the affinity and reaction of the material to the body cells, it can be divided into biocompatibility and biocompatibility polymers [6].

At present, the most widely used biodegradable polymers in biomedical materials include: natural biodegradable polymers such as starch, cellulose and its derivatives, polysaccharides, proteins, lipids, chitin, chitosan, etc. Synthesis of degradable polymers such as polylactic acid (PLA), Polyglycolide-lactide (PLGA), polyhydroxy butyrate (PHB), polyglycolide (PGA), polylactone (PCL), polyhydroxyalkylester (PHA) and their copolymers. As shown in Fig. 11.1, this kind of polymer contains ester, ether and amide bonds, etc., which can be degraded by enzymes or microorganisms in body fluids, acids and alkalis. The main chain of the polymer structure is broken, the molecular weight is gradually reduced, and finally metabolized into water and CO_2, or high hydrophilic monomers. The natural polymer material has low toxicity and good biocompatibility, but the composition is complicated, the difference between each batch is large, and purification is difficult [7]. High purity polymer, easy to control performance, selectivity, but its degradation product toxicity, degradation rate of polymer itself and accumulation in the body are the main problems.

As a new main force of biomaterials, nano-metals and nano-biomaterials made by using nanotechnology have many amazing properties. If the nano metal has low toxicity, its sensing characteristics and elastic modulus can be close to normal natural biological tissue, which can enable cells to grow on its surface, and has the function of repairing diseased tissue [9]. In medicine, the plastic nano sol preparations provided by nanotechnology go beyond the limitations of surgical implantation surgery, so that the implant has the same surface properties and homogeneity with natural materials, such as the natural superstructure of the human body recently developed by British scientists, which has replaced stainless steel materials for human corrective surgery, this substance composed of hydroxy-phosphatic lime lattice, and biological bone. The material composition is the same, the strength and density of natural bone are the

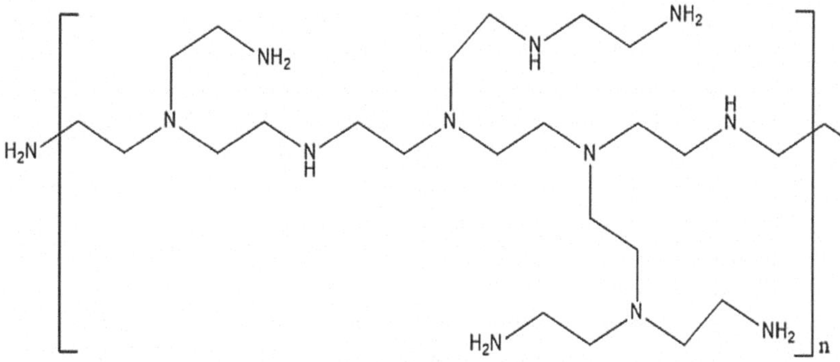

Polyethylenimine (PEI)

Fig. 11.1 Polymers for delivery [8], **a** natural polymer nanoparticles; **b** synthetic polymer nanoparticles

same, and it can be completely fused with the human fracture part, and its biological life period is more than 30 years [10]. It is worth pointing out that some nanomaterials in the field of medicine, such as nanomaterials as drug carriers or diagnostic reagents, are a polydisperse system, or colloidal dispersion system, in which the size of the dispersed phase particles is on the order of nanometers. The use of nanotechnology to make biological materials into nano-sized colloidal particles or into ultra-small devices or nano-devices can be used as drug carriers, medical materials or medical equipment, which brings a new revolution to the field of medicine and pharmacy [11].

11.2 Common Nano Drug Delivery System

The penetration of nanotechnology in the biological field has formed nanobiotechnology, and the research of nanomedical drug delivery system is the focus and hot spot of nanobiotechnology. Countries around the world, especially the United States, Japan and Germany, have focused on nanobiotechnology as a scientific research priority project in the twenty-first century. Priority areas of research funded by the United States include the application of nanotechnology to clinical therapies (drugs and gene vectors). The Japanese government has established a large number of nanotechnology research institutes in national laboratories, universities and companies, and biotechnology is its priority research area. Germany has launched a new round of nanobiotechnology research program, which focuses on the development of nano-missiles that destroy tumor cells and micro-memory that can store data for diagnosis and treatment, and the use of this technology to further develop micro-biosensors for diagnosing the formation of antibodies in infected human blood, treating cancer and various cardiovascular diseases. Research is also underway in the UK, Australia, South Korea and Russia. Compared with the advanced countries, the development of nano biologic technology in China started late, but the 863 programs launched the domestic nano revitalization plan, and also listed nano biologic technology as a special project to give priority support.

At present, the international research of nanobiotechnology in the field of medicine has made some progress. In the research of nano and biomaterials, the research focus and substantial achievements are the drug delivery system [12] and the gene transfer technology of nano particle.

The application of nanotechnology and nanomaterials in the pharmaceutical field has resulted in nanodrug delivery systems, including nanoparticles (NP), nanospheres (NS), nano capsules (NC), nano micelle (NM), nanoliposomes (Nano-liposomes (NL) and nano emulsion (NE), etc. One of the more common is NP.

NP generally refers to solid colloidal particles (1–1000 nm) made of natural or synthetic polymer materials with particle size at the nanometer level. NC is a nanoscale vesicle wrapped with a polymer material, which generally contains an oily liquid phase in the capsule. The main preparation methods of NP and NC are emulsification polymerization, crosslinking polymerization and interfacial condensation polymerization. Recently, nanometer solid micelles with nuclear/shell structure formed by automatic polymerization of diphilic polymer materials (known as self-assembly) have attracted attention. The hydrophilic end of the micelles is outward and the lipophilic end is inward, and the nuclei can be loaded with various drugs, enzymes and genes [13].

As a drug delivery system, NP has the following characteristics:

1) NP is highly dispersed, has a large surface area, and has special surface properties (such as bio adhesion, electricity, affinity, etc.), which is conducive to increasing the contact time and contact area of the drug at the absorption site. In addition, NP has an obvious protective effect on the drug, so it can improve the absorption and bioavailability of the drug [14].
2) Due to the different types and properties of carrier materials (such as molecular weight and proportion, etc.), the different preparation processes of NP and the different structure of NP, the drug release speed can be different.
3) After entering the body, due to the phagocytosis of the reticuloendothelial system, NP can be targeted at liver, spleen, lung and lymphoid tissue with abundant phagocytosis cells, especially small NP can also enter the human bone marrow tissue. NP can actively target the target tissue through antibody-mediated, ligand-mediated or magnetic field mediated.
4) Changing the membrane transport mechanism of drugs NP can increase the permeability of drugs to biofilms, such as increasing the permeability of drugs to the blood–brain barrier and cell membrane, which is conducive to the treatment of some special parts of drugs [15].

NP drug carrier is the most widely studied field of nanotechnology in medicine. Intravenous administration, oral administration and ocular administration of NP as a drug carrier are more common, but there are also studies on intra-nodal administration, subcutaneous administration and transdermal administration. The NP drug delivery system is mainly used for anti-tumor therapy, anti-infection therapy, oral administration of multi-brain protein drugs, treatment of eye diseases and other diseases. With the deepening of research, in addition to ordinary NP, new drug delivery systems such as long cycle NP solid lipid NP, immune NP, magnetic NP, heat sensitive NP and pH sensitive NP have been developed.

At present, the research focus of drug carriers is magnetic nanoparticles, especially paramagnetic or superparamagnetic nano-ferrite particles, which can kill tumors when the temperature rises to 40–45 °C under the action of external magnetic field. For example, magnetic targeting of magnetic doxorubicin albumin nanoparticles in normal liver, distribution in rats and therapeutic effect on metastatic liver cancer in rats, the study results show that magnetic doxorubicin albumin nanoparticles have efficient magnetic targeting, significantly increased aggregation in rat transplanted hepatoma, and have a good effect on transplanted tumors. Most of the foreign research on nanoparticle drug carriers is used in cancer diagnosis and treatment. An external magnetic field is used to locate and fix the drug magnetic particles, and then an alternating magnetic field is used to heat the magnetic particles to destroy the cancer cells.

Another hot spot in the study of nanodrug carriers is polymer biodegradable drug carriers or gene carriers. After the carrier and drug-gene fragments are directed into human target cells through degradation, the surface carrier is biodegraded, and the drug in the core is released to play its effect, avoiding the release of drugs in other tissues.

11.2.1 Nanoparticles Used as Carriers of Cancer Drugs

NP is used as a carrier for the treatment of anti-tumor drugs, mainly to change the distribution and pharmacokinetic characteristics of drugs in the body through the action of NP, improve the targeting of drugs to the tumor site, thereby improving the efficacy and reducing the toxic side effects. As shown in Fig. 11.2, on the other hand, cell permeability at the tumor site is in a hyperactive state, and the longer the existence of anticancer drugs in the blood circulation, the more favorable the treatment of tumors. In this regard, a lot of research work has been done, and the antitumor drugs involved include: doxorubicin, paclitaxel, 5-fluorouracil, methotrexate, mitoxantrone, actinomycin D, aclacinomycin, pervone, cisplatin and carboplatin, etc. Among them, doxorubicin and paclitaxel have more NP studies, such as doxorubicin has been prepared into alkyl cyanoacrylate NP, albumin NP and carboplatin respectively Gelatin NP et al. [16]. From the animal experimental results of various drugs, most of the drugs combined with NP can change the process of the drug in vivo, so that the drug in the body to prolong the residence time, the result is to improve the tumor inhibition rate, compared with free drugs, some can improve the efficacy of more than 50%.The charged nanoparticles can be selectively enriched in the lesion

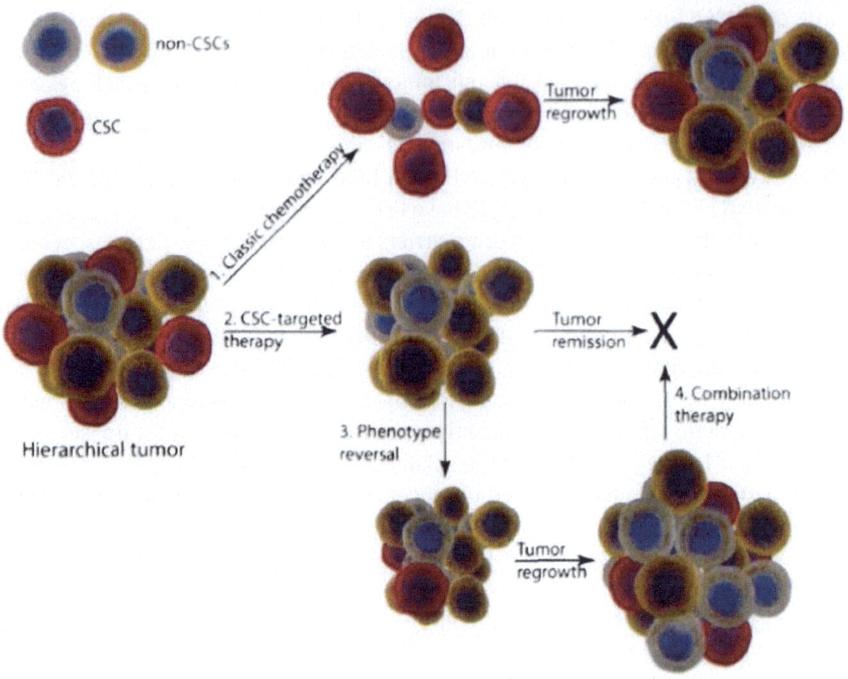

Fig. 11.2 The growth changes of tumor stem cells during tumor therapy [18]

site of lymphatic cancer, which can be used for the treatment of lymphatic cancer: the nanoparticles can selectively cross the blood–brain barrier (BBB), and can also be used for the treatment of brain tumors, such as the use of T80 encapsulated alkyl cyanoacrylate nanoparticles as the brain targeting carrier of Adriamycin [17].

More than 60% of current research and inventions in the diagnosis and treatment of malignant tumors use biodegradable polymer biomaterials as carriers, such as polylactide (PLA), polycaprolactone (PGA), polycaprolactone (PCI), PMMA, polystyrene (PS), cellulose, cellulose-polyethylene, polyhydroxy propionate, gelatin, and in between Copolymers and biological polymers, such as proteins, phospholipids, glycoproteins, liposomes, collagen, etc., use their affinity to combine with gene fragments and drugs to form biological polymer nanoparticles, and then combine with RGD directional identifier, targeting and integrins on the surface of target cells to bind drugs into tumor cells, so as to kill tumor cells or cause tumors The purpose of gene transfection of cells [19].

11.2.2 Nanoparticles Used as Carriers for Anti-infective Drugs

Because of its small particle size, NP can cause inflammation when microorganisms infect the body, and the permeability of cells at the inflammation site is hyperactive, and drug-loaded NP can enter the infection site. If the surface of NP is chemically modified, such as connecting PEG(polyethylene glycol) long chains, NP can circulate in the blood for a long time, and eventually accumulate at the site of infection, which is conducive to the treatment of antimicrobial drugs [20]. A typical example is the development of the antimicrobial fungal drug amphotericin B lipid NP, with an average particle size of 25–50 nm, compared with ordinary amphotericin B, the toxicity to human epithelial cells in vitro is lower, and the renal toxicity to animals is also significantly reduced. This is mainly NP concentrated in the reticuloendothelial system, which reduces the uptake of the drug by the kidneys. The decrease in renal toxicity allows doctors to give patients a relatively high dose of the drug, thereby increasing the therapeutic index of the drug. Recently, there are many researches on the preparation of nano micelles by amphiphilic block copolymers. Amphiphilic block copolymers can be automatically assembled into nano micelles when the molar ratio of hydrophilic and lipophilic fragments is appropriate. This amphiphilic nano micelle with self-assembly properties can solubilize drug control drug release and has the function of long circulation, allowing drugs to be targeted to tissues or organs outside the reticuloendothelial system.

NP can also be used to treat infections by intracellular microorganisms [21]. There are many microorganisms, including all viruses, as well as many common bacteria, fungi and protozoa, that can grow and reproduce in cells. Generally, the ability of antibiotics to enter cells is relatively weak, and antibiotics that enter cells are often unstable under intracellular physiological conditions, or are not sensitive to intracellular microorganisms, or only enter uninfected organelles, etc., so the effect of intracellular chemotherapy is very poor, or even completely ineffective. After combining with NP, antibiotics can enter infected cells in a similar way to microorganisms (such as fusion endocytosis, etc.), and the concentration of intracellular antibiotics is greatly increased after drug release, which can greatly improve the therapeutic effect of intracellular infection [22]. A more successful example is the study of amdinocillin NP. When nude mice infected with Listeria monocytogenes were treated with ampicillin NP, the bacterial count in liver decreased from 105 CFU/ml to 10 CFU/ml and remained there for a week. NP containing 2.4 mg ampicillin was more effective than 48 mg ampicillin. Mice infected with Salmonella typhimurium were given 0.8 mg ampicillin NP, all mice survived, and all mice in the face control group died. Free ampicillin was given 96 mg to have a similar effect. Other antibiotics are gentamicin and amikacin [23].

In addition is also used as a carrier for antiviral vaccines, or for the treatment of Leishmania, or for the treatment of HIV infection, the latter two microbes in the reticuloendothelial system, where NP tends to concentrate. NP can also be used to carry nucleic acids for intracellular therapy.

11.2.3 Nanoparticles Used as Carriers of Polypeptide Protein Drugs

There have been many reports on the use of nanoparticles as oral drug delivery carriers, but the more promising research may be as polypeptide protein drug carriers. Polypeptide drugs are poorly absorbed by mouth, so generally such drugs can only be injected, but for drugs that need to be taken for a long time, injection can bring great pain to patients. With the rapid development of biotechnology, biotechnology drugs (mainly peptide protein drugs) are also increasing, and will become one of the mainstream drugs in the future. Therefore, many large pharmaceutical companies in the world have spent a lot of manpower and resources on the non-injection administration of polypeptide protein drugs.

There are several main reasons for the poor oral absorption of polypeptide protein drugs:

1) Catalytic degradation of acid–base environment in gastrointestinal tract
2) Catalytic degradation of digestive enzymes
3) Biomacromolecule molecules are too large to pass through the mucosa of the gastrointestinal tract
4) After absorption into human blood, there is a transient metabolic effect in the liver

It was found that polypeptide protein drugs were coated or adsorbed with high polymer to make nanocarrier oral preparations, which could significantly improve the absorption of drugs. For example, after oral administration of insulin nano capsules prepared with polycyanoacrylate, the blood glucose of diabetic rats decreased by 50–60% on the second day, and the hypoglycemic effect could last for 6 and 20 days, respectively, when the dosage was 12.6 and 50 μg/kg. In a normal rat model of hyperglycemia caused by oral glucose, administration of the above insulin nano capsules can reduce blood glucose by 50%. The encapsulation rate of cyclosporine A was 88%, and the bioavailability of cyclosporine A was significantly improved after oral administration to animals, and the distribution of cyclosporine A in the kidney was reduced, thus reducing the renal toxicity of cyclosporine A. Cyclosporine A nano emulsion developed abroad has been successfully applied in clinic [24].

The mechanism of nanocarriers improving multi-brain protein drugs may include the following aspects:

First, the nanocarriers can enter the circulatory system through the peyer's node of the small intestine, and the nanocarriers with small particle size can also enter the circulation through the intercellular pathway of the mesentery. Secondly, due to the high dispersion and huge surface area of nanoparticles, the solubility and dissolution rate of insoluble drugs can be increased, and the contact with the gastrointestinal wall can also be increased, thereby increasing the chance of absorption and improving the bioavailability of insoluble drugs. In addition, nanoparticles are more likely to be captured by duodenal microtriche than solvents and remain for a longer time. If nanoparticles are prepared with bio adhesive polymers, the contact time between drugs and cell walls can be further extended simultaneously, and the absorption rate and absorption rate of drugs can be improved. There are also nanocarriers to protect multi-protein drugs from enzymatic or acid–base catalytic degradation.

Another problem with polypeptide drugs is that they usually have a short half-life and poor stability. Due to the short half-term, polypeptide protein drugs generally need to be repeated several times, which brings inconvenience to patients. At the same time, due to the short half-life of polypeptide protein drugs, they are quickly cleared in the body, and cannot give full play to the efficacy of drugs, causing great losses. One way to solve the above problems is to load polypeptide proteins into nanodrug delivery systems (such as nano-liposomes or nanoparticles). It should be noted that biodegradable polymer materials are generally used to prepare nanocarriers, to control the explosive release of drugs, and if it is iv administration, the phagocytosis of the reticuloendothelial system on the NP carrier should be considered. For example, insulin is loaded onto polyisobutyl cyanoacrylate NP, and the hypoglycemic effect can last for 24 h after subcutaneous injection; isohexyl polycyanoacrylate NP containing growth hormone releasing factor was prepared. After subcutaneous administration, it was confirmed that the plasma level of growth hormone releasing factor could be maintained for 24 h, and the bioavailability in vivo was 18 times higher than that of free drugs. The isohexyl polycyanoacrylate carrier material was excreted through urine and feces. The increase of bioavailability is due to the slow-release effect of NP vector and the protective effect of NP vector on growth hormone releasing factor.

11.2.4 Nanoparticles Used as Carriers of Ophthalmic Drugs

Eye drops commonly used in ophthalmology produce a large number of tears in the eye due to irritation and other reasons, which can quickly dilute and remove the liquid, so that the action time of the liquid in the eye is very short. On the other hand, the drugs in the general water-based medium are also difficult to pass through the cornea and into the human eye, resulting in poor treatment effect of eye drops. Drugs in water-based media can also enter the body through the tear duct and produce some toxic side effects. As a carrier of ophthalmic drugs, NP has certain adhesion on its surface, which can prolong its retention time in the eye and reduce the removal of drugs by tears. In addition, the materials of NP are mostly polymer compounds, which have considerable lipophilicity, and are easier to enter the eye through the cornea, increasing the concentration of drugs in the aqueous humor and corneal tissue, which is conducive to the treatment of eye diseases.

Several anti-infective drugs (such as amikacin, acyclovir), anti-inflammatory drugs (such as hydrocortisone, indomethacin), glaucoma drugs (such as pilocarpine, bisoprolol fumarate, indomethacin, etc.) and immunosuppressants (such as cyclosporine A) have been studied in the NP ocular delivery system. Almost all the research results show that NP can effectively increase the penetration of the drug into the cornea, increase the concentration of the drug in the aqueous humor and corneal tissue, improve the efficacy, prolong the action time, and sometimes improve the stability of the drug. For example, amikacin sulfate was prepared into NP by polybutylene cyanoacrylate and compared with amikacin solution group. The concentration of drugs in aqueous humor of NP group was 3 times higher than that of control group, and the concentration of drugs in cornea also increased significantly.

Some people prepared Pilocarpine into 100–200 nm NP with polybutylene cyanoacrylate, and the drug encapsulation rate was 10–18%. After the drug was used in the eyes of animals, the drug concentration in aqueous ether increased by 23%, the drug elimination rate became smaller, and the intraocular pressure (IOP) decreased for a longer period of time. By using interfacial polymerization method, Pilocarpine was prepared into NC from polybutylene cyanoacrylate, and the gel release system was made by adding surfactant. The results showed that it had obvious long-term effect in rabbit eyes. In general, the stronger the hydrophilicity of the drug, the more obvious the improvement in corneal permeability after NP or NC. The therapeutic effect of different carrier materials is different. It has been suggested that polycaprolactone (PCL) is better than isobutyl cyanoacrylate and lactide copolymers as ophthalmic NP materials because PCL has an aggregation process in the conjunctival sac, which results in a longer duration of action. Other researchers have also found that butyl cyanoacrylate can cause cell membrane damage when absorbed by the cornea, which is not a problem when PECL is used as a carrier. Sometimes nano capsules containing the oil phase work better than solid NP because the oil phase protects the drug from the rinsing effect of tears.

Recently, there have been many studies on cyclosporine A ophthalmic NP administration system. For example, chitosan was used to prepare cyclosporine A ophthalmic NP, the encapsulation rate of the drug was 73%, the drug loading was 9%, the surface charge was positive, the surface potential was 37 mV, and the drug release in vitro was divided into fast release and slow-release phases. The concentration of cyclosporin A in the outer eye tissues (such as cornea and conjunctiva) was higher than that in the control group (cyclosporin A solution or suspension) 48 h after administration. The concentration of the drug in the inner eye tissue (such as the iris and aqueous humor) and in the blood was below the detection line. These results suggest that cyclosporine A ophthalmic NP drug delivery system will be very beneficial to improve the therapeutic effect of external eye tissue.

In addition to using NP as a carrier of ophthalmic drugs to make eye drops and administer them in the eye, NP is also used as a carrier of ophthalmic drugs for intraocular administration to treat fundus diseases.

11.3 Special Nano Drug Delivery System

11.3.1 Nanoliposomes

Liposomes, also known as phospholipid membranes (as shown in Fig. 11.3), proposed by Bangham in the mid-1960s, have become a typical model and have been widely used in various scientific fields such as biophysics, therapeutic physics, chemistry, and colloid.

Liposome is widely used as a drug carrier. Liposomes can be divided into three types according to their structure: small single-compartment liposomes (SUV), large single-compartment liposomes (LUV) and multi-compartment liposomes (MLV). When liposomes were used as drug carriers in the 1970s, they were neglected due to their extremely unstable structure and low drug encapsulation rate. With the in-depth research on the physical and chemical properties of liposomes, structural characteristics, and the interaction between liposomes and cells and other basic theories, the problems of industrial production, stability and encapsulation rate have been initially solved. In the early 1990s, several foreign pharmaceutical companies successfully developed economic and efficient anti-fungal and anti-cancer drug liposomes products, and successively put into the market, which greatly promoted the development of liposomes Research and development. Such as daunorubicin liposomes, developed by Nexstar Pharmaceutical Company in the United States, the product name DaunoXome, approved by the FDA after the market, its particle size is about 50–60 nm. Due to its small particle size, it is not easy to be recognized by phagocytes of the reticuloendothelial system after entering the blood circulation, so it can stay in the blood circulation for a relatively long time. At the same time, due to the change of distribution performance, the drug preparation has no obvious cardiac toxicity.

The stability of liposomes includes physical, chemical and biological aspects. By controlling the preparation conditions such as particle size, pH value, ionic strength, antioxidant and complexing agent, liposomes can be stabilized for more than 1 year. If cryoprotectants are added to the liposomes or freeze-dried products, they can

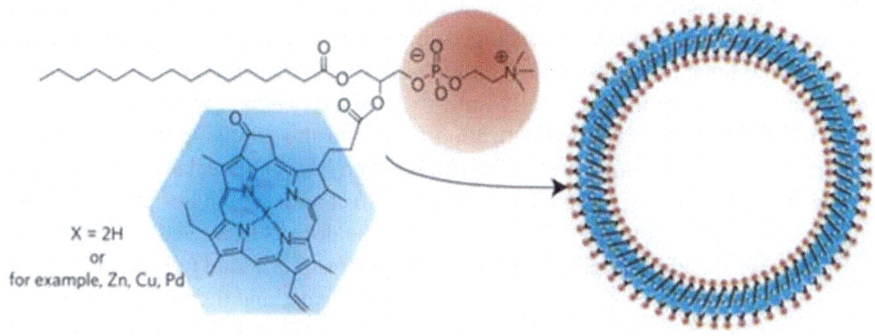

Fig. 11.3 Porphyrin-phospholipid bilayer liposome nanovesicles [25]

be stored for a long time. Now, some special liposomes, such as long-circulating liposomes and Ph-sensitive liposomes and temperature-regulated liposomes, can be obtained by using macromolecular protective colloids, or by adding inert surface molecules for surface modification, or by joining stable branched-chains.

Biodegradable polymers play an important role in the stabilization and versatility of liposomes. The drug loading of liposomes in the range of 50–150 nm increased with the increase of particle size, but the stability of liposomes in the range of 80–200 nm decreased with the increase of particle size, and the film thickness should be 4 nm. The addition of 5 mol/L of 2000 PEG (PEG molecular weight is 2000 Da) and the composition containing cholesterol can significantly enhance the mechanical and elastic properties of the film, and enhance the stability. Potential, isotonic pressure and other factors directly affect the drug loading of liposomes.

The excellent liposomes have long residence time, small distribution volume and high blood concentration in vivo. Conventional liposomes are easily phagocytic by monocytic system (MPS) as foreign particles, mainly Kupffer cells in liver and spleen, so they have passive targeting of lymphatic system. For other systemic diseases, overcoming the phagocytosis of MPS is a necessary condition to ensure the efficacy of drug therapy. The introduction of inert terminal ligands into the lipid membrane of liposomes to make recessive long circulating liposomes or to make active targeting liposomes by grafting specific groups has been a hot topic in the study of liposomes. Similarly, increasing the structure of insoluble drugs (such as amphotericin) and protecting sensitive drug molecules (such as DNA, RNA, antisense oligonucleotides, ribozymes, etc.), promoting intracellular transfer and changing in vivo distribution and pharmacokinetic behavior are also important contents of liposome research.

11.3.2 Solid Lipid Nanoparticles

Solid Lipid Nanoparticles (SLN) are nanoparticles prepared from a solid lipid material. Emulsions, liposomes, microcapsules and nanoparticles are all based on synthetic polymers or natural polymer materials. Due to the problems of physical stability of emulsions and liposomes and the limitation of industrial production, researchers have researched and developed new nano preparations with more application value and hope for industrialization. In the early 1990s, with the development of emulsification technology, especially milk homogenizers After the successful application of High-Pressure Homogennization (HPH) in the emulsion pharmaceutical industry, a new nano dosage form SLN with good physicochemical stability, smaller particle size, low toxicity, targeted slow-release effect and suitable for multi-channel drug delivery has been further developed. First, the drug is dissolved in the lipid at 5–10 °C above the melting point, and then the colostrum is added to the water phase containing surfactants under agitation to make colostrum, and then the colostrum is made into O/W type nano emulsion through a high pressure milk homogenizer, and cooled to room temperature or below the melting point of the lipid, so that the lipid crystallization into solid nanoparticles. According to the different melting point and drug sensitivity of the lipids, two preparation methods can be used: hot high pressure milk homogenization and cold high pressure milk homogenization. The drug in SLN is encapsulated in the membrane material in the form of solid solution, lipoid nucleus and solid nucleus drug. Compared with liposomes and nanoparticles has the following remarkable characteristics:

1) The production process is simple, the application of mature HPH method is suitable for industrial production
2) Low toxicity, good biocompatibility, can avoid the use of organic solvents used lipid materials such as fatty acids are endogenous, bioavailable non-toxic substances, a wide range of sources, low price, in vitro cytotoxicity tests showed that in the same observation period, containing 0.5%PLA or PLA/PGA nanoparticles caused 100% normal cell death, while 80% of 10%SLN remained viable in cell culture medium
3) High drug load, good stability. The drug is dissolved in soluble materials to make colostrum, and finally stored in solid form, which is most suitable for drugs with poor chemical stability; Suitable for multi-route administration. According to medical needs, SLN can be further processed into oral capsules, pellets, tablets or cold injections

11.3.3 Magnetic Nanocarriers

Commonly used magnetic materials are ferric oxide, ferric oxide, iron cobalt alloy and so on. These magnetic materials have good magnetic responsiveness, and can be easily obtained in the nanometer scale with appropriate methods. The smallest average particle size is only a few nanometers, and the particle size is generally normally distributed.

After surface coating, magnetic nanomaterials can be used as superparamagnetic ferric oxide nanomaterials in magnetic resonance imaging and have important applications in disease diagnosis. It can also be used for the preparation of magnetic microspheres, magnetic microspheres have products in foreign countries, and have a wide range of applications in the biomedical field.

Magnetic nanomaterials can also be used for the preparation of drug-carrying magnetic nanoparticles or drug-carrying magnetic liposomes. For example, polycyanoacrylate alkyl NP containing 3H actinomycin D has been prepared by using ferro oxide magnetic nanomaterials of 10–50 nm, and its particle size is 220 nm. The addition of magnetic nanomaterials did not increase the toxicity of NP. After iv administration, a magnetic field was applied to the left kidney of mice (850 mT, held for 10 min), and the radiation activity of the left kidney was 3 times higher than that of the right kidney as the control. In addition, NP containing indomethacin was prepared and injected intravenously into rats, and a magnetic field was applied to the tail of the rats until 60 years after injection at min, the concentration of the drug in the rat tail was 60 times higher than that in the control group. When no magnetic field was added, the concentration of polymethyl cyanoacrylate NP containing indomethacin

Table 11.1 Basic properties of two kinds of magnetic nanoparticle aqueous dispersions sold abroad

Company	Chemicell	Nano-technologiesGBR
pH value	7.4	7.4
Granularity	100 nm	50–150 nm
Magnetic material concentration	50 mg/ml	15 mg/ml
Ionic concentration	30 mg/ml	9 mg/ml
Stabilizer	25 mg/ml	
Nanoparticle concentration	About 10^{10} capsules /ml	About 1.5%
Smell	Neutral	Neutral
Color	Opaque black	Opaque black

was higher in the liver and spleen. These results suggest that magnetically mediated NP enables drugs to effectively target the site of action.

At present, there are many products of magnetic nanomaterials abroad. For example, Chemicell, a company based in Berlin, Germany, produces a magnetic nanomaterial whose surface is coated with a starch polymer with a particle size of 100 nm. Another German company, Nano-Technologies GBR, makes a similar product. Both products are aqueous dispersions of polymer-coated magnetic nanoparticles whose basic properties are shown in Table 11.1.

11.3.4 Long-Circulation Nanocarriers

Common colloidal particles include nanoparticles, nano-liposomes and pharmacosome, etc., which are all nanoparticles. Due to the passive targeting of the reticuloendothelial system, colloidal particles lead to two results: on the one hand, it is beneficial to improve the therapeutic effect of the corresponding RES tissues (such as liver, spleen, lung and bone marrow, etc.) (the toxic side effects on these tissues may also increase correspondingly); on the other hand, the therapeutic effect may decrease for diseases of other tissues other than the reticuloendothelial system. Therefore, to avoid colloid particles being engulfed by RES and improve the active targeting of drugs has become another research goal of nanoparticles. In order to prolong the circulation time of drugs in the body, the affinity of RES phagocytes for drugs must be reduced and the stability of the particle structure must be maintained. The current approach is as follows.

(1) Coating with surfactants mainly enhances the hydrophilicity of NP surface, and may also produce better steric hindrance. It is generally believed that the steric hindrance can be effectively played when the coating thickness is more than 10 nm.
(2) Cross-linking PEG (polyethylene glycol) or polyacrylamide on the surface of NP is generally prepared by linking polymer materials with PEG. PEG has

good hydrophilicity, and the long chain of PEG can also produce good steric hindrance and surface flexibility, which reduces the adhesion of opsonin in blood to colloidal particles, thus reducing the recognition of colloidal particles by RES phagocytes.

(3) Inhibition of the complement system the complement system plays an important role in the uptake of NP by phagocytes, so inhibiting the activity of the complement system with heparin or ganglioside can also prolong the circulation time of NP in the body.
(4) Using the effect of NP surface charge, it is currently believed that negatively charged nanoparticles have a strong affinity for the lymphatic system. On the contrary, particles with cationic and neutral particles on the surface are relatively difficult to be recognized by RES. In conclusion, reducing the adhesion of opsonin to colloidal particles is the main way to weaken the phagocytosis of RES and increase its circulation time in vivo.

11.3.5 Temperature Sensitive Nanocarriers

Thermosensitive liposome (Thermosensitive liposome) is a similar targeting vector. The study of temperature-sensitive NP carriers is the result of the rapid development of new polymer materials in recent years (Fig. 11.4). Dipalmitoyl phosphatidylcholine (DPPC) and distearoyl phosphatidylcholine (DSPC) were mixed in a certain proportion to achieve a specific phase transition temperature (Tm) from condensed matter to liquid crystal. When the liposome reaches the liquid crystal phase transition temperature, its phosphatidyl acyl chain is disorganized, its activity is enhanced, the membrane fluidity is enhanced, and the drug release is increased. Applying this principle, the liposomes with phase transition temperature slightly higher than the physiological temperature can be designed, that is, at the physiological temperature, while the drug can be released in the preheated tumor or inflammatory lesion site. Generally, the degree of curvature in the single-compartment liposome structure is large and the phase transition temperature is low. However, large single-compartment liposomes have high temperature sensitivity and low blood clearance, and can obtain high targeting index (DTI) in the body. By adjusting the ratio of DPPC and DSPC, the rate of drug release in the matrix can be adjusted to meet the needs of design and treatment. It has been reported that hydrophilic polymers are used as lipid anchors to act as temperature valves [26].

The research of temperature sensitive NP carrier is the result of the rapid development of new polymer materials in recent years. Du et al. prepared a thermosensitive liposome (F7) using dipalmitoyl phosphatidylcholine, myristoyl lysolecithin and dipalmitoyl phosphatidylcholine protein-polyethylene glycol 2000 as membrane materials by pH gradient active drug loading method. The encapsulation rate and drug loading of F7 thermosensitive liposomes were determined by HPLC. The phase transition temperature of liposome was investigated by differential scanning calorimeter. The morphology of liposome was characterized by atomic force microscopy. The

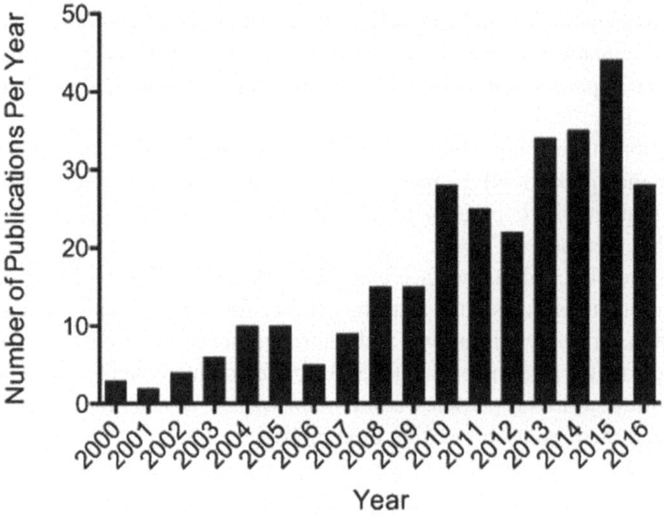

Fig. 11.4 Number of publications of heat-sensitive liposomes since 2000 [27]

heat sensitive release of liposome was examined by dialysis. The particle size and potential distribution of liposomes were investigated by laser particle size analyzer. The cytotoxicity of F7 and F7 thermosensitive liposomes was measured by MTT method. The freeze-drying process of F7 heat sensitive liposomes was optimized using Design-expert software. Results The encapsulation rate and drug loading of F7 thermosensitive liposomes were 97.56% and 1.51% respectively. The phase transition temperature is 39.9 °C, it is electronegative, the potential is − 15.10 mV, the particle size is 86.94 nm, and the polydispersion coefficient is 0.17. Compared with F7 ordinary injection, F7 thermosensitive liposomes showed stronger growth inhibition on lung cancer cells H1299 and breast cancer cells MCF-7 in a concentration-dependent manner [28]. After lyophilization, the liposomes were redissolved in good condition, the encapsulation rate was still 95%, and the particle size was about 130 nm. Conclusion The preparation of F7 thermosensitive liposomes by pH gradient active drug loading method has high encapsulation rate, good stability, and is simple and feasible, which provides a basis for the research and development of post-thermosensitive liposomes.

11.3.6 PH Sensitive Nanocarriers

PH-sensitive nanocarriers are a kind of drug delivery system in which the performance of the carrier changes in a specific acid–base environment in vivo, resulting in increased drug release. With this characteristic, targeted drug delivery to focal tissues can be realized in vivo.

PH-sensitive NPS are often used in oral administration studies, taking advantage of the difference in the pH value of the gastrointestinal tract, using different Ph-sensitive polymer materials (soluble in different pH environments) to prepare NPS, which can deliver the drug carrier to a specific site in the gastrointestinal tract. For example, ethylene pyrrolidone (VP) and acrylic acid (AA) were crosslinked with methylene bisacrylamide (MBA), and fluorescein isothiocyanate-dextran (FITC-dextran) was used as model marker to obtain 50 nm NP, which had high encapsulation rate and good redispersability after freeze-drying. The release of markers from NP is Ph-dependent, which is very slow in acidic media and increases with the increase of pH value. Moreover, the NP carrier system is also temperature sensitive, and the release of markers increases with the increase of temperature. Other PH-sensitive polymer carrier materials include complexes of polyamino acids and dodecanoic acids, such as polyarginine, polyhistidine, polylysine and dodecanoic acid complexes.

When this type of carrier circulates in the body to the tumor, inflammation or infection area, due to the low local pH value, the membrane structure of the liposome can be changed, and a large number of drugs can be released. The conditions for PH-sensitive liposome composition are as follows:

(1) When phosphatidyl ethanolamine (PE) or palmitoyl homocysteine (PHC) was introduced into the lipid structure, the amphiphilic colloid membrane was charged when $pH > 7.4$ and $pH > pK_a$, and the potential difference between the membranes played a stabilizing role. On the contrary, the surface of $pH < pKa$ liposomes is protonated or the charge decreases, and the bilayer cannot be stabilized.
(2) The polyethylacrylate (PEAA) cross-linked on the liposome membrane lipids has a stretchable hydrophilic stereo structure under alkaline or neutral conditions. After acidification, the hydrophilicity decreases, and the structure shrinks into a ball due to hydrophobicity, resulting in the destruction of the lipid structure, the formation of micelles, bulk expansion, and the release of the contained drug.

11.3.7 Immune Nanocarriers

At present, the research and development of monoclonal antibody has entered the practical stage, and many monoclonal antibody products have been marketed. After the monoclonal antibody is linked to the drug, it can deliver the drug to the focal site. However, drugs and monoclonal antibodies are often chemically bonded, which may cause the loss of antibody activity. At the same time, monoclonal antibodies can carry a limited number of drug molecules. The enzyme system present in circulation may also break the chemical bond between the drug and the monoclonal antibody. To solve these problems, scientists came up with nanocarrier systems. If the drug-carrying NP is linked to the monoclonal antibody to form the drug carrier-antibody complex, more drug molecules can be carried, and the antibody to the carrier will not cause the loss of activity of the antigen-binding site on the monoclonal antibody.

At present, monoclonal antibodies are generally adsorbed to the surface of NP to form immune nanocarriers. On the one hand, this NP carrier with immune function can passively target the tissues and organs of the reticuloendothelial system; on the other hand, it may be modified with different antibodies on the surface of NP, so as to have active targeting function and target specific tissues and organs in vivo. For example, anti-osteosarcoma nanoantibodies were adsorbed on the surface of polyhexyl cyanoacrylate NP and then added to the medium containing T24 cell line and human osteosarcoma cell line 788T. The results showed that the above-mentioned immune NP only selectively recognized human osteosarcoma cell line 788T and this specific binding could be maintained at 4 °C, indicating that the adsorption of anti-osteosarcoma monoantibody on the surface of polycyanoacrylate NP is relatively firm.

In recent years, as a vaccine carrier, NP has become a new research hotspot to play an immune role through injection or even oral or nasal mucosal administration. For example, the tetanus toxin vaccine was adsorbate on the surface of a new polyester polymer NP, and then inoculated 6-week mice by oral administration, nasal mucosal administration or intraperitoneal administration, and compared with traditional alumina hydroxide sorbents or solutions, that is, the titer of antibodies in the blood was analyzed and determined by ELISA method at 4 and 6 weeks after inoculation. The results showed that the titer of IgG and IgA increased to 3×10^3 and 2×10^3 after oral and nasal mucosal administration of polyester NP containing tetanus toxin. After oral administration, the small particle size of NP carrier can obtain higher antibody titer than the large particle size of NP carrier, while the medium size of NP carrier can produce antibody after administration of nasal mucosa. The authors suggest that of the drug delivery routes studied, nasal mucosal inoculation may be the most promising.

11.3.8 Photosensitive Nanocarriers

The basic idea of photosensitive nanocarriers is to carry photosensitized compounds on the nano drug delivery system. Photosensitized compounds (such as porphyrin, hematoporphyrin, phycocyanin or naphthol cyanine, etc.) can be absorbed and retained by tumor tissues. After iv administration, red light is irradiated at the tumor site, and the photosensitized compounds can produce new ecological oxygen, the latter of which has a strong effect Tumor cell killing effect. Liposomes containing photosensitive substances have been prepared.

Zhang etc. [29] The photosensitive liposomes (1-2-LP) containing docetaxel (1) and fullerene C_{60} malonic acid derivatives (2) were prepared by thin film dispersion method. The particle size, zeta potential, appearance morphology, drug encapsulation rate, in vitro release, in vitro cytotoxicity, and the expression of reactive oxygen species (ROS) in tumor cells were investigated Effects of levels, tumor cell cycle and apoptosis, and antitumor activity in mice. The results show that the average particle size of 1-2-LP is 170.4 nm, and the zeta potential is − 31.28 mV. Transmission electron microscopy (TEM) showed that the liposomes were spheroidal and evenly distributed. The packet rates of (1) and (2) are 86.38% and 93.81%. The addition of (2) did not affect the release of (1) in 1-2-LP, and the cumulative release rate of (1) was 78% within 50 h. 1-2-LP can significantly inhibit the proliferation of human cervical cancer (HeLa) and mouse melanoma (B16-F10) cells under light irradiation (532 nm laser irradiation for 3 min). The results of ROS detection showed that the fluorescence intensity of HeLa and B16-F10 cells could be significantly enhanced by 1-2-LP under light. Flow cytometry showed that 1-2-LP could change the cell cycle and promote early apoptosis after light exposure. The results of in vivo experiments on B16-F10 tumor bearing mice showed that tumor growth was significantly inhibited (inhibition rate was 85.33%) in 1-2-LP light group, indicating the synergistic anti-tumor effect of co-loaded (1) and (2) liposomes combined with laser irradiation.

11.3.9 Nanocarriers for Gene Delivery

With the development of human genome research, gene therapy will become an important means of disease treatment. There are three important steps in gene therapy. One is to find the faulty gene (the so-called target gene); The second is to obtain normal genes to replace the target genes (the so-called target genes); The third is to have appropriate methods or means to deliver normal genes to the target site to achieve gene therapy. Initially, gene therapy is only a simple correction of gene defects, but now it can also deliver genes that can express therapeutic effects of proteins, and directly or indirectly kill tumor cells in target organs [30]. Whether the treatment is to correct a genetic defect or to produce a therapeutic protein, the DNA molecule needs to be efficiently delivered to the target cell. The delivery of genes usually needs a carrier to complete, and now the commonly used carrier is a virus. The expression level of virus-gene therapy system is high, but there is no specific selectivity, and there are many side effects, such as carcinogenic wild-type virus generation and cytopathological changes [31].

Liposomes used for gene delivery include cationic and anionic liposomes, but cationic liposomes are more commonly used, which can adsorb and carry genes through the action of electric charge. As an exogenous gene carrier, liposome has the characteristics of simple preparation, non-toxicity, non-immunogenicity, large carrying capacity, etc. At the same time, as a gene or nucleic acid carrier, liposome can resist the hydrolysis of nuclease, delay the degradation of genes and other inclusions, and is safe and reliable.

NP prepared with biodegradable polymer materials for gene delivery has many advantages, such as stability, non-toxic, non-antigenicity, biocompatibility, controlled release (can extend gene expression time) and gene protection. The best NP carrier has a particle size of 26 nm, a surface potential of + 31 mV, and a DNA carrier rate of more than 90%. When there is no plasma in the medium, the transfection efficiency of the cells is the highest when the mass ratio of NP to DNA is 10, then the surface of the NP is negatively charged, and when the mass ratio of NP to

DNA is 30, the surface of the cells is positively charged. The DNA-carrying NP has been prepared by using polyglycol lactide and compared with liposomes carrying the same DNA. The results show that this NP system has an obvious slow-release effect and can release DNA sustainably for 1 month in vitro, and the structure and function of the embedded DNA remain unchanged. After 7 days of administration in rats, the transfection efficiency of NP was 1–2 orders of magnitude higher than that of liposome, and it also showed obvious slow-release effect. Similarly, some people wrapped the reporter gene LZ in chitosan nanoparticles and gave it to rats by oral administration. As a result, NP transported the gene to the gastrointestinal tract of rats and obtained the expression of LZ gene in the gastrointestinal tissue [32].

11.4 Applications of Nanomaterials in Biology

11.4.1 Material Assembly and Application of Biomacromolecules

The Crystalline surface layers (S-layers) are a collection of protein or glycoprotein subunits that make up the surface layers of bacterial and palaeobiological cells and are called S-layer protein. S-layer protein exist widely in the envelope bacteria, showing a general paleontological homogeneity. Its lattice arrangement consists of 4000 to 20,000 single-molecule protein and glycoprotein subunits and covers the surface of the cell. In S-layer protein, the subunit lattice planes are arranged in oblique, square and six diamond shapes (Fig. 11.5), and the center of the subunits are symmetrically arranged, forming 2.5–35 nm hollow pores with a thickness of 5–15 nm. The outer surface is smooth, but the inner is cratered, and the porosity accounts for 30–70% of the entire molecular surface. Most S-layer protein are composed of weakly acidic glutamic acid and aspartic acid, with an isoelectric pH of 3–6. The S-layer protein of the organism are separated and arranged in the suspension to form stomatal crystals in single molecules on the surface of liquid interface membrane (including lipid membrane and liposome membrane), solid support material, etc. Among them, the functional stomata with S-layer protein lattice localization and orientation arrangement have important application value in diagnosis, immune vaccine, bionics and molecular biology technology [33].

(1) SUMs of S-layer protein isolated from Bacillus have zeolite-like properties and have sensitive sorting and filtration. The S-layer (SAPs), which is composed of the S-layer protein themselves on cell fragments, has a pore size of up to 1 μm and can selectively bind specific molecules (such as enzymes).
(2) S-layer protein as matrix fixed functional molecules Because S-layer protein mesh has the physical and chemical properties of recognizing various sub nanometer size molecules, that is, it has the ability to fix specific functional structures of molecules. For example, a specific macromolecule is fixed by the

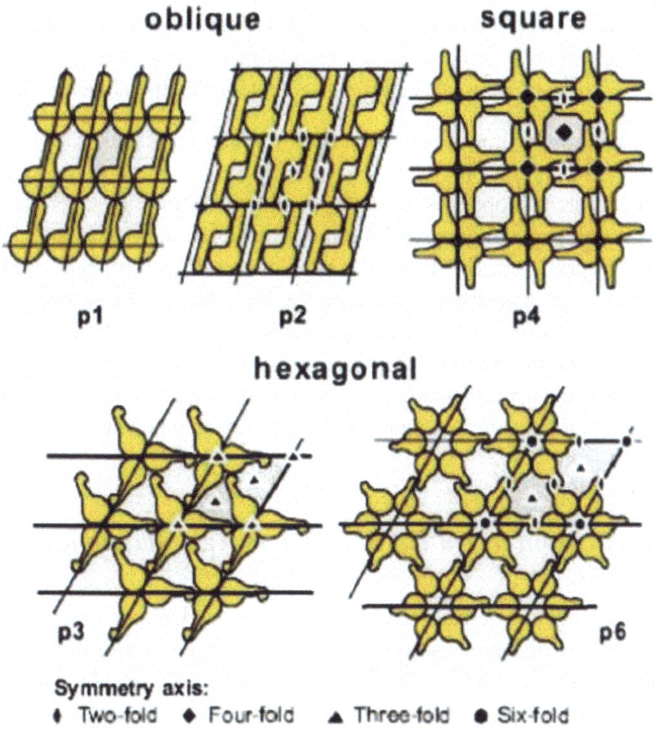

Fig. 11.5 Diagram of different symmetric forms of S-layer protein [33]

end of the carbon chain of the glycoprotein S-layer protein. The self-assembled SAPs obtained from Gram-positive bacteria had a cylindrical structure with an inner diameter of 1 μm and two ends covered by polysaccharide S-layer protein. The different S-layer protein have the properties of fixing corresponding macromolecules, and can be used for monomolecular and multimolecular enzymes, as well as sensors.

(3) SAPs was used as the weak antigen and hapten fixation substrate, and glycosylated or non-glycosylated S-layer reticular protein was used as immune vaccine carrier. Compared with tetanus and diphtheria toxin vaccines, S-layer protein carrier vaccines do not cause intradermal injury and other side effects after subcutaneous or intradermal injection.

(4) As the supporting structure of the functional lipid membrane, S-layer protein is a biofilm made of structural recombination with special functional appendage (ion channel, carrier proton pump, stomatal protein, receptor), which overcomes the disadvantages of poor fluidity, short kinetic lifetime and slightly strong mechanical properties of the general lipid membrane. The most significant application is that the S-layer protein is covalently combined with the phospholipids in the liposome to form molecular appendages, and the S-layer crystalline films

are covered on its surface. The bionic results formed are similar to the shell of paleontological cells and contain a large number of stomatal crystal structures, which provides a broader application prospect for the liposome as a drug carrier.

In short, S-layer protein is the simplest biofilm in biological evolution. It is based on single molecule glycoprotein. The mesh-arranged sub nano pores have specific physical and chemical properties. By improving the technology of S-layer protein mesh-fixed natural/artificial phospholipid membrane, the molecular membrane with characteristics is prepared. By molecular surface modification or as a substrate, it is used to identify, fix pairing and regulate the action of single molecules. On the other hand, S-layer protein can be directly used for gene manipulation, using known functions of multiple brains to reassemble into new congener biofilms or enzymes, ionic electrodes, microparticle carriers, biosensors, vaccines, diagnostics, etc., to provide biocompatibility and biodegradability for tissue renewal.

11.4.2 Application of Antisense Nucleic Acid Technology

In 1978, Paul Zamecnik of Harvard University in the United States first demonstrated the synthesis of Antisense oligonucleotides by inhibiting the replication of Rous Sarcomavirus with the DNA-13 base oligonucleotide complementary to Rous Sarcomavirus RNA oligonucleotide (AONs) can block gene expression.

Because AONs specifically blocks gene expression, it has broad prospects as antisense therapy in anti-tumor and immune diseases. AONs can be used in two ways: one is to hybridize with DNA in the nucleus to form a three-strand helical structure and inhibit DNA transcription; The second is to bind to mRNA in the cytoplasm, which can inhibit mRNA translation, but also to the formed RNA–DNA double bond structure, which is translated by ribonuclease H (RnaseH), interrupting the translation of mRNA by ribosomes. One oligonucleotide molecule can mediate several mRNA molecules that can inhibit expression for purposes. There are two main ways to prepare AONs: the short strand DNA of 15–30 bp is synthesized by DNA synthesizer or expressed by expression vector. The introduction of DNA fragments into the body produced large amounts of antisense DNA in the patient's cells. Current methods of introducing AONs into the body include:

I. Physical introduction methods: such as particle bombardment
II. Chemical introduction: DNA transfection mediated by colloidal dispersed vectors
III. Biological introduction method: virus (retrovirus, Epstein-Barr virus, adenovirus) is used as carrier to introduce it into somatic cells

Because AONs are unstable and easily digested by nucleases in cells in the body, it is necessary to continuously inject AONS into the body to ensure its effective therapeutic concentration for therapeutic purposes. Nano-sized liposomes were prepared from biodegradable polymer materials. Liposomes were used as carriers to regulate the distribution and pharmacokinetic processes in vivo through surface charge, molecular size, structure and morphology. Antisense oligonucleotide encapsulated in liposomes could prevent the degradation of AONs in vitro. At the same time, the similar structure of liposomes is easy to be taken up by dividing cells, so that the therapeutic effect is greatly improved.

Recently, there are many researches using nanoparticles as antisense nucleic acid carrier. Studies have shown that AONs can be easily combined with lipophilic biodegradable polymers NP (preferably with opposite ions), or encapsulated in some hydrophilic systems (such as alginic acid nano systems). After chemical modification to increase the lipophilicity of AONs, it is also beneficial to bind to the polymer NP. After binding with NP, AONs can protect it from enzyme destruction, increase its permeability to the cell, and significantly change the distribution characteristics of AONs in vivo, increasing its distribution in liver and lung. Isohexyl polycyanoacrylate (NP) containing anti-RAS oligonucleoside acid was administered at the same site 48 h after implantation into tumor cells. After 21 days, the animals were killed and the tumor tissues were weighed. The results showed that NP containing anti-oligonucleotide specifically inhibited tumor growth compared with the control group, and the effect was very obvious.

11.5 Application of Nanomaterials in Medical Field

11.5.1 Application of Medical Nanomaterials

Medical nanostructured materials consist of nanoparticles, which are usually assembled from nanocrystals and nanofilms, and their specificity depends on the composition of the two units. Nanoparticles are an important component in the preparation and processing of nanomaterials [34]. Nanoparticles have important applications in cancer monitoring, therapy, cell and protein isolation, targeting and slow-release drugs. In addition, after CPC curing, the microporous structure composed of different arrangements of crystals, reported to be 2–5 nm in diameter, can block the passage of bacteria, but does not affect the passage of ions and small molecules. Using CPC as the carrier, the drug is added into the microcapsule or pellet, and the micropores in the material can be quantified or measured to release the drug. It has been reported that when antibiotics or anticancer drugs are added to the biomaterial CPC in bone repair, the drugs do not affect the cohesion and tissue repair of the material, but also can perform chemotherapy on the damaged local tissues or malignant tumors after resection to prevent infection and tumor recurrence, playing a dual role of tissue repair and drug therapy [35].

Biomaterials and nanotechnology are also advancing tissue engineering. Scaffolds play an important role in tissue engineering. In order to simulate the natural skeleton structure of extracellular collagen tissue, for example, nanofiber skeleton was synthesized using biodegradable polymer materials. This nanofiber material has a three-dimensional porous structure, and the pore size of the fiber is 50–500 nm. This nanofiber material can prevent the entry of foreign substances (such as microorganisms, etc.), and there is no immune rejection of natural collagen, which is very conducive to the inoculation, distribution of cells on the nanofiber material and the growth of new tissues. According to the different chemical composition, molecular weight, structure type, degradation rate and mechanical properties of nanofibers, a series of products can be obtained, which are very useful as scaffolds for tissue engineering [36].

Intelligent materials refer to functional materials that can imitate living systems and have elements such as perception and feedback of corresponding information, and these materials are endowed with biological characteristics. N-ethyl propyl acrylamide (NPAAm) is a kind of hydrophilic gel [37]. When the temperature is above its lowest critical temperature (LCST), its volume expands tens of times in a small area and can absorb water quickly. On the contrary, when the temperature is below its lowest critical temperature (LCST), it can rapidly shrink, squeeze the network structure (nanostructure), and release the drug-containing aqueous solution. NPAAm was grafted onto polyvinyl alcohol (PVA) substrate with higher minimum critical temperature to enhance its mechanical strength and retain its original high physiological reactivity [38]. This intelligent composite material, whose swelling and antiswelling properties can be adjusted by temperature or pH, can be used in new types of sensors and has great potential applications in diagnostic, immunological and targeted drug delivery systems.

11.5.2 Nanorobot

Nanomechanics and nanorobots are the most attractive research directions in nanotechnology. The use of atoms and molecules to directly assemble nanomachines is not only faster and more efficient than conventional machines, but also has a wide range of applications and special functions. Manufacturing nanorobots can save materials and reduce pollution levels, which is unmatched by the manufacturing of existing robots. The use of nano-macromolecular biological components and small molecular inorganic crystal structure combination, the use of nano-electronics control assembly into nanorobots, will bring a profound revolution to human medical science and technology, so that many modern difficult diseases to be solved. These molecular robots use light sensors as a switch, from the glucose and oxygen dissolved in the blood to obtain energy, and according to the programmed program to explore the body objects, with a doctor's pre-programmed program to conduct a whole-body health check to clear blood clots in the blood vessels, remove fatty deposits in heart arteries, devour viruses and tissue fragmentation cells, kill cancer cells, monitor the body's lesions and so on. Nanorobots can also be used for human organ repair work, such as repairing damaged organs and tissues, cosmetic surgery, genetic assembly work, removing harmful DNA from genes or installing normal DNA in genes, so that the body can restore normal function. Tiny memory (ROM) devices made from nano-silicon wafers are implanted in the brain and connected to neural pathways that could be used to treat Parkinson's disease or other neurological disorders.

It is predicted that nano-dentistry will be possible through the use of nanomaterials and the introduction of biotechnology, especially dental nanorobots. In 10–20 years, the first dental nanorobots will be built. This device can precisely control toothache and enable dental replacement with whole teeth cultured in a single clinic visit, or rapid and precise restoration of dental tissue at the nanoscale. This means that it is clinically possible to make repaired or replaced dental tissue as naturally grown, or to achieve permanent desensitization, or to complete orthodontic treatment in a single visit, or to covalently bind jewel-like enamel, or to maintain oral hygiene for a long time.

11.6 Application of Nanomaterials in Biomedicine Field

11.6.1 Application of Cyclosporin a Nano-emulsion

(1) Cyclosporine A(CyA), A metabolite of fungi, is A cyclic polypeptide containing 11 amino acids. Cyclosporine A is a white acicular crystal (Me_2CO) with a melting point of 148–151 °C, $[\alpha]_{20}^{D}$ – 244 °C (c = 0.6, chloroform), Soluble in methanol, ethanol, acetone, ether, chloroform, non-volatile oil, slightly soluble in water.

Cyclosporine A, as a highly effective immunosuppressant, was first developed by Sandoz Pharmaceutical Company in Switzerland, and in 1983, the United States Food and Drug Administration (FDA) approved CyA as an immunosuppressant for kidney, liver and heart transplantation. Its discovery ushered in a new era of organ transplantation. At present, CyA is widely studied in Western Europe, North America, Japan and China, in addition to organ and tissue transplantation, it is also used in the treatment of autoimmune diseases.

CyA selectively inhibits cellular immunity, and its inhibitory effect on thymocytes and T helper cells (T_H cells) is the core of its immunosuppressive effect. Its main effects are: inhibiting thymus factor secretion, blocking the maturation of T cells in the thymus, thus inhibiting thymus function and inhibiting the expression of T cells' grade II major histocompatibility antigen; Inhibit TH cell differentiation and proliferation; Inhibit antigen or mitogen to stimulate the differentiation and proliferation of T cells; Inhibit the secretion of interleukin 2,3,4, ν-interferon, tumor necrosis factor α,β and other lymphatic factors, thereby inhibiting rejection and immune response.

CyA inhibits cell surface major histocompatibility antigens [MHCII (I_a)antigens].

Expression, change prostaglandin E2 synthesis; Inhibition of macrophage production of interleukin-1; Inhibition of monocyte chemotaxis; Block the lymphokine-dependent T-lymphocyte-macrophage interaction, thereby reducing the amplification effect of the immune response.

CyA has an inhibitory effect on antibody production. CYA indirectly affects the differentiation and proliferation of B cells by inhibiting T cells from producing interleukin-2 and other lymphatic factors. On the other hand, CYA also has an inhibitory effect on the secretion of some non-T cell-dependent antibodies. It has been demonstrated that the titer of serum IgM type anti-mitochondrial antibody in patients with primary biliary cirrhosis is significantly reduced after CyA treatment.

CyA has been used in bone marrow transplants since the late 1970s and has been shown to significantly reduce the incidence of acute graft-versus-host disease (GVHD). At that time, nephrotoxicity was more serious, and the severity of nephrotoxicity has been significantly reduced by CyA and methotretin combined therapy today. Due to the success of bone marrow transplantation, 60–80% of aplastic anemia patients and 10–70% of leukemia patients become long-term (>3–5 years) asymptomatic survivors. CyA is now widely used for kidney transplants worldwide. The use of CyA has increased the survival rate after kidney transplantation from 50% to more than 90%. Kidney transplantation has become a better option for the treatment of advanced kidney disease and is considered to be superior to dialysis treatment both in terms of effectiveness and economy.

The use of CyA has led to a 5-year survival rate of more than 75 percent, and 90 percent of heart transplant patients return to normal work. Due to the emergence of CyA, liver transplantation has made great progress in the past few decades, with a 1-year survival rate of 70% and a 5-year survival rate of 60% after transplantation. In addition to the above organ transplants, transplants of lungs, pancreases, and intestines have also been successful in humans.

The bioavailability and pharmacokinetic parameters of cyclosporine A vary greatly from individual to individual, and its main side effects are dose-related. Due to the longer medication time of organ transplant patients, the insufficient dosage of drugs affects the curative effect and the rejection reaction, which is difficult to distinguish from the toxic reactions such as kidney damage and liver dysfunction caused by overdose. Therefore, its pharmacokinetic process has been extensively studied at home and abroad.

Oral cyclosporin A is mainly absorbed in the small intestine and enters the lymphocyte-rich area around the human thoracic duct through the Small intestinal chyle lymphatic vessels. The oral absorption of CyA was poor, the peak time was 1–8 h, the average was 3.87 h, and the absorption rate was 0.6–4.7 h^{-1}. The bioavailability (F) showed great individual differences (5–89%) due to the influence of the transplant organ type, patient age, post-transplantation time, gastrointestinal and hepatobiliary functional foods, and combined drugs. Long-term administration of CyA can significantly increase F, peak concentration, trough concentration and AUC. Because CyA is a fat-soluble drug, bile can help the absorption of CyA, which can be increased when taken with milk and chocolate. Food and liver diseases can affect the absorption of CyA.

After being absorbed, cyclosporin A is initially distributed in the blood, and 60–70% is taken up by blood cells, of which 41–58% are red blood cells, 5–12% are granulocytes, and 4–9% are lymphocytes. The rest is bound to plasma lipoproteins, of which 18–22% are high-density lipoproteins, 11% are low-density lipoproteins, and 2.5 are very low-density lipoproteins other proteins accounted for 2.5–3.5%. The uptake and binding of CyA by red blood cells increased with decreasing temperature, and reached saturation at 3–5 mg/L. However, the binding of lipoprotein to CyA decreased with decreasing temperature in the concentration range of 0.025–3.8 mg/L.

CyA has a high tissue binding rate, and most of the dose is distributed outside the blood with a large apparent volume of distribution (4–8 L/kg), and the volume of distribution is significantly increased in children. Human autopsy shows that CyA is widely distributed in various tissues, and the concentration in tissues rich in white blood cells (such as thymus, spleen, thyroid, sleep fluid gland, lung, skin) is more than 10 times higher than the concentration in blood.

Cyclosporin A is metabolized by cytochrome P450IIIA enzyme system, resulting in lightening, demethylation and cyclization, but the structure of the parent ring remains unchanged. This process mainly occurs in the liver. So far, 30 metabolites have been identified. However, the most active metabolites, their activity is only 10–20% of CyA, and metabolic toxicity is very low.

CyA clearance rates ranged from 0.38 to 3 L/(kg·h). The rate of clearance was rapid in children, decreased in patients with liver damage and low serum LDL levels, and was more likely to decrease in middle-aged and elderly people. CyA clearance in tissues follows linear dynamics with an apparent clearance half-life of 19 h in the blood, but the actual half-life exceeds 24 h due to extensive tissue distribution. Brynskov found that the elimination half-life of homeostasis increased from 7.9 to 11.9 h for a single dose. More than 90% of the dose is excreted from bile, of which < 1% is the prototype; 6% is excreted in the urine, of which only 0.1% is the prototype drug.

The pharmacokinetics of CyA are multicompartment models. According to most reports, both oral and intravenous administration showed a bicatrioventricular model.

(2) Preparations of cyclosporine A Preparations of cyclosporine A can be divided into intravenous infusion preparations, oral preparations, eye preparations and skin preparations according to different routes of administration. Among them, intravenous infusion and oral administration preparations are commercially produced.

1) CyA Oral liquid commercially available, the Mountain Ming oral liquid (Sandimmun, trade name) produced by the former Sandoz Pharmaceutical Company (now merged into Novartis) is represented. Mountain Ming oral liquid is known as emulsion preconcentrate, and its typical prescription is 100 mg CyA, 300 mg Labrafil M 1944CS (a surfactant), 100 mg anhydrous ethanol, and 425 mg olive oil. The preparation method is as follows: at room temperature, the CA is dissolved in the organic solvent anhydrous ethanol, and then the oil solution containing the surfactant is added, and the oil solution is stirred evenly to obtain a clear and viscous oil solution. The oil solution is quite stable at room temperature. Because the preparation contains surfactant and oil phase, the three elements of the emulsifier (surfactant, oil phase, and water) can be met with water or gastroenteric fluid, and the emulsion can be automatically emulsified to

form an emulsion, so this Drug Delivery system is also known as Self-Emulsified drug delivery System, SEDDS). It should be noted that Labrafil M 1944CS is a lipophilic surfactant with an HLB value of 3–4, so the Mountain Ming oral liquid can form a conventional emulsion with an average particle size of 1 μm after self-emulsifying with water. The above preparations are generally taken after taking the prescribed dosage and adding it to drinking water or drinks before immediate use and stirring (emulsifying).

2) CyA soft gel capsule is produced by the former Sandoz Pharmaceutical factory under the trade name Mountain Ming. CyA soft gel is prepared by placing the CyA emulsified concentrate directly into the soft gel. After the gelatin of soft gel was disintegrated in gastroenteric fluid, the CyA emulsion concentrate contacted with gastroenteric fluid to self-emulsify into emulsion, which was helpful to improve the absorption of CyA. Soft capsule preparation is single-dose packaging, accurate dose division, convenient and safer to take.
3) CyA injection commercially available CyA injection is also produced by the former Sandoz Pharmaceutical factory under the trade name also Mountain Ming, which is an intravenous infusion of condensed liquid. Each ml of CyA injection contains 50 mg of CyA, 650 mg of Cremophor EL (another surfactant) and 33% ethanol by volume. The concentrated intravenous infusion of CyA should be diluted with normal saline or 5% glucose at the ratio of (1:20) to (1:100), and then slowly injected intravenously for 2–6 h. A significant number of patients who are given large amounts of Cremophor EL intravenously for a short period of time will experience an allergic reaction, so intravenous administration is only used in patients who cannot take it orally.
4) At present, CyA eye drops have entered the clinical stage and are being evaluated in multi-center clinical trials for the treatment of eye related diseases. CyA eye drops are mostly CyA suspensions in oils, commonly used oils are castor oil and olive oil with concentrations of 0.5%, 1% and 2%. Other formulations of CyA include tablets, topical formulations, microspheres and nanoparticle formulations, liposomes, and solid dispersions, which are currently under laboratory study. CyA tablets can be prepared by direct pressing. It has been reported that each tablet contains 100 mg CyA, 250 mg mannitol, 200 mg microcrystalline cellulose, 20 mg stearic acid and 10 mg sodium dodecyl sulfonate. There was no significant difference in bioavailability between Mountain Ming soft gel by cross administration of 6 dogs (400mg/ capsule).

Cyclosporin A is a cyclic polypeptide with a relative molecular mass of 1202, which is theoretically difficult to penetrate the skin, but it has been reported that under appropriate conditions, human skin permeability can reach 204.19 ng/cm^2. Within 22 h, the CyA concentration in subcutaneous tissue could be maintained at 146.62 ng/ml, which had reached the concentration of effectively inhibiting T lymphocytes.

CyA biodegradable materials microspheres and nanoparticles have been widely reported. For example, CyA polylactic acid (PLA) microspheres were prepared by solvent volatilization: 2 mg CyA and 18 mg PLA were dissolved in 2 ml choline chloride, then added to 7 ml 2% (W/V) PVA aqueous solution,100 W phacoemulsification, and electromagnetic stirring at room temperature (500 r/min 20 h) The choline chloride was removed, the microspheres were collected by centrifugation, washed with distilled water and dried. The average particle size measured by scanning electron microscopy was 260 nm, and the CyA encapsulation rate was 88%-93%. After intramuscular injection, the plasma drug concentration reached the peak at 2–4 h, and the lymph node drug concentration was significantly higher than the plasma drug concentration. The Poly (DL-lactide-co-glycoside, PLGA) microspheres and nanoparticles of CyA were prepared by the same method. The particle size (0.2–30 um) was related to the emulsification rate and polymer molecular weight. In vitro release of PLGA showed that CyA was released when PLGA was dissolved, and the dissolution rate was 40%-80% at 1 h, 75%-95% at 4 h, and more than 80% at 8h. The ideal dissolution curve can be obtained by controlled release particle size, and the researchers hope to develop this preparation into an oral sustained release preparation for clinical use.

CyA liposomes have also been reported. For example, CyA liposomes were prepared by thin film dispersion method: lecithin, cholesterol, dicetyl phosphate and CyA (the molar ratio of which was 1:0.33:0.1:0.7) were dissolved in chloroform, the chloroform was removed by rotary evaporation, so that a film was formed on the wall of the device, phosphate buffer (PBS) was added while stirring, and then ultrasound was performed for 5–10 min, obtained liposomes containing CyA (10 mg/mL). Intravenous administration in dogs (25 mg/kg) showed that liposomes reduced the toxicity of CyA, and there was no significant difference in pharmacokinetic parameters compared with control group.

(3) Cyclosporine An oral nano emulsion Introduction Nano emulsion refers to a special emulsion with the diameter of the milk drop at the nanometer scale. Sandimmun Neoral is a new product based on the CyA self-emulsifying drug delivery system (Sandimmun Neoral). A typical prescription for Sandimmun Neoral is 100 mg CyA, 200 mg 1, 2-propylene glycol, 320 mg refined glycerol-ester transfer corn oil, and 380 mg Cremophor RH40(surfactant).

Because Cremophor RH40 is a hydrophilic surfactant, the HLB value is 14–16. The nano emulsion with an average particle size of about 30 nm can be automatically emulsified when it meets water. The droplets of this nano emulsion are so small and colloidal that they are thermodynamically stable. But more importantly, because the milk drops of the nano-emulsion are so small, the new mountain clearly improves the pharmacokinetic properties of the original mountain, showing the following advantages:

1) CyA nano-emulsion can significantly increase the absorption of CyA. Nano-emulsion can release CyA rapidly in gastroenteric fluid, and has good contact with the gastrointestinal upper cortex. As a result, CyA is easily absorbed and the bioavailability is significantly improved. The average bioavailability of Sandimmun Neoral oral liquid was 34%, and Holt's study indicated that the average bioavailability of Xinshanming was 29% higher than that of Shanming,

an increase of nearly 1 time. Therefore, the dosage and cost can be reduced by using new Sandimmun Neoral.

2) CyA nano emulsions can accelerate the absorption of CyA by 1 h.
3) CyA nano emulsion can improve its pharmacokinetic stability in vivo and reduce individual differences between patients. It has been reported that the variation of bioavailability of Shanming oral liquid during steady-state is from 5 to 89%, the absorption rate constant is from 0.6 to 4.7 h^{-1}, and the peak time is from 1 to 8 h. The individual differences are very large, which brings great difficulties to clinical drug use. The absorption of new Sandimmun Neoral oral liquid was stable, the fluctuation of blood concentration was small, and there was a good linear relationship between drug dose and blood concentration. The correlation between trough concentration and AUC is good, and the monitoring of trough value of blood concentration is more reliable. The difference of peak time was also significantly decreased, ranging from 0.5 to 1 h. Due to the stable pharmacokinetic performance and good predictability of Xinshanming oral liquid after administration, it is relatively easy for doctors to control or adjust the blood drug concentration of patients in clinical practice, and the number of blood drug concentration monitoring and dose adjustment can also be reduced. Smaller individual differences are beneficial to improve treatment effectiveness, reduce side effects and reduce costs.

The study also showed that the absorption of CyA nano emulsions was less affected by the physiological state of the gastrointestinal tract (such as bile, secretory feeding, and fat content in food). In addition to oral liquid, the original Shan Daoist company also developed new mountain Ming soft gel capsules. The prescription of Xinshanming soft gel is the same as that of Xinshanming oral liquid, except that it is filled in the soft gel, and after the gelatin of the soft gel is disintegrated in the gastroenteric fluid, Xinshanming will automatically emulsify into nano-emulsion after contact with the gastroenteric fluid.

In summary, cyclosporine A nano emulsion cleverly uses nanotechnology, significantly increases the effectiveness and safety of medication, and reduces the cost of treatment for patients. At present, Xinshanming oral liquid and soft gel are sold worldwide. Pharmacoeconomic studies have shown that the application of cyclosporine A nano emulsion can reduce the dose by 10–15%, and the comprehensive medical cost can be reduced by about 20%, which can bring great social and economic benefits, and is known as a super drug in the industry. Cyclosporine A nano emulsion has been rapidly popularized in clinic, which shows the great practical application value of this new dosage form, and also provides a very successful example for the application of nanotechnology in the field of medicine.

11.6.2 Application of Magnetic Nanomaterials in Clinical Magnetic Resonance Imaging

The application of Magnetic nanomaterials in clinical Magnetic Resonance Imaging (MRI) mainly refers to nanoscale Superparamagnetic Iron Oxide (Superparamagnetic Iron oxide) application of oxide (SPIO) particles in MRI SPIO is a new type of negative contrast agent for magnetic resonance [39]. Gadolinium diethylenediamine pentaacetate (Gd-DTPA) is the most commonly used contrast agent in clinical magnetic resonance (MR), which can effectively improve the diagnostic level of MR Imaging. With the application and development of Magnetic Resonance Angiography (MRA), Gd-DTPA has become the main contrast agent to enhance MR Angiography. However, with the increase of use, Gd-DTPA has some obvious shortcomings, such as: it is eliminated too quickly in the body (after entering the blood, it can quickly pass through the cell space and be excreted by the kidney). Moreover, the distribution in vivo is not specific, so that the contrast of MR Images cannot be significantly improved, and the corresponding equipment is also required to scan quickly, and its price is more expensive. Therefore, scientists have developed a series of magnetic nanomaterials—superparamagnetic iron oxide, which has high tissue specificity in vivo and good safety characteristics.

(1) General properties of superparamagnetic iron oxide particles. Since the end of the twentieth century, the superparamagnetic iron oxide has been extensively and deeply studied at home and abroad, and various products have been developed on the basis of animal experiments and clinical trials. The basic structure of SPIO is mostly obtained by encasing iron oxide particles with glucose, and the diameter is generally on the order of nanometers. The different particle size will affect the distribution specificity and magnetization characteristics in vivo.

Different SPIO products have larger particle sizes. According to its particle size can be divided into two categories. One is the ordinary superparamagnetic iron oxide particles (SPIO): generally ranging in diameter from 40 to 400 nm [40]. One common formulation is SHU555A (trade name Resovist) developed by Schering. It is composed of Fe_3O_4 and Fe_2O_3, outsourced carboxyglucan, iron oxide core is composed of several single crystals, each single crystal diameter of about 4.2 nm, the diameter of the entire hydrophilic particle is about 62 nm. The most common is AMI-25 (Ferumoxides, trade name Feridex) developed by Advanced Magnetics with a diameter of (50 ± 29) nm and a core iron oxide particle with a diameter of about 20 nm. AMI-25 has completed clinical trials in the United States, Europe and Japan, and has been sold, and is now used clinically in some areas in China. There are also MSM and NSP430 developed by Nycomed. The other category is ultrasmall superparamagnetic iron oxides (USPIOs): the maximum diameter does not exceed 30 nm, such as AMI-227 developed by Advanced Magnetics Ferumoxtran and Nycomed's FeO-BPA, which has an average diameter of 4–6 nm and smaller particles, are among these agents.

(2) Biological properties of superparamagnetic iron oxide the distribution of superparamagnetic iron oxide in vivo has obvious specificity. This is mainly due to the reticuloendothelial system of the human body (RES) is rich in phagocytes, which are part of the human cellular immune system. When superparamagnetic iron oxide nanoparticles enter the human body through intravenous injection, bind to plasma proteins, and are recognized by the reticuloendothelial system under the action of opsonin, phagocytes will ingest the superparamagnetic iron oxide nanoparticles as foreign bodies, thus making superparamagnetic Sexual iron oxide is concentrated in tissues and organs rich in reticuloendothelial cells. Therefore, superparamagnetic iron oxide is a contrast agent for reticuloendothelial system, which can be used for MR Enhancement in tissues and organs rich in reticuloendothelial cells such as liver, spleen, lymph nodes and bone marrow. The phagocyte phagocytosis of superparamagnetic iron oxide particles reduced the signal in the corresponding area, while the tumor tissue kept the signal unchanged because it did not contain normal phagocytosis cells.

The particle size of superparamagnetic iron oxide has a great influence on the site of its entry into the reticuloendothelial system. Generally, SPIOs with large diameter are mainly ingested by the reticuloendothelial system of the liver and spleen. If the plasma half-life of Feridex is less than 10 min, 83% of Feridex is distributed in the liver and 6–10% in the spleen within a few minutes after intravenous injection. 2 h after intravenous injection, the liver content reached the highest value, about 89%. USPIOs mainly enters lymph node tissue and bone marrow tissue due to its small particle size. The weakened action with plasma protein and opsonin affects the uptake of S9 by phagocytes, the blood circulation half-life is more than 100 min, and the longer stay in the blood vessels allows it to pass through the capillary wall, more widely distributed in the tissues, especially through the capillary endings into the bone marrow and can be transported to the lymph nodes through the lymphatic vessels.

(3) Strengthening principle of superparamagnetic ferric oxide contrast agents' Common paramagnetic MR Contrast agents, such as Gd-DTPA, contain unpaired electrons (in the outer orbitals of gadolinium atoms), through proton/dipole-electron/dipole interactions, can enhance local magnetic field fluctuations and enhance proton relaxation, resulting in a shorter relaxation time for tissues. Superparamagnetic ferric oxide can also enhance proton relaxation, but the principle of enhancement is different from that of paramagnetic compound contrast agents. Due to its special chemical structure, superparamagnetic iron oxide particles can produce greater magnetism even in a weaker magnetic field, and the magnetic field quickly disappears after the withdrawal of the external magnetic field, that is, the so-called superparamagnetic. After the distribution of superparamagnetic iron oxide particles in the tissue, the distribution is uneven, resulting in an uneven local magnetic field, which reduces the T2 and/or T1 values of the tissue, and reduces the tissue signal (negative enhancement) or increases the tissue signal (positive enhancement). Due to the large size of SPIOs particles, it mainly promotes T2 relaxation and reduces tissue signal, but

has little effect on T1 relaxation. Due to the small particle size and long plasma half-life, USPIOs can induce a significant T1 contraction effect, which causes a significant T1 contraction while shortening T2 value, and increases tissue signal on T1WI while decreases tissue signal on T2WI.

(4) Clinical application of superparamagnetic iron oxide in liver magnetic resonance imaging

1) Detection and diagnosis of liver cancer. After being ingestion by phagocytes in vivo, superparamagnetic iron oxide is mainly present in normal liver tissues rich in phagocytes, which weakens T2WI signal in normal liver tissues. However, there is a lack of phagocytes in malignant liver tumors such as liver metastases and primary hepatocellular carcinoma, and T2WI signal does not decrease and relatively high signal areas appear, thus increasing the correlation between liver cancer tissues and normal liver tissues Compare. Benign liver tumors containing phagocytic cells or pools of blood (hemangiomas) also cause reduced tumor signal on T2WI, thus distinguishing them from the high signal of malignant tumors.

The clinical study on the detection efficiency of superparamagnetic iron oxide in liver tumors showed that SPIOs could significantly improve the contrast to noise ratio (CNR) between tumors and normal tissues, and more lesion tissues could be detected under different field intensities after contrast agent enhancement. For example, AMI25 was used to perform enhanced scanning on 15 cases of liver metastases at a field intensity of 0.6 T. Compared with MR Images before enhancement, the number of detected lesions increased by 16 times, and the minimum diameter of detected lesions decreased from 1.5 cm before enhancement to 0.3 cm, showing a very obvious effect.

Superparamagnetic iron oxide particles can also be absorbed into tumor cells, and the T1WI signal is typically enhanced at the tumor site, while the T2WI signal is decreased, so it can be distinguished from other benign liver tumors. Although the signal of benign liver tumor is reduced, it is weaker than that of normal liver tissue, and has certain characteristics, so the detection rate of benign liver tumor can still reach more than 90%. USPIOs is more helpful than SPIOs in the detection and characterization of hepatic hemangiomas due to its long plasma half-life, obvious T1 effect, and vascular characteristics of tumors.

2) Diagnosis of cirrhosis, hepatitis and other diseases. Due to the lack of phagocytic cells, the signal of inhomogeneity in cirrhotic sites (proliferative fibrous septa and regenerated nodules) was decreased, and the degree of signal reduction was lower than that in normal liver tissues. Active hepatitis can be characterized by uniform mild signal reduction due to impaired phagocyte function. Normal liver tissue showed uniform and obvious signal reduction on T2WI. Therefore, the application of superparamagnetic iron oxide can improve the sensitivity of MRI in detecting liver diseases such as cirrhosis and hepatitis.
3) Applications in hepatic magnetic resonance angiography (MRA). After three doses of SHU 555A were injected, the portal system imaging was compared

before and after enhancement. The results showed that the display of portal system was significantly improved by intravenous injection of superparamagnetic iron oxide. Moreover, the higher the SHU 555A concentration, the more significant the decrease in the strength of hepatic parenchymal signal and the greater the contrast of portal system, in which the left main branch of portal vein was the most obvious. Some people injected SHU 555A into patients with suspected liver cancer cells in clinic, and then obtained 3D MRA images through computer reconstruction. The results clearly showed liver cancer cells, portal vein and hepatic vein. A clinical trial using SHU 555A in 48 patients showed that SHU 555A significantly improved the display of the portal system, and the display rate of tertiary branches increased from 15.2% before enhancement to 87.0% after enhancement.

As an ideal MRA contrast agent, superparamagnetic ferric oxide particles should have a long plasma half-life, so that the concentration of contrast agent in blood can be kept as constant as possible during the imaging process. To achieve this, the particle size of the iron oxide core can be reduced, and the smaller iron oxide core will lead to a more significant TI shortening effect, making it more suitable for MRA. In this sense, USPIOs may be a more suitable contrast agent for MRA. A clinical trial of 3 doses of AMI-227 was conducted in 16 patients. The results showed that the vascular enhancement effect of all patients was very obvious 45 min after intravenous injection of AMI-227, and there was no significant difference in the degree of enhancement under the three doses. Due to the relatively long plasma half-life of AMI-227, the injection time becomes less important when AMI-227 is used as a blood pool contrast agent, which is more convenient in implementation and can reduce time delay. Similarly, AMI-227 may be an excellent MRA contrast agent due to its long plasma half-life, the ability of a single enhancement to perform multiple imaging and to increase the signal intensity of distorted or slow blood vessels. FeO-BPA, a novel USPIOs blood pool contrast agent with a diameter of less than 20 nm, has a lower R2/R1 value than AMI-227, a stronger TI effect, and may be more suitable for MRA. In animal model, FeO-BPA was applied to MRA of abdominal vessels, and the results showed that FeO-BPA could perform MRA under normal field strength.

4) Toxic side effects of superparamagnetic iron oxide the toxic side effects of superparamagnetic iron oxide mainly come from the content of iron. Iron is a normal component of the human body, but if the amount of iron in the human body is too high, there will be toxicity problems. For example, when the iron content of normal people per gram of liver tissue reaches 200 g, it will develop into cirrhosis and hepatocellular carcinoma. On the other hand, when the total amount of iron in the human body reaches more than 15 g, subacute and chronic toxic reactions will occur. However, from the current clinical application, the clinical application dose of superparamagnetic iron oxide is far less than the threshold of its toxicity. Superparamagnetic iron oxide at conventional clinical doses, if injected slowly, side effects are very small.

In the case of AMI-25, a 70 kg adult needs only about 80 mg to meet diagnostic needs. AMI-25 was applied to a mouse model up to 3000 pmol (Fe)/kg, and no obvious acute toxic reaction was observed. Since its introduction into clinical studies in 1988, AMI-25 has been commonly used in doses of 10–15 pmol (Fe)/kg, and the incidence of mild side effects is 10–15%, among which low back pain is relatively common, the incidence is about 3–4%, and other side effects include flushing, rash and dyspnea, etc., but all are reversible and can recover without treatment. Moderate toxic reactions such as hypotension are rare. The conventional clinical dose of SHU 555A is smaller, and 8 pmol (Fe)/kg can meet the diagnostic needs. Clinical trials have shown that there are no obvious pain or other discomfort symptoms during and after the administration of SHU 555A, and no obvious cardiovascular toxicity.

No serious side effects were observed in AMI-227. Some reported the incidence of side effects of 20%, including tachycardia, arrhythmia, heat sensation, abdominal pain, headache, flushing and hypotension, but are relatively mild and can be recovered within 25 min. In general, SPIOs and USPIOs are two types of MR Contrast agents with high safety.

11.6.3 Application of Immunomagnetic Microspheres Prepared by Magnetic Nanomaterials in Biomedicine

At present, the application of magnetic microspheres in biomedicine is rapidly developing, represented by immunomagnetic microspheres, and there are many successful applications. Magnetic nanomaterials are an important part of magnetic microspheres, so the application of magnetic microspheres is also the application of magnetic nanomaterials.

(1) Preparation and properties of magnetic nanomaterials Commonly used magnetic materials are ferric oxide, ferric oxide, iron cobalt alloy and so on. These magnetic materials have good magnetic responsiveness, and can be easily obtained in the nanometer scale by appropriate methods. For example, take a certain amount of magnetic material dissolved in an appropriate amount of distilled water, filter and mix, add a certain amount of distilled water to dilute, stir evenly, add an appropriate surfactant as a dispersant. At a certain temperature, keep stirring, add the solution to the system at a certain speed, and continue to stir for half an hour after adding the drops. Take it out, place it on the magnet, make the iron oxide particles settle, remove the supernatant, then add the aqueous solution of the dispersing agent, disperse and filter by ultrasonic wave, and obtain the color colloidal solution of the magnetic iron oxide nanoparticles.

During the preparation process, the concentration of lye has great influence on the properties of iron oxide nanoparticles, and the magnetization strength of magnetic materials is low when the concentration of lye is low. The droplet acceleration of lye affects the properties of magnetic materials. The slower the droplet rate, the smaller the particle size of magnetic materials and the lower the magnetization intensity. The reaction temperature of the system also has an effect. When the temperature is high, the particle size is larger and the magnetization is enhanced.

The magnetic nanomaterials obtained by the above method can be properly diluted with water, and electron microscope photos can be taken, and the size and distribution of the particle size can be determined by the image analyzer, or the size of the laser particle size tester can be directly determined, and the nanomaterials of different sizes can be obtained, and the minimum magnetic materials with an average particle size of several nanometers can be obtained, and the particle size is generally normally distributed. The structure of magnetic nanomaterials can also be analyzed by X-ray diffraction analyzer, and their magnetization can be measured by magnetometer.

(2) Preparation of magnetic microspheres by magnetic nanomaterials Magnetic microspheres can be prepared by magnetic nanoparticles and polymer skeleton materials. The polymer materials include polystyrene, silane, polyethylene, polyacrylic acid, starch, dextran, gelatin, albumin, ethyl cellulose and so on. There are natural and synthetic, can be used alone or combined as a skeleton material. These skeleton materials should have stable properties, high strength and no side effects.

The method of preparing magnetic microspheres can be divided into one-step method and two-step method: the one-step method is to add magnetic nanomaterials before forming the ball, and the polymer will be wrapped in it when forming the ball; The two-step method is to first prepare non-magnetic pellets, then make magnetic materials enter them by processing, and finally the magnetic nanoparticles exist in a dispersed form in the skeleton material of the microspheres.

The one-step method was developed earlier, and there are many methods, four of which are described below:

1) The magnetic nanomaterials (such as ferric oxide nanoparticles, etc.) are dispersed in water, the monomer of the polymer is added, and the initiator is added to trigger the polymerization reaction under appropriate conditions, so that the monomer polymerizes around the ferric oxide nanoparticles to form magnetic microspheres. This method is used to prepare magnetic microspheres with synthetic polymer material as the skeleton.
2) The magnetic nanomaterials were dispersed in the aqueous solution of the polymer skeleton material, and the appropriate surfactants were added to emulsify the W/O emulsion in a hydrophobic solvent. The polymer skeleton material was solidified into magnetic microspheres by thermal curing or crosslinking curing. This method is used to prepare magnetic microspheres with natural polymer materials as the skeleton.
3) First, Fe^{2+} and Fe^{3+} are precipitated in alkaline solution to form superparamagnetic iron oxide, and then coated with silane to form microspheres. The silane magnetic microspheres can be dispersed in aqueous medium without precipitating quickly, and can be easily recovered by magnetic field.
4) Using the magnetite itself as part of the oxidation–reduction system, the polymer can be completely wrapped around the magnetite. The polymerization reaction is

initiated by the diffusion of iron ions from the magnetite particles and becomes a free radical through the reduction of persulphuric acid. A water-based gel magnetic microsphere containing acrylic resin can be prepared by this method.

(3) Immunomagnetic microspheres (IMMS), or immunomagnetic beads (IMB), are magnetic microspheres with monoclonal antibodies on their surfaces. Due to the need to bind the appropriate antibody on the surface of the magnetic microsphere, it is required that the magnetic microsphere used can be combined with the monoclonal antibody through the chemical groups on its surface, or has a large surface adsorption force, and can be firmly combined with the monoclonal antibody. The crosslinked polystyrene microspheres have high strength and easy to be chemically modified on the surface, which is an ideal skeleton material for preparing immunomagnetic microspheres.

The connection between the microsphere and the antibody has two forms: adsorption binding and covalent binding. The adsorption binding depends on the non-specific adsorption force of the surface of the microsphere to the antibody, while the covalent binding depends on the covalent reaction of the active group on the surface of the microsphere to the antibody, and the adsorption binding can be relatively strong only when the microsphere has a very large surface area. In this way, for the relatively flat surface of the microsphere, it is necessary to improve its antibody binding force by treating its surface to ensure that there is a sufficient number of antibodies on the IMMS surface. After the surface treatment of magnetic microspheres and monoantibody are cultured in an appropriate buffer, the antibody quickly binds to the magnetic microspheres by physical adsorption, and if there is an active group on the surface of the microsphere, it is then covalently bound to the surface of the magnetic microsphere by a slower chemical reaction.

Immunomagnetic microspheres are mainly used for cell separation. Because it can specifically bind to the target substance and make it magnetically responsive, the immunomagnetic microsphere can be selectively bound to the target substance by antigen–antibody reaction if it is co-cultured with a complex mixture containing the target substance (substance to be separated). When this mixture passes through a magnetic field device, the target material bound to the immunomagnetic microspheres is trapped by the magnetic field, and thus separate from other complex substances. Immunomagnetic microspheres used for cell separation should have the following conditions: chemical stability, no condensation. No nonspecific binding to cells. The binding of magnetic microspheres with antibody is firm, the size of magnetic microspheres is uniform, the magnetic response is good, and the content of magnetic nanomaterials is uniform. The magnetic microspheres are of appropriate size and are not easy to be swallowed by cells.

IMM can be bound to cells in two ways, direct method and indirect method. The direct method means that the antibody is directly attached to the magnetic microsphere and then binds to the target cell. Indirect method refers to the mixing culture of cells and specific antibodies, so that the specific antibodies are bound to the cell surface, and then the magnetic microspheres are added with anti-mouse IgG (secondary antibody). The magnetic microspheres are indirectly bound to the target cells. The direct method can reduce the washing and culture steps, but it is rarely suitable for IgM monoclonal antibodies. Compared with the direct method, the indirect method can use a set of monoclonal antibodies in addition to a wide range of application, so it will obtain better cell clearance effect. However, after multiple washing steps, the specificity will be reduced.

The development of monoclonal antibodies against monocytes made it possible to isolate cells with specific surface markers. There are generally three different types of methods, namely flow cytometry (FACS), heterologous red blood cell wreath technology with a secondary antibody attached to the surface, and Panning technology where antibodies are passively adsorbed on polystyrene tissue culture plates. FACS is expensive and technically complex, and is often plagued by problems such as the volume, activity and sterility of the sorted cells. Red cell garland technology cannot handle a large number of cells, and there is no mature and simple method to conjugate antibodies to red blood cell membranes. Panning technology also has many limitations, it is difficult to enlarge the operation, the procedure is cumbersome, cannot quantify, the target cells are often mixed with non-specific antibodies adsorbed on the bottom of the culture plate other cells. Relatively speaking, the immuno-magnetic microspheres have the advantages of convenient operation, rapid and complete separation, and high cell purity, especially in the simple and convenient operation, which is difficult to compare with other separation methods.

(4) Application of immunomagnetic microspheres in biomedicine in recent years, autologous bone marrow transplantation (ABMT) following high-dose chemotherapy has become an option for some patients with acute leukemia, Hodgkin's disease or non-Hodgkin's lymphoma (NHL), neuroblastoma, and Wilms tumors. ABMT has recently shown encouraging results in small cell lung cancer, testicular cancer, and breast cancer. However, ABMT also has a limitation, that is, tumor metastasis cells may involve the bone marrow. Therefore, in vitro elimination of cancer cells in the bone marrow is necessary to ensure the success of ABMT. Various methods have been developed to purify bone marrow in vitro. Among them, the method of physically removing cancer cells by using immunomagnetic microspheres has the characteristics of safety, good effect and simple operation, and has been widely used. One of the reasons for the development of immunomagnetic microsphere technology is the growing evidence that a subclass of tumor cells can reject complement-mediated cytolysis. IMMS purification of bone marrow does not require specific ISOTYPE antibodies, nor is it affected by anti-complement factors in normal bone marrow, and can eliminate tumor cells with low antigen expression. Since the advent of immunomagnetic microspheres, they have been applied to almost all cancer types requiring bone marrow purification. Such as neuroblastoma, lymphoma, leukemia, myeloma, small cell lung cancer and breast cancer.

IMMS was first used to clear neuroblastoma cells. In 1984, T et al. established a model system for removing cancer cells from bone marrow using IMMS. Magnetic microspheres previously cultured with sheep anti-mouse immunoglobulin were then added. After incubating at 4 °C for 2 h, the mixture was placed on a conveyor and passed through a series of magnetic fields. Neuroblastoma cells are trapped and the bone marrow flows down naturally. The results showed that when 1% of the nucleated bone marrow cells were mixed with neuroblastoma cells (cell line CHP100), the IMMS could eliminate 99.9% of the tumor cells. The damage to normal cells was minimal, and after passing through the separation device, the activity of the cells was more than 98%. Colony analysis showed that BFU-E, CFUC, CFU-GEM and CFU-E were not significantly reduced. Immuno-magnetic microspheres for the removal of cancer cells in bone marrow also have some shortcomings, mainly in the methodology, target cell types, sensitivity of detection methods and other aspects of variability, but it is still a safe, simple and effective method.

The removal of cancer cells in autogenous bone marrow transplantation is only one application of immunomagnetic microspheres, which has a wide range of applications in medicine, such as the separation of other nucleated cells, the separation of organelles, and the application in microbiology and molecular biology. Microorganisms bound to magnetic beads by an immune response usually remain active and can be expanded in large numbers if provided with the right nutrients. The process is generally to mix the immune magnetic beads with the sample, culture for 10–30 min, and extract the magnetic beads with live microorganisms with an external magnetic field. It can be washed before injection into a suitable medium, and both monoclonal and polyclonal antibodies can be applied. The immunomagnetic bead technology has many advantages: the target bacteria can be separated from the environment, concentrated from a large volume to a volume suitable for culture: the growth inhibitory factors in the sample are removed from the bacteria, which is conducive to the growth of bacteria; When there are both pathogenic bacteria and a large number of non-pathogenic varieties in the sample, selective medium is often unable to isolate the target organism, and immunomagnetic beads have great potential to selectively isolate strains with specific genotypic families associated with pathogenicity.

More and more people are combining the immunomagnetic bead technology with PCR, called IMB-PCR technology. One factor that affects the direct application of PCR to clinical diagnosis is that certain components of the sample, such as stool and blood, affect the sensitivity of the Taq enzyme. This problem can be solved by treating it with immunomagnetic beads before applying PCR. The concentration of bacteria in the sample to the appropriate volume also eliminated non-specific inhibitors of Taq enzyme. For some samples, pretreatment can increase the number of target organisms, thereby improving the sensitivity of PCR. In addition, frozen samples may contain dead bacteria that are not suitable for culture, and PCR can be used for detection. The microorganisms detected by IMB-PCR included porphyromonas gingivalis, Yersinia enterocolitis, Salmonella, Helicobacter pylori, etc. In the instrument system, there is an automatic detection system ORIGEN that combines immunomagnetic beads and electrochemical luminescence sensors (IMB-ECL). Firstly, the target substance is captured by immunomagnetic beads, and then the three (bipyridine) chelate complex Ru(bpy)2+3 is connected by sandwich method, The labeled reporter antibodies were concentrated by magnetic field and then stimulated by voltage to produce electroluminescence. Some researchers have used it to detect A variety of biotoxins and bacterial spores, analysis of botulinum toxin type A, cholera beta subunits, toxins, staphylococcal endotoxin B and other detection limits up to femtogram (10–15 g). When analyzing the Bacillus anthracis spores, 100 spores could be detected.

References

1. Y. Wang, Y. Huang, H. Bai, G. Wang, X. Hu, S. Kumar, R. Min, *Biosensors-Basel* **2021**, *11*, 124–127.
2. H. Liu, R. Jian, H. Chen, X. Tian, C. Sun, J. Zhu, Z. Yang, J. Sun, C. Wang, *Nanomaterials* **2019**, *9*, 1165–1169.
3. Y. Huang, D. Wang, X. Zhu, D. Yan, R. Chen, *Polym. Chem.* **2015**, *6*, 2794–2812.
4. X. Chen, Y. Kang, J. Wu, *Chinese Pharm. J* **2017**, *48*, 142–149.
5. J. Mei, J. Zhuang, S. Tang, Y. Wang, L. Ma, L. Mo, Z. Shen, *Mater. Rev.* **2014**, *28*, 139–142.
6. J. Wang, R. Chen, J. Fu, Q. Zhao, W. Wang, S. Zhan, *Plastics* **2021**, *50*, 83–87.
7. H. Idrees, S. Z. J. Zaidi, A. Sabir, R. U. Khan, X. Zhang, S.-u. Hassan, *Nanomaterials* **2020**, *10*, 506–512.
8. M. Harun-Or-Rashid, M. N. Aktar, M. S. Hossain, N. Sarkar, M. R. Islam, M. E. Arafat, S. Bhowmik, S. -i Yusa, *Polymers* **2023**, *15*, 4563.
9. A. Elzoghby, *Curr. Pharm. Design* **2016**, *22*, 3303–3304.
10. J. Liao, C. Ning, Z. Yin, G. Tan, S. Huang, Z. Zhou, J. Chen, H. Pan, *Chemphyschem* **2013**, *14*, 3891–3894.
11. F. J. Stadler, A. Garcia-Penas, *Polymers* **2022**, *14*, 1164–1169.
12. K. Shetty, A. Bhandari, K. S. Yadav, *J. Control. Release* **2022**, *350*, 421–434.
13. J. Zhang, J. Jiao, M. Niu, X. Gao, G. Zhang, H. Yu, X. Yang, L. Liu, *Int. J. Nanomed.* **2021**, *16*, 6497–6530.
14. H. Zhang, T. Fan, W. Chen, Y. Li, B. Wang, *Bioact. Mater.* **2020**, *5*, 1071–1086.
15. S. Yang, C. Xu, J. Yao, *Chen, Polym. Chem.* **2019**, *50*, 11–18.
16. W. Mao, Y. J. Son, H. S. Yoo, *Nanoscale* **2020**, *12*, 14996–15020.
17. Z. Li, S. Tan, S. Li, Q. Shen, K. Wang, *Oncol. Rep.* **2017**, *38*, 611–624.
18. G. Guney, H. M. Kutlu, *NSTI Nanotech. Conference and Expo.*, **2011**, *6*, 400–403.
19. S. Bhunchu, P. Rojsitthisak, *Pharmazie* **2014**, *69*, 563–570.
20. P. Subhaswaraj, A. Syed, B. Siddhardha, *Curr. Drug Discov. Tec* **2020**, *17*, 457–468.
21. S. Menina, J. Eisenbeis, M. A. M. Kamal, M. Koch, M. Bischoff, S. Gordon, B. Loretz, C.-M. Lehr, *Adv. Healthc. Mater.* **2019**, *8*, 321–326.
22. H. Kotrange, A. Najda, A. Bains, R. Gruszecki, P. Chawla, M. M. Tosif, *Int. J. Mol. Sci.* **2021**, *22*, 573–576.
23. V. Karthik, S. Poornima, A. Vigneshwaran, D. P. R. D. D. Raj, R. Subbaiya, S. Manikandan, M. Saravanan, *J. Mol. Struct.* **2021**, *1243*, 903–908.
24. S. K. Nitta, K. Numata, *Int. J. Mol. Sci.* **2013**, *14*, 1629–1654.
25. M. Michelon, R. A. Mantovani, R. Sinigaglia-Coimbra, L. G. de la Torre, R. L. Cunha, *Food Res. Int.* **2016**, *79*, 95–105.
26. A. Ding, P. Xiong, Y. Chen, *Fudan J. Med. Sci.* **2015**, *42*, 262–266.
27. L.-A. Tai, Y.-C. Wang, C.-S. Yang, *Nitric Oxide-Biol. Chem.* **2010**, *23*, 60–64.
28. T. Ta, T. M. Porter, *J. Control. Release* **2013**, *169*, 112–125.
29. Y. Zhang, X. Xu, M. He, Y. Zhang, X. Yang, Z. Zhang, X. Guo, *Chinese Pharm. J.* **2022**, *53*, 335–344.
30. H. Zou, Z. Wang, M. Feng, *J. Control. Release* **2015**, *214*, 121–133.
31. X. Tang, Z. Wang, Y. Zhang, W. Mu, X. Han, *Chem. Eng. J.* **2022**, *435*, 104–107.
32. Y. Li, K. Tang, X. Zhang, W. Pan, N. Li, B. Tang, *Chem. Commun.* **2022**, *58*, 8754–8765.
33. C. Zhu, G. Guo, Q. Ma, F. Zhang, F. Ma, J. Liu, D. Xiao, X. Yang, M. Sun, *Prog. Biophys. Mol. Biol.* **2017**, *123*, 1–15.
34. T. C. Lowe, R. A. Reiss, *Nano SPD,* **2021**, *16*, 6497–6530.
35. A. Liu, K. Sun, P. Zhao, *Mater. Rev.* **2005**, *19*, 17–19,30.
36. R. Z. Valiev, I. P. Semenova, V. V. Latysh, H. Rack, T. C. Lowe, J. Petruzelka, L. Dluhos, D. Hrusak, J. Sochova, *Adv. Eng. Mater.* **2008**, *10*, 315–317.
37. R. Z. Valiev, E. A. Prokofiev, N. A. Kazarinov, G. I. Raab, T. B. Minasov, J. Strasky, *Materials* **2020**, *13*, 1054–1059.

38. L. Mishnaevsky, Jr., E. Levashov, R. Z. Valiev, J. Segurado, I. Sabirov, N. Enikeev, S. Prokoshkin, A. V. Solov'yov, A. Korotitskiy, E. Gutmanas, I. Gotman, E. Rabkin, S. Psakh'e, L. Dluhos, M. Seefeldt, A. Smolin, *Mater. Sci. Eng. R-Rep.* **2014**, *81*, 102–119.
39. Q.-Q. Tang, Y.-F. Zhang, Y. He, M.-L. Peng, G.-H. Zhai, H.-M. Fan, *Prog. Biochem. Biophys.* **2019**, *46*, 353–368.
40. K. P. Loh, D. Ho, G. N. C. Chiu, D. T. Leong, G. Pastorin, E. K.-H. Chow, *Adv. Mater.* **2018**, *30*, 1123–1126.

SPRINGER NATURE

GPSR Compliance

The European Union's (EU) General Product Safety Regulation (GPSR) is a set of rules that requires consumer products to be safe and our obligations to ensure this.

If you have any concerns about our products, you can contact us on ProductSafety@springernature.com

In case Publisher is established outside the EU, the EU authorized representative is:

Springer Nature Customer Service Center GmbH
Europaplatz 3
69115 Heidelberg, Germany

The manufacturer's authorised representative in the EU is Springer Nature Customer Service Centre GmbH, Europaplatz 3, 69115 Heidelberg, Germany. If you have any concerns regarding our products, please contact ProductSafety@springernature.com

Printed and bound by CPI Group (UK) Ltd, Croydon, CR0 4YY
01/10/2026
02225857-0015